Latin America and the Caribbean

A SYSTEMATIC AND REGIONAL SURVEY

Latin America and the Caribbean

A SYSTEMATIC AND REGIONAL SURVEY

THIRD EDITION

BRIAN W. BLOUET
College of William and Mary

OLWYN M. BLOUET
Virginia State University

JOHN WILEY & SONS, INC.

New York   Chichester   Brisbane   Toronto   Singapore   Weinheim

Acquisitions Editor	Nanette Kauffman
Marketing Manager	Catherine Faduska
Senior Production Editor	Elizabeth Swain
Designer	Nancy Field
Manufacturing Manager	Dorothy Sinclair
Photo Editor	Lisa Passmore
Photo Researcher	Ramón Rivera Moret
Illustration Editor	Edward Starr
Cover Photo	Ed Simpson/Tony Stone Images

This book was set in 10/11.5 Sabon by V&M Graphics and
printed and bound by Hamilton. The cover was printed by Phoenix Color.

Recognizing the importance of preserving what has been written, it is a
policy of John Wiley & Sons, Inc. to have books of enduring value published
in the United States printed on acid-free paper, and we exert our best
efforts to that end.

The paper in this book was manufactured by a mill whose forest management programs include
sustained yield harvesting of its timberlands. Sustained yield harvesting principles ensure that
the number of trees cut each year does not exceed the amount of new growth.

Library of Congress Cataloging-in-Publication Data:
Latin America and the Caribbean : a systematic and regional survey /
   [edited by] Brian W. Blouet, Olwyn M. Blouet. — 3rd ed.
         p.   cm.
   Includes bibliographical references and index.
   ISBN 0-471-13570-4 (pbk. : alk. paper)
   1. Latin America. 2. Caribbean Area. I. Blouet, Brian W., 1936–   .
II. Blouet, Olwyn M., 1948–   .
F1408.L3863   1996
972.9—dc20
                                                                        96-9925
                                                                        CIP

Printed in the United States of America

10 9 8 7 6 5 4 3

Preface

The first edition of the present text (1982) consisted of essays on the physical environments, historical geography, transportation, population, agriculture, urban places, and industries in Latin America. In the second edition (1993), regional chapters were included covering Mexico and Central America, the West Indies, Andean America, Brazil, and the Southern Cone.

In this third edition, an introductory chapter on Latin American issues has been added, a generalized overview of the climate and vegetation regions has been inserted into the coverage of the physical environments, and the chapter of Latin America and the World Scene has been expanded. The essay on transportation, written by Peter Rees, is not in this edition, but Peter has expanded the coverage of Middle America by creating separate chapters for Mexico and Central America. One new member joined the team for the present edition. Susan Sargent has taken a large role in revising and rewriting the Southern Cone chapter. All chapters in the book have been revised. In the interests of coverage, the editors have inserted materials into several chapters.

Once again, we want to thank many people for their help in preparing the text. Paul S. Anderson, Illinois State University, Charles Bussing, Kansas State University, Dennis Conway, Indiana University, Robert Layton, Brigham Young University, Robert Mancell, Eastern Michigan University, Mary Lee Nolan, Oregon State University, Francisco A. Scarano, University of Wisconsin-Madison, reviewed material for the present edition. We are still indebted to Jim Curtis, Don Zeigler, James Mulvihill, Greg Ridenhour, James Hughes, Ramesh Dhussa, and Robert McKay. Many of their ideas are still reflected in the present book.

The editors thank the Wiley staff for excellent support. Frank Lyman enthusiastically initiated the revision, and when Nanette Kauffman took over as acquisitions editor, she introduced many new ideas which have helped improve the text. Elizabeth Swain was again the production supervisor, keeping us all steady as deadlines were approached. Edward Starr saw to the revision of the maps and diagrams, and made many improvements. Lisa Passmore undertook the photo searching. Kevin Murphy supervised the new design. Dorothy Sinclair handled the manufacturing. Alice B. Thiede (Carto-Graphics), Eau Claire, contributed several new maps. Elizabeth Leppman, Department of Geography, University of Georgia, did the copy editing and helped resolve a number of place name questions.

Jill Glover Bauserman, School of Education, College of William and Mary, word processed all the chapters, as she did for the previous edition. Her careful work greatly contributed to the revision.

The editors and the contributors want to finally thank all the scholars, ancient and modern, on whose work we have drawn to produce this survey of Latin America.

Brian W. Blouet
Fred J. Huby Professor of Geography and International Education
The College of William and Mary in Virginia

Olwyn M. Blouet
Associate Professor of History and International Studies
Virginia State University

Contents

SYSTEMATIC GEOGRAPHY

Chapter 4

Agriculture
C. GARY LOBB, *California State University–Northridge*

Chapter 5

Population: Growth, Distribution, and Migration
BRIAN W. BLOUET, *College of William and Mary*

Chapter 6

The Latin American City
CHARLES S. SARGENT, *Arizona State University*

CHAPTER 1

Latin American Issues

BRIAN W. BLOUET

INTRODUCTION

Pick up the international section of a good newspaper, browse through a weekly magazine, tune into a cable news network, or plug into the Internet. What Latin American issues are you likely to encounter in the 1990s? Sensational stories about drug cartels in Colombia and Mexico, human interest articles concerning Indian lands and rights in Guatemala or Mexico's Chiapas state, migrants flooding into the United States from Mexico, Central America, or the Caribbean—all these issues might be covered in the media.

The North American Free Trade Agreement (NAFTA) linking Mexico, the United States, and Canada in an economic pact commands attention; as does global environmental concern about rain-forest destruction in Amazonia and elsewhere. Hurricanes and earthquakes are always newsworthy. General issues, such as urban problems, shantytowns, pollution, population pressure, poverty, and crime might all appear in a Latin American context. It is worth noting that, although some of these issues are new or have a new focus, others are long-standing and have not just appeared. However, the pace of change has quickened in recent years, and Latin American issues have assumed dynamic proportions.

If you want to gain an idea of Latin America more than half a century ago, seek out the first edition of Preston James' *Latin America* in your college library.[1] James, a great authority on the geography of Latin America, spent much of the 1920s and 1930s in the region and published his influential book in 1942. In his general introduction, he commented on the issues and characteristics of Latin America in the late 1930s. For example, he mentioned population. Surprisingly to us, James thought the population of the region relatively small, since Latin America comprised 19 percent of the earth's land surface but supported only 6 percent of the world's population. In 1940 Latin America had about 130 million inhabitants, about as many people as lived in the United States at the time.

Today the population of Latin America, approaching 500 million, is nearly twice as large as that of the United States, which stands at 265 million. Few commentators in 1940 foresaw the great surge in Latin American population numbers that was to come. The region was still predominantly rural, but there were popula-

---

[1]James, P. E. *Latin America*. New York and Boston: Lothrop, Lee and Shepherd, 1942.

tion concentrations in cities separated by lightly settled zones. Population clusters focused on urban cores, such as Mexico City, Buenos Aires, and Rio de Janeiro, each with more than a million inhabitants. In these cities, and many others, James thought that urban growth was out of proportion to the hinterlands that the cities served, and he did foresee the growth of huge primate cities. Today many Latin American cities have massive populations. For example, Mexico City exceeds 20 million and Buenos Aires has over 12 million inhabitants.

Preston James also commented on industrialization and attendant problems. By 1940, industrialization was accelerating in the region, and James saw that industrial development would further enlarge cities, widening the gap between urban and rural places. In the rural areas, the organization of society still retained feudal elements. The owners of great estates exercised social and economic power, and were influentially connected to government circles in capital cities. But the vast majority of the members of rural society did not own land and did not wield power. They were peons (unskilled agricultural workers, employed on a daily basis), sharecroppers, or tenants who rented land from estates. The social and economic gulfs were large, and frequently, countries were held together by the army, "the most powerful force in political life" (James, 1942, p. 38). James perceived that the countries of the region lacked established democratic institutions. As industrialization and urbanization acted on the existing social system, leading to further concentration of wealth and power in the hands of a small elite, political stability was unlikely.

Another issue that James identified was environmental destruction. James saw that industrialization would hasten the pace of environmental degradation. Amazonia was already in the news. In the 1930s, the Ford Motor Company, via a Brazilian subsidiary, attempted to establish rubber plantations in part of the Amazon region. Rubber trees in plantations, rather than scattered naturally in the rain forest, were disease prone, and clearing the forest to create plantations exposed the soil to erosion. Eventually the Ford rubber plantations were abandoned.

Like some earlier observers, James realized that the soils of Amazonia were not rich, except in flood plains where annual inundations of silt brought fertility. He saw that little of the region was cultivated. The production of cash crops was almost nonexistent, and only near the mouth of the Amazon (in the region within reach of the city of Belém) were such crops as sugar and cacao produced. Ominously, he noted that when land went out of cane cultivation, it was used for cattle grazing. We shall return to the issue of converting rain forest into cattle runs later.

In sum, James identified several controversial issues that continue to be debated. The problems included urbanization and industrialization, economic and social gulfs in Latin America, and the social structures that prevented the achievement of political stability and democracy. In addition, James recognized environmental problems within Latin America. It has been left to recent scholars to identify possible global implications of some of these environmental issues.

OVERVIEW OF SYSTEMATIC CHAPTERS

This book opens with an examination of themes in systematic geography—the physical environment, historical geography, agriculture, population, urban and economic geography.

Tom Martinson, in Chapter 2, describes the diverse physical environments of Latin America that act as a backdrop to human activity. Latin America extends from Baja California (32½°N) to Tierra del Fuego (55°S) off the southern tip of South America. The range of environmental types is far greater than in Europe or North America. Tropical rain forests, savannah grasslands, high grasslands, thorn scrub, temperate grasslands, coniferous forests, and deserts are all found in Latin

America. The region is prone to natural hazards, such as earthquakes, volcanoes, flooding, mud slides, and in the Caribbean, hurricanes.

In Chapter 3, Robert West outlines the ways in which the pre-Columbian inhabitants of the region used resources, organized socially, and settled in towns, cities, and innumerable rural settlements. The European intrusions brought new crops, new tools, and domesticated animals. Existing economic systems were changed, and new patterns emerged. Intercontinental encounter developed, leading to the creation of a group of Latin American cultures that incorporate Indian, European, and in some regions, African traits. West emphasizes that many characteristics of modern Latin America cannot be understood without reference to the historical background.

In the following chapter on agriculture, Gary Lobb describes aspects of agriculture in Latin America. Although much indigenous agriculture was destroyed or modified by Iberians, many parts of Latin America are still cultivated by traditional shifting cultivation techniques. Lobb describes these traditional patterns and shows how, as population numbers increase, the fallow periods in the system are shortened, and the forest area is reduced because it lacks time to rejuvenate.

Latin America became a producer of tropical agricultural commodities for export in the colonial era, when crops like tobacco, cotton, and particularly sugarcane became the basis of new economic and social systems, often based on plantations and slavery. During the nineteenth and twentieth centuries, the range of export commodities increased as markets expanded in Western Europe and North America. For example, after the development of refrigeration, meat was shipped from the temperate grasslands of Argentina and Uruguay, across the tropics, to Europe. Brazil and Colombia developed as major exporters of coffee. Tropical and subtropical fruits, such as bananas and citrus, began to be exported from the Caribbean and Central America. Recently, Brazil has become a major producer of citrus juices and wine. Chile, in the Southern Hemisphere, is able to grow apples, grapes, and other fruits for sale in northern markets during the winter months, when prices are high. Colombia exports large quantities of flowers (in addition to large quantities of cocaine, derived from coca leaves).

There are exceptions, such as banana growers in Ecuador, but the success in agricultural export markets touches few peasant farmers in Latin America. The majority of them cultivate tiny plots, raise subsistence crops, and sell small surpluses in local markets. The issue of disparity between commercial and peasant agriculture remains. In addition, effective land reform still awaits attention in many regions of Latin America, as Lobb indicates.

Chapter 5 examines issues concerned with the distribution, growth, and migration of population in Latin America. As noted earlier, when Preston James first published his text, the population of Latin America was about 130 million. Today in Latin America and the Caribbean, the total population approaches 500 million. During the same period, between 1942 and 1996, the population of the United States has doubled from about 130 million to around 265 million. Currently, the population of the United States grows at 0.7 percent per annum, and if that rate continues, will double in a century. In Latin America, the population growth rate is 2 percent and at that rate will double in only 35 years, although the rate is slowing. It is worth considering several issues regarding demographic increase. For instance, can the region sustain economic development in the light of rising population numbers? Increased food production and sufficient job creation are two basic necessities. In their absence, migration, especially to the United States, may be a strategy many individuals will try to pursue.

The surge in Latin American population numbers is a result of natural increase. In the 1930s, average death rates in the region began to decline, while birth rates remained high. In the 1940s, 1950s, and 1960s, population growth rates in many countries exceeded 3 percent per annum. Populations then doubled in about 20 years. Overall birth rates have fallen in recent years, although population numbers continue to rise.

As numbers grew rapidly, the age structure of populations altered. There were more young people, and as they reached early adulthood, many migrated from rural to urban areas, swelling already established migratory flows and helping to create the megacities that characterize modern Latin America. Most countries in the region now have urban majorities, meaning more people live in towns and cities than in rural places. In urban areas, although rates of natural increase have recently slowed, population numbers still rise due to in-migration. Latin American cities are overcrowded; housing in squatter settlements is wretched; services do not extend into all parts of the urban area, unemployment and underemployment are widespread.

Urban themes are developed by Charles Sargent in Chapter 6. One of the tools of Iberian settlement in the Americas was the creation of towns. The Spanish and Portuguese gave Latin America, at least in accessible areas, a network of towns that defines the distribution of urban places to the present. The colonial towns, with gracious central plazas surrounded by churches, government buildings, and palaces of prominent officials, have been overwhelmed by traffic problems and pollution. The issue here is whether these cities can function effectively to provide a safe environment, adequate housing, services, transportation, and sufficient employment.

In Chapter 7, Alan Gilbert discusses mining, manufacturing, and service industries. Latin America is rich in mineral resources. Iron ore is plentiful. Copper, tin, silver, lead, gold, bauxite, manganese, and tungsten are exported from the region. But mining is a fickle business. Demand can rise and fall, as evidenced by the decline of the Bolivian tin industry in the 1980s, when tin cans were increasingly replaced by plastic packaging, and frozen foods became widespread. In addition, economies built on mineral exploitation face the possibility of resource depletion. As the best deposits are worked out, extraction costs rise.

Currently, petroleum is in world demand, and the oil industry is profitable. Most major countries in the region produce oil. Many, including Mexico, Venezuela, Colombia, Ecuador, Peru, and Chile, export it. The petroleum industry is, however, capital intensive and uses little labor. Revenues from the industry go largely into the treasuries of governments and multinational corporations. Relatively little reward goes to regions of production. As electric cars begin to make a high-priced appearance, no one can be certain that technological advances will not result in efficient electric vehicles and a decrease in the demand for petroleum to power cars. Along these lines, how long can it be until solar power is harnessed effectively and cheaply to production processes?

Fifty years ago, Latin America was not a major manufacturing region. However, all countries now produce a range of consumer goods, and Argentina, Brazil, Chile, and Mexico have significant manufacturing industries. Brazil, for example, exports cars, planes, steel, railroad equipment, and chemical products.

OVERVIEW OF REGIONAL CHAPTERS

The second section of the book examines countries and regions within Latin America—Mexico, Central America, the Caribbean islands, Andean America, Brazil, and the Southern Cone. Each of the regional chapters raises contemporary issues in a Latin American context. The last chapter looks at Latin America in a global perspective.

Peter Rees begins the regional coverage by focusing on Mexico, where he identifies several distinct regions and points to the difficulties of national integration. Against this background, we can debate how Mexico is to take a place in an international organization such as the North American Free Trade Area. What are the problems and prospects involved in NAFTA in which Mexico, the United States, and Canada are moving toward free trade?

There are many issues to consider. It appears that some types of assembly manufacturing in Mexico will gain under NAFTA agreements—at the expense of operations in the United States and Canada. Yet, in agriculture, when the Mexican market is fully open to U.S. and Canadian grain exports, traditional producers of subsistence crops in Mexico may become uncompetitive, perhaps triggering another wave of migration to urban areas.

In addition, NAFTA provides for the opening of the service sector, including banking and finance. As this occurs, U.S. banks will further penetrate Mexican markets and take over a significant portion of the business of raising capital and financing international trade.

Until recently, many aspects of the Mexican economy have been cocooned from competition, protected and subsidized. The rules on environmental protection, safety in the workplace, and child labor await complete legislation and enforcement.

How the pressures created by NAFTA will play out in the distinct regions of Mexico, which Rees describes, is an issue to watch. For instance, many inhabitants of Chiapas (an economically marginal, predominantly Indian region on the Mexico–Guatemala border) are resisting the forces of integration and competition. Indian leaders want better economic conditions and more voice in politics for their people. They champion the indigenous rights movement that is growing in intensity all over Latin America. Chiapas has preserved itself partly by keeping out external influences. Allowing competitive, modern economic activity into a region of subsistence farmers will not improve living standards for the majority. Peasant farmers will be forced off the land and then exploited as cheap labor. Indian leaders seek to have some input in the terms of interaction and negotiation.

At the northern end of Mexico, the major issue is migration. Since the Bracero program started during World War II, there has been a strong flow of legal and illegal migrants crossing from Mexico into the United States, particularly California. The movement was powered for many years by a labor shortage north of the border. When the California economy experienced a downturn in the early 1990s, efforts were made to cut the inflow of migrants from Mexico.

The development of *maquiladoras* (assembly plants) in Mexican border towns is unlikely to halt migration, because wage rates in the United States are higher. Creating economic growth poles on the Mexican side of the U.S. border attracts more Mexicans (and Central Americans) closer to the United States. In the new factories they acquire skills that will be useful on entering the United States.

Central America is one of the many areas in Latin America that struggles to create political stability, as Peter Rees shows in Chapter 9. The strategic importance of the region brought foreign intervention (particularly from the United States) that helped create internal conflicts. El Salvador, Guatemala, and Nicaragua have suffered civil wars. The Guatemalan regime persecutes its Indian population and covets the territory of Belize. Panama tries to attain democracy in the aftermath of the United States invading the country in 1989 and taking the head of state to Florida for trial and imprisonment. The region suffers from economic dependency and reliance on the export of a few agricultural commodities. Of the countries in the region, only Costa Rica and Belize are politically stable and relatively prosperous. Both countries seek to develop eco-tourism (environmentally sensitive tourism) in an attempt to balance the demands of economic development and environmental prudence.

The Caribbean islands, discussed in Chapter 10, are a mosaic of contrasting cultures—the products of differing colonial histories and diverse ethnic backgrounds. Few indigenous inhabitants remain, but large numbers of people with

African heritage inhabit the region. The French colonies of Guadeloupe and Martinique are *départements* of France and part of the European Union. Most former British colonies are independent and democratic. The Dutch colonies retain ties to the Netherlands, although Aruba is pursuing independence. The Danes sold their islands to the United States in 1917, creating the U.S. Virgin Islands.

The former Spanish colonies present a diverse picture. Puerto Ricans are U.S. citizens, although not directly represented in Congress. Cuba retains a communist dictator and one-party state, long after the East European model has crumbled. The island of Hispaniola is divided between Haiti, with strong African and French influences, and the Dominican Republic, where Hispanic heritage remains powerful.

The Caribbean is a region of relatively small states, many too small to be fully independent economic systems. The well-being of many countries will depend upon the ability of the United States, Canada, and the European Union to allow Caribbean countries access to their markets. The danger is that trade organizations like NAFTA, by creating advantages for larger countries such as Mexico, disadvantage small Caribbean states.

Many current global issues are magnified in the islands because the resource base is small, population-to-land ratios are high, and unemployment is also high. One coping strategy has been migration to the United States, Canada, and European nations with former colonial connections to the Caribbean.

South America is covered in three chapters. The Andean countries of Venezuela, Colombia, Ecuador, Peru, and Bolivia are discussed together. Brazil, by virtue of extent, population, and Portuguese heritage, has a chapter to itself. The Southern Cone countries of Chile, Argentina, Paraguay, and Uruguay are considered together.

Politically, South America is an unstable continent. If the majority of states have elected governments, democracy is a recent innovation, and its institutions are fragile. The military has only recently left the presidential palaces and returned to barracks. The underlying problems, associated with maldistribution of wealth, and the prestige of small, elite classes who control the socioeconomic system, remain. Venezuela experienced a failed military coup in 1992. In Colombia, the president is suspected of financial links to drug cartels. Ecuador is peaceful. Peru has an elected government, but in fighting terrorist groups (such as the Shining Path), abbreviates the rights of citizens. Bolivia is ethnically divided between Indian and mestizo.

Chile is modern, economically expansive, export oriented, and eager to join NAFTA. The country has only recently emerged from a military government headed by General Augusto Pinochet. The military still retains a veto on political action. Argentina is also emerging from a period of military control between 1976 and 1983. The country struggles to understand the era of "the disappeared ones," when citizens went missing and assassination squads were active. The military dictatorship lost power after an unsuccessful attempt to invade the Falkland Islands (Malvinas), where the 2,000 inhabitants were adamant that they did not wish to go under Argentine military rule.

Brazil, by far the largest and most populous country in South America, was under a military regime from 1964 to 1985. It, too, tries to establish democracy in a land rich in resources with the most potential for development, but with vast gulfs between rich and poor and between different regions.

Currently, the economies of Brazil, Argentina, Uruguay, and Paraguay are linked via Mercosur (Common Market of the South). However, international transportation links are poor throughout Latin America, and inadequate road and rail facilities hinder movement of goods.

It is worth pointing out that the countries of South America still have boundary disputes. Peru and Ecuador have skirmishes on their border. Chile and Argentina need periodic mediation concerning their Andean border and rights in the Beagle Channel. Several countries covet territory in Antarctica. These examples may constitute flashpoints in the future.

The concluding chapter looks at Latin America in the global scene, emphasizing economic and international issues as Latin America strives to modernize and attempts to play a role in world affairs.

ENVIRONMENTAL ISSUES AND SUSTAINABLE DEVELOPMENT

Latin America is at the center of global environmental debates. There is good evidence that the rain forests of the region, for example in Amazonia (see Chapter 12, Brazil), are being cleared by processes that destroy the existing habitat. Tropical rain forests have been exploited for centuries as sources of dye woods and valuable hard timber, such as mahogany. In the modern world, the growing market for high-quality furniture and panelling in Europe, North America, and Japan has greatly increased the demand for tropical hardwoods. When the valuable species have been extracted, the remaining rain forest is often burned off. With the forest gone, grasses take the place of trees, and cattle are raised. This process is common in Amazonia, where soil fertility rapidly declines because nutrients are stored in the vegetation, not the soil. When the forest no longer gives a rich supply of organic matter, through leaves and boughs falling to the forest floor, fertility is lost.

Satellite imagery shows that in many areas of Amazonia the forest is simply being burned without prior extraction of timber, as land is turned into low-grade grazing pasture. The innumerable acts of clearance—can we call it eco-vandalism—are often made without regard to the customary usages that Indian communities have practiced in the forest environment. Traditional Indian groups occupied and used the forest on a communal basis. Individual title to land was unknown, and communities rarely had clear title, although Indian reserves have recently been created. As the forest is destroyed, the environment in which the communities hunted, trapped, fished, and gathered is disturbed and destroyed.

The issue of rain forest destruction may have global climatic implications. Amazonia is a huge reservoir of water in its rivers, soil, vegetation, and highly humid air. The forest grew as a response to a well-watered climate. The trees transpire and help create the high humidity that feeds the rainfall of the region. Remove the forest from extensive areas, and precipitation may decrease. The drier spells may become longer, placing stress on trees in marginal areas.

The trees in the forest absorb carbon dioxide (CO_2). Burn the forest, and the CO_2 is released into the atmosphere. Globally, the amount of CO_2 has been increasing in the atmosphere, at least since the nineteenth century, partly as a result of the burning of fossil fuels by industrial societies. Some observers believe that the build-up of CO_2 in the atmosphere is contributing to a "greenhouse effect," in which gases like carbon dioxide help to retain more heat in the atmosphere and contribute to global warming. Others think that there are natural warming and cooling trends in the earth's climate and point to historical meteorological records that illustrate fluctuations in past weather patterns.

The impact of atmospheric warming, whether caused by human activity or natural cycles of cooling and warming, is not easy to predict. A warmer climate would give higher temperatures, but it might also be wetter, and it is difficult to know whether desert and semiarid areas would expand or contract. Everyone agrees that if there is an increase in heat energy in the atmosphere, there will be more severe weather events, such as hurricanes. In the Caribbean hurricane season (June to November) of 1995 there were more than 20 hurricanes (storms with winds over 74 m.p.h.) and tropical storms. In the twentieth century, only 1931 has recorded more hurricanes and tropical storms.

The effects of clearing the Amazon rain forest may be felt beyond Latin America, and the region's environmental problems have therefore attracted global interest. In 1992, the "Earth Summit," or United Nations Conference on Environment and Development, was held in Rio de Janeiro. It drew official delegations from around the world to discuss habitat destruction and the resulting extinction of plant and animal species. Siting the conference in Rio was related to the need to raise awareness of environmental issues in a region with a history of environmental plundering.

Habitat destruction is not unique to Latin America, but Amazonia is the largest and richest of tropical rain forests. It is estimated to contain at least 20 percent of the world's plant varieties and bird species. The elimination of plant species may have dietary, pharmaceutical, and material implications.

Linked with the issue of preserving habitats—and the species that occupy them–is the idea of *sustainable development*. In the environmental sense, sustainable development means the use of resources in a manner that passes to future generations an environment that has not been degraded by species destruction, soil erosion, pollution, or any other means. Values included in the idea of sustainable development include biological diversity, protection of indigenous people, environmentally sensitive production of agricultural commodities and minerals, watershed protection, and global climatic stability. Some commentators have suggested that sustainable development should include efforts to produce a more equitable distribution of wealth within countries and between the developed and less developed regions of the world. In the case of most Latin American states, the maldistribution of income and wealth inside individual countries is a major problem, and any transfer of wealth from North America, Europe, or Japan to Latin America will likely be absorbed in government machines that are more committed to preserving the privileges of elites than improving the living standards of the poor.

READING ABOUT LATIN AMERICA

Latin America is in the news every day, and major newspapers have coverage of events in the region. *The Christian Science Monitor, Los Angeles Times, The New York Times, The Wall Street Journal*, and *The Washington Post* all carry frequent articles on Mexico, Central America, the Caribbean, and South America. All these newspapers are indexed and microfilmed. *The Miami Herald* is reported to have the best coverage of Caribbean issues of any newspaper. Reference to a good atlas is essential to the study of Latin America. *Goode's World Atlas* (19th ed., Chicago: Rand McNally, 1995) will help you locate the places, physical features, and regions of Latin America. The Foreign Broadcast Information Service (FBIS) publishes a *Daily Report: Latin America*, which contains transcriptions of broadcast material in Latin American countries. Many university libraries have this publication.

Below follows a concise list of textbooks, journals, handbooks, and statistical sources regarding Latin America.

► *Textbooks*

Bethel, L. ed. *Cambridge History of Latin America*. Cambridge, England: Cambridge University Press, 1984–1992.

Butland, G. *Latin America: A Regional Geography*, 2d ed. New York: John Wiley, 1966.

Burns, E. B. *Latin America: A Concise Interpretive History*, 6th ed. Englewood Cliffs, NJ: Prentice Hall, 1994.

Caviedes, C., and G. Knapp. *South America*. Englewood Cliffs, NJ: Prentice Hall, 1995.

Cubitt, T. *Latin American Society*; 2d ed. New York: John Wiley, 1995.

James, P. E. *Latin America*. New York and Boston: Lothrop, Lee, and Shepard Co., 1942.

_____. *Latin America*, 4th ed. New York: The Odyssey Press, 1969.

James, P. E., and C. W. Minkel. *Latin America*, 5th ed. New York: John Wiley and Sons, 1986.

Pendle, G. *A History of Latin America*. Harmondsworth, Middlesex, England: Penguin, 1963.

West, R. C., and J. Augelli. *Middle America: Its Lands and Peoples*, 3d ed. Englewood Cliffs, NJ: Prentice Hall, 1989.

Williamson, E. *The Penguin History of Latin America*. London: Penguin, 1992.

Winn, P. *Americas: The Changing Face of Latin America and the Caribbean*. Berkeley: University of California Press, 1992.

► Scholarly Journals

Bulletin of Latin American Research. Elsevier, Oxford UK., 3 times a year.

Caribbean Geography. Mona, Jamaica, University of West Indies, semi-annual.

Hispanic American Historical Review. Durham, N.C.: Duke University Press, quarterly.

Journal of Development Studies. Frank Cass, London, six times a year.

Journal of InterAmerican Studies and World Affairs. Coral Gables; Florida, quarterly.

Journal of Latin American Studies. Cambridge, England: Cambridge University Press, 3 times a year.

Latin American Research Review. University of New Mexico; quarterly.

Latin American Perspectives. Thousand Oaks, Calif.: Sage, quarterly.

Yearbook of the Conference of Latin Americanist Geographers. Austin: University of Texas Press, annually.

► Handbooks

Caribbean Islands Handbook. Bath, England: Trade and Travel Publications, Ltd., 1995.

Mexico and Central American Handbook. Bath, England: Trade and Travel Publications, Ltd., 1995.

South American Handbook. *Bath*, England: Trade and Travel Publications, Ltd., 1995.

Collier, S., H. Blakemore, and T. Skidmore. *Cambridge Encyclopedia of Latin America and the Caribbean*. Cambridge, England: Cambridge University Press, 1985.

Handbook of Latin American Studies. Austin: Texas University Press, 1936– .

Steward, J. H. *Handbook of South American Indians*. Washington, D.C.: U.S. Government Printing Office, 1946–1959.

► Statistical Sources

Standard yearbooks, such as *The World Almanac*, are useful as sources of basic information on population, resources, industry, language, religion, and political systems of Latin American countries.

The Statesman's Year-Book (Library of Congress reference JA 51.S7), has been published annually since 1864. The yearbook provides a wide range of information, including the addresses of embassies. Many embassies have a publicity department that provides official information about a country. Your library may have a long back run of *The Statesman's Year-Book*, and it can be used as a source of economic and demographic trends in a country over a period of years. Remember that in earlier decades the figures given may be government estimates.

An excellent source for historical statistics is B. R. Mitchell, *International Historical Statistics: The Americas 1750–1988*, 2d ed., New York, Stockton Press,

1993. The source allows comparisons between countries, and if there are problems with statistics, they are noted. Mitchell can be used to track the rise and fall of mineral and agricultural commodity production in Latin American countries.

Standard statistical sources for Latin America include:

Statistical Yearbook for Latin America and the Caribbean. New York: United Nations Economic Commission for Latin America and the Caribbean, 1991.

Wilke, J. W., ed. *Statistical Abstract of Latin America.* Los Angeles: UCLA Latin American Center Publications, 1995.

Some sources on population data are referred to in Chapter 5 of the present text.

► *Journalistic Sources*

All the standard weekly periodicals cover Latin America if there is a newsworthy story. *The Economist* has coverage of Latin American politics and economics on a regular basis. Latin American Newsletters, London, produces regular reports on the region including *Latin American Weekly Report, Caribbean and Central America Report,* and *Mexico and NAFTA Report. The Latin American Times* is published 10 times a year by World Reports Limited, London. *The Caribbean Monthly Bulletin*, published in English and Spanish by the Institute of Caribbean Studies, University of Puerto Rico, discusses many aspects of Caribbean life.

► *Bibliographies*

Brunnschweiler, T. "Geography: Bibliography." In *Latin America and the Caribbean: A Critical Guide to Research Sources,* edited by P. H. Covington. New York: Greenwood Press, 1992, pp. 277–290.

Delorme, R. L., ed. *Latin America, 1983–1987: A Social Science Bibliography.* New York: Greenwood Press, 1988.

Fenton, T. P., and M. J. Heffron. *Latin America and Caribbean: A Directory of Resources.* Maryknoll, N.Y.: Orbis Books, 1986.

Grieb, K. J., ed. *A Bibliography of Latin American Bibliographies.* Metuchen, N.J.: Scarecrow Press, 1968.

Martinson, T. L., ed. *Geographic Research on Latin America; Benchmark 1990; Proceedings of the Conference of Latin Americanist Geographers.* Auburn, Ala.: Conference of Latin Americanist Geographers, 1991.

Martinson, T. L., and S. Brooker-Gross, eds. *Revisiting the Americas: Teaching and Learning the Geography of the Western Hemisphere.* Indiana, Pa: National Council for Geographic Education, 1992.

McNeil, R. A., and B. G. Valk, eds. *Latin American Studies: A Basic Guide to Sources.* Metuchen, N.J.: Scarecrow Press, 1990.

Mikesell, M. W., ed. *Geographers Abroad: Essays on the Problems and Prospects of Research in Foreign Areas.* Chicago: University of Chicago Department of Geography, Research Paper No. 152, 1973.

CHAPTER 2

Physical Environments of Latin America

TOM MARTINSON

INTRODUCTION

The diverse physical environments of Latin America have been perceived and used in many different ways. Amerindians, Europeans, enslaved Africans, travelers, and contemporary inhabitants have all developed images of the region.

CHANGING PERCEPTIONS OF THE ENVIRONMENT

► Environment as God

Native Americans revered the environment. The high cultures of the Aztecs in the Valley of Mexico, the Mayas of the tropical lowlands of the Yucatán, and the Incas of highland Peru all attained their status, at least in part, by successful relationships with the environment.

> . . . in Mesoamerica, cultural development was most likely affected by a world-view carried down from hunting-gathering times. Fate (symbolized by the reed) seems to be a concept central to Mayan thinking. Also, the cyclical view of existence that formed an important part of later religion probably took its lead from nature, where seasons come and go, and life follows death. Nature also suggests a unity, and the idea may have existed, too, that all elements of life are an intricate part of the whole. The bug on the leaf and the passing cloud are brothers, in this sense, parts of the same absolute identity. This view is close to the pantheistic idea that the world is God. (Nelson, 1977, p. 13)

The idea of the environment as God is not dead. Indians in isolated mountain and forest communities are still animists who believe that features in the environment have spiritual significance.

► Environment as Prey

Iberian conquerors saw the Americas as resource rich. The Indians were overcome by European technology, and the land yielded to the European pick and domesticated animals. Resources, wealth, and prestige flowed to the conquistadors and on to Spain and Portugal, firing the spirit of adventure. The Spaniards loved the adventure of the quest and found enterprises like the Portuguese commercial colonies in Asia distasteful. To own land, to achieve eminence, to become noble, and to wield influence were the real reasons why the conquistador wanted gold (Picon-Salas, 1963, p. 33).

To gain wealth by plundering the environment remains a theme in contemporary Latin America. Whether the perpetrator has been soldier, landowner, foreigner, or native, the region's water resources, landforms, soils, and vegetation have suffered over the centuries. Currently the Amazon rain forest is being plundered, and some of the damage is irreparable.

► Environment as Illusion

At the beginning of the age of encounter, the New World appeared as a mirage, too good to be true. As Carl Sauer writes in *The Early Spanish Main*:

> *The beauty of the islands moved Columbus greatly . . . The trade-wind shores he found as tropical nature at its best, and he reveled in praise of their charm and bounty. The perfume of the trees and flowers, he said, was carried out to the ships at sea. The islands were lands of perpetual spring. Birds of many forms and colors sang sweetly in a vast garden of innocent nature . . . (Sauer, 1966, p. 29)*

Christopher Columbus saw the value of Hispaniola, which he thought abounded in spices, gold, and other metals. It possessed mountains of great beauty, extensive plains, and fruitful fields well suited to tillage, pasture, and settlement (Major, 1961).

Of course, hurricanes and other environmental upheavals could disrupt this image of the Americas as a Garden of Eden, revealing some of the difficult conditions of the region. Much of the apparent fertility of the region proved to be illusionary. As the botanist Gardner (1849, p. 9) pointed out, the vegetation was rich, but the underlying soil was poor.

► Environment as Laboratory

During the eighteenth century, Latin America had several universities and was home to a cultural and scientific elite connected with the Enlightenment, or Age of Reason. The environment became a laboratory in which scientists and artists from the Americas and Europe observed and explored nature.

Many distinguished scientists, explorers, and artists visited Latin America in the nineteenth century. For example, the German geographer Alexander von Humboldt (1769–1859) wrote rich descriptions of the places he observed in Mexico and South America during his expedition of 1799–1804 (Humboldt, 1829). His work helped to counteract scientific opinion, current in Europe at the time, that the Americas were inferior to European environments (Gerbi, 1973). Humboldt admired artists, including Johann Moritz Rugendas, and Comte de Clarac. The latter traveled to Latin America to record the region's landscapes and his *Intérieur d'une Forêt du Brésil* was first exhibited in Paris in 1819. Rugendas, trained in reportorial art, was artist to Baron Landsdorff's expedition to Brazil in 1821. Later Rugendas painted in Mexico and elsewhere in Latin America. Prince Maximillian (1820) traveled in Brazil from 1815 to 1817. This is the same Maximillian who explored the Missouri valley in the 1830s accompanied by Karl Bodmer, another artist admired by Humboldt.

Both Darwin (in Moorehead, 1969) and Wallace (1853) worked in the region prior to developing theories of evolution. By mid-century, artists from the United States, including George Catlin and Frederick Edwin Church, who painted the Ecuadorian

Charles Comte de Clarac, *Interieur d'une Forêt du Brésil*. The painting had an impact on Humboldt and Darwin.

volcano *Cotapaxi* repeatedly (Manthorne, 1989), were painting the region. Other distinguished observers include Louis Agassiz (1868) and Sir Richard Burton (1868), both of whom wrote on Brazil. There is, of course, an extensive literature on Latin America in the nineteenth century, written by external observers. Most college libraries have a selection of material, which often provides a good starting point for term papers (Welch and Fijueras, 1982).

► *Environment as Obstacle*

There are difficult physical environments in Latin America. Several writers have gone so far as to suggest that the environment has determined the region's development. For example, Herring wrote:

> The mountains and their high plateaus have largely determined the political, economic, and social patterns of much of Latin America. The Andes' high barriers established the boundaries of Chile and Argentina. The plateaus and inaccessible valleys furnished refuge to aboriginal peoples, delayed their conquest by European invaders, and helped to preserve their ancient ways (Herring, 1961, p. 6)

Most geographers now reject the idea of environmental determinism, preferring to outline the problems posed by the environment and indicate how people have responded to the challenges. In the view of Preston James:

> If a group of people is to remain for a long time in any area, some kind of workable connection must be formed with the earth resources. Not even city

Frederic Edwin Church, *Cotopaxi*, 1855. Cotopaxi (19,347 ft) is one volcano among many in the Andes of Ecuador. Cotopaxi lies just south of the equator and Quito, but even in the tropics there are permanent snow fields at high altitudes.

people are free from this necessity. From the land must come the fundamental needs of human life—food, clothing, and shelter; and from the land, also, must come the material means by which a human community can raise its standard of living above the minimum essentials of existence. To the problem of making a living, however, or of creating wealth from the resources of the earth, the earth itself remains essentially passive and indifferent. (James, 1959, pp 41–42)

MAJOR CLIMATE AND VEGETATION REGIONS

Let us start our review of the physical environment by dividing Latin America into four regions: South America east of the Andes; the Andes; the west coast of South America; and the Caribbean, Central America, and Mexico. An explanation of processes in the physical environment comes later.

► *South America East of the Andes*

A great part of South America east of the Andes lies in tropical latitudes. The equator extends from the mouth of the Amazon across the continent to the Pacific Ocean (Figure 2.1). The equatorial region of the continent of South America is warm and moist, and there is sufficient precipitation to support rain forest, as can be seen by comparing the climate and vegetation regions shown in Figure 2.2 and Figure 2.3.

North and south of the equatorial zone are regions that are still tropical. Frosts are unknown, except in the Andes. Plant growth is not interrupted by low temperatures, but there is a seasonal pattern to the rainfall regime: the summer

Johann Moritz Rugendas (1802–1858), *Distant View of Orizaba, Mexico.*

The Orinoco River in the rainy season. The water level in the river is high. Clouds are forming for the next rain storm. Much of the Orinoco basin is grassland, but along the river bank, conditions are wet enough to support forest.

Figure 2.1   Major climate and vegetation regions, South America.

months are wet, the "winter" months are dry. These tropical wet and dry climates are indicated in Figure 2.2 by the letters Aw in the Köppen system of climatic classification. The tropical wet and dry zone is a land of tropical grasslands: green and lush in the wet season; dry, dusty, and brown in the time of drought.

All of South America north of the equator lies in the tropics. Southward, however, the continent extends beyond the Tropic of Capricorn into subtropical, warm temperate and, in Patagonia and southern Chile, into cool temperate regions. If we were to start southward from the Tropic of Capricorn (23½°S), in the center of the continent, in the Gran Chaco region, we would pass through a semiarid region with drought-resistant *quebracho* trees. The area is far from the oceans and receives less than 20 inches of rainfall a year. Agriculture is possible only in favored areas.

South of the Chaco is the Pampas, a temperate grassland region with a warm summer and a cool winter. Rainfall decreases to the west as distance from the Atlantic Ocean increases and the higher plains fall into the shadow of the Andes, which cut off storms from the Pacific. Comparison of the rainfall data for Buenos Aires (Figure 2.4) and Mendoza (Figure 2.5) reveals the decline in precipitation as we move westward across the Pampas to the slopes of the Andes. The weather stations referred to in this chapter are located in Figure 2.6.

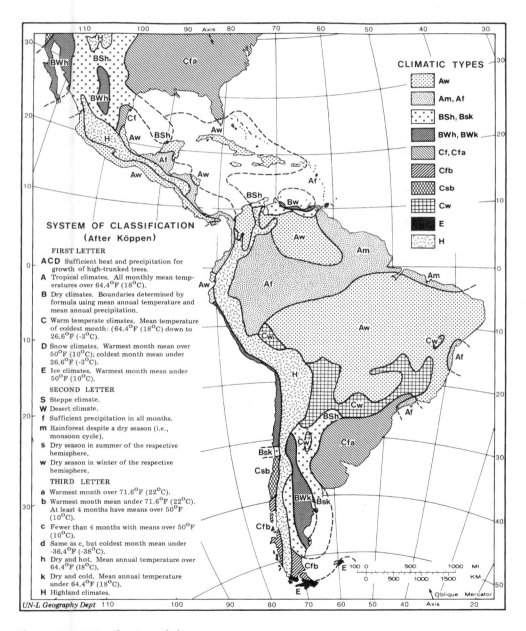

Figure 2.2   Distribution of climates.

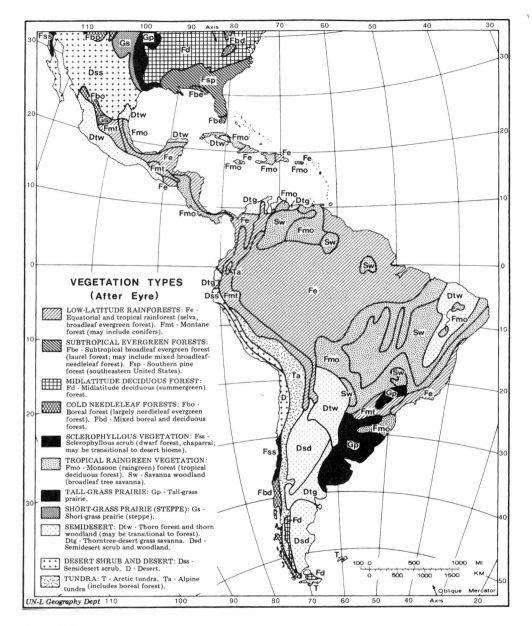

Figure 2.3 Vegetation types.

South of the Pampas is Patagonia, windswept and treeless except in sheltered valleys, a land of grassland and grazing for sheep and cattle, although in the incised valleys that cut into the Patagonian plateau and run west to east from the Andes to the Atlantic, there is arable farming and orchards. The climate data for Rio Gallegos (Figure 2.7), near the southern end of Patagonia, shows that the warmest months have mean temperatures in the low 50s fahrenheit (about 12°C). The rain-shadow effect of the Andes keeps rainfall totals low (around 10"), and agriculture in the sheltered valleys depends on water taken from the rivers.

► *The Andes*

The Andes, like any major mountain range, display altitudinal zonation of climate and vegetation. As we go up in the atmosphere it cools, 3.7°F, on average, for every thousand feet ascended. In tropical areas, there is a well-marked zonation described by Humboldt (Stadel, 1992). Vegetation and crops undergo changes as the mountains rise (Figure 2.8). Within the tropics, places from sea level to just over 3,000 feet (1,000 meters) lie in the *tierra caliente* (hot/warm land); from 3,000 to 6,000 feet within the *tierra templada* (temperate/moderate land); from 6,000 to 12,000 feet within the *tierra fria* (cold land); and, in the highest elevations, *tierra helada* (freezing land) prevails (on Figure 2.8, 1 meter equals 3.281 feet).

South of the tropics, grasslands predominate in the mountains, except on the Pacific coast of southern Chile where the westerly winds, blowing off the ocean, bring in ample rainfall, associated with cyclones, and provide sufficient rainfall to support coniferous forests. There are some environmental similarities between the

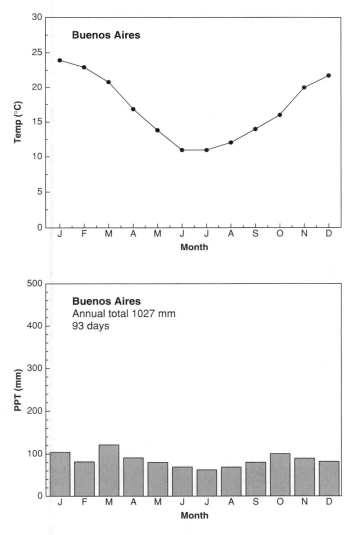

Figure 2.4   Buenos Aires, temperature and rainfall.

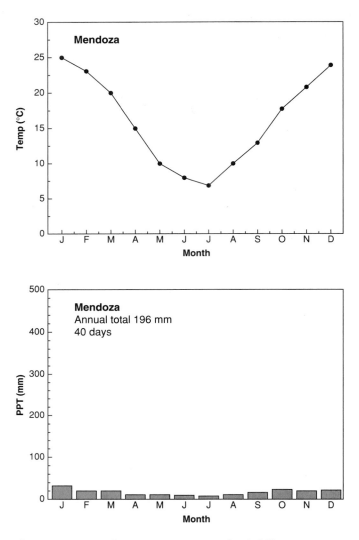

Figure 2.5   Mendoza, temperature and rainfall.

Pacific northwest of North America and southern Chile, but the Chilean region is colder than Oregon or Washington.

► *West Coast of South America*

The climate of the west coast of South America is heavily influenced by the character of the offshore currents and the air masses that lie on the ocean. North of the equator, the Pacific Ocean, on the coast of South America, is a warm, tropical sea. The winds that blow from sea to shore are warm and moist, and when they are forced upward on the western flanks of the Andes, heavy rainfall results. Vegetation is predominantly tropical rainforest.

South of the equator, the west coast climate of the continent is dominated by the north-flowing Peru, or Humboldt, current. The current flows north from Antarctic waters and, together with upwelling currents from the ocean depths, brings cold water into temperate and tropical latitudes.

In southern Chile, temperatures are lower than at comparable latitudes on the west coast of North America. The Central Valley of Chile has a Mediterranean-type climate with dry, sunny summers and a cool, moist winter (Csb on Figure 2.2), but temperatures are not as warm as found in European Mediterranean climates.

North of the Central Valley of Chile, the influence of the westerly winds that bring moisture to southern Chile and the Central Valley in winter is not felt. The sub-tropical high-pressure system that lies over the Pacific at around 30°S (Figure 2.9) is present throughout the year. In this stable air system, the influence of the Peru current is increased. The air above the ocean current is cooled as a result of contact with the cold water. When the cool air comes ashore, it encounters a land surface that is warmer, and the air starts to warm. Warming air absorbs moisture and is unlikely to yield rainfall. Along the coast of South America, from northern Chile, through Peru, into Ecuador and the Gulf of

Figure 2.6 South America, weather stations mentioned.

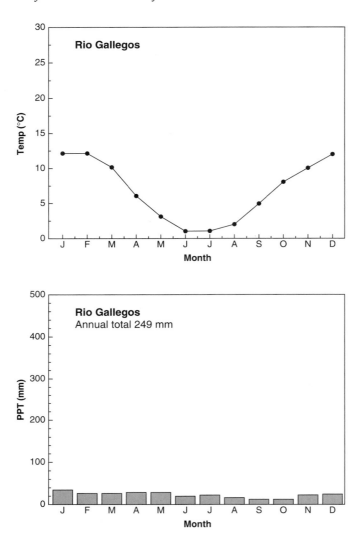

Figure 2.7   Río Gallegos, temperature and rainfall.

Guayaquil, south of the equator, is the Atacama desert. This is an arid land interrupted only by the verdant valleys of streams flowing east to west from the Andes to the Pacific.

The Peru current is rich in fish, particularly anchovies, which support large populations of seabirds and fishing boats. The seabirds deposit their droppings as guano on offshore islands. The fishing boats sweep the sea of shoals of anchovies that are turned into fishmeal and fertilizer. Every few years, the north-flowing Peru current is displaced along the desert coast by south-flowing, warm water from equatorial regions. This warm current, known as *El Niño*, brings storms and flooding to the desert coast. The anchovies leave with the displaced cold water, the seabirds die from lack of food, and the fishing fleets and fertilizer plants close down for lack of fish to catch and process. *El Niño* usually occurs after a rise in sea surface temperatures in the Pacific equatorial zone. The effects of this warming also influence climates in North America and across the world.

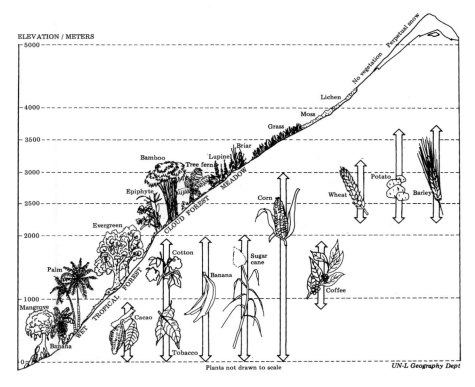

Figure 2.8   Vegetation and altitude.

THE CARIBBEAN, CENTRAL AMERICA, AND MEXICO

In Middle America, the same general principles apply regarding climate. Most of the region, except northern Mexico, is in the tropics, and is proximate to the oceans, although the Sierra Madre Oriental and the Sierra Madre Occidental (Figure 8.1) shield much of central and northern Mexico from Caribbean and Pacific maritime influences (i.e., the mountains shield interior areas from rainfall brought in by winds blowing off the sea). Those regions (mostly coastal) open to maritime tropical air coming off the Caribbean and the Pacific have warm and moist climates. As we ascend the mountains of mainland Middle America, temperatures are lowered and there is altitudinal zonation similar to that described for South America. Mountain ranges have high rainfall on slopes facing prevailing winds coming off the Caribbean. There are rain shadows as the prevailing winds descend and warm on the opposite slope. This is common to many Caribbean islands, such as Jamaica (Figure 10.4).

In Mexico, we leave the tropics north of a line that runs roughly from Cuidad Victoria on the Caribbean side to Mazatlán on the Pacific coast. In winter, cooler northern air penetrates in the form of *nortes* (north wind) that can bring frost. In summer, the pool of hot, dry air that helps create the desert southwest of the United States heats and expands. Summers are hot, drought prevails, the landscape is semi-arid, and agriculture depends upon irrigation.

This sketch of the climates of South and Middle America is greatly simplified. There are many exceptions to the general patterns described, and little has been said in this section about the processes acting in the physical environment. We now introduce these processes.

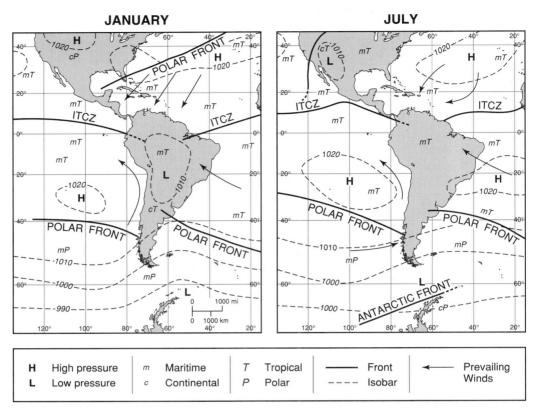

Figure 2.9   Seasonal distribution of air masses.

CLIMATE PROCESSES

▶ *Temperature Characteristics*

Solar energy received at the earth's surface, insolation, fuels the climate machine. Temperature, a measure of solar energy, varies with the nature of earth–sun relationships. The sun produces a constant supply of energy, and the amount captured by the earth varies little, but the concentration of energy changes from season to season because of the variation in the angle at which the sun's rays strike the surface of the earth as it orbits the sun. The rays of the sun are most direct between the Tropic of Cancer (23½°N) and the Tropic of Capricorn (23½°S). At higher latitudes the sun's rays are oblique and diffused over a wider area. Latin American land and water areas that lie within the tropics are subject to more intense heating over the year than locations outside the tropics where marked seasonal temperature changes occur. The atmosphere receives little heat directly from the sun's rays. The atmosphere is heated by contact with the earth's surface, and this heat fuels meteorological processes. The amount and rate of energy transferred to the atmosphere depends on the season, latitude, slope, and aspect, together with the physical nature of the surface material. Over water, heating and cooling occurs more slowly than over land. Air masses overlying water tend to be warmer in winter and cooler in summer than those over land.

Local variations in this broad pattern of temperature distribution result from several factors. Some solar energy, especially that which strikes the high latitudes where snow and ice are found, is reflected. Elevation affects temperature, with high locations having lower temperatures than places at low elevations at the same latitude. An example of the impact of altitude on temperature is Quito (9,300 feet), in the Andes, with an average annual temperature of 54.6°F (Figure 2.10). Guayaquil, on the Pacific coast (at about the same latitude as Quito), has an average annual temperature of 78.2°F, a difference of 23.6°F.

Notice that Quito, on the equator, has little variation of mean monthly temperatures because the receipt of solar radiation varies little. International meteorological statistics are recorded on the metric system. Temperatures are in °C, and precipitation is recorded in millimeters in Figure 2.10. Thirty-two degrees fahrenheit equals zero degrees centigrade. One inch of rainfall equals 25.4 mm.

The air circulation system (Figure 2.9) redistributes heat energy. The major air circulation elements over Latin America are the trade winds that blow toward the equator from the subtropical high-pressure belts at about 30°N and S latitudes, and the westerlies that prevail from about 30°N and S latitudes to the subpolar lows at

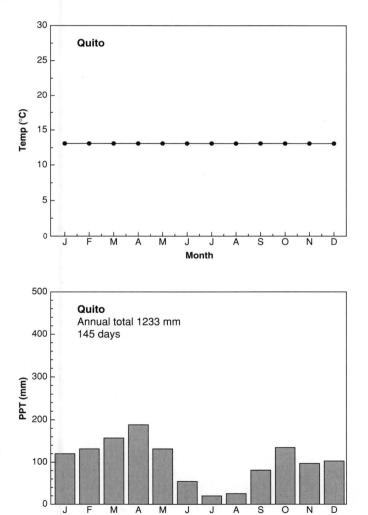

Figure 2.10   Quito, temperature and rainfall.

about 60°N and S latitudes. The trade winds converge on the equatorial region and, in the intertropical convergence zone (ITCZ), unstable air tends to rise and create showers and storms. Weather activity can be intense when "waves," or disturbances, form in the ITCZ. Rainfall totals are high in equatorial regions, particularly where unstable air is raised by passage over mountain chains. When rainfall results from the cooling of air that has been forced upward over higher ground, we speak of orographic lifting and orographic precipitation. Rainfall totals of 200 inches per year are recorded on the eastern flanks of the Andes where the trade winds are forced over higher ground. Orographic uplift causes cooling, condensation, cloud formation, and precipitation.

► *Moisture Characteristics*

The passage of water through the hydrologic cycle reflects the availability of moisture and solar energy. The three stages in the hydrologic cycle are:

1. evaporation;
2. condensation; and
3. precipitation.

1. Evaporation occurs if there is sufficient energy to change water to a gas: water vapor. In high latitudes and high altitudes there is little precipitation and little evaporation because less energy is available. At low elevations in the tropics, precipitation and average temperatures are higher, and evaporation is maximized. In southern South America, the air over Patagonia is dry compared to the high relative humidities experienced in the Amazon Basin. Once evaporated, water vapor can be carried long distances by the prevailing winds. The tropical eastern flanks of the Andes owe their high rainfall totals to the importation, by the trade winds, of moist air from warm Atlantic ocean surfaces.

2. Condensation, the second stage in the hydrologic cycle, is the precursor of precipitation. Condensation results from the cooling of moist air by convective, cyclonic, or orographic lifting. The water vapor changes to a liquid when air is cooled by lifting to the dew point, the temperature at which the relative humidity is 100 percent.

3. Precipitation takes many forms and includes rain, snow, hail, dew, and mist. Rainfall may result from convective, cyclonic, and orographic processes. The daily convective showers that characterize many tropical lowlands result from local heating, uplift, and adiabatic cooling of air. Orographic rainfall results when moist air is forced upward over higher ground, cooling and condensation result in precipitation. Cyclonic precipitation, resulting from the uplift of warm, moist air along a boundary with cooler, drier air, occurs in the Latin American middle latitudes that come under the seasonal influence of the polar front, where polar air meets tropical air. Parts of Mexico are subject to spring and autumn showers because of invasions of northern polar air into buoyant tropical air. Frontal storms, on the boundary between polar and tropical air, may travel hundreds of miles into Mexico, bringing rainfall. As the front passes, cold northerly winds (*nortes*) arrive, accompanied by temperature drops of 20°F. *Nortes* penetrate as far south as Costa Rica or cross the Caribbean and invade Cuba. The cold air may bring frost and damage citrus and coffee crops.

Cold polar air penetrates from the south in the Southern Hemisphere winter, and temperatures drop markedly in Patagonia, as can be seen in the Río Gallegos temperature statistics (Figure 2.7). The temperature drop is not so marked by the time polar air reaches the Pampas of Argentina, but the region is invaded by cold winds from the south accompanied by squalls (*Pamperos*) (Figure 2.4). After the squalls pass, the weather cools as temperatures drop beneath clear skies. The polar air from the south can penetrate into the tropics and equatorial regions. In the 1850s the British pioneer geographer and naturalist Henry Bates (1825–1892) lived on the Rio Tefé, a southern tributary of the Amazon, and recorded the impact of the air of polar origin in the rain-forest region as temperatures dropped into the low 60s. He wrote:

> *There is, in the month of May, a short season of very cold weather, a most surprising circumstance in this otherwise uniformly sweltering climate. This is caused by the . . . cold wind, which blows from the south over the humid forests . . . the temperature is so much lowered that fishes die in the river . . . the wind is not strong; but it brings cloudy weather, and lasts from three to five or six days in each year . . . the period during which this wind prevails is called the* tempo da friagen, *or the season of coldness. (Bates, 1930, pp. 293–294)*

When the modified polar air invades the more southerly, subtropical region around São Paulo, it can cause frost and damage the coffee crop.

Dry lands are found in continental interiors far from maritime sources of moisture and where subtropical high pressure prevails, about 30°N and S latitudes. The Sonora Desert of Mexico and the Atacama Desert of northern Chile and coastal Peru are at least partially caused by subtropical high pressure systems. The seasonally dry *chaco* of interior Argentina–Bolivia–Paraguay results from its position far from the oceans. The Atacama Desert, one of the driest in the world, owes aridity to a combination of factors including subsiding air within a stable, subtropical, high-pressure system centered offshore and the upwelling of cold water associated with the north-flowing Peru Current. The cold water causes fogs and low stratus clouds, but the air is stable at low elevations and rainfall is rare. These conditions are reflected in the temperature and rainfall graph of Lima (Figure 2.11). Lima and its port of Callao are infamous for their thick drizzling mists that dampen clothes but rarely lead to rain.

An anomalous arid region exists on the Caribbean shore of Venezuela and Colombia. Here descending, stable air that is warming and tending to absorb moisture is a cause of aridity. The annual rainfall of La Guaira, on the Caribbean shore of Venezuela, is 11 inches, and rainfall is surprisingly low on the Gulf of Maracaibo (Figure 2.12).

The thorn scrub *caatinga* zone of northeast Brazil (Dtw on Figure 2.3) suffers from periodic droughts that have caused out-migration of farmers to other parts of the country. The Brazilian author Euclides da Cunha, in *Rebellion in the Backlands*, has written vividly of the situation, as has Graciliano Ramos in *Vidas Sêcas*, available in English translation as *Barren Lives* (Ramos, 1992).

► *The Distribution of Climates*

The distribution of climates can be described in several ways. We can talk about the air masses that prevail at different seasons of the year and the manner in which they control climate, or we might try to define climate regions using various criteria.

Four major air masses control temperature and moisture conditions in the region.

1. Cold, dry continental polar (cP) air masses that originate over the continental high-pressure cells of the polar latitudes, that is, Antarctica and the Canadian north.

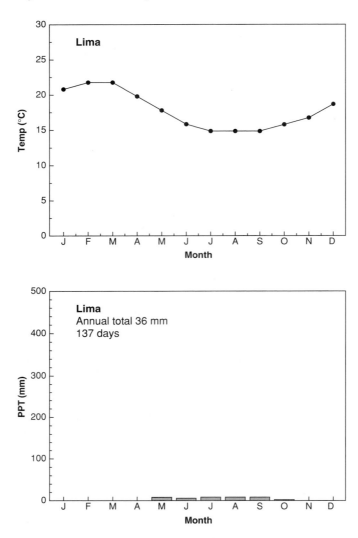

Figure 2.11   Lima, temperature and rainfall.

2. Warm, moist maritime tropical (mT) air masses, formed over tropical and subtropical seas.

3. Hot, dry continental tropical (cT) air masses that develop over dry land-masses in the subtropics.

4. Cool, moist maritime polar (mP) air masses that form over the oceans of arctic and subarctic regions.

The three most common air masses over Latin America are 2, 3, and 4; the warm moist (mT), hot dry (cT), and cool moist (mP). Figure 2.9 shows the seasonal distribution of air masses. In January—the Northern Hemisphere winter—the polar front can penetrate to the Caribbean, but maritime tropical air still controls temperature and moisture conditions over the Caribbean area and the center of the South American continent. In the south, the cool, moist, maritime polar air of

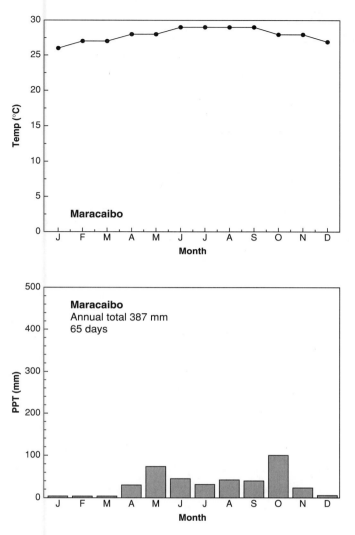

Figure 2.12   Maracaibo, temperature and rainfall.

the South Atlantic meets the hot, dry (cT) air of central Argentina along a polar front that is well south in the Southern Hemisphere summer.

In July—the Southern Hemisphere winter—the polar front moves north of the Río de la Plata. In North America, the polar front—the leading edge of cool, Canadian air—has retreated off the map. The whole Caribbean region and much of tropical South America are bathed in maritime tropical (mT) air. On Figure 2.9, notice the seasonal variation in the position of the intertropical front, or ITCZ. In the Northern Hemisphere summer, it moves north, bringing with it a zone of clouds and rainfall and creating a rainy season in the Caribbean and the Orinoco basin. In the northern winter, it shifts south, bringing a rainy season to the Aw climate zone (Figure 2.2) of Brazil, south of the equator, as demonstrated at Cuiabá (Figure 2.13).

Figure 2.2 displays the climate types of Latin America after a system devised by the German climatologist Wladimir Köppen (1846–1940). Köppen's system was based upon average data, and averages conceal weather surprises. The characteristics of each of Köppen's climate types are described in the key to Figure 2.2.

The lowlands at the mouth of the Amazon, lying in Köppen's Am zone (Figure 2.2), are warm all year, but rainfall varies considerably from month to month, as can been seen in the data for Belém (Figure 2.14). Mexican border towns expect hot, dry cT air in July and usually cool, dry conditions in January. In the Argentine west, Mendoza experiences warm, dry conditions associated with cT air in January and cool, moist maritime polar air in July (Figure 2.5).

Santiago (Figure 2.15), in the Central Valley of Chile, has a Mediterranean-type climate, with warm, dry summers, and cool, moist winters. These conditions illustrate the role of air masses in influencing climate. During the Southern Hemisphere summer, the Central Valley is dominated by continental tropical air associated with the subtropical high that helps create the Atacama Desert of Northern Chile. In the Southern Hemisphere winter, the subtropical high shrinks in

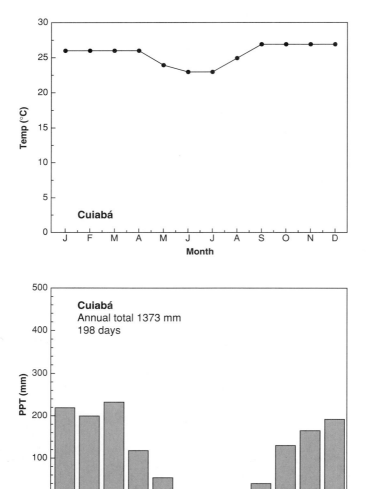

Figure 2.13   Cuibá, temperature and rainfall.

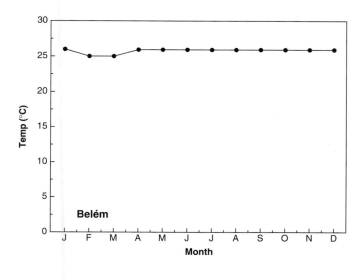

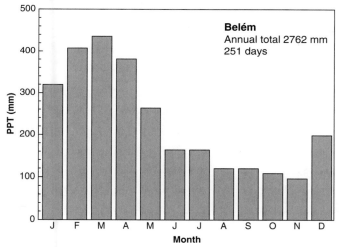

Figure 2.14 Belém, temperature and rainfall.

size, and the Central Valley comes under the influence of the westerlies and maritime polar air that gives rain from May through September. Offshore of Chile is the cool, north-flowing Peru Current, which keeps temperatures at a lower level than are found in a true Mediterranean-type climate.

WATER

Some of the world's lushest rain forests and most arid deserts are found in Latin America. These environments are reflections of water surplus and water deficiency extremes on the continuum of water supply in the region.

► *Rainfall*

In regions where convectional, cyclonic, and orographic precipitation occur, rainfall totals are high, as on the Andean rim of the Amazon Basin. Some generalizations can be made about the distribution of precipitation in Latin America.

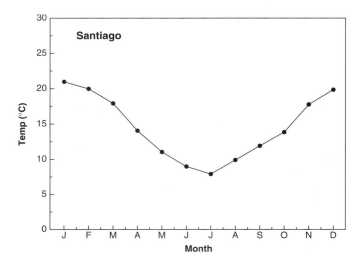

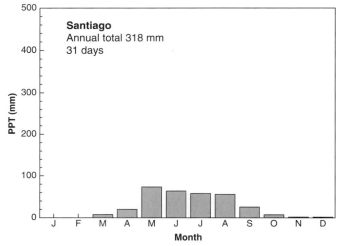

Figure 2.15   Santiago, temperature and rainfall.

   1. There is an equatorial zone of maximum precipitation, related to the convergence zone of the trade winds, the ITCZ.

   2. North and south of the equatorial zone are regions of rainfall seasonality in which there is a wet season in the summer and a dry season in the "winter" months.

   3. Storms in the Latin American middle latitudes are related to incursions of relatively cold air into areas usually dominated by warm, moist air.

   4. Snow is commonly confined to the high latitudes and high altitudes.

   5. The dry zones of Latin America are found in areas dominated by high pressure and on the leeward sides of mountains barriers. Examples of dry zones include coastal Peru, northern coastal Venezuela, northeast Brazil, and the Balsas depression of Mexico.

   6. Orographic precipitation results when adiabatic cooling is induced by the rise of warm, moist air over higher ground. Mountain ranges in Latin America have orographic precipitation belts on their windward slopes. Orographic processes cause heavy precipitation in valleys that face toward the northeast trades blowing off the Caribbean.

SURFACE WATER

On reaching the ground, precipitation is redistributed by runoff, percolation, and evaporation. Alternatively, it may be stored for short terms as ice and snow or for long terms in chemical combinations with other substances.

Where rainfall is plentiful, water occurs on the landscape as lakes, streams, and rivers. This surface water accomplishes most of the erosion in Latin America. Controlled surface runoff that remains in established channels or is impounded poses no problems for human activities, but uncontrolled runoff that results from deforestation or urban waterproofing can threaten agriculture and settlement.

The water that percolates into the subsurface may accomplish solution and remove materials. In limestone areas, the subsurface rock dissolves, and subsurface solution creates sink holes down which streams disappear into underground caverns. Sink holes, called *cenotes* in the Mexican Yucatán Peninsula, contain archeological sites that hold Mayan relics.

Erosion by the runoff of surface water can be serious where rainfall is intense or prolonged. Farmers have evolved agricultural practices to deal with the threat of water erosion. Terracing (or *tablón* agriculture as it is known in Guatemala) reduces topsoil loss and permits irrigation during a dry season. The Inca terraces of Peru were built to facilitate irrigation and to offer a flat surface for planting crops.

Many present-day agricultural practices depend on the elimination of surplus water from the soil. Drainage ditches are common in tropical lowlands, especially where commercial agriculture is practiced. The ancient ridged fields of the Amazon, Orinoco basin, and Middle America were attempts to solve the problem of excess soil moisture.

In comparison with other large continental masses, not a great deal of water is retained on land surfaces in Latin America. The South American landmass tapers as it approaches the South Pole. There is no great "continental" influence. The possibilities for snow accumulation, glacier development, and ice formation are small when compared with North America. The high altitudes of Latin America offer some opportunities for snow accumulation and glacier growth, as the glaciers of southern Chile illustrate.

▶ River Systems

In areas dominated by continental tropical (cT) air masses, such as interior northern Mexico, permanent streams are often absent because evaporation exceeds precipitation. The only major rivers originate in wetter places and pass through deserts on their way to the ocean. An example of an exotic stream is the Río Bravo del Norte (Rio Grande in the United States). The river provides irrigation water for the production of fruit and vegetables grown for United States markets.

In high-altitude locations stream flow may be limited by the winter freezeup. Summer snowmelt produces runoff. Andean summer meltwater benefits agriculture in Peruvian coastal valleys and the Central Valley of Chile, where irrigation is necessary to produce crops.

Regions under maritime polar (mP) air mass conditions throughout the year receive rainfall in all seasons. The river systems of southern Chile and Patagonia flow throughout the year. The mP regions of Latin America are small compared with the mP zones in counterpart latitudes in the Northern Hemisphere that foster rivers such as the Mississippi or the Rhine.

Hot, wet conditions in the maritime tropical (mT) air mass regions promote extreme river development. Large annual rainfall totals generate surface waters that are discharged into major river systems such as the Amazon. Henry Bates wrote in 1863:

> *We beheld . . . a sea-like expanse of water, where the Madeira, the greatest tributary of the Amazons, after 2000 miles of course, blends its waters with those of the king of rivers. I was hardly prepared for a junction of waters on so vast a scale as this, now nearly 900 miles from the sea. Whilst travelling week after week along the somewhat monotonous stream, often hemmed in between islands, and becoming thoroughly familiar with it, my sense of the magnitude of this vast water system had become gradually deadened; but this noble sight renewed my feelings of wonder. (Bates, 1930, pp. 162–163)*

The northern tributaries of the Amazon contribute the greatest volume of water in the Northern Hemisphere summer when the ITCZ lies north of the equator and over the northern headwaters of the Amazon. The tributaries entering from the south have maximum flow at the opposite time of year, in the Southern Hemisphere summer.

Water levels on the Amazon and its tributaries vary on a seasonal basis. At Manaus, which is on the Rio Negro close to the confluence with the Amazon, a floating dock allows ships to tie up although the level of the Rio Negro varies by over 40 feet during the course of the year.

Water levels in the Orinoco River are seasonal. The Orinoco basin is far enough north to lie in the tropical wet and dry zone—Köppen's Aw climatic region, which has no monthly mean temperature below 64.4°F (Köppen's definition of a

Tropical rain forest. For miles, the large line of a pipeline construction crosses the tropical rain forest in the Amazonas region. The trees were pulled down by bulldozers.

tropical climate) and a "winter" dry season (Figure 2.2). In the Northern Hemisphere summer, as the ITCZ migrates north, there is a rainy season, and rivers in the Orinoco system overflow their banks. When the ITCZ moves south in the "winter" months, there is a dry season, stream levels fall, and the landscape becomes dry and parched.

▶ Ocean Water

Oceans influence the distribution of climate by acting as heat and moisture reservoirs. As storage fields for heat they contribute energy to the Caribbean hurricanes, as attested in the deadly hurricane season of 1995, when more than 20 hurricanes and tropical storms passed through the Caribbean Sea.

Hurricanes and tropical storms form off the coast of West Africa before traveling across the Atlantic and tracking into the Caribbean. The storms strengthen on their journey. Hurricanes are slow-moving storms formed around an "eye" of very low pressure. Within the "eye," low pressure pulls up the surface of the sea so that, as a hurricane comes on shore, it brings with it a wall of water in addition to winds in excess of 74 m.p.h. and heavy rainfall. The economies of small Caribbean islands can be heavily damaged by the passage of a tropical storm. In 1831, the sugar economy of Barbados was brought to a halt by a major hurricane that destroyed many plantation houses and sugar mills. Barbados possesses few seventeenth- and eighteenth-century buildings as a result of the storm. In 1995, many homes and hotels were destroyed in the Caribbean hurricane season, affecting power and utilities for months.

There are some rich fisheries in the oceans adjacent to Latin America. For example, the cold Peru Current provides a hospitable environment for plankton and the fish that feed on them. Large fish catches, especially of anchovies, enter world trade from Chile and Peru as fish meal to feed chickens in the United States and Europe.

Water provides the opportunity for trade, and much local, national, and international commerce passes over water routes. Landlocked countries such as Bolivia and Paraguay are disadvantaged by poor access to international commerce. El Salvador does not have a Caribbean shore and cannot participate directly in the sea trade of the Gulf of Mexico.

▶ Human Intervention in the Hydrologic Cycle

Deforestation and the seasonal burning of tropical grasslands alter the hydrologic cycle. These human activities result in a loss of plant surfaces for evapotranspiration and contribute to the supply of nuclei for condensation. The burning of the rain forest, which approaches conflagration proportions in the pioneer zones on the eastern flanks of the Andes and along the Amazon penetration roads in Brazil, is depleting one of the world's few remaining stands of virgin timber.

Vegetation clearing reduces the water storage capacity of the soil, resulting in quick and sometimes destructive runoff. Urbanization changes the temperature and moisture characteristics of local areas by introducing new heat sources and encouraging precipitation with the accumulation of dust and other forms of pollution that act as hygroscopic nuclei. In addition, the weatherproofing of cities retards absorption of water by the soil and reduces cooling by evaporation because the rainfall is quickly directed away through drainage channels.

Some large cities have outgrown local supplies of water. Mexico City, located by the Aztecs on an island in Lake Texcoco, grew to occupy a larger area, and the lake was drained. Removal of the groundwater for urban uses as the Spanish city grew caused shrinking of the underlying soil. The result in contemporary times has been an unstable base to build on and dependence on distant water supplies.

LANDFORMS

► *Tectonic Forces*

The earth's crust consists of plates that float on a plastic layer. The major plates that influence South and Middle America are outlined in Figure 2.16. More detail

Figure 2.16   Plates, trenches, and fracture zones.

is provided in Figure 2.17, which displays the major structural regions of Latin America and the fracture zones that mark the margins of the plates.

The core of the continent—the South American plate—is represented in Figure 2.17 by the Gondwana shield areas of the Guiana highlands, the Brazilian highlands, and Patagonia. Gondwanaland was an ancestral continent. The South American plate broke off from Africa and drifted westward.

Along the margins of plates, earthquake activity is caused as the plates move against each other. The Middle America Trench and the Peru-Chile Trench are associated with subduction zones in which the margins of plates are pushed down and

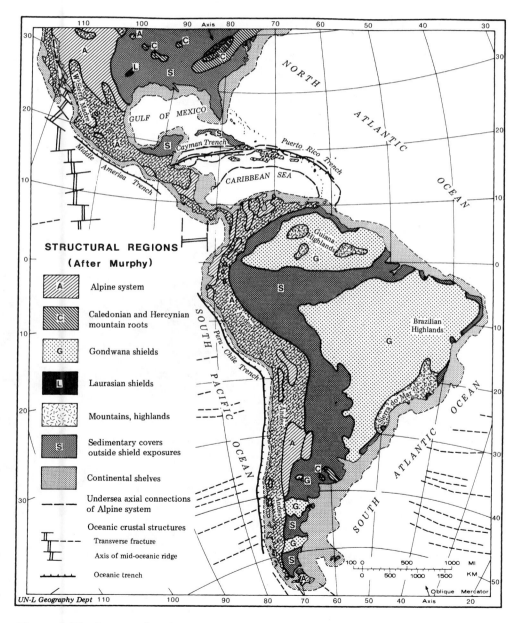

Figure 2.17 Structural regions.

there is volcanic activity on the landward side. A chain of volcanoes is found on the West Coast of Latin America from Mexico and Central America into the Andes. The area is particularly prone to seismic activity and destructive earthquakes.

Volcanic activity is a persistent theme in Middle American and Andean American history. For example, in February 1943, a fissure opened in a field near the Mexican village of Paricutín, Michoacán, through which a stream of molten lava poured. Volcanic ash, thrown over the countryside, killed crops and trees, and made the land useless for cultivation. Eventually, the ash will become fertile soil, but meanwhile the population makes a living guiding tourists who come to observe the solidified lava flows through which partially buried buildings protrude (Simpson, 1961, pp. 1–2).

In 1902, Mt. Pelée exploded on Martinique. The eruption was preceded by ashfalls and a flood of boiling mud. On May 28, a black cloud of superheated steam and dust issued from the volcano, obliterating the city of St. Pierre, and killing 30,000 inhabitants within a few seconds. Ships in the harbor capsized or burned to the waterline. Only two inhabitants of the city survived, one as a prisoner in an underground cell. Mt. Pelée continued erupting for several months, finally stopping on August 30. Destroyed were St. Pierre, the "Paris of the West Indies," and five other towns on Martinique.

The largest volcanic disaster since Mt. Pelée was the November 13, 1985, eruption of Nevado del Ruiz in west-central Colombia near Manizales. Large explosions dislodged a lake, melted a glacier, and produced enough volcanic debris to inundate several towns in the vicinity. At least 25,000 lives were lost.

Enormous destruction and loss of life occurred on September 19, 1985, when a major earthquake, 7.8 on the Richter scale, struck Mexico City. It was followed

Earthquake destruction, Mexico City, 1985. In the foreground, the floors of buildings have collapsed one upon another, crushing people to death. Some buildings have not failed structurally. What factors might account for the differing incidence of damage?

by another, 36 hours later, which measured 6.5. The quakes caused the deaths of at least 5,000 people (some say 10,000). It is estimated that 100,000 homes and 200 public buildings were destroyed. Such devastation may occur again, because of Mexico City's location near major fault zones and its site atop an ancient lakebed. Human actions such as erecting buildings and concentrating populations near active faults increase the potential for destruction.

► *Weathering, Erosion, and Climate*

In areas dominated by cold, dry cP air most of the year, as in southernmost South America, low temperatures in winter cause ice and snow accumulation. Mechanical weathering by freezing and thawing and ice erosion prevail. In summer, temperatures may increase enough to allow chemical weathering and erosion by water.

In areas in mP air for much of the time, chemical weathering and water erosion are experienced the year round. Temperatures are cool, and there is enough rainfall to produce extensive river systems.

The deserts of Latin America are found in areas dominated by continental tropical, cT, air masses. There is little or no rain, and in the arid climate, the wind picks up sand. The blowing sand is an agent of erosion that sculpts landforms.

In regions that lie under maritime tropical (mT) air for most of the year, temperatures are warm, the air humid, and rainfall heavy, at least in some months of the year. In this climate, chemical weathering predominates, creating a deep waste mantle. Soils are often saturated and mass movement is common.

Landforms are not solely the product of climatic influences. Underlying geology plays a role, and in zones of recent crustal movement, stream profiles are steeper, valleys are V-shaped, and faults may create steep slopes (Figure 2.17).

SOIL

Soil is related to climate, water resources, landforms, and vegetation in the environmental complex (Figure 2.18). Human beings have altered soil processes as they have interfered with other physical processes in the region.

Both organic and mineral soils are found in Latin America, but mineral soils are better known, cover a wider area, and are more important agriculturally than organic soils. Mineral soils contain only about 5 percent organic material.

Mineral soils in Latin America may be analyzed in terms of their composition and properties as they have developed in response to environmental controls, especially climate. Soil develops slowly on sloping land where water erosion is prominent. Erosion is accelerated where the natural vegetation cover is removed, as in cultivation. Serious erosion occurs on arable land that has been planted to row crops without considering slope.

Climate exerts a dominant influence on the development of organic material in the soil. In general, as temperature increases, the amount of organic soil material decreases. Increased organic material leads to higher soil fertility and greater crop productivity. The high humus content of the loess soils of Argentina's humid Pampas, for example, is responsible for their fertility.

Although climate is the most important control of soil development, other factors are important. For example, in a young soil that has not lain in place long enough to be radically changed by exposure to the elements of climate, properties are determined by parent material.

Soils developed from parent materials newly transported by water, ice, or wind have not lain in place long enough to be severely leached, that is, to have water flowing through and removing minerals. The floodplains of major streams in

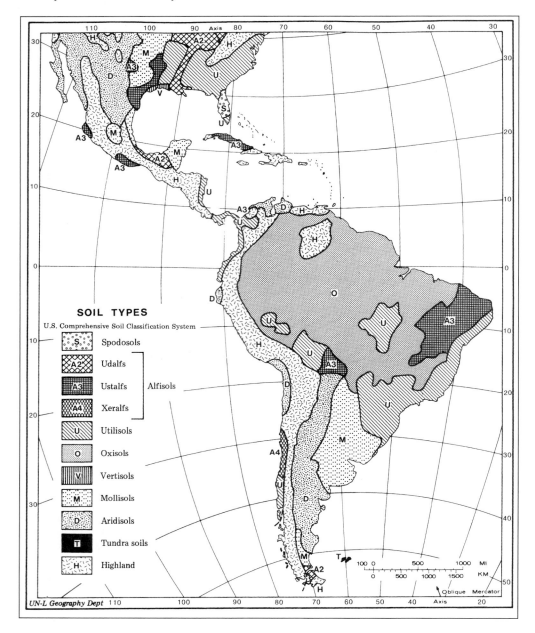

Figure 2.18 Soil types.

Latin America are prized for agriculture because soils are "new" and contain more plant nutrients. Seasonal flooding may deposit organic matter and minerals. The banks of the Amazon and its tributaries are more densely populated than interior locations due to the ease of transportation, the availability of fish, and soils, that, if properly drained, are more fertile and productive than leached soils on interfluves.

Soil properties are influenced by vegetation. As plants grow and die, they are incorporated into the soil's organic material or humus. Where vegetation is sparse, as in arid regions, little humus accumulates. The wet lowland tropics of Latin

America, when cleared of forest, have humus-poor soils, because high temperatures lead to rapid decomposition of organic debris. Where environmental conditions are moderate, as on the pampas, humus and nutrients accumulate, and more fertile soils result.

► *Soil Orders*

Mineral soils develop common properties under similar controls and may be grouped into soil orders. There are three major soil orders: zonal, intrazonal, and azonal.

Zonal soils occur over a broad area and owe their characteristics to the climate in which they develop. The soils have recognizable layers, the A, B, and C horizons, that form the soil profile. The layers may be distinguished on the basis of color or constituent material. The top, or A, horizon, for example, may be darker in color than the lower horizons because more humus has accumulated. The B horizon may be lighter due to a lower humus content and deposits of easily dissolved minerals (often lighter-colored minerals such as calcium carbonate) that have been leached from the A horizon. The A and B horizons, constituting the *solum*, overlie the C horizon, the parent material.

Zonal soils in Latin America result from at least four processes: gleization, podzolization, laterization, and calcification. One of these processes dominates each environmental realm.

Gleization, the accumulation of organic materials in the top layers of soil, occurs primarily in the cP realm. Here, under cold, dry conditions, the soils tend to be shallow, acidic, and humus-rich. Gleization is the soil-forming process that dominates in the southern reaches of South America as well as in high elevations. The resulting soils are not particularly fertile; this, together with the climatic extremes of the areas in which they develop, precludes intensive agricultural use.

Podzolization, the accumulation of silica in the top layers of the soil, is dominant in the cool, wet mP realm. Cool temperatures slow biochemical activity, and there are concentrations of organic matter in middle-latitude mixed forests. Over time, seasonal rains percolate through the organic debris of the forest, forming weak acids that leach out the easily dissolved minerals and leave gray silicon oxides. The soils are fertile and agriculturally productive.

Laterization, the accumulation of iron and aluminum oxides in the top layers of the soil, occurs in the hot, maritime tropical climates. Organic debris decays through bacterial action, allowing rapid absorption of nutrients by plants. This leads to the accumulation of nutrients in the vegetation instead of the soil, and under the hot, wet conditions, nutrients not absorbed by plants are carried away by the large volumes of water. The end product is an infertile soil covered by lush vegetation. If the forest is cleared, agricultural yields are poor, except on floodplains, because few nutrients are available in the soil.

Calcification, the accumulation of carbonates in the top layers of the soil, is found in the hot, dry cT realm. The absence of water and reduced leaching produce characteristically whitish soils rich in dissolved salts and low in humus. The white color results from evaporation at the surface and the deposition of minerals there. Irrigation allows these soils to be productive, as in the oases of northwestern Mexico.

Intrazonal soils are different from zonal soils in that they develop horizons under the influence of local climate, water conditions, or parent material. They form pockets of distinctive soil within more widely distributed zonal soil orders. They take on local characteristics because of their environments, becoming hydromorphic if waterlogged, halomorphic if subjected to salt concentrations, or calcimorphic on calcium-rich parent materials.

Azonal soils, like intrazonal types, are limited in extent. They lack well-developed profiles because of limited time or specialized topographic conditions. Lithosols are surface deposits lacking soil horizons. Alluvial soils and aeolian soils are of agricultural significance on river floodplains and in the Pampas, respectively, because they are young and contain more plant nutrients.

► Acculturated Soils

Acculturated soils are soils altered by human action. Agriculture influences soil development, even when traditional techniques of cultivation are employed. For example, fire is often used to clear fields for shifting cultivation, and heat may bake the top layer of the soil, changing its physical structure and eliminating microorganisms. Replacement of natural vegetation by cultivated crops may draw plant nutrients from the soil more quickly and deplete fertility. Consequently, farmers must move to new plots as yields decline. Although shifting cultivation can be an environmentally sensitive method of farming in the tropics, population pressures now force agriculturalists to use more land, leaving less time for fallowing. The conflict between environmentally acceptable and economically desirable uses of the soil will grow.

VEGETATION ASSOCIATIONS

Vegetation is an indicator of environmental relationships and human activities. Natural vegetation mirrors the distribution of climate, water resources, landforms, and soils and is the first factor of the environment influenced by human activities (Figure 2.3).

► Environmental Influences

The most important environmental factor influencing the distribution of natural vegetation is climate, as reflected in temperature and moisture conditions. Solar energy is the vital force behind plant growth because it supplies the power for photosynthesis. Biochemical activity in plants increases with higher temperature when moisture is adequate. As a result of variations in temperature and moisture conditions there are several recognizable natural vegetation associations.

Specially adapted plants occupy each environmental realm. Hydrophytes (water plants) are common in lakes, swamps, and marshes. Xerophytes (desert plants) populate arid regions. Usually a great number and variety of plants comprise a vegetation association, so the association is named after the dominant vegetation. For example, trees are the dominant plants in a forest association even though there may be many other plant types there, such as scrubs, grasses, and epiphytes.

Forest Associations

On the basis of temperature and moisture conditions, four principal types of forest can be distinguished: broadleaf evergreen forest, broadleaf deciduous forest, needleleaf evergreen forest, and needleleaf deciduous forest.

The broadleaf evergreen forest, or tropical rain forest, is luxuriant. It is found in tropical lands where mean monthly temperatures exceed 64.4°F., rainfall is plentiful, and there is no pronounced dry period, allowing growth to occur without interruption.

The structure of the tropical rain forest may consist of at least three layers of forest trees and several other plant components. At the top is a discontinuous layer of crowns as high as 90 to 120 feet. The second set of crowns, at 50 to 90 feet, tends to be a continuous canopy, shading the lower vegetation. The lowest set is much shorter, between 15 and 50 feet in height, and less dense. Thousands of different tree species may be found over limited areas. When sufficient light penetrates the canopy, herbs, vines, saprophytes, epiphytes, and parasites are found.

Observers comment on several characteristics of tropical rain forest, including the gloom caused by the canopy of foliage shutting out the sun, the quiet, the accumulation of rotting vegetation on the forest floor, the absence of ground flora resulting from a lack of sunlight, and the large number of lianas climbing the high trees to the sun (Hudson, 1904; Tomlinson, 1912; Wallace, 1853). If the forest is disturbed or cleared, other forms of vegetation appear as sunlight reaches what was the dark forest floor.

Forests at the tropical margins and in the middle latitudes are dominated by broadleaf deciduous or needleleaf evergreen trees. Where rainfall is seasonal or temperatures are lower or change with the seasons, deciduous trees are common. Needleleaf trees survive in places too cold or dry for deciduous trees; their thick, needlelike leaves reduce moisture loss by transpiration, their thick bark and sugar-rich sap resist the cold, a shallow root system enables them to grow in thin soils, and they are tolerant of highly acid conditions. The middle-latitude forest has a single story and little undergrowth because of lower temperature and moisture levels.

Grassland Associations

Grasslands are composed of perennials and annuals that vary in height from a few inches to several feet, depending on climate and soil. Trees are usually confined to watercourses.

Tropical grassland or savanna such as the Llanos of Venezuela and Colombia is found at the margins of the forest. Some observers believe that Llanos has been formed by repeated burning by people or is due to local soil conditions that promote rapid percolation. However, the dry season is pronounced, and if trees did grow widely in the Llanos, the species mix would have been different from that found in the tropical rain forest of Amazonia.

In temperate lands, or at higher elevations, grassland prevails, ranging from tall prairie to short steppe. The Pampas of Argentina was originally covered with tall prairie grass. Meadow grass is found above the tree line on high mountain slopes or in the high basins of the central Andes in Colombia, Ecuador, Peru, and Bolivia.

Desert Associations

As dryness increases, grassland gives way to desert. The vegetation of both cool deserts and warm deserts is sparse and specialized. All plants are xerophytic—adapted to a limited water supply. The qualities ensuring the survival of desert plants include a short growth period among the perennials, a quick life cycle among the annuals, the ability to store moisture over extended periods of time, small size, special root structure (extending to deep water sources or spreading out over an extensive area near the surface), and seeds designed to withstand prolonged drought. In spite of these adaptations, vegetation is nearly absent in the Atacama, one of the driest deserts in the world. The northern Mexican deserts seem lush by comparison, but rainfall is seasonal and must be supplemented by irrigation if the deserts are to be farmed.

► *The Distribution of Vegetation*

Natural vegetation mirrors other environmental factors. Tropical rain forests are found in maritime tropical climates where temperature averages and moisture totals are high. Associated with the rain forests are extensive rivers, chemical weathering, and transportation of soil nutrients.

In the deserts associated with continental tropical air are xerophytic plants. Physical weathering is dominant, and soils tend to be calcified. Physical weathering, wind erosion, and occasional water erosion after storms produce angular landforms.

Needleleaf evergreen and broadleaf deciduous trees predominate in the mP environmental realms. Adequate moisture and cool temperatures encourage chemical weathering, and podzolization processes produce spodosol soils associated with forest conditions.

In the cP environmental realms are some hardy trees, mountain grasses, and low, xerophytic plants. Physical weathering is dominant in this realm, and gleization processes have produced thin soils.

Temperate grasslands occupy areas between the cold desert of the cP environment and the needleleaf evergreen–broadleaf deciduous forest of the mP environmental realm. The altitudinal counterpart of the tundra is meadow grass vegetation (*páramo*), found above the tree line. In the middle latitudes prairie grasses form the boundary zone between the tropical rain forests and the subtropical deserts.

SUMMARY

Over the course of space and time people have interacted with the environment to project at least five images: environment as God; environment as prey; environment as illusion; environment as laboratory; and environment as obstacle. There are still many people in Latin America whose relationships with the environment are based on ritualized reverence, and many views of the region are illusions based on incomplete evidence or biased perception. Today there is an adversarial relationship between people and environment. This is apparent in the destruction of the Amazon rain forest (Caufield, 1991). Technology and increasing population menace the environmental base of Latin American life and may threaten world environmental stability.

FURTHER READINGS

Bates, H. W. *The Naturalist on the Amazons.* New York: Dutton, 1930. (Originally published 1863).

Martinson, T. L. "Interrelationships Between Landscape Art and Geography in Latin America: First Response to a Challenge." *Geographic Research on Latin America: Benchmark 1980.* Muncie, Ind.: Conference of Latin Americanist Geographers 1981, pp. 347–356.

———— ed. *Geographic Research on Latin America: Benchmark 1990.* Auburn, Ala.: Conference of Latin Americanist Geographers 1992, pp. 347–356.

Parsons, J. J. "The Northern Andean Environment." *Mountain Research and Development* 2(1982): 253–262.

Place, S. E. ed. *Tropical Rainforests: Latin American Nature and Society in Transition.* Wilmington: Scholarly Resources, 1993.

Stadel, C. "Altitudinal Belts in the Tropical Andes: Their Ecology and Human Utilization." *Benchmark 1990*, Conference of Latin Americanist Geographers, Vol. 17/18 (1992): pp. 45–60.

Sternberg, H. O'R. "Aggravation of Floods in the Amazon River as a Consequence of Deforestation?" *Geografiska Annaler Series A: Physical Geography* 69 (1987): 201–219.

Townsend, J. "*Magdalena, River of Colombia.*" Scottish Geographical Magazine 97 (1981): 37–49.

Welch, T. L., and M. Figueras. *Travel Accounts and Descriptions of Latin America and the Caribbean*, 1800–1920: *A Selected Bibliography.* Washington, D.C.: Organization of American States, 1982.

West, R. C., J. P. Augelli, et al. *Middle America: Its Lands and Peoples*, 3d ed. Englewood Cliffs, N.J.: Prentice-Hall, 1989.

West, R. C., ed. *Natural Environment and Early Cultures: Handbook of Middle American Indians*, Vol. 1. Austin: University of Texas, 1964.

CHAPTER 3

Aboriginal and Colonial Geography of Latin America

ROBERT C. WEST

INTRODUCTION

Latin America has the longest continuous record of European occupation in the New World. It is also the locale of the aboriginal civilizations of America—the Aztec, Mayan, and Incan cultures, whose way of life in some respects equaled or excelled that of sixteenth-century European society. The aboriginal and colonial

Mayan Ruins, Chichen Itza, Mexico.

heritage of Latin America is strong today, and many aspects of the contemporary scene are difficult to comprehend without a knowledge of the past 500 years.

This chapter points out the salient features of aboriginal and colonial life that have influenced the land and people of Latin America. The aboriginal scene on the eve of European conquest is considered, followed by the story of the European invasion and its influence on native inhabitants. In dealing with the historical geography of the colonial period, emphasis is placed on the exploitation of human and natural resources by the three main groups of European settlers—Spaniards, Portuguese, and North Europeans.

ABORIGINAL PATTERNS ON THE EVE OF CONQUEST

The aboriginal background is basic to understanding European colonial development in Latin America. In most of Spanish America in particular, the aboriginal element was as significant as the European in the evolution of colonial life and landscape. Today, the people of southern Mexico, Central America, and the central Andean countries are still considered to be largely Amerind in character, although modified by European culture.

On the eve of European conquest, Mexico, the West Indies, and Central and South America were occupied by aboriginal groups that varied in culture; they ranged from primitive bands who practiced Stone Age culture to civilized states such as the Aztec and Inca. Generally, the influence of aboriginal groups on European colonial development varied with their degree of cultural attainments and population densities. The civilized, densely peopled states of Mexico and the central Andes fashioned the pattern of Spanish colonial settlement and economic development in those areas differently and more thoroughly than the simple forest tribes influenced the course of Portuguese occupation of Brazil.

Prior to European contact, two fundamentally different patterns of livelihood prevailed in Latin America: (1) *nonagricultural economies,* characterized by gathering, hunting, and fishing, mainly in the extratropical zones of the southern and northern peripheries of Latin America, and (2) *agricultural economies,* which were found mainly in the tropical highlands and lowlands (Figure 3.1).[1] Because the farming areas contained the great bulk of the aboriginal population, including the two centers of civilization (Mexico and the central Andes), they were far more significant in European colonization than the nonfarming sectors.

NONAGRICULTURAL ECONOMIES

Low population densities, impermanent settlement, and rudimentary technology characterized most of the nonfarming people. Many inhabited isolated refuge areas where they may have been pushed by invaders of a higher culture. Among the most primitive people were the Fuegian shellfish gatherers of the cold, wet southern end of South America. Probably never more than 15,000 in number, these Stone Age people inhabited one of the least desirable parts of the continent. Europeans delayed effective contact with them until the nineteenth century, and since then the Fuegians have become practically extinct. In the West Indies a similar shellfish-gathering group, the Ciboney, who once occupied all of the Greater Antilles, had been almost wholly replaced by farming peoples at the time of Spanish contact.

---

[1]Other categories have been suggested for the aboriginal economies of Latin America. For example, Lockhart and Schwartz used the classifications sedentary, semisedentary, and nonsedentary, which correspond to the terms civilization, simple farming, and hunting and gathering (1983, p. 35).

Figure 3.1 Aboriginal economies.

North of the Fuegian peoples of southernmost South America there extends a large area that was inhabited by small bands of hunters, the "*pampean* Indians." The region includes most of Argentina and all of Uruguay. Physically, it is composed of the arid, bush, and grass-covered Patagonia plateau and the grassy plains of the Pampas. Little is known of the Indians who occupied the area until after they had acquired certain European traits, such as the horse for riding, in the middle of the seventeenth century. It is known that they hunted the guanaco (an American camel related to the domesticated llama of the Andes), the rhea (or

American ostrich), and several small mammals. Spears and bolas were the main weapons, and the *pampeans* apparently moved from place to place seeking animals and plants for food. After they acquired the horse, these hunters became mobile and warlike, their hunting techniques became highly specialized, their population grew, and their social organization became complex. Aided by Indian horsemen who had migrated from Chile, they evolved into a formidable aboriginal military force that helped delay European occupation of the fertile Pampas of Argentina. Not until the 1850s were the last of the *pampean* bands subdued and placed on reservations. Parallels can be drawn between the *pampeans* of Argentina and Uruguay and the Plains Indians of the United States and Canada. Both hunted large mammals in grassland environments; their cultures were changed by the adoption of the horse at about the same time; and the relations between native and European evolved in a similar fashion.

Throughout the vast forested and grass-covered Brazilian highlands, small groups of hunters and gatherers formed enclaves within a large area of farming economy. Examples include the Bororó and the Gê-speaking bands who occupied much of the Brazilian plateau and were hunters and gatherers who occasionally farmed. Many of the present-day inhabitants of the headwaters of rivers within the Brazilian plateau are descendants of the once widespread Gê-speaking bands.

Hunting and gathering peoples occupied a large territory on the northern periphery of Latin America, especially in the deserts of northern Mexico and the peninsula of Baja California. The inhabitants of the peninsula were the least numerous and culturally the least advanced of any Mexican aboriginal group. After Spanish missionary contact in the seventeenth century, the Baja Californians disappeared because of culture shock and disease. The desert nomads of north-central Mexico were more numerous and more culturally advanced than the Californians, particularly in social organization. Warlike and mobile, especially after the acquisition of the horse from the Spaniards in the seventeenth century, these people, collectively called Chichimecs, became a partial barrier to northward expansion of Spanish settlement from central Mexico. The hunting and gathering economy of northern Mexico can be considered a southern extension of a similar culture that covered much of the arid portion of western North America, which included such groups as the Apache, Paiute, and Shoshone. Hunting and gathering bands also extended to the Pacific coast of North America, where the "Digger" Indians of southern California and the acorn gatherers of central California formed a relatively dense population that was missionized during Spain's political expansion into upper California in the late eighteenth century.

AGRICULTURAL ECONOMIES

Prior to the conquest, more than two thirds of Latin American territory was occupied by farmers. In contrast to the nomadic hunters and gatherers, most farmers were sedentary people who settled in hamlets or villages ranging in size from a dozen to several thousand inhabitants. With a more abundant food supply through plant cultivation, the densities of farming populations were higher than those of nonagricultural people. By far the greater part of the land devoted to agriculture was occupied by farmers whose cultivation techniques were rudimentary, at least one-half of whose food came from hunting and fishing, and whose social organization was based on tribal affiliations. Use of advanced agricultural practices and nearly complete reliance on domesticated plants and animals for food were most characteristic of aboriginal civilizations in highland Mexico and the Andean area.

► *Simple Farmers*

Most of the simple farmers used a tillage system called slash-and-burn, or swidden, an ancient farming technique. In the Americas it was practiced on steep, wooded slopes in the highlands and on both slopes and level land in the lowland forests. Slash-and-burn farming involves the following procedures. Small plots are cleared within the forest and the brush is burned. Seeds or tubers are then planted in holes punched into the ash-covered soil with a sharpened stick, or dibble. After 2 or 3 years of cultivation, yields usually decline because of loss of soil fertility and weed competition. The plot is then abandoned for perhaps 10 to 20 years to permit soil rejuvenation and the reestablishment of second-growth forest. After that time, the same plot may be recleared and the cycle repeated. Meanwhile, in the surrounding forest, the farmer has cleared new plots that go through the same cycle of cropping and abandonment. Thus, the farmer is continually shifting small fields, and occasionally an entire village may be moved to a new site when land around it has become overcropped. Today this aboriginal method of shifting cultivation plots is found in subsistence farming from Mexico through Amazonia to Chile.

► *Tropical Forest Farmers of South America*

The simple farmers who inhabited the tropical forest of Brazil and northern South America cultivated starchy root crops, such as manioc (both the bitter and sweet varieties), sweet potatoes, and arrowroot as staples. Except for the peanut, few protein-yielding crops were grown, and in most areas maize (corn) was secondary or absent. Crops were raised in clearings along riverbanks, where the more fertile alluvial soils occur. Interfluves, characterized by leached, infertile soils, were rarely cropped. Protein foods consisted of fish and game obtained from the river and forest.

Because they were riverine people and expert canoers, tropical forest people migrated long distances. Long before Portuguese contact, the Tupi-speaking bands of the lower Amazon Basin had spread southward along the Atlantic coast and inland as far as southern Brazil, and the related Guaraní had occupied the fertile lands of eastern Paraguay. The Arawaks of the upper Amazon settled along the western tributaries of the river and, by the beginning of the Christian era, had migrated into the Greater Antilles as far as Cuba and the Bahamas, taking their manioc–sweet potato farming complex with them. Similarly, bands of Caribs from the Guianas and Venezuela had island-hopped into the Lesser Antilles by A.D.1400. Moreover, groups that spoke languages related to Chibchan pushed northward by land from Colombia into the tropical forests of southern and eastern Central America, carrying South American life-styles into that area. As a result of such migrations, Amazonian-type farming based on slash-and-burn cultivation of root crops became the most widespread agricultural complex in aboriginal America. Manioc and other tubers are still the basic foodstuffs throughout the region.

Aboriginal population densities in areas of root crop cultivation were much above those of nonagricultural peoples. Because of an abundant and stable food supply from land and water, the riverbanks and sea coasts had densities up to 30 persons per square mile (Denevan, 1966a, 1992). The account by the Spanish explorer, Francisco de Orellana of his journey down the Amazon from Peru to the Atlantic in 1542 mentions the large and almost continuous villages along it. The Tupi settlements on the Brazilian coast were probably just as numerous. The Greater Antilles, however, was the most densely settled area occupied by natives of Amazonian culture. The island Arawak developed a sedentary type of farming characterized by the construction of permanent, raised beds (*conuco*) for planting that produced prodigious crops. At Spanish contact (1492), Hispaniola (Haiti and the Dominican Republic) may have had more than 1 million native inhabitants, giving it an overall population density of 35 persons per square mile (Sauer, 1966). Cook and Borah (1971) estimated the population of Hispaniola, on the eve of contact, at 7 to 8 million people. Lovell (1992) draws attention to the range of population estimates for Hispaniola which run from 60,000 to 8 million.

A minor area of simple farming was located in central Chile, well outside the tropics, forming the southernmost extent of agriculture in South America. Inhabited by Indians of Araucanian speech, the northern, sparsely wooded part of the area was under Incan rule, while the southern, heavily forested sector from the Maule River to Chiloé Island was held by independent tribes at the time of Spanish contact. Using slash-and-burn methods, the Araucanians cultivated maize and potatoes as the main staples, supplemented by beans, squash, and local grains. In contrast to the docile tropical forest farmers of Brazil and the Greater Antilles, the Araucanians were warlike and succeeded in blocking Spanish settlement in their forested land until the nineteenth century. Their modern descendants, called Mapuches, still practice slash-and-burn farming on reservations in south-central Chile.

► Simple Farmers of Mexico and Central America

A major area of simple farming covered the mountain slopes and adjacent lowlands of northern Central America and southern Mexico, with an extension into northwestern Mexico. In these areas the slash-and-burn farmers cultivated seed crops, the most important of which were maize, beans, and squash. Root crops, such as the sweet potato and sweet manioc, were secondary. The crop triad of maize, beans, and squash affords a fairly well-balanced diet; partly for that reason, hunting and fishing were less important among these people than among the forest Indians of South America. Maize furnishes starch and is rich in oil and certain proteins; beans provide most of the protein component; and squash offers a variety of essential vitamins. The Indians cultivated all three crops together in the same plot, as many do today. Aboriginal maize foods, such as tortillas, tamales, and atole, still form the basic food staples throughout Mexico and most of Central America.

Although they are classed as simple farmers, most of the lowland inhabitants of southern Mexico, such as the Huastec and Totonac of the Gulf coast, were descendants of older civilizations or had been engulfed by Aztec conquests at the time of European contact. Slash-and-burn farming was the rule in the northern part of the Yucatán Peninsula, even during the height of Mayan civilization several centuries before the coming of the Spaniards. It is still the rule among Mayan subsistence farmers throughout the peninsula.

The simple farmers of northwestern Mexico practiced slash-and-burn cultivation on hillslopes and along stream bottoms. Within the Sierra Madre Occidental, the Tarahumar and Tepehuan formed the largest groups, while near the coast, tribes such as the Yaqui and Mayo planted the river floodplains, utilizing the moist alluvial soil when the annual floods receded. Farther north in Sonora, the Ópata and Pima groups had some knowledge of irrigation, and the Pueblo Indians of the upper Rio Grande had developed a sophisticated system of canal irrigation. Having a sizable population, such groups would later attract Spanish conquest and settlement into what is now U.S. territory.

► Aboriginal Civilizations

At European contact, two main areas of aboriginal civilization had evolved in the New World. The northern area, *Mesoamerica*, included southern Mexico and the northern part of Central America. The southern area, *Andean America*, covered the central Andean highlands and adjacent Pacific coast of South America. At least three quarters of the total aboriginal population south of the Rio Grande was concentrated within these two areas, which comprised less than one fifth of the land surface of Latin American territory.

The inhabitants of Mesoamerica and Andean America had developed several cultural attributes, including (1) a stratified social class system, in which a small group of nobles and priests held tight control over a large proletariat for public labor, tribute, or civic duties; (2) the concept of a political state, forged by organized military operations; (3) intensive cultivation techniques, which could produce an abundant and stable food supply; (4) the use of metals, primarily gold and silver for ceremonial ornamentation and secondarily some of the lesser metals (copper and bronze) for utilitarian purposes; and (5) the growth of true cities, characterized by large concentrations of population, functions such as manufacturing and public services, and the presence of monumental architecture. The attributes of an organized and subservient labor force and the presence of precious metals attracted European conquest and gave rise to the two great centers of Spanish exploitation and settlement in the New World: Mexico and Peru.

► *Mesoamerica*

By A.D. 1500, two strong military states existed in the highlands of central Mexico: the Aztec realm in the east and the Tarascan state in the west. The Aztecs established the larger and more populous of the two states and, when the Spaniards arrived in 1519, had extended political hegemony from the Pánuco River in the northeast to the modern frontier of Guatemala in the south. The Aztec domain was not an empire in the modern sense; it was a tribute state. All conquered towns paid tribute in the form of products of the land. Political power emanated from the Aztec capital, Tenochtitlán (in the Valley of Mexico), a city of perhaps 100,000 inhabitants. Thomas (1993, pp. 612–613), provides a useful review of European estimates of the population of Tenochtitlán.

Although the Tarascan state was not comparable in size and wealth to the Aztec state, it was a cohesive political unit that corresponded roughly to the present state of Michoacán in west-central Mexico (Stanislawski, 1947). The Tarascan state was more of a political empire than that of the Aztec. The Tarascans exacted tribute from conquered areas and colonized people along the frontiers; by this means, they spread their language and culture through methods not used by the Aztecs. Tarascans and Aztecs had strong armies, and military clashes occurred along the frontier between their realms.

Southernmost Mesoamerica was organized into small independent chiefdoms instead of states. Despite high attainments in religious art, science, and architecture, the Maya of the Yucatán Peninsula never developed a true political state, and at Spanish contact, their territory consisted of 18 small chiefdoms. A similar political structure characterized the Guatemalan highlands and the Pacific lowlands of Central America as far south as northwestern Costa Rica.

During preconquest times, Mesoamerica formed one of the most densely settled areas of the New World. Population estimates for preconquest Mexico range from 4 to 30 million inhabitants (Thomas, 1993, p. 609). Rural population densities in the Valley of Mexico and other highland basins in the eastern part of the Aztec state may have been as great or even greater in A.D. 1500 than they are now. The inhabitants of Central America may have numbered 5 million, and the Yucatán Peninsula, 800,000 (Denevan, 1992a, p. 291; Lovell and Lutz, 1992, pp. 127–129).

Sustaining the large population required the adoption of farming techniques that were more productive than the slash-and-burn method of the simple farmers. Sophisticated techniques of cultivation had evolved in the highlands. Among these techniques were permanent fields that were fallowed every two or three years; terraced fields on hillslopes; and, in places, irrigation. Irrigation was employed as far south as the Pacific lowlands of Nicaragua, especially for tree crops. The most productive cultivation method, however, was the *chinampa*, or the so-called floating garden, still used in the southern part of the Valley of Mexico. A type of land reclamation technique, *chinampas* are raised, elongated plots constructed along the margins of shallow, freshwater lakes. Irrigated and heavily fertilized, a *chinampa* plot can produce as many as three crops annually. This technique was used mainly in the Valley of Mexico, but it was also practiced in other parts of east-central Mexico (Armillas, 1971; West, 1970; Wilken, 1969). Few cultivation systems in the world can match the productivity of *chinampas* (Whitmore and Turner, 1992, p. 409).

Like other Mesoamerican farmers, the Aztecs and their neighbors relied on maize, beans, and squash as staple foods; other crops included the chili pepper, a grain called amaranth, cotton, maguey (from which the native alcoholic drink, pulque, is made), and many tropical fruits. Cacao, or chocolate bean, was the most important tree crop; it was widely used as currency and formed an important item in long-distance trade.

► *Andean America*

At the beginning of the sixteenth century, Andean America was dominated by the Inca Empire, the largest aboriginal state in the New World. Expanding outward from the high Andean valley of Cuzco (11,000 feet elevation), Incan armies, within a period of 90 years (1438–1526), had conquered a vast area that extended 2,800 miles from southern Colombia to central Chile. The empire included the central Andean highlands (Peru, Ecuador, Bolivia, and northwestern Argentina), as well as the productive oases of the Peruvian coastal desert and the tropical coastal lowlands of Ecuador.

Unlike the Aztec tribute state in Mexico, the Incan state was a true empire. It was tightly controlled by the Incan nobility through a highly organized sociopolitical system. Theoretically, all natural resources, including farmland, belonged to the state. The state language, Quechua, and the state religion of sun worship were imposed on conquered peoples. (The widespread use of the Quechua language today in Ecuador, Peru, and Bolivia stems from ancient Incan policy.) Production and distribution of agricultural

Machu Picchu, Peru.

products were organized according to a tripartite system whereby farmers of each village produced crops for themselves, the Incan nobility, and the state priesthood. Surplus production was stored and distributed to areas in need, precluding food shortages or famine. A well-maintained road system and the use of llama pack trains insured food distribution and effective military control throughout the empire (Figure 3.2).

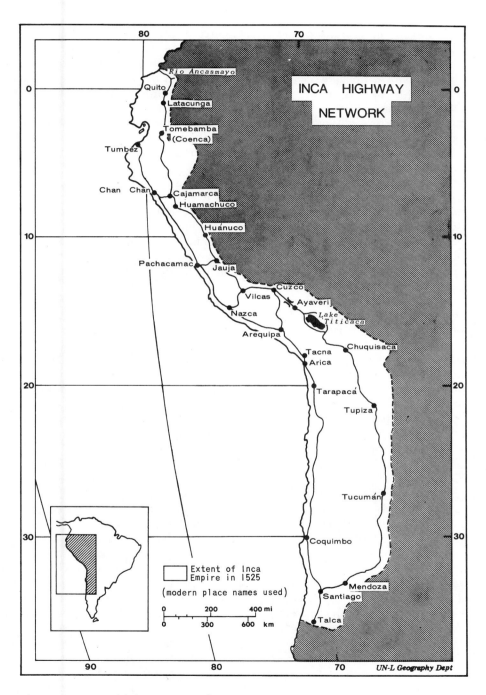

Figure 3.2 Inca highway network.

Moreover, through a labor system called the *mita*, all able-bodied males were obliged to serve a few weeks of each year in the army or on public works. After the conquest, the Spaniards continued the *mita* labor system, extending it particularly to the mining industry, much to their advantage.

Cuzco, the center of Quechua nobility, was the chief administrative seat of the Inca. Early in the sixteenth century, a second capital was established at Quito to control the northern part of the empire.

In the Andes, north of the Inca Empire, at least one other aboriginal culture was sufficiently advanced to be classed as a civilization. This was the Chibchan domain in the eastern cordillera of present-day Colombia. At conquest, several chiefdoms of Chibchan-speaking Indians living in the highland basins had confederated to form a rudimentary political state. Unlike the Inca, however, the Chibchans built no monumental architectural structures and showed little tendency to expand territorially outside the eastern cordillera. Nonetheless, Chibchans had established a well-organized trade with the simple farmers living in the Magdalena and Cauca valleys to the west, exchanging their salt and their cotton cloth for the lowlanders' gold and gold ornaments.

Andean America was less densely populated than Mesoamerica. Estimates of total population at Spanish contact vary. Rowe (1946) suggested a population of 6 million. Smith (1970), using various models to calculate population numbers, argued that 12 million people lived within the Inca Empire. Cook suggested a maximum population of 8.6 million and a minimum population of 3.1 million for Peru in 1520 (Cook, 1981, pp. 93–97). No more than 500,000 inhabited the Chibchan domain (Eidt, 1959). Because of aridity, high altitude, and mountainous terrain, the central Andean environment is harsh. There are relatively few valleys in the mountains, and many of these, being over 10,000 feet in elevation, are subject to killing frosts eight months of the year. In the high Bolivian *altiplano* (11,500 feet elevation), the best arable land clusters around Lake Titicaca, whose shores were densely settled in Incan times, as they are today. Other small spots of dense population occurred in the river oases in the Peruvian coastal desert. As in parts of Mesoamerica, population in most of the Inca Empire probably had reached maximum densities in terms of available arable land and the agricultural technology developed at the time of the Spanish conquest.

The Indians of Andean America developed advanced farming techniques and a large crop complex. In addition, they domesticated more animals than any other aboriginal group in the Americas. Root crops were dominant, as they were in most of South America. In the highlands, the potato was the staff of life. Frost-tolerant tubers, including the potato, were preserved by drying, forming the light, easily transportable *chuñu*, which lasts years without spoiling. Native plants, such as quinoa, could be grown up to 14,000 feet. Maize was a secondary crop in most highland areas. Introduced into the Andes from Mesoamerica as early as 3000 B.C. (MacNeish, 1970), maize never replaced the native tubers as the main staple, because it does not mature well in areas over 10,000 feet in elevation. In pre-Spanish times, maize was used in Andean America chiefly for making *chicha*, a ceremonial beer. In the coastal oases of Peru, the dominant crops were tubers such as sweet manioc and sweet potatoes (both introduced at an early date from the Amazon Basin), and maize, beans, and squash. Animal domesticates of the Andean area included the guinea pig, raised for its meat (and thus the main source of protein among the highlanders); the llama, reared for its wool and as a beast of burden; the alpaca, a close relative of the llama, bred for its fine fleece; the muscovy duck, probably introduced from the Amazon Basin; and varieties of dogs.

Irrigation reached its height of development in aboriginal America in the coastal oases of Peru where water was diverted from streams and carried in canals to fields far down the valleys. In the highlands, stone-faced terraces were constructed on steep mountain slopes to increase the amount of cultivable land; water from springs and small streams was diverted by canals onto the terrace surfaces. In many parts of the Peruvian highlands, Incan terraces are still maintained for cultivation. The most widespread farming technique in the highlands, however, was the *era*, or permanent ridged fields on hillslopes, made for the cultivation of potatoes and other tubers. The ridges were constructed with the Andean foot-plow, or *taclla*, and are made in the same way today as in pre-Spanish times (Gade and Rios, 1972).

► *Raised Fields*

In many areas of Latin America, there are remains of irrigation works, terraces, and raised fields that date from the pre-Columbian era. Doolittle (1992, p. 393) tells us that "raised fields are agricultural landforms constructed by piling earth . . . into hills, platforms, or ridges." Such agricultural artifacts are found widely in the region (Denevan, 1992b, Figure 1). Some, like the chinampas, are still in use. The cultivation of Arawak *conuco* is understood from historical accounts. However, in the San Jorge basin of northern Colombia, on the Gulf coast of Mexico, in the western Orinoco basin, the Guiana shore, the upland basins of Andean Ecuador (Knapp, 1991, pp. 147–162), the Guayas Valley of Ecuador, the coastal desert of Peru, and the lands around Lake Titicaca, the fields are relict features whose operation is not fully understood.

Ridged and raised fields can serve a variety of functions in marginal environments. In poorly drained lands, raised fields prevent water logging of root systems. At high altitudes in the Andes, raised fields facilitate the drainage of cold air from fields. In regions of tropical wet and dry climates, such as the Orinoco, ridged fields would drain excessive moisture in the wet season, and if the depressions between the ridges were choked with earth by cultivators at the end of the rainy season, they might be a means of carrying more ground water into the dry season.

Claude Lévi-Strauss points out that in Amazonia and surrounding areas in Bolivia and Colombia there is evidence of defunct but formerly intensive agricultural systems:

> *Over tens, at times hundreds of thousands of hectares of flood-land, man-made embankments several hundreds metres long, and separated by drainage canals, guaranteed year-round irrigation while protecting fields from rising floodwater. Here the Indians [of Amazonia] practiced an intensive type of agriculture based on tubers which, combined with fishing in the canals, could support more than a thousand inhabitants per square kilometer. (Lévi-Strauss, 1995, p. 20)*

EUROPEAN INVASION

The European conquest and settlement of Middle and South America began in the late fifteenth century with the landing of Christopher Columbus in the West Indies. Spain and Portugal were the major conquerors. The Spaniards quickly managed to occupy the populous areas of aboriginal civilization in Mexico, Central America, and western South America—all of which were prodigiously wealthy in people, land, and minerals. The Portuguese, on the other hand, slowly gained a foothold on the Brazilian coast. They established sugar plantations among the tropical forest Indians, whom they enslaved, eventually obliterated, and replaced with enslaved Africans. A third European force, composed of North European nations, including England, France, and Holland, entered the Latin American scene first as pirates and later as settlers and entrepreneurs in parts of the West Indies and in the Guianas. Like the Brazilian Portuguese, they established sugar plantations and brought in African slave labor.

Thus, three spheres of European influence evolved in colonial Latin America. The Spanish sphere was the largest, potentially the richest, and the most diverse culturally and physically. The Spaniards tended to look inland toward the highland areas of dense aboriginal population and mineral wealth; except for a few ports and the dry littoral of Peru, they generally neglected the coastal zones. In the interiors the Spaniards developed the semifeudal, *hacienda*-type agricultural estate, using aboriginal serf labor to supply food to a local market—mainly the Spanish towns and mining centers. In contrast, both the Portuguese in Brazil and the North Europeans in the West Indies and the Guianas developed the capitalistic plantation, using black slave labor to produce sugar for an overseas market. The interiors of both Brazil and the Guianas were neglected. To this day, a coastal orientation dominates the economy and culture of most of the West Indies and the Guianas. In Brazil effective settlement of the immediate interior did not occur until the eighteenth century. Occupation of the far interior lies in the future, despite a recent spate of road building and political slogans to encourage frontier settlement.

► Centers of Conquest and Settlement Spread

Particular focal areas emerged within each of the colonial spheres of influence. Once firmly held by Europeans, they became centers or "hearths" for new conquests and settlement of adjacent areas. The motives for the establishment of such centers and for the movements from them varied. For Spaniard, Portuguese, or North European, the pervading reason to conquer and settle was economic—the desire to gain quick wealth. The Spaniards in particular were attracted by precious metals. The Portuguese and North Europeans also sought gold and silver, but most of them settled for more prosaic agricultural pursuits such as sugar, tobacco, or indigo production. After initial conquest, the spread of Christianity by missionaries became another motive for conquest and settlement, particularly by Spain and Portugal, less so by the North European nations. Finally, political considerations led to frontier expansion, such as the Spanish settlement of upper California to thwart the Russian influence from Alaska and the occupation of the Banda Oriental (Uruguay) to meet the southward expansion of Portuguese from Brazil.

► Spanish Centers

The Spaniards established several centers for territorial expansion (Figure 3.3).

Conquest of Middle America

The West Indian island of Hispaniola, "discovered" by Columbus on his initial voyage, was the first center of Spanish expansion. Attracted by the island's gold deposits and dense native population of Arawaks, the Spaniards in 1496 founded the city of Santo Domingo on the southern coast as the first permanent European settlement and administrative center in the New World. For almost half a century (1493–1530), mining gold with native labor was the main Spanish activity on the island. Rapid decline of the Indian population, however, led the Spaniards to adjacent islands, the Gulf coast, and the mainland of northern South America to recruit Indian labor by force for the Hispaniola mines. Gold deposits were also discovered in Cuba and Puerto Rico; for that reason, Spanish settlement and culture in those islands became well established. Hispaniola was an early proving ground for many Spanish social and economic institutions, as well as crops and livestock, that were

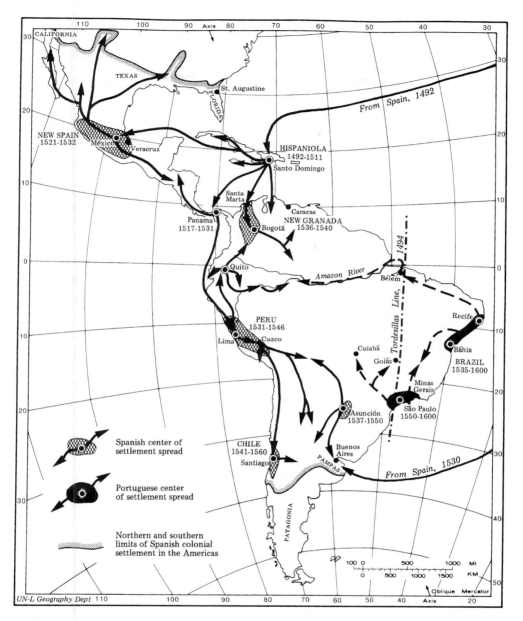

Figure 3.3 Spanish and Portuguese conquest and settlement of the New World.

later transferred to mainland Latin America. Hispaniola functioned as a stepping-stone for conquest and settlement of the mainland.

By the first decade of the sixteenth century, the Spaniards on Hispaniola were familiar with the north coast of South America, where they sought Indian slaves to replenish the declining population of the island. Most of the Colombian and Venezuelan coast was then called Nueva Andalucía, and only a few impermanent settlements were made along the coast to serve as bases from which to raid the interior for slaves, around the coastal islands of Margarita and Cubagua to exploit pearl oyster deposits, and at various points inland to mine gold. One of the inland sites was Santa María la Antigua del Darién, founded in 1510 near the present Panama–Colombia border. From this small settlement Vasco Nuñez de Balboa crossed the isthmus and discovered the Pacific, or South Sea, in 1513.

The Spaniards realized the strategic value of the Panamanian isthmus as a transit zone between the Atlantic and Pacific oceans. For that reason the city of Panama was established in 1517 as the capital of Castilla del Oro, as the isthmus was then called. Panama became the second major point of dispersal for Spanish conquest and settlement in the New World. From the isthmus the southern part of Central America was explored and settled. But, more important, Panama was the departure point for exploration and conquest by sea to the coast of Peru, where the Spaniards discovered the vast Inca Empire.

The Spaniards established a third major center of conquest and settlement in central Mexico, which was the focus of Mesoamerican Indian civilization and the locale of the Aztec and Tarascan tribute states. Landing in the vicinity of Veracruz after a short voyage from Cuba in 1519, the Hernán Cortés party quickly discovered the extent and wealth of the Aztec realm. The fall of Tenochtitlán, the Aztec capital in the Valley of Mexico, occurred two years later; on its ruins the Spaniards slowly built Mexico City. This city became the administrative center of New Spain, the name given to colonial Mexico and most of Central America.

Spanish expansion from the Valley of Mexico occurred in two phases, each encompassing a definite geographical area. The first phase was characterized by an easy and rapid overrunning of heavily populated Mesoamerica, which included central and southern Mexico and the northern third of Central America. Within 10 years after the fall of the Aztec state, Cortés and his men had taken control of the native people by means of the *encomienda*, a system of tribute[2]; they had staked out claims to gold and silver deposits and some of the choice agricultural areas; and they had founded several Spanish towns, or *villas*, in strategic spots to control the Indians.

The second and later phase of expansion involved the slow conquest of northern Mexico, an arid region inhabited by warlike and nomadic Chichimecs. In contrast to the rapid overrunning of Mesoamerica, the occupation of northern Mexico took 200 years. The discovery of silver deposits, beginning at Zacatecas in 1546, drew the Spaniards northward. Beyond Zacatecas, one line of northward advance followed a series of silver deposits along the western side of the Mexican plateau, through Durango and Chihuahua, where grassy plains afforded forage for livestock. Continuing farther north, farmers and stockmen in 1599 made settlements among the Pueblo Indians in the upper Rio Grande Valley, within present U.S. territory. Another route of advance led northeastward from Zacatecas across the interior desert to the mines of Mazapil, Saltillo, Monclova, and finally into Texas, where San Antonio was founded in 1718 to thwart French expansion from Louisiana. In the seventeenth century, a third line of northward movement was made into Sonora in northwestern Mexico by Jesuit missionaries, who were followed by Spanish stockmen and miners. From Sonora this line of settlement reached into southern Arizona. The last northward thrust of Spanish settlement into what is now U.S. territory was the occupation of upper California by Franciscan missionaries, stockmen, and the military in order to counter Russian advances from Alaska.

---

[2]For definition and discussion of the *encomienda*, see section on Colonial Economy of Spanish America.

Conquest of South America

The Spanish occupation of South America is dominated by the conquest of the Inca Empire. In Panama, the Spaniards heard tales of a high civilization to the south, and exploratory trips were made along the rain-drenched Pacific coast of northwestern South America beginning in 1524. Finally, in 1531, a party led by Francisco Pizarro and Diego Almagro landed in northern Peru, where the coastal desert begins.

The story of this conquest, including its fortuitous timing with civil strife within the Inca Empire, is well known. By 1533, both Cuzco, the capital city of the Inca, and Quito, the seat of the northern part of the Incan realm, had fallen to the Spaniards. Almost invariably the Spaniards built their highland towns on the foundations of the looted Incan cities, among which Cuzco, Quito, Cajamarca, and Jauja were the most important. In Peru, however, the Spaniards did not choose an interior highland site within an area of dense Indian population for their main administrative city. Instead, Francisco Pizarro founded as the capital Lima, or Ciudad de los Reyes, in one of the coastal oases called Rimac, or Limac, at the terminus of the easiest route to the sea from Cuzco. Such a choice may have been dictated by the Spaniards' tenuous hold over the potentially rebellious Incan nobles and by the danger of committing the main administrative seat to a vulnerable place far into the mountainous interior. Lima became the political center of the extensive viceroyalty of Peru and one of the most opulent cities of the Spanish colonies. With its port of Callao, it was also the main collecting point for the mineral wealth of highland Peru.

In Peru as in Mexico, the search for precious metals was the motivation for much of the Spaniards' efforts at conquest and settlement. However, the looting of the gold and silver artifacts from Incan temples and the debilitating civil war between the Pizarro and Almagro factions in Peru may have delayed for a decade the discovery of rich silver deposits in the *altiplano* (Upper Peru) south of Cuzco. Although small Incan silver mines such as Porco continued to be exploited after Spanish conquest, not until 1545 were the rich Potosí mines discovered in the cordillera east of the altiplano. The findings of many other deposits in the central Andes followed. Thereafter the mines in the Altiplano and its surrounding areas became the main focus of Spanish economic activity and settlement in South America.

Both before and after the discovery of silver in Upper Peru, two highland cities within the conquered Incan realm—Cuzco and Quito—were departure points for further Spanish conquests of western South America. With these conquests, Spain acquired a vast area that was placed under the jurisdiction of the viceroyalty of Peru. Chile was occupied by settlers led by Pedro de Valdivia, who left Cuzco in 1540 and in the following year founded Santiago as the capital of the province. During the remainder of the century, other Spanish towns and gold placer camps were laid out in the fertile central valley of Chile. But permanent settlers pushed no farther south than the Bío-Bío River, which was the beginning of dense, middle-latitude forests and the locale of warlike Araucanian Indians. For nearly 250 years the Bío-Bío marked the southern frontier of Spanish settlement in South America. Santiago de Chile itself became a point of departure for settlement when in 1561 a group of Chilean farmers crossed the Andes over the Uspallata Pass (12,650 feet) to found the vineyard towns of Mendoza and San Juan in what is now Argentina.

Most of northwestern Argentina was occupied by stockmen and adventurers from Cuzco, such as those who in 1553 founded the towns of Santiago del Estero in the lowland plains bordering the eastern foothills of the Andes. From this settlement other Spanish towns, such as Tucumán, Córdoba, and Salta, were formed during the last third of the sixteenth century to furnish livestock and other supplies to the mines of Upper Peru.

From Quito, the former northern Incan city, Spanish control expanded in many directions. Guayaquil on the Pacific coast was founded in 1537 as the port of Quito. In 1536 Sebastián de Benalcázar established Popayán and Cali in present-day Colombia. From Cali a part of the group pressed far down the Cauca Valley, where gold deposits were found.

Centers of Spanish settlement in the northern Andes were approached mostly from the Caribbean. The major penetration was carried out by the expedition of Jiménez de Quesada which, in 1536, left the small port of Santa Marta to seek gold and conquer the Chibcha Indian domain in the fertile highland basins within the Cordillera Oriental (in present-day Colombia). Two years later Jiménez established the Spanish town of Santa Fé de Bogotá as the capital of the New Kingdom of Granada. Instead of rich gold deposits, the Spaniards found a dense population of Indian farmers whom they appropriated in *encomiendas*. Later in the sixteenth century a series of towns, including Tunja, San Cristóbal, and Mérida (the latter two in present-day Venezuela), were founded in the highlands, all of which became centers of farming and stock raising on large *haciendas* worked by native labor. Westward from the New Kingdom of Granada, the province of Antioquia was opened up in the 1540s by gold seekers from Cartagena and other Caribbean ports. Exploitation of rich placer mines on the Cauca River and its tributaries made the area the foremost gold-producing district of the Spanish colonies. In 1546 Santa Fé de Antioquia was established near the Cauca River as the administrative center of the province. Late in the seventeenth century, the administrative function was moved to Medellín, today one of Colombia's largest cities.

The Caribbean coast of South America was at first controlled mainly for the purpose of exporting Indian slaves to Hispaniola and elsewhere. The only permanent Spanish settlements were those of Cumaná (founded in 1516) and Coro (1527) along the Venezuelan coast, and Santa Marta (1525) and Cartagena (1533) along the coast of present-day Colombia. Most of the settlements began as slaving stations, and the continuation of slaving for many decades resulted in the depopulation of the coastal region. Not until the last half of the sixteenth century were the highland basins of northern Venezuela opened to Spanish settlement, and the capital of the province was moved from Coro to Caracas (founded in 1567). Cattle ranches were established in highland centers such as Valencia, Barquisimeto, and Tocuyo from which the stock-raising industry moved into the plains (*Llanos*) of the Orinoco, bringing that isolated region under Spanish control. The province of Venezuela was the only large area of Spanish South America that fell outside the direct administration of the viceroyalty of Peru. Until the eighteenth century its political ties were with the *audiencia*[3] of Santo Domingo (Hispaniola), a part of the viceroyalty of New Spain.

From nearly all the main Andean centers of South America the Spanish crown encouraged expansion eastward into the hot lowlands of the Amazon and Orinoco basins. There were three reasons for this expansion: to search for precious metals, to spread the Christian faith, and to stem Portuguese movement westward into areas claimed by Spain. Spanish penetration into the eastern Andean slope and adjacent Amazonian and Orinoco lowlands during colonial times is reflected today in

---

[3]Within the Spanish colonial political system, an *audiencia* was a tribunal composed of a president and generally four judges, who had administrative and judicial jurisdiction over a particular geographical area. Such an area was also known as an *audiencia*. The tribunal was subordinate only to the viceroy and the Council of the Indies in administrative and judicial matters.

the eastern boundaries of Colombia, Ecuador, Peru, and Bolivia. During the six-teenth century, several expeditions set out from Cuzco to explore the southwestern tributaries of the Amazon, and for 100 years (1668–1768), the Jesuit order mis-sionized among the Mojos Indians who lived in the wet lowlands of northern Bolivia (Denevan, 1966b). In 1539 an expedition left Quito (Ecuador) to seek gold in the forest-covered eastern slopes of the Andes. In 1542 one of the members, Francisco de Orellana, continued down the length of the Amazon River, claiming for Spain a vast eastern realm that was later occupied mainly by the church orders. One of these eastern areas settled from Quito was the province of Quijos (along the Napo River), famed for its cotton production in the sixteenth and seventeenth cen-turies. Another was the mission province of Maynas on the upper Amazon; here, forest Indians were indoctrinated in 80 missions founded by the Jesuits in the sev-enteenth century. Santa Fé de Bogotá (Colombia), seat of a large archbishopric, was another departure point for Jesuit missionaries who, in the seventeenth century, moved eastward into the Orinoco lowlands, where they founded 42 mission towns along the Meta, Casanare, and upper Orinoco rivers. Despite vigorous missionary activity, the eastern Andean slopes and adjacent lowlands attracted few Spanish set-tlers, and most of the missionized Indians succumbed to disease. Only in the last few decades have farmers from the highlands penetrated these areas.

The southeastern part of South America lacked mineral wealth. Chiefly for that reason, until the eighteenth century, the Paraná–Paraguay River Basin and the Pampas that border the La Plata estuary were lightly settled by the Spaniards and largely ignored by the Spanish crown, save for the famous Jesuit missions of Paraguay. As early as 1536, the Spaniards attempted to found a colony called Buenos Aires on the La Plata estuary in order to open a direct route from the east to the newly conquered Inca realm in the Andes and to forestall Portuguese designs to expand southward from Brazil. Indian hostilities defeated this venture. But in 1537, the Spaniards succeeded in ascending the Paraguay River for a distance of 800 miles from the sea. On the high east bank of the river, within a rich agricultural zone densely inhabited by Guaraní Indian farmers, they established the town of Asunción as the capital of the La Plata area.

Asunción was a center for further exploration and settlement of Paraguay and northern Argentina. In 1573 Spaniards from Asunción founded the stock-raising center of Santa Fé on the northern edge of the Pampas, and in 1580 they reestab-lished Buenos Aires as a small port. Thereafter the northern Pampas became a source of dried meat and hides for the mines of Upper Peru, and Buenos Aires degenerated into a smuggling station through which contraband merchandise found its way to Potosí and even Lima. The rolling grasslands of the Banda Oriental (Uruguay), directly across the La Plata from Buenos Aires, were not occupied by Spaniards until the eighteenth century, mainly to stem Portuguese penetration from the north. In 1680 the Portuguese had established the port of Colonia on the La Plata across from Buenos Aires as a point for shipping meat and hides to their sugar plantations along the Brazilian coast. In 1726 the Spaniards founded a garri-son town called Montevideo (the present capital of Uruguay) farther eastward on the estuary and succeeded in expelling the Portuguese. Most of the Pampas of Ar-gentina and the desolate scrub- and grass-covered plateau of Patagonia that stretches for 1,000 miles to the south were not effectively occupied until the middle of the nineteenth century.

In contrast to the slow Spanish colonization of most of southeastern South America, the province of Paraguay in the fertile Paraná–Paraguay Basin was quickly overrun by *encomenderos* exploiting the large Guaraní population for food and labor. To protect the remaining Indians, the Jesuits in 1610 began to settle the Guaraní in mission towns, hoping eventually to form a utopian "republic." The first missions were founded in the Guairá area within the present Brazilian state of Paraná and along the lower Paraná and Uruguay rivers within the present Paraguay–Argentine borderland (Caraman, 1976). Because of Portuguese raids, the Jesuits and their charges retreated southward from the Guairá area, but by 1730 30 missions containing 140,000 Guaraní in what is now southern and eastern Paraguay made up the thriving christianized Indian community, tightly controlled by the Jesuit fathers. After the Spanish crown expelled the Jesuits from the New World in 1767, the mission towns disintegrated, but by then the native culture had been sufficiently protected to preserve many of the Guaraní folkways. Today, Guaraní is the language spoken in rural areas and among many urban families in Paraguay.

► Portuguese Centers

Portuguese claims to New World territory were based on the Treaty of Tordesillas of 1494[4] and on Pedro Alvares Cabral's chance landing on the Brazilian coast in 1500. However, occupied with its Asian commitments, Portugal did not colonize Brazil for three decades. During that interval, French and Portuguese adventurers ranged along the Brazilian coast, trading with Indians and cutting brazilwood (hence the place-name, Brazil), a source of dyestuff for use in Europe. In 1532 permanent Portuguese settlements first appeared on the coast. The crown had previously divided the territory into 12 large proprietary land grants called *capitânias*, each of which was given to a *donatario* (a private individual of noble rank) who was responsible for its development at his own expense. Each *capitânia* fronted onto 100 to 400 miles of coast and extended inland for an undetermined distance. Finding no precious metals on their lands, the *donatarios* turned to cutting brazilwood and cultivating sugarcane and tobacco along the tropical and fertile coastal lowlands. Only a few of the original *capitânias*, however, were successful, and in 1549 all of them were transferred to the crown. Among the successful ones were Pernambuco and Bahia in the northeast and São Vicente in the south. These two areas later became hearths from which Portuguese conquest and settlement spread to other parts of Brazil (Figure 3.3).

Northeastern Center

For nearly two centuries, Portuguese settlement in the northeast was confined chiefly to the coastal plain. The most prosperous sugar plantations and tobacco farms of Brazil were in the *capitânias* of Pernambuco and Bahia. There the coastal forest was completely cleared for agriculture and stock raising. The native Tupi Indians were at first enslaved, but they either fled into the interior or died; as a result, black slaves were imported from Africa to provide labor. From Pernambuco sugarcane cultivation spread into other *capitânias* as far north as that of Maranhão. But in those *capitânias* of Ilheus and Porto Seguro, south of Bahia, invading Indians in 1560 wiped out the thriving plantations, and Europeans did not succeed in resettling that area until the nineteenth century. Two main port cities developed in the northeast: Salvador (Bahia) and Pernambuco (Recife). Because of its wealth in sugar and its large harbor, for more than 200 years Salvador was the political capital of all colonial Brazil, a rank it retained until 1763, when the capital seat was shifted to Rio de Janeiro. In the northeast the Brazilian plantation, or *fazenda*, based on black slave labor and the export of sugar, reached its apogee. The rich plantation owners formed a social aristocracy from which came many of the political leaders,

---

[4]The treaty gave to Portugal exclusive rights to regions of the New World east of a north-south line drawn 370 leagues west of the Cape Verde Islands and to Spain all lands west thereof. This line intersected eastern South America from approximately the mouth of the Amazon to the São Paulo coast, Brazil.

writers, and entrepreneurs of colonial Brazil. For these reasons the northeastern center is sometimes regarded as the "culture hearth" of the Brazilian nation (Schmieder, 1929) (Figure 3.4).

So lucrative were the northeastern sugar plantations that, for a short period (1624–1654), the Dutch took the *capitânia* of Pernambuco from the Portuguese and operated the sugar industry with efficiency and profit. The Dutch introduced improved methods of refining sugar and developed the port city of Olinda (near Recife), which still reflects Dutch influence in its colonial architecture.

There were at least two periods of significant emigration from the densely populated coastal lowlands of the northeast in colonial times. First, early in the seventeenth century, many Portuguese and mixed people began to settle the higher and drier backlands—the *sertão*. Most of these people engaged in stock raising on large ranches located no more than 200 to 300 miles from the coast. A second movement occurred on a much larger scale in the eighteenth century, after gold was discovered in the backlands of Minas Gerais, 200 miles from the coast. For most of that period, gold vied with sugar as Brazil's most valuable export.

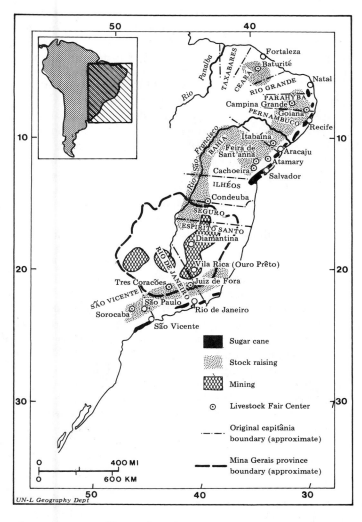

Figure 3.4 Brazilian colonial economy, 1650–1750.

During the seventeenth century, a third movement out of the northeastern center resulted in Portugal's territorial claims and tenuous occupation of a large part of northern Brazil. Permanent port settlements were made in the northernmost *capitânias* and beyond: Fortaleza in Ceará (1609), São Luis de Maranhão (1612), and Belém at the mouth of the Amazon (1616). Belém was a base for occupation of the Amazon Basin. Portugal used the expedition of Pedro Teixeira (1637–1639) up the Amazon River to claim this vast forested area, deep into Spanish territory west of the Line of Tordesillas (Phelan, 1967, p. 32). Moreover, for more than a century (1637–1755), various religious orders based in Belém had established missions among the Indians along the Amazon and its lower tributaries, further consolidating Portugal's claim on the territory (Rippy and Nelson, 1936, p. 237). Despite missionary activity, the Portuguese used the Amazon area mainly as a source of Indian slaves for the coastal sugar plantations. Slave raids continued there well into the eighteenth century, forcing the Indian population to flee into the upper tributaries.

Southern Center

During the sixteenth century, the plantations established on the narrow coastal plain of São Vicente, the southernmost of the early *capitânias*, were second only to those of the northeast in sugar production. The significance of the southern center for Portuguese expansion, however, did not derive from sugar but from the rise of a large group of mixed Portuguese-Indian stockmen on the plateau of São Paulo, immediately inland from the coast. Called Paulistas, or *bandeirantes*, these people of mixed ancestry became more adept at enslaving Indians than at raising livestock. They wandered far into the interior, capturing Indians for export to the coastal plantations, especially during the first half of the seventeenth century (Morse, 1965). After the expulsion of the Dutch from Pernambuco in the 1650s, slaves from Africa once more became available, Indian slavery declined, and the Paulistas turned to seeking deposits of gold and diamonds in the interior. They discovered the goldfields of Minas Gerais in 1693, Cuiabá in 1718, and Goiás in 1725, and found the diamond deposits of Minas Gerais in 1729. These discoveries resulted in Portuguese settlements of widely separated points within the south-central part of the colony—the first successful penetration of the Brazilian interior. Other than mining camps, the Paulistas made few permanent settlements. Their role was that of pathfinders, not settlers.

▶ *North European Centers of Settlement*

As settlers, the English, French, and Dutch came to Latin America late. They were first attracted to the Caribbean in the middle of the sixteenth century as pirates preying on Spanish galleons richly laden with gold, silver, and other products from Mexico and Peru. The more successful pirates, abetted by their rulers, established bases on a few unoccupied islands within the Caribbean.

The French and English, for example, operated from Tortuga, off northern Hispaniola; later, the English used Jamaica, especially Port Royal, for privateering; and, during the eighteenth century, the Bahamas, lying close to the main Spanish homeward route through the Florida Strait, were bases for various North European pirates. In general, the areas that the North Europeans claimed and later settled in Latin America were those that initially were of little interest to Spain or Portugal. Devoid of gold and inhabited by warlike Caribs, the lesser Antilles held little attraction for Spaniards, but these small islands would become England's and France's prized possessions by the end of the seventeenth century. Although Spain considered all of Hispaniola as its rightful territory, there were few Spanish settlements in the western third of the island, which in 1697 was transferred by treaty to France as the colony of St. Domingue (modern Haiti). Spain occupied Jamaica in 1509, but England captured the island in 1655 and later developed it into one of its most valuable possessions in the Americas. Spain kept a strong hold on Santo Domingo (the eastern two thirds of Hispaniola), Cuba, and Puerto Rico, where Spanish lifeways had been entrenched during the period of gold mining in the early sixteenth century. On the northern coast of South America, between Venezuela and the mouth of the Amazon, the hot, forested Guiana coast was a no-man's-land until colonized by the Dutch, English, and French in the seventeenth and eighteenth centuries.

Permanent English and French colonization of the Lesser Antilles began in 1624 on St. Christopher (St. Kitts) with the settlement of small farmers who grew tobacco and cotton for export. From there English settlement spread to other islands of the Leeward (northern) group, such as Nevis and Anguilla in 1628 and Antigua and Montserrat in 1632. Farther south the English colonized Barbados in 1627. French colonists occupied Martinique and Guadeloupe as well as several islands of the Windward (southern) group and, in the 1640s, began settlement of the western part of Hispaniola, long before it became legal French territory. The Dutch colonized some of the Leeward islands and Aruba and Curaçao in the southern Caribbean. In the 1640s sugar cultivation and black slavery were introduced from Brazil into the North European holdings of the Caribbean. Sugar became the prevailing cash crop of these small islands, making them France's and England's most lucrative eighteenth-century possessions in the New World. On Jamaica, where grazing land was available, the raising of livestock was an important activity (Morgan, 1995).

Outside the sugar islands of Jamaica and the Lesser Antilles, English-speaking whites, mulatto freedmen, and black slaves occupied the Bahamas, several small islands in the western Caribbean, and some parts of the Caribbean shores of Central America that were ineffectively held by Spain. The British colony of Belize (British Honduras) had its origin in English exploitation of dyewoods along the coast during the seventeenth century. The influence of English smugglers was so strong along the Caribbean coast of Honduras and Nicaragua that until the middle of the nineteenth century much of that area (the Mosquito Coast) was a virtual protectorate of Great Britain.

The first successful colonizers of the Guiana coast were the Dutch, who early in the seventeenth century established small settlements on some of the rivers of present-day Guyana (former British Guiana). The English followed in the 1650s when peasant farmers from Barbados, squeezed out by the sugar barons, settled along the Suriname River. After several attempts, the French finally established a small colony on the Cayenne River in 1674. Ravaged by disease and based on the cultivation of tobacco and indigo, none of these initial settlements was economically successful. In the early eighteenth century the Dutch began to move to the coast, where swamps were drained and sugar plantations were established with some success on rich organic soils. As in the Caribbean islands, the Guiana colonies changed hands between the North European powers several times, and final claims were not established until the beginning of the nineteenth century.

The Spaniards acquired a sphere of influence quickly, having accomplished conquest and settlement of the richest, most heavily populated, and most civilized parts of the New World by the middle of the sixteenth century. The Portuguese, having occupied the coastal fringe of Brazil about the same time, waited until the late seventeenth century to expand inland. The North European countries, starting late with what Spain and Portugal considered to be the less favorable of New World lands and beset by continual international wars, finally consolidated their respective spheres in the Caribbean and Guianas early in the nineteenth century.

EFFECTS OF EUROPEAN CONQUEST
ON THE ABORIGINAL POPULATION

The European conquest of Latin America affected the native peoples both physically and culturally. Spaniards, Portuguese, and North Europeans introduced their lifeways, including food crops and habits, farming methods, language, social organization, and religion. These they imposed on the aboriginal peoples and landscapes with varying degrees of effectiveness. In turn, the indigenous cultures influenced the Europeans, so that in many parts of Latin America elements of Old and New World peoples and cultures fused to form an amalgam that characterizes the human scene (Viola and Margolis, 1991).

A far-reaching effect of European conquest was the drastic reduction of the Indian population following European contact. Another was miscegenation, or race mixing, which increased progressively during the colonial period and continues to the present. These changes have greatly reduced the aboriginal element in Latin America as a whole in both relative and absolute terms. Indian peoples and cultures still persist, however, in some areas of the ancient civilizations (for example, Guatemala, Peru, and Bolivia) or in isolated sections (for example, parts of interior Brazil).

► *Decrease of the Indian Population*

Table 3.1 presents examples of the abrupt decline of the aboriginal population, based on estimates. The most disastrous collapse of any indigenous New World population occurred on the island of Hispaniola, the locale of the first European settlement in the Americas. By 1518, 26 years after Spanish contact, the native population had dropped to about 12,000. Thus, even if we accept a low estimate of the preconquest population such as Rosenblat's 100,000 (Rosenblat, 1992, p. 59), the indigenous population was down to about one eighth of its former level within a

TABLE 3.1   Decline of the Indian Population in Three Selected Areas of Latin America

	Estimated Population at European Contact	1570	1650	1800
Hispaniola	1,100,000[a] 1,000,000[j] 100,000[k]	100[b]	?	0(?)
Central and Southern Mexico	25,000,000[c] 20,700,000[j]	2,600,000[d]	1,500,000[e]	1,500,000[f]
Central Andean Area	12,000,000[g] 11,696,000[j]	1,500,000[h]	?	600,000[i]

[a]Columbus census of 1496, cited in Sauer, 1966, p. 66, and in Cook and Borah, 1971, pp. 388–389.
[b]Cook and Borah, 1971, p. 401.
[c]Borah and Cook, 1963, p. 88.
[d]Borah and Cook, 1963, p. 4.
[e]Dobyns, 1966. (Gerhard, 1972, p. 24, gives 1 million for 1650.)
[f]Gerhard, 1972, p. 24.
[g]Smith, 1970.
[h]Lipschutz, 1966.
[i]Patch, 1958.
[j]Denevan, 1992.
[k]Rosenblat in Denevan, 1992.

quarter century of contact. By 1570, only 100 Indians were left (Cook and Borah, 1971, pp. 397–401). The rapid decline of labor for gold mines in Hispaniola prompted the Spaniards to import Indian slaves from the neighboring islands, and they too, died. By the close of the sixteenth century, most of the West Indies had lost their native population—a case of near extinction of a large human group within a century. A similar pattern of population decline occurred among some simple farmers, such as the Quimbaya of northern Colombia, who were gone by 1570 (Friede, 1963).

The conquest took an appalling toll of Indians in Mesoamerica and the Andean area. Within 50 years after contact, the natives of both central Mexico and the central Andes were reduced to one tenth of their original numbers (Lipschutz, 1966). The low ebb in the central Mexican population came at the beginning of the seventeenth century, when about 1 million Indians are estimated to have survived (Borah and Cook, 1963; Whitmore, 1991); thereafter the Indian population began a recovery. The nadir of the Andean peoples, however, came about a century later before an increase set in. The conquest decimated the civilized Indians, but it did not obliterate them as it did the Arawak of the West Indies.

Various factors contributed to this dreadful mortality. The conquest took many native lives. Enslavement of Indians contributed to death and dislocation of populations. The Spaniards made traffic in Indians a lucrative business in the Caribbean, Mexico, and Central America until the practice was outlawed in 1542. In Brazil enslavement of Indians occurred on a large scale for much of the colonial period and may have been the prime cause for the disappearance of the Tupi population from the eastern part of the colony. Morse (1965, p. 24) estimates that during the sixteenth and seventeenth centuries the Paulista slave hunters of southern Brazil took more than 350,000 Indian prisoners, most of whom were trekked to the coastal sugar plantations, where they died. Aside from enslavement, other factors such as harsh treatment, disruption of normal food supply, and psychological despair contributed to the death rates. The *encomienda*, a system of tribute and forced labor instituted by the Spaniards, contributed to such conditions. Sauer (1966, p. 203) believed that disruption of the native economy by Spanish demands for Indian labor caused the demise of the aborigines of the Caribbean Islands.

The leading cause of Indian mortality was Old World diseases. The killers were influenza, smallpox, measles, and typhus, against which the American Indians had no immunity. Columbus' second voyage transmitted influenza to Hispaniola (Lovell, 1992, p. 428). The first smallpox epidemic broke out in Hispaniola in 1518, spread to Mexico with the initial Spanish conquest, and by the late 1520s had become pandemic over much of Middle America and western South America (Crosby, 1967, 1986). Subsequent epidemics of Old World diseases occurred throughout Mesoamerica and the Andean area at frequent intervals for the rest of the colonial period, but the worst ones came in the sixteenth century, causing the precipitous decline of the native population. The Indians gradually developed partial immunity to smallpox and measles, and by the end of the colonial period, their numbers had begun to increase. Little is known of other Old World diseases that Europeans and their African slaves carried to the Americas. Malaria probably took its toll in the tropical lowlands of Mexico and the coast of Peru, where in many sections the native population practically disappeared during the sixteenth century.

N.D. Cook (1981, pp. 59–62) studied demographic collapse in the population of Peru where between 1520 and 1635, there were outbreaks of smallpox, measles, typhus, influenza, diphtheria, mumps, and possibly bubonic plague. As the indigenous population had not been exposed to these diseases, mortality was high. In Europe, an outbreak of smallpox killed approximately 30 percent of those infected. In unexposed New World populations, the death rate was closer to 50 percent. In

Peru a smallpox epidemic ran from 1524 to 1526. Measles arrived in the early 1530s and killed 25 to 30 percent of the remaining population. Typhus broke out in the 1540s, and in the next decade simultaneous epidemics of influenza and smallpox hit. Many communities did not survive successive epidemics killing off 25 to 30 percent of their population. In the coastal region, Indian communities were decimated. Alchon in a study of the impact of diseases in the highlands of north central Ecuador in the colonial era, documented outbreaks of smallpox, measles, plague, typhus, and influenza. The 1520 precontact population of Ecuador may have been 1.08 million people. As a result of preconquest civil wars, the conquest, and disease, population numbers had been reduced to 105,000 by 1598 (Alchon, 1991, pp. 18–47).

Melville studied population collapse in the Valle del Mezquital, a region on the Tula River north of Mexico City. Melville (1994, p. 43) calculated that between 1519 and 1600, population numbers declined by about 90 percent. The major cause was successive epidemics. Whitmore (1991) simulated population collapse in the basin of Mexico and concluded that a 90 percent decline in numbers was common throughout the region. Brooks (1993) argues that the preconquest estimates of Mexican population are too high, and that the death rates from smallpox are overstated.

The decline of the Indian population had serious cultural and economic consequences, especially among the native civilizations. The ability of the Aztecan, Tarascan, and Incan states to resist Spanish invasion was weakened by the first smallpox pandemic. After the conquest, the declining population brought about decreased tributes and a scarcity of labor so severe that, according to some authors (Borah, 1951), a long period of economic depression ensued in Mexico and Peru during the seventeenth century. Moreover, the Europeans imported large numbers of African slaves into areas that suffered nearly complete extinction of the native population, as in the Caribbean Islands and portions of the Brazilian coast. Descendants of Africans have contributed distinct cultural and racial attributes in those areas to this day.

► *Racial Mixing*

A lasting biological effect of the conquest on the native population was the gradual dilution of Indian blood with that of Europeans and Africans. Interbreeding has produced the present mixed groups that predominate in much of Latin America. In Spanish America, the mixing of Europeans and Indians began slowly. By law the Spanish crown discouraged mixed unions, and compared to the large native population, there were relatively few Spaniards in the colonies during the sixteenth century (Table 3.2). In Mexico and Central America a sizable mestizo population did not exist until the beginning of the eighteenth century. Thereafter the number of mestizos increased rapidly, accounting at the close of the colonial period for about 35 percent of the population.

During the colonial period, large numbers of enslaved Africans were imported into Middle and South America (Mellafe, 1975). Black slavery had been a tradition among Iberian peoples for centuries. Moreover, in the tropical lowlands of the New World blacks proved to be resistant to disease, especially malaria, and they survived longer than the Indians at hard labor. Most of the Africans brought to Latin America were shipped to Brazil (Miller, 1990), the West Indies, and the Guianas, where Indian labor had disappeared or was greatly depleted. From 1650 to 1830, slave traders transported between 3 and 4 million enslaved Africans to the British and French West Indian islands; from 1540 to 1860 the Portuguese may have shipped nearly 4 million to the Brazilian coast (Curtin, 1969, Table 10.3). By the close of the eighteenth century, blacks dominated the population of these two areas, and a large group of mulattoes from black and white unions had evolved. Moreover, Portuguese in Brazil bred with Indian women in the back country, giving rise to the *mameluco*

TABLE 3.2   Changes in the Racial Composition of the Population in Selected Areas of Latin America During the Colonial Period

| | Mexico and Central America | | |
	1570	1650	1825
Whites	60,000 (2%)	120,000 (6%)	1,000,000 (14%)
Mixed	36,000 (1%)	260,000 (14%)	2,500,000 (35%)
Indians	3,000,000 (97%)	1,500,000 (80%)	3,600,000 (51%)
Total	3,096,000	1,880,000	7,100,000

| | West Indies (North European and Spanish Possessions) | | |
	1570	1650	1825
Whites	7,500 (12%)	80,000 (13%)	482,000 (17%)
Blacks	56,000 (85%)	400,000 (66%)	1,960,000 (69%)
Mixed	—	124,000 (21%)	400,000 (14%)
Indians	2,000 (3%)	(nearly extinct)	(nearly extinct)
Total	65,500	604,000	2,842,000

| | Brazil | | |
	1570	1650	1825
Whites	20,000 (2%)	70,000 (7%)	920,000 (23%)
Blacks	30,000 (4%)	100,000 (11%)	1,960,000 (50%)
Mixed	—	80,000 (8%)	700,000 (18%)
Indians	800,000 (94%)	700,000 (73%)	360,000 (9%)
Total	850,000	950,000	3,940,000

Source: Based on data modified from Rosenblat, 1954. Vol. 1.

mixed blood; on the coast, black and Indian interbred to form the *cafuso* element. More racial mixture occurred in Brazil than in any other area of Latin America.

Enslaved Africans were imported into Spanish America in smaller numbers (about 1.5 million), but they were used in nearly every Spanish area of the New World. Most were sold in the Greater Antilles, but many were taken to Mexico and Central America, where they worked in mines, on stock ranches, in port towns, in sugar mills, and at other pursuits thought to be too arduous for Indians (Palmer, 1976). Others were used as farmhands in the oases of the Peruvian coastal desert, which had been depleted of its aboriginal population by disease. A large contingent of blacks in Spanish America was sent to New Granada (present-day Colombia) to work in the lowland alluvial gold deposits because the natives quickly perished under Spanish rule or were uncontrollable.

Black slaves in Spanish America interbred with Indians and whites, in spite of regulations to the contrary. Many blacks escaped to Indian villages, and other run-aways formed isolated settlements, taking Indian wives with them. Black and mulatto freedmen also mixed with the other races, so that a complex racial caste system had evolved in Spanish society by the close of the colonial period. Except in Colombia, Venezuela, and northern Ecuador, however, blacks in Spanish America have been almost completely absorbed into the mestizo element. The dominant black population along much of the Caribbean coast of Central America is not a Spanish colonial development but stems from more recent migrations from the West Indies.

Because of the high death rate and miscegenation, the percentage of the native population declined almost everywhere in Latin America during the sixteenth, seventeenth, and eighteenth centuries. Since the close of the colonial period, Indian groups have been reduced to pockets of varying extent in southern Mexico; southwestern Guatemala; the highlands of Ecuador, Peru, and Bolivia; south-central Chile; and a few areas in the Amazon and Orinoco basins. Even the "Indians" of these areas have today been "mestizoized" to varying degrees.

COLONIAL ECONOMY

European activities in the New World that most thoroughly affected the geographical landscape were economic in nature. Spaniards, Portuguese, and North Europeans exploited human and natural resources within their respective realms of influence in various ways. In the early formative years of occupation, the Spaniards emphasized the exploitation of the native population through institutions such as the *encomienda*, labor levies, and enslavement of rebelling Indians. Formal establishment of industries based on the exploitation of natural resources such as mining and agriculture followed. The Portuguese in Brazil and the North Europeans in the West Indies were less concerned with native peoples, who were far less numerous and civilized than those in the Spanish realm. As indicated previously, Portugal and the North European countries relied on slaves imported from Africa to exploit the land.

The colonial exploitation of natural resources in Latin America was dominated by two industries: (1) the mining and processing of precious metals, and (2) the cultivation and manufacture of sugar. Mining, localized in the mountainous interior, formed the economic core of Spanish America. The sugar industry, best developed on fertile soils along tropical coasts, was the main economic basis for the Portuguese in Brazil and the North Europeans in the Caribbean. The products of these two industries furnished the most valuable exports shipped from the Latin American colonies to their respective ruling countries. Mining and sugar were the center of most of the economic activities within the colonies, including staple food production, stock raising, and transport. Both industries required a large, cheap labor force, supplied by native Indians in the case of mining and by imported African slaves in the case of sugar. In short, within their respective areas, the two industries were instrumental in shaping the form of the geographical landscape of colonial Latin America, including settlement and land use patterns, as well as distribution and racial composition of the population. Much of this landscape is retained today, and the present export orientation of most Latin American countries stems from the nature of the colonial economy.

The geographical limitation of mining to the Spanish realm and sugar to the Portuguese and North European areas was not absolute. Gold mining was important in colonial Brazil, but not until the eighteenth century. Sugar was widely grown in Spanish America, but mainly for internal consumption, except in Cuba, which became a major sugar exporter late in the eighteenth century. Moreover, colonists exported other products, such as dyestuffs, hides, and tobacco, but in terms of value, these were comparatively insignificant in the export economy.

Colonial Economy of Spanish America

Encomienda.   Although mining and agriculture dominated the colonial economy of Spanish America, the *encomienda* was Spain's most effective economic and social institution in the New World during the initial period of settlement in the early sixteenth century. The *encomienda* system arose in Spain during medieval times; in the Americas it was modified to suit local conditions. In brief, groups of Indians—

usually several contiguous villages—were "commended" to a deserving Spaniard (the *encomendero*), who was obligated to instruct his wards in the Spanish language and the Catholic faith. In return, the *encomendero* had the right to exact tribute in the form of economic products or labor from his Indians. In theory the *encomienda* did not entail landownership. *Encomienda* grants were made throughout the Spanish realm, but they were successful mainly in New Spain and Peru, where the native population was numerous and civilized. In 1549 the rendering of labor service by *encomienda* Indians was abolished. Thereafter, by law, the system became exclusively tribute payments by natives to their *encomendero*, and Indian labor for agriculture and mining was supplied mainly through levy systems that the Spanish authorities imposed on Indian villages. By means of their *encomiendas*, many Spaniards in the New World were able to survive during the early sixteenth century because tributes consisted of agricultural products, gold, cloth, or any other product that the *encomendero* could consume or sell. The drastic decline in the native population in the late sixteenth century contributed to the demise of the *encomienda*. By the seventeenth century, tributes had so decreased that few Spaniards found their *encomiendas* profitable, and the larger ones had reverted to the crown.

Mining Economy.   Following the rapid exploitation of alluvial gold deposits in the Greater Antilles, the Spaniards continued the search for quick wealth on the mainland, where they found gold and silver in abundance. From Mexico to Chile, gold was encountered as dust and nuggets in the sands and gravels of lowland rivers that drained crystalline rock mountains. Because alluvial deposits were quickly depleted, requiring the frequent shifting of labor gangs along stream courses, their exploitation rarely led to the establishment of permanent settlements. Silver and some gold, however, occurred as complex ores in veins that penetrated deeply into mountain flanks within the highlands. Since vein mining and ore refining were slow processes and required large investments in machinery, reagents, and labor, such operations usually resulted in the founding of permanent mining camps, some of which grew into large, opulent cities. So great was the demand for labor, food, clothing, draft animals, and raw materials in the large mining centers that these settlements became the "dynamos" that generated much of the economic activity within colonial Spanish America.

Mexico and Upper Peru (Bolivia) were the foremost producers of silver during the colonial period, and New Granada (Colombia) was the source of much of the gold in Spanish America. Other producing areas were minor by comparison: the Greater Antilles (gold), Honduras (silver and gold), Ecuador (gold), and Chile (gold and copper). Although Taxco, founded southwest of Mexico City about 1531, was the first silver district to be developed by the Spaniards in the New World, large-scale production of the metal did not begin until after the opening of the Zacatecas mines in 1546 and the subsequent exploitation of rich silver lodes north and southeast thereof within the "Silver Belt" of New Spain (Parral, Guanajuato, and Pachuca). At about the same time (1544), the great silver deposits of Potosí were discovered in the high Altiplano of Upper Peru. Later finds in the same area and in the central Andes (Oruro, Castrovirreyna, and Cerro de Pasco) gave the viceroyalty of Lima even more wealth than Mexico in the sixteenth and seventeenth centuries. Between 1580 and 1670, for example, Upper Peru accounted for 65 percent of all American silver shipped to Spain, whereas by the late eighteenth century, Mexico's share was 67 percent. During most of the colonial period, Zacatecas and Potosí were the largest and most productive mining centers in Spanish America. Zacatecas produced about one third of Mexican silver and was considered the second city of New Spain; it possessed 5,000 inhabitants in the middle of the seventeenth century.

At that time, Potosí was much larger, having 160,000 inhabitants. It produced two thirds of Peruvian silver and was considered the leading metropolis of South America (Bakewell, 1971; Brading and Cross, 1972).

Mine Labor.   The colonial mining industry in Spanish America could not have operated without a plentiful labor supply. Fortunately for the Spaniards, the silver-producing districts of Mexico and Peru lay within or near areas populated by civilized Indians. In both areas the Spaniards applied a system of forced labor for the mines. In Mexico it was called the *repartimiento*, and in Peru it was the *mita*. Both derived from aboriginal systems of obligatory public service. Both were labor levies whereby Indian villages were required to furnish able-bodied males to work for various periods in the mines or in the ore-processing plants. In Mexico forced labor was largely replaced by free wage workers by the middle of the seventeenth century, especially in the northern mines, far from the densely populated central part of the colony. In Peru, however, the *mita* continued until its abolition in 1812, but it was considered an oppressive form of servitude that caused the death of thousands of Indians. A total of 13,300 Indians were obliged to work at four-month intervals in the Potosí mines alone, some having to walk from the Cuzco district, a distance of 600 miles (Vázquez de Espinosa, 1942, p. 624). The decline of the native population through disease and other factors gave rise to frequent labor shortages. In New Granada, for example, Indians who were forced to work in the lowland gold placers through the *encomienda* system quickly died off and were replaced with enslaved Africans.

Refining Techniques and Reagents.   In 1556 a new method of refining silver, called the *patio*, or amalgamation, process, was introduced into Mexico. It revolutionized the mining industry in Spanish America and caused a sharp increase in silver production. Smelting, effective for refining high-grade silver ores, was previously the sole method known to the Spaniards. Amalgamation, a simple, inexpensive means of refining large quantities of low-grade ore, involved finely grinding and puddling ore and applying mercury, salt, and copper pyrite to extract the silver. Strictly controlled by royal monopoly, mercury was shipped in large quantities from Spain to Mexico for this process. Because of the distance and cost of mercury shipments, amalgamation was not introduced into Peru until 1574 after the discovery of a large deposit of mercury at Huancavelica in the central Andes, 800 miles northwest of the Potosí mines. Thereafter silver production at Potosí boomed (Bakewell, 1977). In both Peru and Mexico, maximum silver output occurred between 1580 and 1630, but for the rest of the seventeenth century, production slumped badly, mainly because of mercury shortages. With the restoration of regular shipments of this reagent and improvements in mining technology in the late eighteenth century, silver production began to exceed that of any previous time (Brading and Cross, 1972; Fisher, 1975).

The need for large quantities of salt in the amalgamation process created an adjunct industry to that of silver mining. In Peru and Mexico salt was obtained by scraping the surfaces of dry lake beds. The Peruvian supply originated in the large salines on the high, arid Altiplano, whereas the salt for the Mexican refineries came from the central desert, north and east of the Silver Belt, and from tidal lagoons along the Pacific coast. Since copper pyrite was found near most of the mining centers, its supply posed few problems for the silver refiners.

Wood was also needed for many operations in the mining centers. The lumber business developed to supply beams and planks for buildings, mine supports, and machinery, and charcoal was manufactured for use as fuel in smelting ore and in various operations of the amalgamation process. For miles around every mining center in Mexico, woody vegetation was seriously depleted by woodcutters and charcoal makers; the effects of this devastation are seen to this day. But in the high, treeless country of Potosí and adjacent areas, lumber and charcoal were hauled great distances, from the forested eastern slopes of the Andes and even from Ecuador. Peruvian refiners were often obliged to use llama dung and bunch grass to fuel smelters because of the lack of charcoal.

Agriculture. Throughout the colonial period, production of food for subsistence remained primarily in the hands of the Indians and mixed bloods, who retained native crops and farming methods. Commerical agriculture, however, was directed mainly by the Spaniards, using Old World plants, animals, and techniques. The development of commercial farming and stock raising was closely tied to the mining industry, since the mines were the main markets for grains and animal products. The Spanish towns were secondary markets. Moreover, large numbers of pack animals, especially mules, were required for land transport to connect ports and food-producing areas with mines and towns. Except for dyestuffs and hides, few agricultural products were exported overseas. Production for local consumption was the rule.

In the early years of conquest, the Spaniards relied on Indian foods, but they soon introduced Old World plant and animal domesticates into the Americas. Wheat and barley became important crops in the tropical highlands of Mexico and Peru and in the middle-latitude areas of Chile. Other significant Spanish crops that were introduced included grape vines, olives, and figs, which were grown chiefly in the desert oases of northern Mexico and coastal Peru. Such crops did poorly in the tropical lowlands, where Old World plants such as plantains, citrus, and sugarcane were successfully grown and even adopted by Indian farmers. The introduction of European domesticated animals resulted in fundamental changes in land use. The Indians had no large domesticated animals except for the Andean llama and alpaca, and a true aboriginal herding economy was unknown. The range animals—cattle, sheep, horses, and mules—became the mainstay of the livestock industry, which was established in most parts of the Spanish realm. The animals reproduced so rapidly that in the tropical savannas of the West Indies and lowland Mexico and the grassy Pampas of Argentina, herds went wild and were annually "harvested" for hides by hunting expeditions until well into the seventeenth century.

In the Valle del Mezquital, Mexico, sheep flocks increased rapidly until they exceeded the carrying capacity of the environment. After 1570 sheep numbers dropped, but damage had been done, and the overgrazed lands declined in quality (Melville, 1994, pp. 54–55). In the Andes, sheep seem to have been adopted successfully into indigenous farming (Gade, 1992, p. 467).

In New Spain by the 1530s the Spaniards had established wheat farms in the basins of Mexico and Puebla on the central plateau, and cattle, sheep, and mules were herded in the surrounding hills that were unoccupied by Indians (Gibson, 1964). At that time wheat flour and meat were supplied mainly to Mexico City and other Spanish towns. However, with the opening of the Silver Belt north and southeast of Zacatecas during the following decades, the main market for agricultural produce shifted to the mines. New wheat-farming and stock-raising areas such as the fertile Bajío of Guanajuato and the Valley of Guadalajara in the western portion of the central plateau were opened to supply Zacatecas, Guanajuato, and smaller mining centers nearby. As Spanish settlement progressed northward from Zacatecas along the Silver Belt, irrigated wheat and maize farms sprang up along river valleys near every mining center, and stock ranches were formed in the surrounding grass-covered plains. Thus, in northern Mexico the mining community, the stock ranch, and the grain farm, being economically interdependent and geographically proximate, combined to form the colonial ranch-mine settlement complex, traces of which persist today. Using Indian labor, the farmers supplied wheat and maize to the mines; the stockmen furnished mules (for mine work), beef (for food), hides (for ore sacks), and tallow (for candles, the chief means of illuminating the dark shafts and tunnels).

In the viceroyalty of Peru, the agricultural pattern as it related to the mining industry was more complex than that in New Spain. The main market for agricultural produce was the mining center of Potosí and the surrounding satellite communities in Upper Peru. Products from as far away as New Granada and Quito in the northern Andes, central Chile, the Argentine Pampas, and Paraguay converged on Potosí, the major economic magnet for the entire viceroyalty. Most of the staple foods consumed in the Potosí area, however, came from nearby Indian lands and Spanish estates, especially (1) in the fertile, well-watered valleys of Cochabamba, Chuquisaca (La Plata), and Tarija on the eastern slopes of the Andean highlands (maize and wheat), and (2) on the Altiplano (source of the native *chuño*, the main food consumed by Indian mine laborers). Food was also brought in from the irrigated farms along the Peruvian desert coast, where land was taken over by Spanish farmers after the Indians had perished in the smallpox epidemics of the early sixteenth century. The most valuable exports of the southern coastal oases, such as Pisco, Ica, and Arequipa, were wine and brandy, products of the vineyards introduced by Spaniards and worked by black slave labor. Consumed in large quantities in the mining centers, wine and brandy also came from the vineyards of Mendoza in western Argentina. Today, these vineyards, as well as those of the Peruvian oases, are still among the major wine-producing districts of South America. In the Peruvian mining centers and administrative towns, the Spaniards and mestizos consumed most of the wine and brandy, while the Indian workers drank the native *chicha*, or corn beer, made locally from maize brought in from the eastern valleys. In some of these valleys, the Spaniards had taken over aboriginal coca farms and, with Indian labor, produced this stimulant for sale in the mining centers (Cobb, 1945, 1949; Helmer, 1950).

The Potosí mining district was also the major market for the products of stock ranches: meat, hides, and tallow as well as draft and pack animals. Such products were supplied mainly from the large ranches established in the last half of the sixteenth century in northwestern Argentina and the northern edge of the Pampas. Probably the most renowned aspect of the stock-raising industry to develop through the stimulus of the Potosí market was the mule trade of northwestern Argentina, where each year during the seventeenth and eighteenth centuries 30,000 to 60,000 animals from the ranches around Córdoba and Tucumán were sold at the mule fair of Salta and driven to Upper Peru. Mules were also driven to the Peruvian mines from the Cúcuta pastures of northern New Granada, 2,500 miles from Potosí (Vázquez de Espinosa, 1942, p. 323). In addition, native llamas in Peru were used as pack animals and for food; 100,000 were consumed each year in Potosí at the beginning of the seventeenth century (Jiménez de la Espada, Vol. 1, 1965, p. 308).

In other areas of the viceroyalty, mines as well as port cities and administrative centers stimulated farming and stock raising. In central Peru, Spaniards established grain farms and cattle ranches in the high Andean basins, such as Huancayo, Jauja, and Huamanga (Ayacucho), in order to supply food for the mines of Cerro de Pasco and Huancavelica and the city of Lima. The same pattern developed in the highlands of Quito (Ecuador), where wheat, barley, maize, and cattle were shipped to the Zaruma gold mines and the port of Guayaquil. Similarly, the Spanish grain and cattle estates in the highlands of New Granada around Bogotá and Tunja depended on the gold placers of Antioquia and the upper Cauca for their main markets.

Agriculture and Land Tenure. A fundamental aspect of commercial farming and stock raising in colonial Spanish America was the development of large landed estates, variously called *haciendas, estancias,* or *hatos.* The land tenure problems of many Spanish American countries since independence stem partly from adverse social and economic conditions associated with the large estates owned by elite families.

During the colonial period, the Spaniards normally obtained private lands by means of the *merced*, a municipal or royal grant. Single grants for raising livestock were large, some more than 5,000 acres in size. Through the accumulation of several contiguous grants by purchase or otherwise, some Spaniards came into possession of immense estates, particularly in northern Mexico and the Argentine Pampas, where the Indian population was sparse and nomadic. Original land grants were smaller in areas of dense native population, as in central Mexico and the Andean highlands, where a mixed grain-livestock economy evolved and where tracts occupied by Indians were legally protected from usurpation. Nonetheless, as the number of Indians declined, their abandoned lands gradually were attached to adjacent Spanish holdings, creating large *haciendas*.

The church, especially orders like the Jesuits, Dominicans, and Augustinians, acquired large *haciendas* that were operated more efficiently than the lay estates. Beginning with the expulsion of the Jesuits from the New World in 1767, church lands were gradually absorbed by private estate owners.

The focal point of the *hacienda* was a compact cluster of buildings called the *casco*. It included the "big house" of the owner or overseer, the huts of the workers, often a chapel, and the corrals and granaries. By the end of the eighteenth century, the *hacienda* and its *casco* had become the dominant rural settlement form in much of Spanish America. *Hacienda cascos*, some of which are in ruins and others in use, are still a conspicuous feature of the countryside in many Spanish American countries.

Inadequate labor and uncertain markets posed constant problems for the estate owners. In the early colonial period, the *haciendas* within areas of dense native population were often supplied with hands through government-controlled work levies, such as the *repartimiento* in Mexico. In Peru and Chile some *encomenderos* who managed to obtain land grants near the villages of their Indian wards probably used the male population for farmwork long after the abolition of *encomienda* labor services in 1549 (Morner, 1973). However, after the epidemics of the sixteenth century had reduced the Indian population, most *hacienda* owners used extralegal means to get sufficient labor. In Mexico and Central America, Indians were induced to leave their villages to settle on the estate by offers of relatively high wages paid in kind. By advancing them credit, the owner was able to keep his workers in perpetual debt (debt peonage), which bound them to the *hacienda* (Chevalier, 1963). In the Andean area, Indians were bound to the estates by being permitted to cultivate a portion of *hacienda* land for subsistence in return for their labor in tending the owner's crops and herds (Vargas, 1957). Bondage proved to be one of the most pernicious social aspects of the *hacienda* and helped to precipitate agrarian revolutions and reforms in various Spanish American countries during the twentieth century.

The Spanish American *hacienda* produced for a local market (mines and towns) rather than for overseas trade. Because of frequent fluctuations in metal production and population in mining centers, high costs of land transport, and occasional shortages of farm labor, *hacienda* owners normally underproduced to maintain high prices, and relatively small amounts of capital were invested to improve the land. Consequently, large tracts of potentially productive land were purposely retired from agricultural production, a practice that has continued into modern times despite the increase of population and demand for food. Furthermore, in Spanish American society the mere accumulation of large landholdings enhanced the social prestige and political power of elite families.

Although the Spanish American *hacienda* is usually defined according to the attributes described here, its nature varied from place to place, so that nowhere did there exist a "classical" form. Moreover, the *hacienda* as both a major institution and a landscape element evolved slowly during the colonial period. In some places, such as Bolivia, it did not reach full development until the beginning of the twentieth century (Morner, 1973).

Manufacturing Industries.   To protect industry in Spain, the Spanish crown discouraged large-scale manufacturing in its American colonies. Besides processing agricultural products such as sugar, grapes, and dyestuffs, only one colonial industry—textile manufacturing—grew to sizable proportions in various parts of Spanish America. Again, the mining centers consumed the greater part of textile products, although Spanish towns and Indian communities were also markets. The weaving of cheap, coarse cotton and woolen cloth formed the basis of the colonial textile industry, which developed in many areas inhabited by Indians who were weavers by tradition. Three textile-making areas, however, produced most of the cloth that entered local trade: (1) east-central Mexico, especially the cities of Puebla, Tlaxcala, and Mexico; (2) the highland basins of Ecuador, where weaving was carried on in Indian towns around Quito, Riobamba, and Latacunga; and (3) the districts of Córdoba, Santiago del Estero, and Tucumán in northwestern Argentina. In all three areas, cloth was woven by Indians who either paid tribute in cloth or worked as forced laborers in small Spanish-controlled factories called *obrajes* (Greenleaf, 1967; Super, 1976).

The greatest cloth production occurred in the Ecuadorian highlands, where large herds of merino sheep grazed and abundant cotton was obtained from the hot lowlands of the Amazon Basin and the Pacific coast (Phelan, 1967; Vargas, 1957). From Quito textiles were shipped to all parts of the viceroyalty of Peru, but mainly to the Potosí mining district, 1,500 miles away. Potosí also obtained large amounts of cotton cloth from the textile towns of northwestern Argentina. Mexican textiles were sold mostly in the northern mines, where wages were often paid in pieces of cotton or woolen cloth. Since the cloth and other textiles produced in the Americas were of poor quality and were manufactured primarily for Indian workers in mines and *haciendas*, the weaving guilds in Spain raised few objections to the colonial industry as long as they could export Andalusian velvets, silks, and brocades for sale to rich miners and *hacendados* in Peru and Mexico.

Of the agricultural products raised in Spanish America for export overseas, only dyestuffs and hides were of significance. Two dyestuffs native to America were cultivated for export and used in the European cloth industry. One was cochineal, a crimson dye extracted from insects raised on the leaves of the prickly pear cactus by the Indians of New Spain. In Europe cochineal commanded fabulous prices; by 1600 it ranked in value second only to silver among Mexican exports to Spain. The other dyestuff was native indigo, cultivated on Spanish estates by black African labor mainly along the Pacific lowlands of Central America (MacLeod, 1973). In the seventeenth and eighteenth centuries, it became the main economic product of that area and was produced in El Salvador as late as 1920, when it was finally eclipsed by synthetics. Although most of the colonial Spanish American stockmen found their main markets for livestock in the local mines and towns, in some areas cattle were raised or hunted chiefly for hides that were exported to Spain for the leather industry. The half-wild herds in the Spanish West Indies, the Gulf lowlands of Mexico, and the Argentine Pampas were the main source of these hides. According to Chaunu (1955, Vol. 6, p. 105), between 1562 and 1620, by weight and volume, hides were the most important export to Spain from the American colonies. By value, however, hide exports were insignificant when compared to precious metals.

Colonial Economy of Portuguese America

Whereas the colonial economy of Spanish America tended to revolve around the mining of gold and silver in the mountainous interior, that of Portuguese

America was based largely on the production of sugar along the Brazilian coastal plain. Not until the eighteenth century did mining become significant in the Brazilian interior, and thereafter the nature of Spanish and Portuguese colonial economies became more similar.

The Brazilian Sugar Plantation. The decision of Portuguese *donatarios* to turn to sugar cultivation in Brazil in the middle of the sixteenth century was prompted by rising sugar prices in Europe. Since 1520, the price had soared, mainly because the Ottoman Turks had occupied Syria and Egypt, which at that time were the main sources of sugar for Europe. Already versed in sugar production on their Atlantic islands of Madeira and São Tomé, the Portuguese readily turned to that activity in Brazil. By 1580 Brazil was the leading sugar-producing area of the world, a rank it held for more than a century until surpassed by the West Indies around 1700.

Along the wet, forested Brazilian coast, plantations were established on large tracts of land, called *sesmarias*, each averaging about 12 square miles in size. Such tracts were allotted to colonists, who were obliged to begin cultivation and make other capital improvements, including forest clearing and mill construction. Because such improvements required investment of considerable capital, only the wealthier colonists became successful planters. By 1570 the northern *capitânias* of Pernambuco and Bahia had become the major sugar areas and the politico-economic center of the colony.

The sugar plantation was the major settlement form along the coast; except for ports and trading centers, the Portuguese did not establish formal towns as did the Spaniards. Each plantation usually consisted of four parts: the cane fields, in actual cultivation or in fallow; pastureland for grazing oxen; the *roça*, or land cultivated for food; and the woodland, to provide fuel and land for future cultivating. The grinding mill and cooking vats for processing sugar formed the plantation center, around which were the huts of the workers and the big house of the owner or overseer. The Brazilian sugar plantation was physically similar to the Spanish *hacienda*, but it was quite different in function, producing a single processed product for export to overseas markets.

Along with sugarcane and the tropical plantation concept, the Portuguese brought over oxen to power the mills and to draw cane carts and plows (Table 3.3). Oxen were so valuable as draft animals that cattle were rarely raised for meat on

TABLE 3.3 Sugar Production in the Americas for Export, 1580–1875 (in tons)

Region	1580	1650	1700	1770	1800	1825	1850	1875
Brazil	4,800[a]	28,500	22,000	20,400	21,000[b]	99,000[c]	138,000[d]	102,500
English and French West Indies	—	7,000	31,400	168,200	161,400	183,000	134,600	260,700
Guianas	—	—	4,100	9,200	6,900	47,700	38,900	91,100
Spanish West Indies	—	—	—	10,000[e]	28,400	54,700	273,200	790,100

Source: Compiled from Deerr, 1949, Vol. 1.
[a]All figures rounded to nearest 100.
[b]1796.
[c]1823.
[d]1853.
[e]1770–1778 average.

the sugar plantations. Until the seventeenth century much of the dried meat and hides consumed on the plantations were imported from Portugal, the Atlantic islands, and even the Argentine Pampas. Because of the tropical climate, the cultivation of wheat and other Mediterranean crops was never successful in Brazil. The Portuguese colonists adopted Indian crops, such as manioc and sweet potatoes, as food staples for themselves and plantation slaves.

The Portuguese operated their holdings almost wholly with slave labor. As indicated previously, they used both Indian and African slaves but, because the docile, forest-dwelling Tupi died quickly in servitude, black slaves comprised most of the workforce on the plantations. Each large plantation needed 150 to 200 slaves, most of whom were purchased at stations along the Guinea coast of Africa and in Angola, a Portuguese colony. Relations with Brazilian natives, however, were not confined to slavery; in the southern *capitânias* an active trade developed between the Portuguese and the Tupi, who supplied manioc to the plantations and to the transatlantic slave ships in exchange for iron tools (Cortesão, 1956).

Although dominant, sugar was not the sole product of plantation agriculture in colonial Brazil. In the seventeenth century, especially during and after the Dutch occupation of Pernambuco, tobacco became an important export crop on some of the smaller holdings in the northeast. Moreover, many plantation owners with land near the coast continued to cut brazilwood for export as a secondary activity well into the seventeenth century. Exports decreased sharply after 1650, mainly because of the gradual destruction of the coastal forests (Mauro, 1960). Brazilwood has long since disappeared from commerce, but sugar and tobacco, together with cacao, are still major commercial crops of the coastal plain of the Brazilian northeast.

Pastoral Activities.   In Brazil the range cattle industry first evolved to meet the demand for animal products on sugar plantations. In the northeast, colonists who lacked capital to establish plantations obtained large land grants to run cattle, especially in the dry scrub forests of the *sertão* from Pernambuco northward to Ceará. Cattle raising was expanded in the 1630s, when the tropical savannas of the plateau were penetrated by stockmen fleeing the Dutch occupation of Pernambuco. By 1700 there had been a remarkable development of stock raising in the São Francisco Valley, which drains much of the northeastern plateau. In 1711, 1.3 million head of cattle grazed on the São Francisco ranches, some of which were 40 square miles in size (Poppino, 1949). Ranchers from the northeastern interior supplied livestock, dried meat, tallow, and hides to the adjacent coastal sugar plantations in return for raw sugar, cane brandy, food, and other necessities. Such transactions took place in a series of market centers, or fairs, established on the coastal edge of the *sertão* (Figure 3.4). A similar pattern developed on a lesser scale in southeastern Brazil, where large cattle ranches on the São Paulo plateau supplied the coastal sugar plantations of São Vicente and Rio de Janeiro. In the Brazilian northeast, this simple symbiotic relationship between coast and interior continued into the twentieth century (Lasserre, 1948). Through both the sugar plantation and the cattle ranch, large holdings became the dominant form of land tenure in colonial Brazil.

Mining.   In 1693 the accidental discovery of gold in the headwaters of the São Francisco and Doce rivers greatly expanded the colonial economy of Brazil. During the eighteenth century, Brazil became the world's leading producer of gold, and for a time, the precious metal surpassed sugar as the colony's most valuable export. The initial discovery in Minas Gerais, as this highly mineralized plateau area came to be called, precipitated the world's first major gold rush. Thousands from the Brazilian coast, Portugal, and other parts of Europe flocked into Minas Gerais. Discouraged by falling sugar prices, many plantation owners of Bahia and Pernambuco liquidated their lowland holdings and moved their slave gangs into the interior to mine gold. It is estimated that during the eighteenth century gold mining attracted 800,000 Portuguese to Brazil.

Minas Gerais gold was found mainly in alluvial deposits in streams and on terraces or, less frequently, in weathered soils on hillsides (Boxer, 1962). Thus, most settlements were ephemeral mining camps, which were shifted to a new location after depletion of a particular deposit. The few permanent towns in the mining district were administrative centers, such as Vila Rica, later called Ouro Prêto.

Food for the mining crews was imported or raised on small plots in the area. The demand for dried meat in the camps gave the stockmen of the São Francisco Valley and the São Paulo plateau an expanded market and cattle fairs, similar to those that served the sugar plantations were established in various parts of Minas Gerais (Webb, 1959). Moreover, the small town of Rio de Janeiro on the southeastern coast became the main port for the mines. Because of the growing importance in the Brazilian economy and the relative decline of sugar in the northeast, in 1763 the colonial capital was moved from Salvador to Rio de Janeiro. The subsequent discovery and exploitation of gold farther inland (Goiás and Cuiabá) and of diamonds in various parts of the plateau added to the growth of Brazil's mining industry and to the token settlement of scattered points in the far interior.

Thus, during the eighteenth century, the colonial economy of Brazil revolved around the exploitation of precious metals, paralleling the pattern developed earlier in Spanish America. Brazil's "golden age," however, was short-lived; by the early nineteenth century the major gold placers had been depleted, and other boom activities, such as coffee growing, were beginning in the São Paulo area. Nonetheless, the tradition of placer mining still persists among the *garimpeiros*, the folk miners who wash gold and diamonds from stream gravels in the Brazilian interior.

Colonial Economy of the North European Possessions

Like the Portuguese in Brazil, the English and French in the West Indies developed sugar as the main facet of the colonial economy. Except for their plantations in the Guianas, the Dutch concerned themselves primarily with shipping, and their small islands in the Antilles functioned mainly as trading centers. Lacking deposits of precious metals in their islands and coastal lowland possessions, none of the North European nations engaged in mining.

The English and French devoted the first two decades of colonizing the Lesser Antilles (1620s–1640s) to the settlement of farmers, who used white indentured servants from the homeland to cultivate small holdings planted with various tropical crops marketable in Europe. In the English islands, indentured servants were mainly youths between the ages of 15 and 25 who contracted to serve from four to six years; such labor made up the bulk of North European immigration to the Leeward Islands and Barbados until the middle of the seventeenth century. Crops grown for export included, first, tobacco and cotton, and, secondarily, indigo and ginger (Deerr, 1949, Vol. I).

Individual farmers as well as trading companies in London and Paris subsequently saw the possibilities of establishing sugar plantations on the islands, as the Portuguese had done earlier on the Brazilian coast. Large-scale sugar cultivation began on Barbados, Guadeloupe, and Martinique in the 1640s. Through Dutch influence, landowners on those islands copied Brazilian methods, including plantation organization, black slave labor, and reliance on an overseas market. It was on Barbados that the planter class, composed of rich sugar barons, first took shape in the Caribbean (Dunn, 1972). The sugar estate was gradually introduced into other English and French islands, so that by the first decade of the eighteenth century the center of sugar production in the Americas had shifted from Brazil to the Caribbean (Table 3.3).

As in Brazil, natural conditions in the West Indies—lack of frost, plentiful rainfall with a dry period for harvest, and fertile soils—favored sugar cultivation. The volcanic and coral limestone soils of the Lesser Antilles and the alluvial plains of Jamaica and St. Domingue (Haiti) were highly productive. The paucity of arable land in the mountainous islands, however, limited the extent of sugar cultivation, and in most cases (flattish Barbados excepted), plantations occupied the narrow coastal rims close to the sea and the ports.

The plantations, acquired through government grants or by purchase, were smaller than those in Brazil, but they were similarly organized. Cane fields occupied the best land; plots tilled for food, primarily manioc and sweet potatoes, were relegated to mountain slopes; and the plantation house, the mill, boiling house, and workers' quarters formed the settlement center. Mills were powered by water, wind, or animals. On the mountainous islands swift streams were often harnessed to turn the mill wheels; on Barbados and other flatter islands the trade winds turned huge, Dutch-type windmills for grinding cane. Oxen, however, formed the main power source, and these were raised on the plantations or imported from the North American colonies. Cattle and dried meat were imported from the North American colonies and Venezuela. Unlike the Brazilian case, an extensive range cattle industry never developed on the islands to supply meat and other animal products to the plantations. Instead, most protein food, primarily in the form of dried fish, was imported from North America to feed the slave labor. To this day, despite the decline of the plantation, some West Indians prefer dried cod from maritime Canada or New England to local fish, and most of the smaller islands still rely heavily on imported food and clothing.

The West Indian planters were even less successful than their Brazilian counterparts in enlisting native labor; the rebellious Caribs of the Lesser Antilles would not work, and the natives of larger islands had long since vanished. Thus, the North Europeans imported black slaves, whose descendants today comprise most of the West Indian population outside the Spanish-speaking islands. After the abolition of slavery in the 1830s, the English imported contract workers from southern Asia (mainly India). The descendants of these indentured workers still live in some West Indian islands (chiefly Trinidad and Jamaica) and Guyana.

Using the plantation system and black slaves, the French and English West Indies became the leading sugar-producing areas during the eighteenth century. The largest producers were St. Domingue (Haiti), settled by France in the last half of the seventeenth century, and Jamaica, ceded to England by Spain in 1655 (Table 3.4).

TABLE 3.4   Sugar Production in the North European West Indies, 1770 (in tons)

Saint Domingue (French)	60,000[a]
Jamaica (English)	36,000
Martinique and Guadeloupe (French)	19,600
Barbados (English)	8,600
Leeward Islands (English)	28,800
Windward Islands (English)	12,370

Source: Compiled from Deerr. 1949, Vol. 1.
[a]All figures rounded to nearest 100.

Partly because of the black rebellion in Haiti (1793), the emancipation of slaves in 1838 (Britain) and 1848 (France), and the increasing competition from other cane and beet sugar producers, the colonial plantation system was declining in the West Indies by the middle of the nineteenth century. Although sugarcane continued to be cultivated on the British and French islands, by 1850 major production had shifted to the Spanish islands of Cuba and Puerto Rico. By 1900, with the influx of U.S. capital and the development of the large mechanized sugar factories, the Spanish-speaking islands, including the Dominican Republic, had become the leading growers of sugarcane in the New World.

► *Population Migration in the Colonial Era*

Economic activity in the colonial era led to the foundation of large numbers of towns, as Charles Sargent describes in Chapter 6. Towns, mines, and *haciendas* attracted migrants, just as new industries and cities do today in the region. Throughout the colonial era, there was internal migration in Latin America. Some migration was *forced*. Indians relocated as Iberian colonists occupied territory. Indigenous inhabitants were coerced into occupying settlements as *reducciones* were created to place Indians in nucleated settlements. Indians were forced to move to provide labor in mining camps. Priests, organizing *misiones*, coerced Indians to reside in new settlements. Many examples of forced migration are indicated in this chapter.

Less well known are the high levels of *voluntary* migration that existed in the colonial era as Europeans, *blancos*, mestizos, mulattoes, blacks, and Indians moved to take advantage of opportunities (Robinson, 1990). Modern censuses postdate the colonial era in Latin America, and the censuses are often unreliable. Precensus historical demography has developed rapidly in recent decades. Using parish records of births, marriages, and deaths, French historical demographers have recreated the demographic characteristics of seventeenth- and eighteenth-century communities in France. Schofield and Wrigley (1981) projected population totals, birth rates, and death rates for parishes in England and Wales for periods long before the first census in 1801. Often Latin American communities have source material that can be used to reconstruct population numbers and vital statistics. However, as Robinson (1990, p. 12) points out, migration data are not included in parish records of births, marriages, and deaths, for there is no record of entries and exits from communities other than births and deaths. Other population records from the colonial era may allow the reconstruction of elements in migration patterns. Evans (1990), in a study of migration processes in Upper Peru in the sixteenth and seventeenth centuries, compared the population distribution of the region as indicated in the *Visita General* (1575) with the *Numeración General* (1683–1686). The enumerations were done on different bases, but in broad terms, they reveal the distribution of population in Upper Peru at the times they were taken. Comparison shows there was a redistribution of people between 1575 and 1686, as population moved south to mining areas and east into the Yungas.

TRANSPORT PATTERNS OF COLONIAL LATIN AMERICA

Europeans introduced their own modes of travel into the New World and established a transport pattern that, in some places, has lasted until the present time. Spaniards, Portuguese, and North Europeans depended on the sailing ship to transport goods overseas. Land transport, however, with the use of pack animals and wheeled vehicles, was far more important to the Spaniards than to the Portuguese or North Europeans, mainly because of the Spaniard's interior orientation.

► Sea Transport and Trade

Transatlantic sea routes and trade, connecting the colonies with the ruling nations and the large markets of Western Europe, were equally significant to all three groups. Like the trade policies of the Portuguese and North Europeans, the Spanish crown established rigid commercial controls, forbidding its colonies to trade with foreign powers. Moreover, early beset by piracy, Spanish ships for most of the colonial period were forced to sail in convoy, carrying manufactures, mercury, and various luxuries to the colonies and returning to Seville with precious metals, hides, and other New World products. Sailing southwest from Spain, the convoys (*flotas*) picked up the trade winds, which carried them to the Caribbean and Mexico (Figure 3.5). To reach the Pacific ports of South America, goods were transshipped across the Panamanian isthmus by mule train and at Panama City were loaded on ships for the voyage southward along the coast. This crossing point was one of the most strategic spots of the Spanish Empire because most of the Peruvian silver was shipped to Spain through Panama. Largely for that reason the province of Panama fell under the political jurisdiction of the viceroyalty of Peru for most of the colonial period. On the return trip to Spain, ships from Mexico and the Caribbean ports assembled in Havana to form a convoy. Controlling the Florida Strait, the main exit for ships heading northward to catch the westerly wind, Havana was another highly strategic point for Spanish transatlantic shipping.

Far to the south of the normal Spanish American sealanes another transatlantic route linked the La Plata estuary with Europe, largely through illegal trade. After its refounding in 1580, Buenos Aires became notorious for its clandestine commerce with smugglers of various nationalities. Flemish cloth and other North European goods were exchanged for silver from Potosí; at least one tenth of Potosí's production was filtered illegally through Buenos Aires (Chaunu, 1955, Vol. 6). In terms of distance and ease of transport, Buenos Aires was the most logical port for Spanish trade into the Potosí mines, but royal edict required that all overseas contact with Upper Peru pass through Panama for security reasons.

Rarely troubled by pirates, Portuguese and North European sugar ships did not sail in convoy (Mauro, 1961). Vessels plied directly between Lisbon and Brazil, taking Portuguese wine and other luxuries to Recife, Salvador, and Rio de Janeiro and returning with sugar and tobacco. In addition to carrying sugar directly to Europe, English shippers in the West Indies took molasses to New England in exchange for livestock and foodstuffs; New England rum, processed from the molasses, was shipped to West Africa, where it was traded for slaves, who were then carried to the West Indies.

Besides its dominant transatlantic commerce, Spain also developed a small but prosperous transpacific trade between the west coast ports of Mexico and the Far East. Beginning in the 1580s, the famous Manila galleon plied annually between Acapulco, Mexico, and Manila in the Spanish-occupied Philippines, bringing in oriental silks and spices in exchange for Mexican silver. With the decline of the Manila trade at the end of the colonial period, Acapulco lost its role as one of the leading Pacific ports of the Americas.

In most of colonial Latin America, coastwise sea routes were probably more significant than trails on land in local trade. Maritime travel was more rapid, comfortable, and economical than transportation overland. This was particularly true in Brazil, where few roads existed between coastal settlements and local commerce was perforce by sea. Similarly, the island character of the West Indies required sea travel for local trade. In Spanish America maritime travel was especially important along the Pacific coast, where products from both the coast and the interior moved more by sea than by land, resulting in the development of several ports, large and small. Most of the Potosí silver, for example, was carried overland to Arica, the

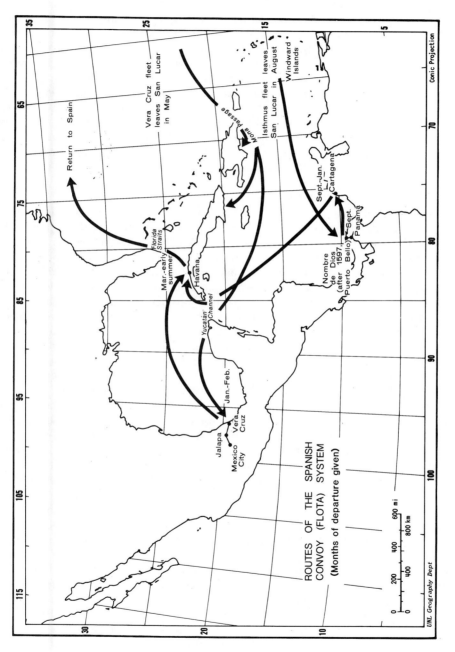

Figure 3.5 Routes of Spanish convoy *(flota)* system.

nearest Pacific port, which was also the main entry for imports into Upper Peru. Again, instead of paying the high cost of overland transport, government officials sent mercury from Huancavelica destined for Potosí to the small port of Chincha, whence it was shipped by sea to Arica and thence over the short Andean trail to the mines (Vázquez de Espinosa, 1942, p. 472). Most of the agricultural products from the Peruvian coastal oases also went by sea to Lima, Panama, or Arica through small local ports, many of which are used today to ship out cotton and sugar. Callao, near Lima, where goods in coastwise trade were officially registered, was the largest South American Pacific port. In size and importance it was followed by Guayaquil, the main shipbuilding center within the viceroyalty of Peru.

In the sixteenth century a sizable trade by sea was established between New Spain and Peru (Borah, 1954). Textiles and luxury items were shipped to Peru from the Mexican port of Huatulco; from the port of Realejo in Central America came cacao for making chocolate and pine pitch for coating Peruvian wine jugs. In return, the merchants of Lima sent silver bullion, mercury, and wine. Because of the increasing restrictions on interviceregal commerce, this trade ceased in the 1630s.

► Land Transport and Trade

The development of land transport in colonial Latin America was closely associated with mining and commercial agriculture, especially in Mexico, the central Andes, and the province of Minas Gerais in Brazil. Mule and donkey trains formed the most common type of transport over rough mountain trails, and in Upper Peru the native llama was used extensively as a pack animal throughout the colonial period. In two areas of Spanish America—Mexico and Argentina—large two-wheeled carts pulled by oxen were employed to haul cargoes to and from mining centers and towns over roads constructed on level terrain. Land transport was arduous and costly. Colonial reports are replete with statements to the effect that to haul a unit of goods for a relatively short distance overland by pack train or cart cost several times more than to bring it by ship across the Atlantic. This large cost differential continued until the advent of the locomotive and automobile, and it persists in isolated areas of Latin America where colonial-type transport is still used.

In the New World, the Spaniards early established land routes, many of which are followed today by railroads and highways. In Mexico three major routes led northward from the capital to the mining districts and the settlement frontier. The western route eventually arrived at Tucson, the eastern route led to San Antonio and the central route followed the main silver belt through Zacatecas to the upper Rio Grande and on to Santa Fe (Figure 3.6). The most frequented routes in Mexico led from the ports of Acapulco and Veracruz to the capital. On these two trails at any given time during the dry season, thousands of mules were used to haul silver bullion and other products to the ports, returning with imported goods. Pack trials also led southeastward from the Valley of Mexico to Central America but, with the exception of the transisthmian route across Panama, Central American trails carried little traffic. Lack of frequent communication and trade between the isthmian provinces may have led to the political separatism that resulted in the formation of the Central American nations after independence.

In the Andes, the mule trails that led from the highlands to the Pacific ports carried a large amount of traffic, such as silver bullion from the mines and luxury goods imported from Spain (Figure 3.7). Examples of highland–coast roads include those connecting Quito and Guayaquil and Potosí and Arica. Though frequently made, the journey between Potosí and Lima through Cuzco required four months of travel by mule over one of the most difficult land routes in Latin America (Coni, 1956, p. 38). Travel between Potosí and the plains of Argentina was easier. By the end of the sixteenth century, a cart road had been built between Buenos Aires and

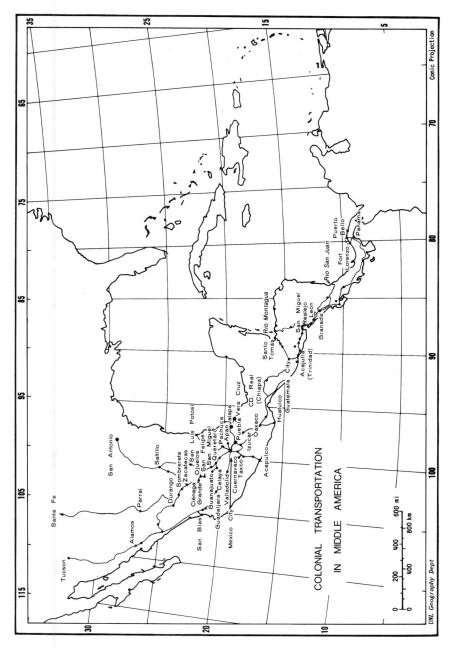

Figure 3.6 Colonial transportation in Middle America.

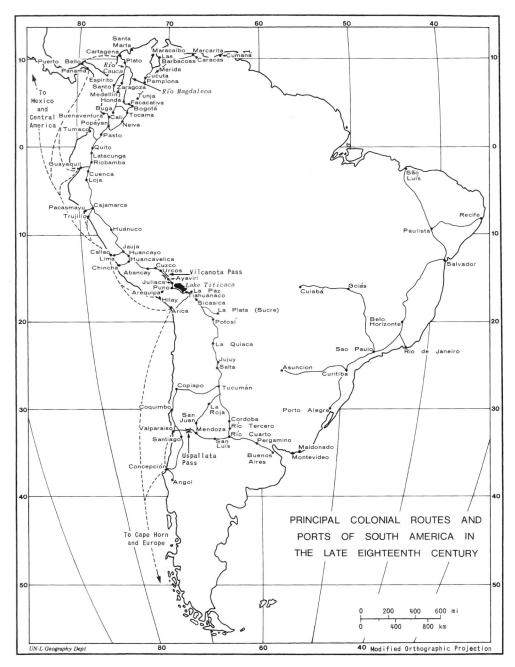

Figure 3.7   Principal colonial routes and ports of South America in late eighteenth century.

the valley towns of Salta and Jujuy in northwestern Argentina, with branches to the vineyards of Mendoza and to Paraguay. Huge two-wheeled carts, each pulled by six to eight yokes of oxen, hauled hides, wine, and smuggled imports to Salta and Jujuy, where the cargo was transferred to mule trains destined for Potosí (Coni, 1956, p. 80; Vázquez de Espinosa, 1942, pp. 671, 691).

In Brazil cattle trails between the *sertão* and the coast formed the only well-established land transport pattern until the discovery of gold in the interior at the close of the seventeenth century. Thereafter Portuguese merchants established a series of donkey trails among the mining camps of Minas Gerais, the port of Rio de Janeiro, and the food supply center of São Paulo (Webb, 1959). The main trail north from Rio de Janeiro up the plateau escarpment was later followed by both a railroad and a highway to Minas Gerais. Today it is one of Brazil's most heavily trafficked thoroughfares.

► River Transport

During the colonial period, neither Spaniards nor Portuguese used rivers for transport and trade to any extent. There were few navigable streams within the main areas of economic activity, and river navigation was not an outstanding Iberian trait. Within the Spanish realm only the Magdalena River in northern South America became a major route of communication, connecting the interior of New Granada (Colombia) with the port of Cartagena on the Caribbean. The Magdalena retained its role as Colombia's main trade artery until the 1960s, when truck roads were completed to the coast. Other rivers that the Spaniards occasionally used for transport included the San Juan in Nicaragua, the Guayas in Ecuador, and the Paraná–Paraguay system in Argentina and Paraguay. The Paraná–Paraguay, however, did not become Paraguay's major commercial outlet until the independence period. Similarly, the Amazon River system in northern Brazil and eastern Peru was little used for commerce until the rise of rubber collecting in the Amazon rain forest during the late nineteenth century.

SUMMARY

Many aspects of modern Latin America, including population distribution, racial composition, and economic and social patterns, have roots in the aboriginal and colonial eras. Colonial Spaniards occupied and developed mainly the parts of the New World that were densely settled by civilized natives and were rich in precious metals: Mexico and the Andean highlands. Today these two areas form the major centers of Spanish culture and of what is left of aboriginal life in the New World. In both areas miscegenation created the dominant mestizo element. In both areas mining was the mainstay of the economy until recently. Lands that lay outside Mexico and the Andes held scant interest for the Spaniards; for example, southern South America and the northern borderlands of Mexico were settled late, more for political than economic reasons, and the development of both has been largely postcolonial by non-Spaniards.

In contrast to the Spaniards' inland orientation toward mineral deposits and concentrations of Indian population in the mountainous interior, the interests of the Portuguese in Brazil and the North Europeans in the Caribbean lay on the coastal fringes. There fertile soils could be exploited for the production of sugar, tobacco, and other agricultural products for export. In Brazil and the Caribbean, native peoples were quickly diminished or destroyed by disease and maltreatment. They were replaced by the importation of enslaved Africans to work the plantations. This is reflected today in the dominant black and mulatto population of the Caribbean and coastal Brazil. Moreover, sugar is still important in the Caribbean economy, and plantation agriculture, though modified from its colonial form, remains dominant in Brazil and in some West Indian islands. In type, the colonial Brazilian economy approached that of the Spanish areas only in the eighteenth century, when gold was mined in the immediate interior. Brazil now relies on both mining and agriculture, but its population is still concentrated near the coast, and most of its far interior is as undeveloped today as it was in colonial times.

In the effectively settled parts of colonial Latin America, landholding was a dominant feature of the rural scene: the livestock-grain *hacienda* in Spanish America, the sugar and cattle *fazenda* in Brazil, and the sugar estate in the West Indies and the Guianas. The agricultural village in Latin America derived largely from Indian tradition. Today the enforcement of land reforms has affected the large landholdings in many Latin American countries, but the agricultural village of peasant farmers is still a significant form of rural settlement. The pattern of dispersed rural homesteads, so familiar in the North American scene, rarely developed in Latin America.

The urban life of modern Latin America has its roots in the colonial past. Most of the Spaniards who migrated early to the New World were town dwellers, and the main type of initial Spanish settlement in the Americas was the *villa* (town) or *ciudad* (city), established as a political, military, and religious control point (Morse, 1962). Port towns, of course, had the additional functions that attend overseas and coastwise shipping. Some inland administrative towns later became trading and manufacturing centers for local markets, but only the mining towns processed goods (bullion) in large quantities for overseas trade. Both Spaniards and Portuguese attached social prestige and political advantage to urban living. For example, the large landowner, in addition to his *hacienda* or *fazenda* mansion, usually owned a townhouse in the nearest provincial seat or even in the viceregal capital, where he spent time close to royal officials. In Brazil most urban centers were port towns; only in the eighteenth century did mining and market centers develop inland from the coastal fringe. During the nineteenth and twentieth centuries, urbanism in Latin America greatly intensified because of migration into the cities and larger towns. The viceregal and many provincial capital cities of colonial days have maintained their primary character as the national capitals of the modern Latin American countries.

FURTHER READINGS

Butzer, K. W. "The Americas Before and After 1492: An Introduction to Current Geographical Research." *Annals of the Association of American Geographers* 82 (1992): 345–368.

Collier, S., T. E. Skidmore, and H. Blakemore, eds. *The Cambridge Encyclopedia of Latin America and the Caribbean*, 2d ed. Cambridge: Cambridge University Press, 1992.

Cook, N. D., and W. G. Lovell, eds. *"Secret Judgements of God": Old World Disease in Colonial Spanish America*. Norman: University of Oklahoma Press, 1991.

Covington, P. H. *Latin America and the Caribbean: A Critical Guide to Research Sources*. New York: Greenwood Press, 1992.

Denevan, W. M., ed. *The Native Population of the Americas in 1492*. 2d ed. Madison: University of Wisconsin Press, 1992.

Galloway, J. H. *The Sugar Cane Industry: An Historical Geography from its Origins to 1914*. Cambridge: Cambridge University Press, 1989.

Robinson, D. J. *Migration in Colonial Spanish America*. New York: Cambridge University Press, 1990.

Sauer, C. O. *The Early Spanish Main*. Berkeley and Los Angeles: University of California Press, 1966.

Viola, H. J., and C. Margolis, eds. *Seeds of Change*. Washington, D.C.: Smithsonian Institution Press, 1991.

West, R.C. and J. P. Augelli. *Middle America, Its Lands and Peoples*. 3d ed. Englewood Cliffs: Prentice-Hall, 1989.

CHAPTER 4

Agriculture

C. GARY LOBB

INTRODUCTION

In spite of rapid urbanization and the growth of new types of employment, agriculture provides the largest single form of employment throughout Latin America. Agriculturalists live in diverse physical and cultural environments. Some farmers engage in production for subsistence, and others produce commercial crops for market. There is no one type, or model, of Latin American agriculture.

The agriculture systems in Latin America can be analyzed from environmental and human perspectives. Environmental factors may have influenced the selection of certain plants and animals and certainly influence the seasonality and intensity of agricultural work. The structure of agricultural systems and the timing of agricultural labor are often related to solar energy, moisture availability, and nutrient resources in the soil. But there is also the human or cultural dimension to be considered in relation to Latin America. In many traditional societies farming and other productive activities are seen as a part of the social fabric. Some aspects of agriculture can be explained only in terms of cultural traditions instead of the type of input-output accounting procedures used in relation to most U.S. farming operations, where agriculture is viewed as predominantly an economic activity.

In Latin America, as in other parts of the world, traditional modes of cultivation are being transformed into more commercial types of agriculture as new technologies become available. This is not to suggest a complete lack of technical innovation in the region, as the pre-Columbian use of irrigation, terracing, fertilizing, and specialized adaptations such as the *chinampa*, or floating garden, of central Mexico illustrate. However, during the last four decades the pace of technological innovation has accelerated as commercial fertilizers, genetically improved crops, and tractors have become available throughout Latin America. Most Latin American countries today pursue policies aimed at applying modern technology to help increase gross agricultural production. Transformation in the agricultural sector of the region has been particularly significant, especially in countries such as Brazil, Chile, and Mexico, where commercial agriculture is export-oriented. Brazil exports coffee, soybeans, and citrus. Chile produces grapes for wine and fresh fruit for export, and Mexico is a major supplier of fruits and vegetables to the U.S. market. A discussion with the fresh produce manager at the supermarket will reveal how widespread are the sources of the fruits and vegetables consumed in the United

States. Latin America is a major source area. Make a list of the vegetable products in your supermarket that come from Latin American countries.

The earliest form of agriculture in the region employed a very basic technology and was tied closely to locally available environmental resources. Traditional methods, ecologically similar to the earliest forms of cultivation in the New World, are still in use in many parts of Latin America. Although it is not certain that the earliest agriculture in the New World emerged in the tropics, the traditional systems in use today have a wide distribution within the tropical portions of the Western Hemisphere. The impediments to productivity of the humid tropical environment—plant and animal pests and low levels of soil fertility—are managed in much the same way as they were 3,000 years ago.

THE ECOLOGY OF TROPICAL AGRICULTURAL PRODUCTION

Much of Latin America, almost 5.5 million square miles (8.8 million square kilometers), or 82 percent of the region, is tropical (Figure 4.1). In this area the traditional systems that use a limited technology have persisted.

The fertility of soils in tropical areas is often poor. Infertile, acid soils occupy more than 80 percent of the tropical areas of Latin America. The soils quickly deteriorate if the existing vegetation cover is removed. Strategies to cope with this problem have been in use for thousands of years in what is now Latin America. Basically, a variety of swidden, or shifting, cultivation methods are employed. Vegetation is cleared and the ground cultivated for a few years; when yields begin to decline, the cultivators move on and clear another plot.

In Latin America there is a lexicon of terms identifying swidden, or shifting, agriculture. The term *roza* is widely used in Spanish-speaking areas from Mexico to Paraguay. In Brazil the Portuguese equivalent is *roça*. There are more localized terms

Roza Plots, Guatemala.

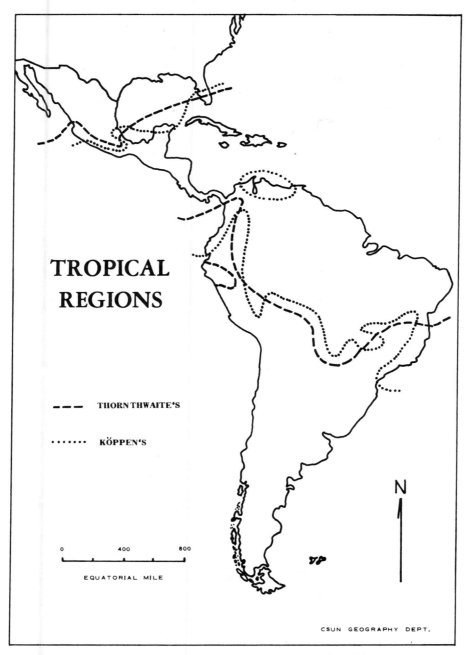

TROPICAL REGIONS

– – – THORNTHWAITE'S

· · · · · · KÖPPEN'S

0   400   800

EQUATORIAL MILE

N

CSUN GEOGRAPHY DEPT.

Figure 4.1 The tropics defined by the climatic classifications of Köppen and Thornthwaite. See Figure 2.2 for Köppen classification.

such as *conuco*, used in Venezuela and the Spanish-speaking area of the Caribbean, and *chaúa*, used in the Amazonia region of Peru. *Milpa*, a term used to identify traditional agriculture in many parts of Mexico and Central America, is sufficiently specialized and productive to be thought of as ecologically different from swidden.

Roza cultivation is small in scale, with cleared plots seldom exceeding 2.47 acres (1 hectare) (Harris, 1972). The plots may be planted in one regionally preferred staple such as maize or manioc (a starchy, vegetatively reproduced root crop), or they may be planted with a variety of useful crops, all grown together in a plot.

Another characteristic of *roza* is that it is necessarily land-extensive. For ecological reasons the plots are cultivated only for short periods of time—1 to 3 years on average—before being abandoned for longer periods. These periods of abandonment are fallows, when natural vegetation is allowed to regenerate on the vacated *rozas*. This long fallow feature provides the most convenient definition of swidden: an agricultural system in which the periods of fallow exceed the periods of cultivation.

To many North Americans the large areas of fallow may seem wasteful. Indeed, if the fallow land is considered a part of the total agricultural land, the long-run yields per unit area are incredibly low (Lobb, 1974). On the other hand, the system can be highly productive in terms of yields per unit of labor expended and has been known, when measured in kilocalories, to give returns in excess of 16 to 1 on the energy invested in agriculture (Rappaport, 1971) (Figure 4.1).

The length of fallow periods varies from 2 to 6 years among the modern Kekchí- and Quiché-speaking Indians of Guatemala (Carter, 1969). The Ka'apor of Amazonia fallow some land for 10 years (Baleé, 1994). Kuikuru and Akanaio of Brazil and Guyana, respectively, use fallows of 25 years or more (Carneiro, 1961; Butt, 1970).

Abandoning one cultivation plot and moving to another involves the additional effort of clearing and burning the natural vegetation. But the move is made necessary by the decline of yields and the increasing difficulty of keeping weeds under control. In fact, the problem of weed invasion may be as important as that of soil nutrient rundown in influencing the time of abandonment. During the fallow, perennial weeds are shaded out by taller vegetation (Seavoy, 1973).

THE SPATIAL AND CULTURAL PARAMETERS OF SWIDDEN

An example of *roza* agriculture in tropical Guatemala may help to illustrate the human inputs and their timing associated with the manipulation of the environment factors. *Roza* cultivation is widespread on the north slopes of the Cuchumatanes Mountains in northwestern Guatemala around the community of Todos Santos Cuchumatán (elevation 7,600 feet). The limestone soils of the Cuchumatanes foothills are of better-than-average fertility considering the tropical nature of the environment. Under such conditions, in addition to the nutrients stored in the living vegetation, the soils are also able to provide some nutrient support for cultivated crops. Temperatures are warm enough to sustain plant growth throughout the year, and the seasonality of production is regulated by the availability of moisture. There is a pronounced dry season from late November to late January. The timing of most major human activities associated with crop production is related to this period of dryness.

Even though the soils are not altogether leached of their nutrients, Todos Santos cultivators depend on the nutrients contained in the living vegetation, which are released through burning. Fully developed forest vegetation is cherished as an environment in which to practice *roza* cultivation because there is more biomass to burn. But well-developed forest ecosystems are becoming rare in this part of Guatemala, as they are in much of Latin America. Many years of *roza* cultivation in Todos Santos have resulted in the destruction of most of the primary, fully developed rain forests because they have been systematically cleared for *roza* plots and fallowed for periods too short to allow the primary vegetation to regenerate. The forestland that presently exists is considered secondary forest, which means that it is a forest ecosystem that has reestablished itself on abandoned *roza* plots. Unless such reestablished, secondary forests are more than 50 years old, they most likely will not contain the variety of forest species or the volume of biomass typical of the original forest, and therefore, they are not as important a source of nutrient support for agriculture. Secondary forestlands and primary forests, where they exist, are referred to locally as *montaña* lands. Todos Santeros consider *montaña* lands a valuable resource because they are potentially the best lands for successful *roza* cultivation.

In the 1970s, only half of the villages in the Cuchumatanes foothills had access to *montaña* lands. Increases in population in the last 20 years have greatly increased the pressure on available agricultural lands, and the result has been that the forests have been cut for *roza* plots at a faster rate than they have been able to regenerate themselves. The lands most commonly being used now for *roza* cultivation are second-growth ecosystems in an early stage of regeneration. These bushlands, called *huatál* lands by the native population, are now the most widespread. They are poor in species composition and biomass content compared with the *montaña* ecosystems and are therefore less valuable as potential agricultural lands. It is now common to use *huatál* land for cultivation, but ideally it should be allowed to grow to *montaña*.

If the peasant cultivator has *montaña* land to utilize for *roza* plots, he will begin to clear the wild vegetation from a selected plot in the middle of November, at the onset of the dry season. First, the underbrush is cut with a large machete, and the large trees are then felled with an axe or power chain saw. Branches are chopped off the trees, and the mass of cut vegetation is allowed to dry. The debris is collected into piles with clearings surrounding them so that the fire will not spread out of control. The dry forest debris is then burned. The amount of work required for this phase of the agricultural cycle varies with the size of the trees and their numbers, but it usually ranges from 15 to 30 days per worker per acre. After a successful burn, the ashes are scattered over the plot to be cultivated and the *roza* may be planted within a few days, "after the ground has cooled," it is said.

The *huatál* land (bushland), which is now the kind of land most available for the new *roza* plots, is cut later in the dry season than the *montaña* because it dries in less time. It is cut in the middle of December with a machete only and is planted a few days after a successful burn. A serious problem confronting *roza* cultivators in this part of Guatemala is the invasion of coarse perennial grasses into the environments where *huatál* lands are cut. The open, treeless, shadeless areas are ideal habitats for the successful growth and reproduction of perennial tropical grasses. After repeated burns and intermittent *roza* croppings, the environment may become largely a grassland and is no longer valuable as a potential *roza* plot. There is not sufficient biomass to provide nutrients for the cultivated plants, and the growth of the grass is never retarded sufficiently to eliminate it as an aggressive weed that smothers domesticated plants. Many square miles of once-forested countryside in the Cuchamatanes foothills have been converted into *pajonales*, or tropical grasslands, as a result of the management practice just described.

Land under cultivation is *rastrojo*, or cornstalk land. If the *roza* plots are cultivated for more than one year in succession, which is the case in this part of Guatemala, last year's *rastrojo* land is prepared for another year of cultivation by collecting cornstalks in piles and burning them, thereby releasing any nutrients tied up in the remains of last year's crop.

Seeds are planted in February, at the onset of the rainy season. Seeds are carried into the cleared fields in a *henequen* shoulder sack; a deep digging stick (*pachán*), is used to open a hole 3 inches deep, and five grains are dropped in. Holes are made at intervals of 16 to 20 inches, and loose soil is mounded over the opening, creating a little hill called a *montón*. If the rains begin on schedule and conditions are average, the maize starts to sprout within 3 weeks.

The removal of annual and perennial weeds becomes more of a problem each successive year of cultivation. The first year after a *montaña* plot has been converted into a *roza* field, competition from weeds is not serious. In the second year, weed

House site and typical dooryard garden in Petén region of Guatemala.

invasion threatens productivity. During the years the *roza* plot is in use (usually 3 years in Guatemala), the number of cultivations to remove weeds varies from one to four per growing season. Four cultivations a year would be typical of a *roza* plot in its final stages of use before abandonment. The usual number of cultivations is two, the first in the middle of April and the second in June or July. The hoe is the only agricultural tool used in the weeding operation. At the time of the first cultivation (*limpiada*), leaves are selectively removed from the corn plants to prevent shading of the lower leaves and thus to enhance productivity. Most of the leaves are hoed into the soil along with the weeds, where they decompose and contribute to soil fertility.

In September most of the green leaves are stripped from the corn and tassels are removed. This reduces photosynthesis, the sugar in the grains converts to starch, and the kernels harden. In October the stalks are all bent over below the lowest dried ear, a process known as doubling (*doblando*), which seems to prevent water from entering the ear and reduces damage from predatory birds, particularly parrots. The harvest takes place in November as the dry season approaches once more.

Shifting cultivation, when practiced on recently cleared *montaña* lands, can be employed for as long as 5 years before the plot is abandoned and allowed to return to forest fallow. The average maize yields are about 446 pounds per acre during 5 years of cropping. However, if the *roza* plot is established on recently cleared bushlands, now the most common practice, the plots can be cropped for only 2 or 3 years, and the yields average from 28 to 46 kilos per hectare, after which the plots must be fallowed for 5 years to allow *huatál* lands to regenerate. If the *roza* plots are cultivated for more than 3 years after the clearing of *huatál* lands, yields decline and the plots become clogged with annual and perennial grasses. When the field is abandoned, it becomes grassland (*pajonál*). If left unburned and uncultivated, these *pajonál* lands return to bushland in 15 to 20 years.

The tropical cropping system described here, which is widely practiced with some variations throughout Central America and southern Mexico, should be thought of as an example of traditional agriculture involving the cultivation of seed plants such as maize and beans. Traditional systems of tropical cultivation share many ecological characteristics, but a wide range of crops are used. In the

Caribbean, the Amazon lowlands, and other areas, root crops such as manioc, sweet potato, arrowroot, and yams are the staple crops.

Swidden agriculture (*roza*) is seldom the only means of support. It is common for *roza* to be used in conjunction with intensive, small-scale plots—dooryard gardens—that are in perpetual tillage and yield a wide variety of food products, including yams, climbing beans (*habeas*), sweet potatoes, chili peppers, bananas and plantains, coffee bushes, and tree crops such as citrus, avocado, papaya, and mango. Dooryard gardens are usually adjacent to dwellings and are fertilized with household organic wastes.

While representing similar strategies regarding the problems of agricultural production in the tropics, the *roza* systems of the seed cultivators of Mexico, Central America, and the sub-Andean regions of South America are distinct from the root cultivators of tropical South America and the circum-Caribbean region. The two traditions of cultivation, the seed culture (or corn) and vegeculture (or root crops), make different demands on the soil. The protein-rich seed crops require larger supplies of nutrients in the ash, litter, and soils than the starch-rich root crops of vegeculture. This difference is accentuated during harvest, when the seed-crop cultivator removes highly concentrated nutrients from the *roza* for consumption in the form of beans and corn. The vegeculturalist removes a smaller fraction of available fertility in the harvest of tubers and roots.

Maize is the mainstay of the seed-culture areas of tropical Latin America. Manioc is the staple of traditional vegeculture societies. Most varieties of manioc are poisonous and require special preparation before they can be eaten. In one form or another, manioc makes up as much as 85 percent of the diet of many cultivators in the Amazon Basin. The timing of the agricultural work being done in a cycle of vegeculture (referred to as a *roça* cultivation in Brazil) is closely related to that of a seed-culture *roza* cycle. Shortly after the end of the rainy season, the forest vegetation is cut and left to dry for several weeks. Just before the next rainy season it is piled up and burned, and planting begins about the time of the first rains. A hole is opened with a dibble, and 4 to 10 cuttings are planted in low mounds made of loose soil. Plots are weeded one or more times a season, depending on the age of the cultivated field. Manioc tubers develop to a harvestable size 5 or 6 months after the cuttings are planted; if they are allowed to remain in the soil for a longer period, say 18 to 20 months, the tubers are considerably larger and attain their highest starch content of about 30 percent (Schery, 1963). Among the vegeculturalists, too, dooryard gardens are common.

Although closely related in many ways to *roza* cultivation, the short-fallowing of permanent fields, as opposed to the abandonment of fields in the swidden system, is common in parts of Mesoamerica and the sub-Andean region of South America. The permanent field system involves the initial clearing and burning of the vegetative cover with 3 or 4 years of successive cropping followed by a short fallow of equal or less duration. Such short-fallow agriculture is made possible by a combination of the following environmental and cultural factors: (1) relatively high levels of soil nutrients common in the cool highlands, particularly around the fertile volcanic and lacustrine basins on the Mesa Central of Mexico, in southwestern Guatemala and Chiapas, and in sub-Andean Colombia; (2) no severe competition from perennial weeds; and (3) the availability of the European plow and draft animals to allow deep plowing. The aboriginal crops of maize, beans, and squash, together with chili, are the principal food plants of the *milpa* in short fallow and are the predominant subsistence foods of most Mexicans and Central Americans. Average maize yields from *milpas* are less than 64 kilos per hectare, but in certain locations the yields are impressive. For example, the average long-term yields for the Quezaltenango Basin in Guatemala are 625 pounds per acre, with two years of cropping being followed by a one-year fallow.

AGRICULTURAL SYSTEMS OF
HIGH-ALTITUDE ENVIRONMENTS

The agricultural systems of the high-altitude environments of the Andean region of South America are different from those of the tropical lowlands. The Andes are an important center of plant domestication, and domesticated plants from Middle America reached the highlands at an early date. The Andes became a center of intensive short fallow agriculture in pre-Columbian America. The agricultural system protected soils from erosion and used fertilizers to maximize yields.

Traditional agricultural practices in the Andean area were intensive and included the use of terraces and irrigation, although many areas cultivated in the pre-Columbian era have been abandoned. These methods are characteristic of the valleys between 7,000 and 12,000 feet in elevation and are also found in the cultivated areas of the cool desert coast of Peru. Andean valleys are characteristically deep and narrow. There is little flat land, and runoff after rains is rapid. The solution to these problems has been the construction of terraces, and most valley sides have been terraced and equipped with stone channels to distribute irrigation water. A dry season makes irrigation desirable almost everywhere. Plows pulled by oxen are used on flat surfaces. On steeper cultivated slopes the Inca foot plow (*tacilla*) is used along with a hoe. A variety of Inca and Old World food plants is presently cultivated. Even in basins over 3,962 meters in elevation eight species of domesticated plants can be raised. The potato is the staple, but other root crops such as the tubers oca (*Oxalis tuberosa*), ulluco (*Ullucus tuberosus*), and ysaño (*Tropaeolum tuberosum*) are cultivated. An important crop in the high valleys is quinoa (*Chenopodium quinoa*). Quinoa is an annual herb that produces small seeds. The seeds can be boiled like rice or ground up to make meal. Quinoa leaves are cooked like spinach.

Market day in a Peruvian village.

Cañihua (*Chenopodium pallidicaule*) is a related plant cultivated at high levels along with hardy varieties of introduced wheat and barley.

In valleys of lower and intermediate elevation wheat and maize replace quinoa and cañihua, but potatoes, chili peppers (*ají*), and squash remain important.

Potatoes are in widespread use throughout the Andean region. At least 13 cultivated species are known. Potatoes are often stored in a dehydrated form known as chuño, and oca is also dehydrated for preservation. In the humid lower valleys, which drain from the Andes to the Amazon and the valleys of the desert Pacific coast, cotton, beans, chili peppers, sweet potatoes, tomatoes, and manioc (yuca) are cultivated. Since the time of the Incas, the products of the eastward draining valleys and the coast have been linked to the economy of the highlands. Coca (*Erythroxylum coca*) is cultivated in the warm lower valleys that drain eastward to the Amazon. Dried coca leaves are transported to the highlands, where they are chewed as a stimulant. Cocaine can be extracted from coca leaves. Bolivia and Peru are major sources of illegal cocaine (Chapter 11).

The dry season in central Andean South America runs from April until November, and the rainy season is from November to the end of March. Potatoes are planted early in the rainy season, in December, and are harvested the following June. Maize and quinoa are sown in September and October and harvested in May. Agricultural work during the rainy season is the weeding of fields, bird scaring, and control of pests.

THE HISTORICAL GEOGRAPHY OF LATIN AMERICAN AGRICULTURE

The earliest systems of agriculture that developed in what is now Latin America must have been ecologically similar to the modern forms of *roza* cultivation. The place, or places, of origin have been identified as the areas in which domestication

TABLE 4.1   List of the Plants First Domesticated in Mesoamerica

Seed Crops	*Legumes*
Maize (*Zea mays*)	Common bean (*Phaseolus sps.*)
Chia (*Salvia hispanica*)	
Amaranth (*Amaranthus cruentus*)	*Tubers*
Huauhtli (*Amaranthus leucocarpus*)	Yautia (*Xanthosoma sp.*)
Chiagrande (*Hyptis suaveolens*)	Arrowroot (*Maranta arundinacea*)[1]
	Sweet potato (*Ipomoea batatas*)[1]
Vegetables	Jícama (*Pachyrhizus erosus*)
Squash (*Cucurbita moschata*)	Manioc (*Manihot esculenta*)[1]
Pumpkin (*Cucurbita pepo*)	
Chilcayote (*Curcurbita ficifolia*)	*Others*
Chayote (*Sechium edule*)	Agave (*Agavaceae sp.*)
	Cotton (*Gossypium sp.*)[1]
Fruits	Cacao (*Theobromacacao*)
Chili peppers (*Capsicum annuum*)	Indigo (*Indigofera suffruticosa*)
Pejibaye palm (*Guilielma utilis*)[1]	
Avocado (*Persea americana*)	
Capulín (*Prunus serotina*)	
Guava (*Psidium quajava*)[1]	
Spodilla (*Achrassapota*)	
Zapote (*Calocarpum mammosum*)	

[1]Possible domestication in both Mesoamerica and South America.

of wild plants took place. The archeological evidence for early agriculture in Mesoamerica is substantial and is becoming more convincing. Mesoamerica has the characteristics of a center of agricultural origin, including an impressive list of domesticated plants (Table 4.1). Other centers have been identified as possible independent areas of origin. The Russian ethnobotanist N. L. Vavilov, writing in

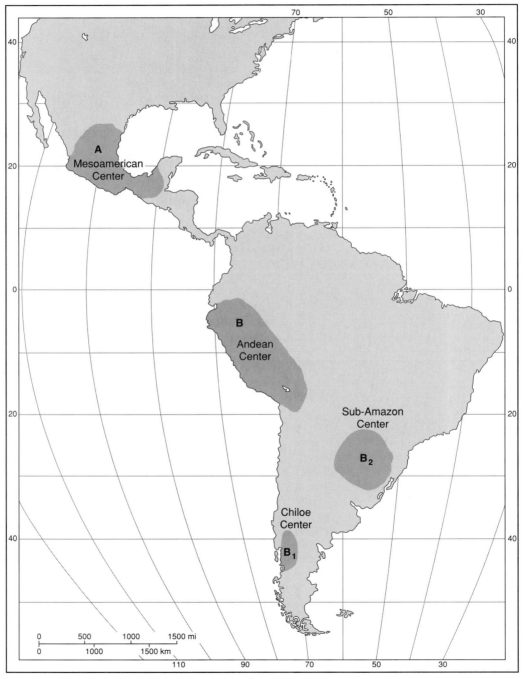

Figure 4.2 Centers of agricultural origin.

The Historical Geography of Latin American Agriculture

the 1920s, suggested a South American center with two subcenters in addition to the Mesoamerican hearth (see Figure 4.2).

The South American locations were considered centers of root crop domestication. Within the Andean center (Center B) certain species of tetraploid potatoes were first cultivated along with other tubers such as oca, ulluco, and ysaño and the herb quinoa. Early agriculture in the Andean highlands was based on plants that reproduced vegetatively by division of the plant parts and not by seeds. In commenting on early Andean agriculture, Carl Sauer (1956, pp. 518–519) noted:

> It is possible to assign to the potatoes the leading role in the agricultural colonization of the Andes, except for the one fact of the existence, side by side with the potatoes, of the lesser tuberous crops. . . . It seems to me therefore, that we have in these minor tuber crops the remnants of the oldest Highland agriculture; that long before potatoes were bred to grow on the bleak reaches of altiplano and páramo, these microthermal native tubers had made sedentary life possible by supplying starch food and had been made into domesticated plants. Also, the storage problem had been solved by inventing chuño making and this was transferred later to potatoes, when these became developed for puna climate cultivation.

Vavilov identified two subcenters of domestication activity. In addition to a central Andean site, he suggested another subcenter of potato domestication in south-central Chile, particularly on the island of Chiloé (Center B1), and another subcenter for the domestication of tropical root crops in sub-Amazon Brazil (Center B2).

Recent studies suggest that domestication may have taken place in many locations throughout South America and that no one center emerged as the dominant site of early domestication (Table 4.2). To quote Harlan (1971, p. 472):

TABLE 4.2 List of Plants First Domesticated in South America

Seed Crops
Quinoa (*Chenopodium quinoa*)
Cañihua (*Chenopodium pallidicaule*)
Mango grass (*Bromus mango*)
Madí (*Madia sativa*)
Jataco (*Amaranthus caudatus*)

Vegetables
Zapallo (*Cocurbita maxima*)
Sicana (*Sicana oderiferal*)
Achoccha (*Cyclanthera pedata*)
Bottle gourd (*Langenaria sp.*)

Fruits
Pepino (*Solanum muricatum*)
Tomato (*Lycopersicon esculentum*)
Chili peppers *ají* (*Capsicum sp.*)
Pejibaye palm (*Guilielma utilis*)[1]
Guava (*Psidium guajava*)[1]
Pineapple (*Ananas comosus*)
Soursop (*Anona muricata*)
Cashew (*Anacardium occidentale*)
Papaya (*Carica papaya*)
Cherimoya (*Anona cherimolia*)

Legumes
Common bean (*Phaseolus sp.*)
Peanut (*Arachis hypogaea*)
Chocho (*Lupinus mutabilis*)

Tubers
Manioc (*Manihot esculenta*)
Arrowroot (*Maranta arundinacea*)[1]
Sweet potato (*Ipomoea batatas*)[1]
Yampee (*Dioscorea sp.*)
Jícama (*Pachyrrhizus ahipa*)
Achira (*Canna edulis*)
Yacón (*Polumnia sonchifolia*)
Potato (*Solanum sp.*)
Oca (*Oxalis tuberosa*)
Ulluco (*Ullucus tuberosus*)
Ysaño (*Tropaeolum tuberosum*)
Maca (*Lepidium meyenii*)
Arrachacha (*Arracacia xanthorrhiza*)

Others
Cotton (*Gossypium sp.*)[1]
Coca (*Erythroxylon coca*)
Tobacco (*Nictiana tabacum*)

[1]Possible domestication in both Mesoamerica and South America.

> *Wild peanut and wild ulluco are found in Jujuy (Argentina) and the adjacent mountainous portion of Bolivia. Wild beans stretch for 5000 kilometers, from Argentina to Venezuela, and . . . Bean domestication seems to have taken place along a band 7000 kilometers long. How can one speak of "a center of origin" for* phaseolus vulgaris *(the common bean)? Its domestication was not even confined to one continent. Sometimes centers exist, sometimes they do not.*

Wherever the places of domestication may have been, the Amerinds must be credited with significant contributions to modern agriculture. Of the 15 principal food crops utilized in the world today, 6 are of American origin: maize, potato, manioc, sweet potato, common bean, and peanut (Harlan, 1975).

Pre-Columbian systems of agriculture, some of which were highly productive and supported high-level civilization, closely resembled traditional subsistence methods such as *milpa* and *roza*, which are still widely used in Latin America. More intensive forms—associated with terracing and irrigation—which contributed greatly to the support of many societies, are not as widespread as at the Columbian contact. The use of the *chinampa* is no longer as widely practiced in the Valley of Mexico, nor are raised fields utilized in the tropical lowlands of South America and the Caribbean, although this agricultural strategy may have contributed to high yields in the areas where it was used. Abandoned, relic raised fields are common in Bolivia, Peru, Colombia, Venezuela, the Yucatán, and Caribbean islands. The fields once supported dense populations and highly productive agriculture in areas where farming is now regarded as marginal or impossible (Chapter 3).

Wherever agriculture originated in the Western Hemisphere, it had spread widely over both North and South America by the time of European contacts, even spreading into the high-altitude environments of Mesoamerica and South America. Much tropical lowland was settled by agriculturalists practicing *roza* cultivation. Groups of nonagricultural hunting and gathering people inhabited the Amazon Basin. Nonagricultural peoples occupied the northern and southern limits of what

Cultivation terraces, Pisac, Peru.

is today Latin America. Except in the Casas Grandes Valley, agriculture was not practiced in the dry, intermontane basins of northern Mexico, the temperate grasslands of the Pampas and Patagonia, and the humid southwest coast of Chile and Tierra del Fuego. Between these areas, the aboriginal inhabitants of what is now Latin America engaged in agriculture to one degree or another.

► Colonial Agricultural Land Use

The most important event in the history of land use in the Americas, apart from the initial development of agriculture, was the introduction of European crops and farming practices beginning in the early sixteenth century (Bray, 1994). A *criollo* peasant agriculture emerged in the hands of Iberian and *mestizo* populations; it represented a blend of native American cropping practices with certain plants, animals, and implements associated with European agriculture. Wheat and barley were spread into the upland environments, and wheat was being grown, for example, by Spanish and *mestizo* cultivators in the Quezaltenango Basin of Guatemala by 1571. Other crops such as onions, garlic, turnips, carrots, radishes, and peas also diffused widely and contributed to the development of *criollo* agriculture. Of great importance were Asian and African tropical plants introduced into the New World by the Spaniards and Portuguese, including bananas and plantains, sugarcane, yams, mangoes, citrus fruits, and rice. These plants spread widely into tropical and subtropical environments and were often incorporated into the *roza* systems of Indian cultivators.

The dooryard gardens associated with indigenous *roza* agriculture were enriched by the addition of bananas, plantains, mangoes, and citrus fruits. Maize remained the staple grain crop among the Indian peoples of Mesoamerica and, in fact, became even more widespread as a result of Spanish and Portuguese colonial activity. Maize cultivation diffused throughout what is today Brazil as a result of Portuguese and Spanish contact with native peoples such as the Tupi and the Guaraní.

The diffusion of exotic crops among the native Indian population was largely made possible by the Catholic church. Around every mission settlement in Latin America European priests, frequently Jesuits, cultivated fruit and vegetable gardens and introduced Indian farmers to new crops and methods of husbandry.

► Commercial Farming During Colonial Times

Production of crops for the use of townspeople was important among pre-Columbian societies such as the Aztecs, Mayas, and Incas. However, it was the Spaniards and Portuguese, with money-based economies, who developed commercial agriculture in Latin America. Not all colonial agricultural institutions were commercial. Large landholdings (*latifundios*) became the basis of a system of social domination of the masses by an ethnically differentiated minority. The landowner was usually motivated by a desire to acquire large properties and live a life-style characteristic of an Iberian *caballero* (Furtado, 1970). Indians often came to occupy the role of laborers or sharecroppers; a class of tenant farmers known variously as *inquilinos, moradores colonos,* or *huasipunqueros,* formed the small-scale (*minifundio*) stratum of the colonial hierarchy. Throughout colonial times and into the twentieth century the prevailing agricultural system was the *latifundio-minifundio* complex, characterized by large estates that controlled many small units worked by tenant farmers.

Commercial production was developed in response to the demands of mining settlements and urban centers. Some types of agricultural production for export were discouraged. For example, sheep-rearing interests in Spain were successful in limiting wool exports from the New World to the Old World. Some produce for which there was excess demand in Spain, such as hides and tallow, was exported from areas such as northern Mexico, Argentina, and Uruguay.

Agriculture was the mainstay of the earliest phase of colonization in the Portuguese territories. Brazilian commercial agriculture was based on the *fazenda de açucar* (sugar plantation), an institution established along the coast of tropical Brazil from the beginning of the sixteenth century. The sugar estates depended on export trade.

Throughout most of the colonial period subsistence agriculture was mainly in the hands of Indians, mestizos, and mulattoes. Traditional methods of production such as *roza* and *milpa* were employed. Some European crops and techniques were incorporated into the indigenous systems. Commercial farming was controlled by the Spanish and the Portuguese. Regions producing precious metals and the imperial administrative centers were the chief markets during the colonial period. The demand for food, rough textiles, and draught animals generated by these centers, and transportation between them, resulted in the organization of tributary economies. For example, in Peru the coastal regions were organized into *latifundios* to supply produce to the urban markets of Lima, Guayaquil, and the interior mines.

Cattle ranching proved to be particularly lucrative as a commercial enterprise in parts of Mexico, Argentina, Uruguay, and southern Brazil. Grassland environments within these countries were an especially suitable basis for a successful export economy involving hides, tallow, and dried or jerked beef (*charque*). Of course, all these products were used in mines: hides in the form of ropes, tallow for candles, and jerked beef as food. Jerked beef was also a foodstuff on sugar plantations in Brazil and the Caribbean. Sugar, tobacco, and indigo remained the most important commercial crops in Brazil and the Caribbean throughout the colonial period.

► Postcolonial Agriculture

At the end of the colonial period, political and economic power in the new republics passed into the hands of substantial landowners and the proprietors of mines. Both groups, together with the merchant class, were in a position to benefit from freer trade by acquiring wider markets for traditional products and by developing new forms of agricultural activity and mineral exploitation. In fact, the impetus for development frequently came from Western Europeans and North Americans seeking new outlets for their technology, capital, and entrepreneurial skills. The colonial period linked Latin America to the Iberian peninsula, which by the late eighteenth century, was one of the most economically backward regions of Europe. Trade between the Iberian peninsula and Latin America tended to be concerned with higher-value commodities: precious metals, sugar, tobacco, and indigo.

During the nineteenth century, the Latin American republics were connected economically with the industrializing areas of Europe and North America, and demands emerged for bulk commodities: wheat, meat, and cheap fruits to feed industrial workers; guano to fertilize fields; nitrates to make gunpowder for modern armies; rubber to provide tires for transportation; and nonferrous ores needed as industrial raw materials.

None of this was immediately apparent. The end of the Napoleonic wars (1815), which had allowed the emerging republics to sever links with the Iberian peninsula, resulted in a sustained slump in world trade. Nevertheless, the British established a network of consuls in Latin America in the 1820s, and the groundwork was laid for the strong trade links that came to exist in the second half of the nineteenth century between Britain and Latin America.

► The Opening of the Pampas

British commercial interests took particular notice of the temperate lands of Argentina, Chile, and Uruguay. In 1825 Woodbine Parish (1797–1882) negotiated a treaty of friendship and commerce with Argentina, and Britain formally recognized the independent existence of Argentina. Argentina thus became the first newly independent Latin American republic to be recognized by an external power.

Parish wrote commentary on Argentina (Parish, 1839). He and others saw the potential of the Pampas temperate grasslands, which contained herds of feral cattle and horses. Before the potential could be exploited, however, much had to be done to improve stock and bring land into cultivation, to say nothing of establishing reliable and regular ocean transport to market areas. Parish thought that the feral cattle were lean, scrubby beasts and doubted if wild dogs would touch the meat of the local sheep. The wool was so coarse that it was hardly worth the expense of cleaning it.

However, it was the export potential of wool that was first developed, since this commodity was not easily perishable and would stand transportation to Europe and North America. Improved breeds of sheep were introduced in the 1830s; wool exports built up until the flocks were diminished in the latter part of the century and pushed onto poorer lands to the west and south, and the more profitable beef production took over the better lands of the humid Pampas.

In the first half of the nineteenth century cattle remained sources of hides, tallow, and *tasajo*. This last product was made at *saladeros*, where the meat was preserved in strips by salting and drying in the sun. *Tasajo* was a low-quality meat product that could not be sold in quantity to European consumers. If export markets were to be gained, a means had to be found to process meat into more palatable forms. In 1847 Justus von Liebig (1803–1873), the agricultural chemist, published a method of producing a meat extract; in 1865 a large plant was set up at Frey Bentos, Uruguay, to utilize the process. Although profitable, the market for extract was relatively small, and attempts were made to can and preserve meat. None of these experiments was wholly satisfactory.

Although the principles of refrigeration and chilling were understood, it took a considerable time to develop systems that would allow meat to be transported from North America, Argentina, Australia, and New Zealand to European markets. The transport of frozen meat in refrigerated chambers across the Atlantic dates from 1877. A system of refrigeration was not enough to insure the success of the trade. As had been pointed out from the beginning, it was no use exporting

Traditional cattle herding by *Gauchos*. The *Gaucho* is the cowboy of the Argentinian grasslands. The heritage of the *Gaucho* includes plains indian and hispanic influences. The techniques of ranching were introduced into the New World from the Iberian Peninsula, but native techiques, such as the use of the *bola*, were retained.

low-quality beef and hoping that it would sell in European markets. Cattle breeds need to be improved. British shorthorn cattle had been introduced in small numbers since the 1820s, but their use had been limited, partly because the *saladeros* preferred a thick-hided animal that was not fully fleshed. In 1876, the first Aberdeen Angus beasts were imported into Argentina, and the process of stock improvement gathered pace. While this was going on, the *saladeros* continued to produce *tasajo* for traditional markets. Many of the improved animals were shipped live to British markets, and it was not until after 1900, when the United Kingdom banned the import of live animals in an effort to control hoof-and-mouth disease, that frozen products came to dominate the La Plata meat trade.

These technical, marketing, and legislative changes, many of them made far away from South America, had a dramatic impact on the landscape of the Pampas. Until the middle decades of the nineteenth century the area still had a colonial look. In 1850 Buenos Aires was a city of fewer than 100,000 inhabitants. Overland transportation was mainly by slow-moving ox carts that used tracks leading to the smaller cities such as Bahía Blanca, Rosario, Santa Fé, and Córdoba. Around the urban centers were farms and gardens, but a great part of the Pampas was roughly divided into *estancias*. The animals looked after themselves to a large extent, and there was no attempt at selective breeding. The flocks and herds were lightly culled for meat and other products.

During the second half of the nineteenth century, however, the Pampas was extensively developed, with resultant changes in land use and settlement. Incidentally, there are many parallels between the development of the Pampas in the second half of the nineteenth century and the settlement of parts of the U.S. Midwest. Both regions are temperate grasslands marked by increasing uncertainty of rainfall toward the west. Drought can be a problem. For example, between 1827 and 1832 there was an extensive drought in Santa Fé and Buenos Aires provinces, huge dust storms developed, and there was heavy loss of livestock. Both areas were occupied by indigenous inhabitants with whom wars were fought for possession of the land. For both regions railroads and European immigrants played a major role in the settlement and agricultural exploitation of the land.

The motives for opening up the Pampas for settlement were various and included: the need to populate territory in order to confirm sovereignty; the desire to push up land values by bringing estates into cultivation; the need of railroads to build up traffic; the demand for locally grown foodstuffs; the need to provide feeds such as alfalfa and grain for livestock being prepared for the meat trade; and the desire of external financial interests to profit from opening up a new area of commodity production.

Railroads played a significant role in opening the region. In 1862 the Argentine government granted a concession to the Buenos Aires and Great Southern Railway, and this company eventually built lines to Bahía Blanca. In 1865 the Central Argentine Railway was given a concession and in 1870 completed a line from Rosario to Córdoba.

The major source of immigrants to the Pampa was Italy, although many other ethnic groups were represented. Immigration was slow between 1850 and 1879, but the 1880s saw a rapid buildup in numbers; in 1889 over 200,000 migrants entered Argentina. Migration collapsed dramatically during the worldwide financial crisis of the early 1890s but recovered to average around 250,000 per annum in the years prior to World War I (1914–1918).

In Santa Fé province, many colonies were set up for migrants, the first of which was established at Esperanza in 1856 (Jefferson, 1926). The colonies were settled farming communities; the migrants either were tenants or were able to purchase the land from the entrepreneurs who organized the settlements. Such colonies were particularly common in central Santa Fé, where cereals were the dominant crop. In southern Santa Fé sheep rearing on more extensive holdings was important, and there were few migrants and agricultural towns in the region. In recognition of these agricultural and ethnic distinctions, the central region was known as the *pampas gringas* and the south as the *pampas criollas*. The immigrant farmers of the *pampa gringas* were politically active and given to populist activity, as in the rural protest movements of 1893 (Gallo, 1977). On the *pampa criollas* the traditional rural *caudillos* continued to dominate political processes.

The pattern of agricultural development in Buenos Aires province was different and did not result in the establishment of as many independent farmers as in the *pampa gringas*. If stockmen in the province of Buenos Aires were to increase production and the quality of animals, the traditional pattern of extensive grazing had to be modified and fodder crops grown. The system employed initially was to offer land to migrants for a few years (usually 3 to 6 years), with the stipulation that alfalfa would be planted before the farmers moved on. The advantage of this scheme to the landowner was that it got land into a fodder crop without the expense of cultivation. When migrant demand for this type of arrangement built up, the landowners eventually asked for a share of the migrant farmers' crops in the years prior to the planting of the alfalfa.

Although some attempts were made to establish the legal equivalents of the U.S. Homestead Act (1862) and the Kinkaid Act (1904), the efforts did not succeed in placing many independent farmers on the land. In 1876 a land law was passed in Argentina that was intended to provide free land for immigrants and promote settlement. Abuses of this law by speculators were great, and the legislation never performed its function. In 1884 a homestead act became law, but this was limited to grazing lands south of the Río Negro in Patagonia, and abuse was again common (Scobie, 1971).

By the end of the first decade of the twentieth century, the Pampas had been transformed. Buenos Aires was no longer a colonial, provincial city, but the first South American city with 1 million inhabitants—more akin to Chicago than to other Latin American cities (Sargent, 1974). Much cultivable grassland on the Pampas had been plowed, and the area of tilled land in the republic as a whole had increased from 373 square miles in 1865 to around 150,000 square miles in 1910 (Jefferson, 1926). In 1850 Argentina was a net importer of grains; by 1900 it was one of the world's leading exporters of wheat. Traditional activities had disappeared or had been pushed outward from the La Plata area. Sheep raising was found in the semiarid lands of the west and south. The production of *tasajo* and extract had moved to less central locations as a result of competition for land from profitable forms of meat processing such as freezing and chilling (Crossley, 1976). If the traditional landowning classes had retained a hold on the wealth and political power during the development process, they had to take notice of the needs of large corporations, frequently capitalized from Britain, that owned railroads, shipping, banks, meat-processing plants, and land.

► Agricultural Exports

As has been illustrated with reference to Argentina, agricultural exports from the temperate lands of Latin America were mainly wheat and animal products. But within the tropical regions of Latin America sugar and tobacco were the most important commercial products until the last quarter of the nineteenth century, when the demand for coffee, cacao, cotton, bananas, and rubber in the European and U. S. markets increased the involvement of Latin America in world trade. In some areas, notably the coffee region of the states of São Paulo and Paraná in Brazil, tropical export agriculture played an important role in general economic development. But in relation to other commodities, for example, rubber in Amazonia, although fortunes were made by individuals, the export economy did little to improve technology, transportation, and general living conditions in the region of production.

In some cases the area of commodity production was owned, organized, and operated by a large international corporation. The United Fruit Company started in a small way in the late nineteenth century as a shipper of bananas from Jamaica. Eventually it owned banana plantations in Jamaica, Guatemala, Honduras, Panama, Costa Rica, and Colombia. The budget of the company exceeded that of several of the states in which it operated, and not unnaturally, its size and dominance of economic life came to be resented. At the opposite end of the scale, small farmers opened up banana cultivation in Ecuador and made that country the world's largest banana exporter (Preston, 1965).

By the beginning of the twentieth century, Latin American export agriculture was well integrated into world markets, and some regions had become dependent on the sale of particular products—Brazil on coffee, Central America on coffee and bananas, Colombia on coffee, Cuba on sugar, and Argentina on meat, wool, and grain.

The agricultural export economies that developed through integration into world markets came to an abrupt end with the Great Depression of the 1930s. Demands for all kinds of raw materials fell drastically, bringing down the price of most of the region's agricultural exports. The most important exports—coffee, cacao, sugar, and bananas—suffered an average decline in real price of about 60 percent between the peak years of the late 1920s and the low point of the Depression in the 1930s. During a time of diminishing incomes in the industrial societies, Latin America found itself the producer of the world's "desserts," and the demand for such luxury imports declined sharply.

The crisis of the 1930s produced a variety of responses in Latin America. In the short run there were attempts to diversify agricultural exports as prices of the various commodities fluctuated. In Brazil, for example, there was a considerable transfer from coffee to cotton as the price of coffee fell drastically and remained low while cotton prices were maintained (Furtado, 1965).

During the Depression years, efforts were made to control the marketing of particular commodities. International agreements covering the production of sugar, tea, wheat, and rubber were signed between February 1931 and January 1936. Schemes were devised to maintain the prices of agricultural exports. A major compensatory scheme involved the destruction of accumulated coffee stocks in Brazil. The government bought up and destroyed 4 to 5 million tons of coffee, about one third of world production during the 1930s. Between the late 1930s and the end of World War II, 550 million coffee trees were destroyed. These measures finally restored some balance in the coffee market, but they were not followed by other producing countries, with the result that Brazil lost a large share of the world market. Brazilian production of coffee increased after World War II, and vast acreage in western São Paulo and Paraná was brought under cultivation.

The crisis of the 1930s stimulated efforts to substitute domestic manufactured goods for imported goods. This process of import substitution became a major stimulus to industrial development; in the years since 1930, the most dynamic sector of the Latin American economy has been manufacturing industry. While the growth of the industrial sector has been impressive, approximately one third of the income derived from exports was earned by agricultural products. The traditional products of coffee, sugar, bananas, and cotton remain leading generators of foreign exchange earnings, although soybeans have been of increasing importance in the last decade. Four traditional export crops (coffee, sugar, bananas, and cotton) occupy approximately 15 percent of the arable land in Latin America. The rise in world prices of most of the region's exportable commodities in recent years has increased the overall value of agricultural exports.

In recent decades, the relative importance of agricultural products as a percentage of total exports has diminished. Even though manufactured exports are more and more important, agricultural exports will remain significant for most countries in the region. As a generator of domestic capital, however, the agricultural sector has been diminishing steadily. The contributions of agriculture to the overall gross domestic product (GDP) have fallen significantly in most Latin American countries since 1960 as economies have diversified.

The production and marketing situation in the commercial sector of agriculture has changed considerably in recent decades. New techniques have altered production prospects considerably. In some countries, such as Mexico and Venezuela, government intervention in the agricultural economy has substantially raised the average yields by providing irrigation infrastructure and technical assistance in plant breeding and fertilizer application. Over one half of the value of Mexican crop production presently comes from irrigated land, but to accomplish this, 10 percent of the federal budget has been spent on irrigation development each year since 1941. There is some question as to whether irrigation has been overextended, particularly in the dry Mexican northwest. Problems with water cost and water quality have been plaguing farmers there recently, and there is evidence of soil loss and desertification in marginal areas. Nevertheless, it should be noted that the average wheat yield obtained on the irrigated land of the Tres Ríos and Colorado Delta regions of northwestern Mexico is higher than in the United States and Canada, where the crop is dry farmed. Average cotton yields are high in Mexico and Peru when they are grown on irrigated land, and maize yields increased from 14.5 pounds per acre to 22 pounds per acre in Argentina and from an impressive 12.6 pounds per acre to 30 pounds per acre in Chile during the period from 1982 to 1987.

Manioc yield more than doubled in El Salvador, Cuba, Ecuador, and Venezuela over the same 20-year period. During the same two decades, maize yields in the United States increased from 22.2 pounds per acre in 1951 to 48.6 pounds per acre in 1972, and wheat yields increased from 10 to 20.3 pounds per acre (Ruddle, 1972).

► Contemporary Food Production

With the exception of the temperate agricultural regions of the Argentine humid Pampas, the central valley of Chile, and the irrigated oases of northwestern Mexico, Latin American agriculture has not been noted for its productivity. Although the traditional farming systems, such as *roza* and *milpa*, proved ecologically stable and attractive from the point of view of yield per unit of labor, they have proven to be inadequate in supplying the increasing human population with nutritious food. It is significant that even in countries such as Mexico, where impressive yields are obtained in the commercial sector, traditional subsistence farming is still the predominant form of agriculture. For example, all commercial acreage combined does not equal the acreage in beans, a traditional subsistence or semicommercial crop.

In the past few decades there has been an expansion of the internal demand for agricultural products. Population growth, rapid urbanization, and the rise in the purchasing power of a part of the population, have increased domestic demand. The agrarian sector has not responded with increased production. Only in exceptional cases has agricultural production exceeded population growth, and demand has grown much more rapidly than the population. The realities are that most Latin American countries have to meet food production deficiencies with imports.

Population growth in the rural areas of Latin America has exerted pressure on land tenure and agricultural systems. Rising agricultural populations in the areas where traditional swidden methods are used result in shorter fallows or prolonged periods of cropping, both of which may be ecologically disastrous and usually result in a malfunction of the swidden cycle. In areas of more intensive *milpa* cultivation the pressure results in untenable partitioning of landholdings into tiny parcels called *minifundios*. Resulting rural refugees usually end up in slums and squatter settlements in large cities and tax the ability of the urban system to absorb them.

Although Brazil is expanding industrial development, and for many years optimistic writers on world food prospects have alluded to the potential of Brazil as an important source of food, it is doubtful whether Brazil will be able to keep pace with its own food needs. The problem is not so much that Brazil has been unable to expand its food production; it has, and at an impressive rate. The agricultural sector grew at over 4 percent per annum during the decade from 1980–1990. The population grows at just over 2 percent per annum, and economic growth has been impressive. Together these two sources of growth in demand are increasing food needs by some 4 percent per year. This means that to be self-sufficient, Brazil needs to increase output far more rapidly than any major country has succeeded in doing.

Since 1980, increases in total food production have occurred in every Latin American country except El Salvador, Guyana, Haiti, Jamaica, and Nicaragua. However, increases in production were offset by population increases in many countries. In populous and important countries such as Argentina, Mexico, and Venezuela the growth of the agricultural sector did not keep pace with the rate of population growth.

► Modernizing Agriculture

Virtually no country in the world has succeeded in fully modernizing its agricultural system. This is not to say, however, that there are no modern agricultural systems in the world. Agriculture in the United States is conducted mostly on large, mechanized units, and farming activity is linked, by efficient communication systems, with the economic life of the country. However, the U.S. system was born relatively modern. When the Midwest and the West were opened up, their economies were extensions of the modern industrial system that existed in the East and in Western Europe. "Pioneer farming" was developed on the basis of substantial landholdings, cash crops, commodity markets, railroads, and agricultural machinery. The pattern of events was somewhat similar on the Argentine Pampas. But the Pampas and the Great Plains are exceptional cases in the history of world agriculture and should not be used as models for other regions.

Overall, in the last 400 to 500 years there is little evidence that Latin America has suffered severe regional food shortages on the basis of traditional agricultural systems, which seem so inefficient to North American eyes. In fact, many of the local systems are adjusted to environmental conditions and the needs of the communities they serve. However, as Latin America has urbanized, the population has left rural communities, and the traditional farming system is not well adapted to supplying large quantities of cash crops to distant markets. Much of Latin American agriculture is small-scale and not capable of absorbing the type of inputs that have improved the efficiency of farming in the United States. If agriculture is to meet the needs of the growing urban communities, traditional, rural life-styles and value systems will undergo alteration. If farming is to modernize, so must the communities in which traditional farmers live. This type of change is not achieved in a generation. Modernizing agriculture will involve employing the techniques of "green revolutions" such as crop breeding to provide higher-yielding varieties of crops, as in Mexico. Yet because the new high-yielding varieties thrive only with chemical fertilizers and irrigation, some 80 percent of Mexico's farmers have been bypassed by the Green Revolution. Above all, a more productive agriculture sector will require additional educational and informational institutions in rural areas, together with storage and communication systems, that will allow surpluses to be marketed when they are available.

► Land Tenure Problems

A major problem is the structure of landholdings. The problem of the concentration of landownership is, as West suggests in Chapter 2, a heritage of the colonial period, when large estates emerged. Many peasants held land from the *haciendas* and *fazendas* and paid rent with produce or labor services. Historically, the *haciendas* and *fazendas* were large units of land cultivated by laborers variously referred to as *peones, colonos, lavradores,* or *inquilinos,* who often obtained small properties on which to practice subsistence farming. Although there were many variations, this became the typical Latin American system of land ownership: large commercial, *latifundia,* and small subsistence *minifundia*. In Mexico, *peones* worked the land under the direction of a manager (*miordomo*) and supported themselves on small plots of land either on or near the *hacienda*. In Brazil, after the end of slavery, sharecropping by former slaves became the typical form of landholding in the sugar areas of the Northeast. In the south, large coffee *fazendas* were also worked by sharecroppers. What was conspicuously absent was the medium-sized family farm that characterized much of Europe and North America.

In 1960, two thirds of the agricultural land in Latin America was occupied by holdings of 1,000 hectares or more, which made up only 1.4 percent of all holdings. On the other hand, three quarters of all holdings were less than 20 hectares, but occupied only 3.7 percent of the agricultural area (Chonochol, 1965, pp. 79–90). The number of *minifundia* has actually been increasing due to subdivision of existing large properties and as a result of small property distribution in areas of agricultural colonization. Since the 1960s, two distinct sectors have emerged in the agricultural economy, one consisting of predominantly small holdings (*minifundia*), with a larger subsistence element and mainly traditional farming methods, and the other modernized, high-energy, large-scale systems. The *minifundia*, however, account for a significant proportion of food crops, since so much of *latifundia* is in nonfood crops.

> Thus in Brazil family farms—those without hired labor—produce 80 percent of the staple foods. . . . In Mexico small farms produce two-thirds of the maize and beans. In Perú small farms, with only 15 percent of the total agricultural area, produce half the cereals and three-quarters of the root crop. The split is most marked in El Salvador, where the larger farms produce coffee and farms of less than 5 hectares grow 60 percent of the maize and beans (Grigg, 1993, p. 196)

During the present century, there have been more and more cries for land reform policies, involving the breakup of large estates. The land question was a major issue that fueled the Mexican Revolution, and the peasantry was eventually given access to land under the *ejido* system in which communities were allocated land. Although land is generally worked by individuals, *ejido* land cannot be sold or rented out. By 1980, about one-half of Mexico's agricultural land was in *ejidos*. Other countries, including Peru, Venezuela, and Bolivia, have broken up large landholdings and given small farmers title to land. Such policies may be a move toward social justice, but they do not necessarily promote more efficient agriculture or better rural living standards.

In the 1960s, land reform was a growing political issue, but effective action was rare. Military governments, as in Brazil, assumed power and protected landowning elites from land reform. Demographically, there was migration out of rural areas to the cities by the younger and more ambitious segments of the population.

The distribution of population in Latin America since World War II has changed from being predominantly rural to being predominantly urban. When countries acquired urban majorities, the political pressures to bring about land reform were reduced. The urban majorities wanted cheap food, and in implementing policies to bring this about, governments often disadvantaged the rural sector. Urban populations, however erroneously, often associated high food prices with agrarian reform (Thiesenhusen, 1989, p. 485).

Many state-sponsored land reform schemes were cosmetic political exercises which involved relatively few farmers and failed to address broader rural development issues. Most countries in the region experienced rapid inflation during the period between 1960 and 1990. Land was a financial hedge against inflation, and landowners were reluctant to sell real assets to agencies redistributing land.

In the 1990s, the issue of land reform is beginning to attract attention once again. A problem that frequently finds its way into newspaper reports is the issue of gaining secure title to land. Even when a small farmer has acquired or been allocated land, it is often difficult to register the title, or it turns out that other parties also have title claims.

► *Agricultural Settlement and Colonization*

Despite attempts at land reform, the number of landless peasants continues to grow, and the subdivision of already small agricultural units is creating more *minifundios*. In many parts of Latin America the settlement of essentially unoccupied terrain, particularly in the humid tropical lowlands or disputed border regions, has been used as a diversion from tenure and modernization problems by many governments and an escape valve for overpopulated highland regions. In the same category are the less directed, more spontaneous settlement activities brought about by opening new roads or new irrigation projects. Newly settled tropical lands have added significantly to agricultural output in the past 30 to 40 years. Expansion of agricultural production continued to depend more on the extension of areas under cultivation than on increased yields.

In most tropical countries the flow of agricultural settlement has been from traditional highland centers of population to the lowlands, which remained sparsely settled during historic times. Important areas of settlement activity have been on the plains and valleys of Mexico's Gulf coast, the tropical interior and coastal lowlands of Central America and interior Venezuela, the Beni and Santa Cruz de la Sierra area of Bolivia, the Guayas River Basin of Ecuador, and the upper Amazon in Peru. The vertical movement is well represented by the general eastern extension into the Amazon lowlands from the Andean foothills by nationals of Colombia, Ecuador, Peru, and Bolivia. The Brazilians, too, are promoting agricultural developments in the Amazon Basin. These developments are dependent on ambitious highway projects such as the Carretera Marginal de la Selva in Peru, Ecuador, and Colombia, and the Cuiabá-Acre, Cuiabá-Santerém, and Trans-Amazonia Highways in Brazil. Along the Cuiabá-Acre Highway (BR364), there has been a remarkable development of commercial soybean production during the last 15 years. In the dry lands of northwestern Mexico wealthy agricultural interests, some with U.S. backing, and peasant farmers working through cooperatives (*ejidos*), have developed lands for agriculture.

SUMMARY

The results of settlement projects have been generally disappointing. There have been problems in processing land titles and protecting property rights. Remoteness from markets has resulted in the persistence of subsistence patterns of agriculture. Frequently mentioned are problems with agricultural credit and extension assistance, and sometimes complete ecological malfunction occurs, particularly in tropical lowland areas. Nelson (1973, p. 5) states:

> It is evident that experience in the development of new lands in the humid tropics has been very mixed. There is no clear understanding of why some areas seem to have prospered and others have stagnated. Only in Brazil has any attempt been made to assess how new land in the humid tropics contributes to the overall development process. There is no record of what has happened as a result of previous land development investments and policies nor of their economic, social, and political objectives and the extent to which these objectives have been achieved. We have only fragmented knowledge about the requirements for capital, managerial talent, and foreign exchange and about how the agrarian structure affects the results of investments.

Perhaps more emphasis should be placed on improving production in humid temperate zones, many in areas above 5,000 feet, instead of new agricultural endeavors on the soils of the humid tropics. For all their dramatic ring, the settlement schemes have not been effective in modifying the traditional land tenure structure, and the *latifundio-minifundio* disparity remains one of the striking characteristics of agrarian life in modern Latin America.

FURTHER READINGS

Adams, W. M. "Sustainable Development?" In *Geographies of Global Change,* edited by R. H. Johnson, P. J. Taylor, and M. J. Watts. Oxford: Blackwell, 1995.

Cubitt, T. *Latin American Society,* 2d ed. London: Longman, 1995.

Hecht, S., and A. Cockburn. *The Fate of the Forest.* New York: Harper Perennial, 1990.

Thiesenhusen, W. C. *Broken Promises: Agrarian Reform and the Latin American Campesino.* Boulder: Westview Press, 1995.

———, ed. *Searching for Agrarian Reform in Latin America.* Boston: Unwin Hyman, 1989.

CHAPTER 5

Population: Growth,
Distribution,
and Migration

BRIAN W. BLOUET

INTRODUCTION

In 1914, a distinguished observer of South American affairs estimated that the population of the continent, excluding Middle America, would be about 150 million persons in the year A.D. 2000 (Bryce, 1914, p. 563). Today, the population of Brazil alone exceeds that figure, and the population of all of South America is passing 325 million. When we include Mexico, Central America, and the Caribbean, the total population of the region is approaching 500 million people (Population Reference Bureau, 1995). Early in the twentieth century, the rate of population growth in Latin America was low, except in Argentina, Chile, and Uruguay—countries that were attracting migrants from Europe. In the middle decades of the century, from the 1930s to the 1970s, population growth rates for the region as a whole surged and exceeded 3 percent per annum. Today, the growth rate for Latin America is 1.9 percent per annum, and it is falling.

In Chapter 3, Robert West examined population change in the colonial era. The present chapter reviews changes in population numbers and distribution in the modern era (Figure 5.1). Migration patterns are also discussed.

POPULATION GROWTH

Population growth may result from two causes: (1) an excess of birth rates over death rates; and (2) an excess of immigration over out-migration. In the first half of the nineteenth century, with the disintegration of the Spanish American empire, natural increase was low and there was little migration to the region. Population growth was relatively slow.

By the mid-nineteenth century, Latin America was being drawn into the world economy. Economic development created demand for labor and attracted European migrants to southern Brazil, Uruguay, Chile, and the Argentine Pampas. The result

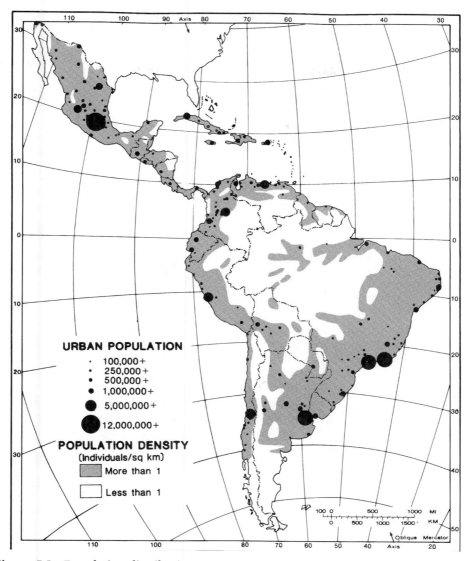

Figure 5.1 Population distribution.

was a large population increase in those areas. In the period 1850–1930, only the
United States, drawing migrants voraciously, had a faster rate of population growth
(Sánchez-Albornoz, 1986). From 1940 until 1980, owing to a high birth rate and a
declining death rate, Latin America as a whole had faster population growth than
any other major world region. During the 1980s Sub-Saharan Africa overtook Latin
America and assumed the problematical distinction of having the fastest population
growth rate in the world.

The surge in Latin American population numbers that began in the second half
of the 1930s is partly related to the beginning of World War II in Europe in 1939.
The demand for foodstuffs and raw materials increased before the war started. After
Germany overran Western Europe in 1940, Britain continued to import heavily from
Latin America, particularly Argentina and Uruguay, long-standing trade partners.

TABLE 5.1   Population Growth Rates for Selected Countries, 1920–1995

	Percentage Annual Growth					
	1920–1925	1940–1945	1960–1965	1980–1985	1990	1995
Total for Latin America	1.9	2.2	2.9	2.4	2.1	1.9
Argentina	3.2	1.7	1.6	1.2	1.3	1.3
Bolivia	1.1	1.8	2.3	2.7	2.6	2.6
Brazil	2.1	2.3	2.9	2.3	1.9	1.7
Chile	1.5	1.5	2.5	1.7	1.7	1.7
Colombia	1.9	2.4	3.3	2.1	2.0	1.8
Ecuador	1.1	2.1	3.4	3.1	2.5	2.2
Guatemala	1.1	3.4	3.0	2.9	3.1	3.1
Mexico	1.0	2.9	3.5	2.9	2.4	2.2
Peru	1.5	1.8	3.1	2.8	2.4	2.1
Venezuela	1.9	2.8	3.3	3.3	2.3	2.6

Source: J. Wilkie et al. (1988), *Statistical Abstracts for Latin America*, vol. 26, p. 109, UCLA: Latin American Center Publications; Gilbert (1990, 1994), *1990 and 1995 World Population Data Sheets*, Population Reference Bureau, Washington, D.C.

When the United States was forced into the war in December 1941, the demand for strategic raw materials increased and prices rose. Latin American economies began to grow, living standards improved, death rates fell, rural-to-urban migration increased, and manufacturing and service sectors in the larger cities expanded. The pace of economic diversification and population growth quickened. The number of people in the region increased rapidly between 1940 and 1970 (Table 5.1). In the 1970s, the rate of population growth began to slow (Dickenson, 1972). Population numbers continued to grow, of course, but at a declining percentage rate.

THE DEMOGRAPHIC TRANSITION MODEL

A general description of the changes in the rate of population growth is contained in the demographic transition model. The model is derived from study of the historical demography of Western countries and has four phases.

Phase 1: Birth rates and death rates are high, and there is little population growth.

Phase 2: Death rates drop, birth rates remain high and may rise. Total population numbers grow rapidly.

Phase 3: Birth rates drop and the rate of population increase slows down.

Phase 4: Birth and death rates are low, and population growth decreases to less than 1 percent per annum.

There is no general agreement on the forces that bring about changes in birth and death rates. In Europe, the fall in the death rate, beginning in phase 2, triggered population growth. The fall in the death rate was brought about by improved living standards, particularly nutrition, which increased the ability of populations to survive disease. The fall in the death rate was also associated with attitude changes as traditional societies emerged from fatalistic belief systems and began to apply elementary rules concerning health and hygiene.

The demographic transition model is not necessarily universally applicable. The model does not fit all parts of Asia, although it has been argued that China experienced the four phases of the transition between 1949 and 1980 (Hornby and Jones, 1993, pp. 90–93).

Latin American countries do seem to be following the transition pattern in a general way, but there are important differences in the pace of change. For example, population growth rates, associated with Phases 2 and 3 of the transition, have been at much higher percentage rates than those experienced in Europe in the nineteenth century. In phase 2, in many Latin American countries, not only has the death rate dropped, but the birth rate has increased, producing rapid population increase.

We will now examine the four phases in more detail and suggest how the demographic transition may be relevant to the Latin American experience.

► *Phase 1*

In premodern, traditional societies, birth rates and death rates are high, population growth is slow and often interrupted by increased death rates brought on by famine, warfare, or disease. Some Latin American regions had both high birth and high death rates during the early decades of the twentieth century, for example, Mexico (Figure 5.2). Today, no Latin American country is in the first phase of the demographic transition. In fact, no country in the world displays the high birth and death rates associated with Phase 1. High birth rates are easy to find in Third World countries. However, even in Africa, where death rates are relatively high (over 20

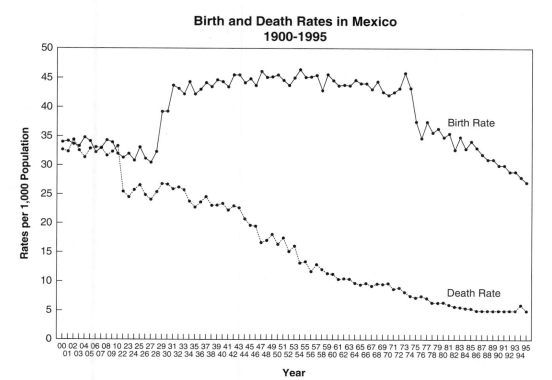

Figure 5.2 Birth and death rates in Mexico.

deaths per annum per thousand inhabitants) mortality rarely reaches half the birth rate. For example, in West Africa, Ivory Coast has a birth rate of 50 per thousand, and a crude death rate of 15 per thousand. Within Latin America and the Caribbean, Haiti has a crude birth rate of 35 per thousand and a crude death rate of 12 per thousand. The gap between the high birth rate and the falling death rate indicates that Haiti is no longer in Phase 1.

► *Phase 2*

In the second phase, birth rates remain high and, in Latin America, often increase. Death rates fall substantially as diseases are controlled. Population numbers rise rapidly. A number of Latin American countries are in this phase, including Bolivia, Peru, Venezuela, and Guatemala (Table 5.2).

All the countries listed in Table 5.2 have had high birth rates for some time. In the case of the Andean countries, birth rates and population growth rates have fallen from even higher figures in the 1960s. When disease and epidemics are under control, high birth rates produce rapid population growth. Rapid population growth produces a youthful population. In a youthful population, the proportion of people dying from old age is low, and the crude death rate figure declines as a function of natural increase. For this reason, crude death rate figures often give rise to anomalies. For example, the crude death rate in Sweden is 11 per thousand. The crude death rate in Peru is 8 per thousand. Sweden has an aging population; Peru has a youthful population. The figures cannot be used to suggest that public health is better in Peru than Sweden.

During Phase 2, population numbers for Latin American countries doubled rapidly. The doubling time is the number of years it takes for a population to double in size. Doubling time is calculated using the rule of 70. Divide the percentage rate of increase into 70. A population growing at 1 percent per annum will double its size in 70 years. A 2 percent growth rate will double a population in 35 years, and a 3 percent rate will result in doubling after 23.3 years. In the 1950s, 1960s, and early 1970s, several Latin American countries, including Mexico, Venezuela, and Peru, were growing at over 3 percent per annum and did double their populations in just over 20 years.

Obviously, the doubling time figure will change if the percentage rate of increase rises or falls. In the 1940s, when the percentage rate of population growth in Latin America was increasing, the doubling time figure was greatly underestimating

TABLE 5.2   Crude Birth and Death Rates, 1995

Country	BR	DR	PGR	% Under 15/65+
Bolivia	36	10	2.6	41/4
Ecuador	28	6	2.2	38/4
Paraguay	33	6	2.7	40/4
Peru	29	7	2.2	36/4
Guatemala	39	8	3.1	45/3
Honduras	34	6	2.8	47/4
Nicaragua	33	6	2.7	46/3
Venezuela	30	5	2.5	38/4

BR = crude birth rate per 1,000; DR = crude death rate per 1,000; PGR = percentage growth rate per annum.
Source: Population Reference Bureau, *1995 World Population Data Sheet* (Washington, D.C.: Population Reference Bureau, 1995).

the time it would take to double the population in Latin America. Today, the population of the region is growing at 2 percent per annum and, if the percentage rate stays constant, the population of Latin America will double in 35 years. In fact, the percentage rate of population increase is declining, and it will take longer than 35 years for the population of the region to double.

► *Phase 3*

In the third phase of the demographic transition, crude death rates stabilize, and the crude birth rate declines substantially. Population numbers continue to grow, but the percentage rate of growth slows. Brazil and Mexico entered Phase 3 in the 1970s. Neither country has completed the phase although crude birth rates continue to drop (Table 5.3). In 1970 Mexico had a classic phase 2 population pyramid with a broad, splayed base indicating large numbers of young people. By 1990 the base of the pyramid had narrowed, indicating a decline in the birth rate. In 1990 there were fewer Mexicans in the 0–4 age group than there were in the 15–19 age group (Figure 5.3). Table 5.3 shows that while crude birth rates are still high in Mexico, they have dropped in the last quarter of a century. Death rates are now low and can drop little further. In fact, as the population ages, crude death rates will rise.

The slowing of population growth rates in Phase 3 results largely from declines in the birth rate. The birth rate is reduced by a number of factors, including:

1. As death rates drop, infant mortality decreases and birth rates cease to be driven by the high death rate among children. Parents no longer need to have large numbers of children to ensure survival of the family.

2. A modernizing society urbanizes. Urbanization produces economic and social pressures that result in fewer births and declining family size. Premodern, rural economies have a large subsistence element; people produce much of what they consume. Urban economies tend to be cash economies and most things have to be paid for in cash. The larger the family, the greater the cash need.

3. The availability of information concerning contraception has helped birth rates to drop rapidly in Phase 3. The dissemination of information is easier in an urban setting than in rural areas, and literacy rates tend to be higher in towns.

4. It has been suggested that "in societies of every type and stage of development, fertility behavior is rational, and fertility is high or low as a result of economic benefit to individuals, couples, or families" (Caldwell, 1982, p. 152).

TABLE 5.3   Birth, Death, and Growth Rates, 1995

Country	BR	DR	PGR	% Urban	% under 15/65+
Brazil	25(39)	8(11)	1.7	76	32/5
Colombia	24(43)	6(26)	1.8		33/5
Mexico	27(45)	5(11)	2.2	71	36/4
Venezuela	30(42)	5(8)	2.6	84	38/4

Figures in parens are 1965 birth and death rates. BR = crude birth rate per 1,000; DR = crude death rate per 1,000; PGR = percentage growth rate per annum.
Source: World Development Report, 1990; Population Reference Bureau, *1995 World Population Data Sheet*

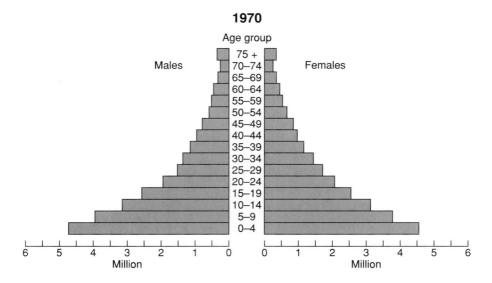

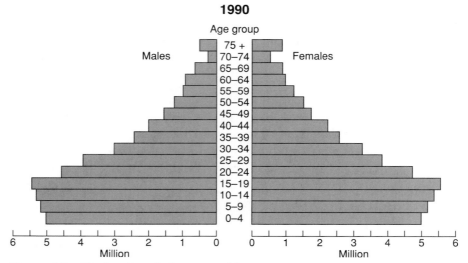

Figure 5.3   Mexico: Population pyramid.

► *Phase 4*

In the fourth phase, birth rates and death rates are low and population numbers grow slowly. For example, in the United Kingdom, the crude birth rate is 14 and the crude death rate is 12 per thousand. The population grows by 0.2 percent a year and will double in three centuries. For Italy, the birth rate is 10 and death rate 9. For Spain the birth rate is 11 and death rate 8. In Latin America and the Caribbean, only Uruguay (birth rate 17, death rate 10), Cuba (birth rate 14, death rate 7), and Barbados (birth rate 16, death rate 9) can be viewed as fourth-phase countries. However, Argentina and Chile are approaching the fourth phase (Table 5.4).

The demographic transition model approximates the experience of a number of European countries during the nineteenth and twentieth centuries. In Latin America the model can best be applied to those countries that have experienced large-scale migration from Europe in the nineteenth century, particularly Argentina and

TABLE 5.4   Crude Birth and Death Rates, 1995

Country	BR	DR	PGR	% Urban
Argentina	21(23)	8(9)	1.3	87
Barbados	16	9	0.7	38
Chile	22(34)	6(11)	1.7	85
Cuba	14	7	0.7	74
Uruguay	17(21)	10(10)	0.8	90
United States	15(19)	9(9)	0.7	75
Canada	14(21)	7(8)	0.7	77

BR = crude birth rate per 1,000; DR = crude death rate per 1,000; PGR = percentage growth rate per annum. *Source:* Population Reference Bureau, *1995 World Population Data Sheet*; 1965 data (in parens) *World Development Report 1990.*

Uruguay. The demographic transition scenario may be less applicable to countries (Guatemala and Bolivia, for example) that have large Indian populations. It is worth noting that some countries have marked regional differences. In Peru, the demographic transition will take place later in the Andean region, dominated by Indian communities, than in the more rapidly modernizing and urbanizing *costa* region.

THE DEMOGRAPHIC TRANSITION MODEL AND THE LATIN AMERICAN EXPERIENCE

Let us look at some histories of population growth in Latin America during the twentieth century and see whether or not the countries selected fit the demographic transition model and, if not, what differences emerge.

▶ *Mexico in the Twentieth Century—Changing Birth and Death Rates*

Examination of Figure 5.2, Birth and Death Rates in Mexico, 1900–1995, suggests that Mexico has gone through a transition from high birth rates, high death rates, and unstable population numbers, to a situation of low death rates, falling birth rates, but still increasing population numbers. The rate of population increase is slowing, but Mexico has not completed the demographic transition.

In the period between 1900 and the outbreak of the Mexican Revolution in 1910, birth and death rates were high, and infant mortality was rising, suggesting that living standards were falling. In some years, deaths exceeded births and population numbers fluctuated. High birth and death rates, together with fluctuations in population numbers, are characteristic of phase 1 of the demographic transition model.

In the period from 1910 to 1922, there are no data, indicating the disruption of life during the revolution. In the 1920s, when data again become available, it is apparent that death rates were falling and birth rates were rising. We should be suspicious of the jump in birth rates that apparently took place in the late 1920s because it may be partially a result of the more complete recording of births as the administration of the country became more efficient in the postrevolutionary era. Even if there is a statistical quirk in the late 1920s, notice that the general trend of the birth rate is upward from the early 1930s to the 1950s. In the same period, death rates trend downward. If the period from the early 1930s through the 1960s represented phase 2 of the demographic transition, Mexico conforms to the general pattern in that death rates fall. The strong upward trend in birth rates, however, is anomalous, but not without precedents. Commentators noticed a tendency for birth rates to rise in some rapidly growing industrial regions in Britain during the nineteenth century (Wrigley, 1983).

Mexico is currently in phase 3 of the demographic transition. Death rates are low. Birth rates are falling, but remain high enough to sustain population growth at 2.2 percent per annum. A growth rate of 2.2 percent is high, but it is significantly down from the 3.5 percent annual rate recorded in the early 1960s. The marked drop in the birth rate in the early 1970s coincides with the years in which the Mexican government dropped pro-natal policies and began to promote birth control.

► *Venezuela, Colombia, Peru*

Mexico is not an isolated case in having a marked rise in the birth rate in phase 2 of the demographic transition. Examination of the graph for Venezuela (Figure 5.4) shows a similar pattern. In the 1930s, birth rates increased, and they peaked in the 1960s. Meanwhile, death rates established a strong downward trend and total population numbers grew. In the 1960s, the population of Venezuela grew at over 3 percent per annum, a rate that would double the population of a country in about 20 years. And the population of Venezuela did double between 1960 (7.4 million) and 1980 (15.0 million).

The demographic statistics for Peru are poor, but trends are the same. In 1940, the birth rate in Peru was about 27 for every 1,000 inhabitants. By 1950, the crude birth rate had risen to 30, in 1960 it was in the high thirties, and it peaked at 46.4 in the mid-1960s. Thereafter, the birth rate trend has been down, but the crude birth rate in Peru today, at 29 per thousand, is still above the figure of 27 per thousand recorded in 1940.

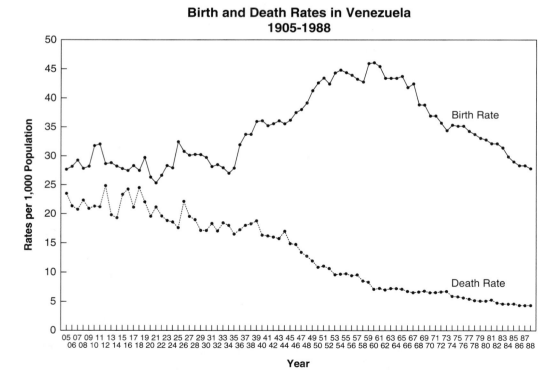

Figure 5.4   Birth and death rates in Venezuela.

There are many inaccuracies in the birth and death rate figures for Latin America. However, the overall trend of rising birth rates in a group of countries that includes Mexico, Colombia, Venezuela, and Peru in phase 2 of the demographic transition—when the standard model suggests they should be stable—is difficult to dispute. Collver indicates that there are widespread inaccuracies in Latin American birth and death statistics. He calculates, "Uruguay was the only country of Latin America with a birth rate below 40 in 1900" (1965, p. 163). In his analysis, birth rates were considerably underrecorded in the early decades of the century and therefore the steep rises in birth rates, characteristic of the period 1930 to 1960, although real, were not as great as suggested by the graphs compiled from official statistics.

In the European experience, the accepted view is that the population increase in the nineteenth century, associated with the demographic transition, was the result of falling death rates. In the Latin American countries we have examined in the twentieth century—Mexico, Colombia, Venezuela, Peru—the increase in population numbers was due to both a falling death rate and a rising birth rate.

▶ *Argentina and Uruguay*

The birth and death rate experience of Argentina and Uruguay was different from that of Mexico (Collver, 1965). The countries occupying the temperate grasslands of South America drew large numbers of migrants from Europe in the second half of the nineteenth century and the early decades of the twentieth. We might expect that the demographic history of Argentina and Uruguay would conform to the demographic transition model.

Examination of Figure 5.5, Birth and Death Rates for Argentina, 1910–1988, shows several contrasts with Mexico. Crude birth rates did not exceed a rate of 40 live births per thousand inhabitants. Except for brief periods immediately before

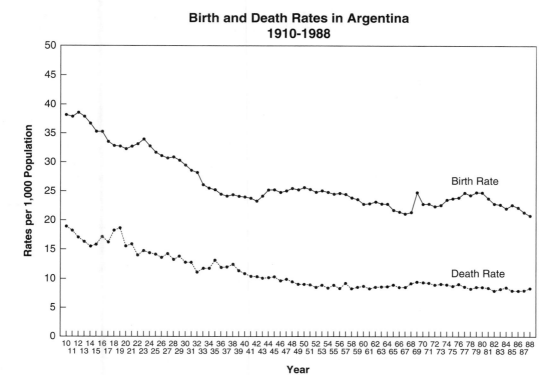

Figure 5.5   Birth and death rates in Argentina.

and after World War I, rates of natural increase did not exceed 2 percent per annum. Like the United States, Argentina was drawing huge numbers of migrants from Europe in the years before World War I. Newly arrived immigrants are not counted as part of natural increase. However, migrants are predominantly young adults and contribute to the creation of high birth rates.

► *Explaining Differences*

Can we explain the differences in this century between the birth and death rate experiences of Argentina and Mexico? Clearly there are cultural distinctions. Argentina is largely a temperate grassland settled by migrants from Europe. Many of the migrants were literate and remained in urban places after arrival. In Mexico, the heavily rural population was largely illiterate at the beginning of the century. There are economic distinctions, too. Argentina, with exports of wheat, livestock, and animal products, was linked into the world economy in the second half of the nineteenth century, earlier than most Latin American countries. Urbanization was rapid and early in Argentina, and Buenos Aires contained more than 1 million inhabitants by 1910. In the same year, Mexico City had a population of around 350,000.

In Mexico, economic reorganization did come in the wake of the Revolution (1910–1917). It is tempting to correlate the emergence of a new economic structure, for example the *ejido* system, which provided land to peasants, with the rising birth rates of the 1930s, as optimism spread across the land and the marriage rate surged.

Rural-to-urban migration accelerated in the 1930s in Mexico, and the pace of this migration increased into the 1960s. From the 1930s through the 1960s, there was economic expansion, optimism rose, family formation accelerated, rural-to-urban migration increased, cities expanded. In the rapid transformation, the attitudes of the traditional rural areas often moved out of the countryside with migrants and into the towns along with premodern ideas on family size, life expectancy, child mortality, hygiene, and health care. Birth rates in urban areas reflected these attitudes until the 1970s, when urbanization, education, and the cessation of pro-natal policies began a rapid decline in birth rates.

INFANT MORTALITY

Infant mortality figures for Latin American countries are unreliable, and the unreliability increases as we go back in time. Nevertheless, the general trends for the region are clear (Table 5.5). Infant mortality rates were very high during the early decades of this century and, while they have dropped rapidly, rates are still significantly above Western European and U.S. levels. In the case of Peru, the rates of infant mortality increased during the 1980s as the country was hit by economic recession and living

TABLE 5.5   Infant Mortality: Selected Countries

Country	1900	1940	1970	1980
Jamaica	174	112	32	12[e]
Mexico	287	126	69	39
Canada	187	58	19	10
Chile	340 (1901)	217	52	33
Colombia	N.D.	142	50	38[e]
Peru	N.D.	128	65	101

e = estimate; N.D. = No Data.
Source: Mitchell (1993).

standards declined. High, though declining, levels of infant mortality are characteristic of phases 2 and 3 of the demographic transition. However, many Latin American countries in phase 3 of the demographic transition are finding that birth rates and infant mortality rates are falling relatively slowly. This may be explained because infant mortality and birth rates are linked. Significant levels of infant mortality and child mortality before the age of five result in pressures to produce more children than would be the case if there were few deaths among children. The policy implication is clear: To reduce the birth rate, attack infant mortality and child mortality.

ECONOMIC DEVELOPMENT AND POPULATION GROWTH

The relationship between economic development and population growth is not simple. In parts of Europe, for example Britain, there appears, at first sight, to be a relationship between industrialization and the onset of the demographic transition. Elsewhere, as in Austria, the demographic transition commenced before the onset of industrialization, suggesting that a change in attitudes and behavior preceded economic transformation.

In general, once economic development starts to create employment opportunities, young men and women find it easier to get jobs. In traditional societies, scarcity of farmland often slows family formation, which reduces fertility rates. In modernizing societies experiencing economic growth, young people migrate from rural to urban areas, marry, and establish families earlier. The result is that in phase 2, rather than remaining steady, the birth rate may rise. This occurred, as we have seen, in a number of Latin American countries including Mexico, Colombia, and Peru. In the 1960s, population in Latin America was growing at over 3 percent per annum. By the early 1990s the rate of increase had dropped to 2 percent a year. The rate of increase will continue to decline due to increasing urbanization, birth control, and changing social values. However, as a result of "demographic momentum" total population will grow as young populations, created by previously high birth rates, have their own families (Gilbert, 1990, p 18).

By the end of the century, the population of Latin America will be around 500 million. Latin America and the Caribbean will add nearly 10 million inhabitants a year even as the annual percentage rate of increase declines. Too few jobs are being created, and it will be difficult to absorb job seekers into the economy. Ironically, a region that has often imported labor now has widespread unemployment.

PROBLEMS ARISING FROM A RAPID RATE OF POPULATION GROWTH

The high rates of population increase displayed in the region give rise to problems. If the population of a country grows at the rate of 2 percent per annum, the gross domestic product of the country must grow at a rate in excess of 2 percent per annum if living standards are to be maintained. In Latin America few countries grew economically in the 1980s. Most economies shrank in real terms, and meanwhile, increased population numbers meant an increasing labor supply that depressed wages and led to a decline in living standards.

In societies with rapid rates of population increase, a large percentage of the population will be found in the younger, predominantly consuming age groups. For example, in Bolivia and Mexico over 40 percent of the population is under 15 years of age. This is a common characteristic of Latin American countries (see Table 5.2). Attempting to provide health, social, and educational facilities to the young can consume a large part of available development funds. At the family level, large numbers of children limit savings and investment, and living standards may be reduced.

THE DISTRIBUTION OF POPULATION

It is apparent from Figure 5.1 that the population of Latin America is unevenly distributed. Some areas are heavily populated; others are sparsely inhabited. The present distribution of population has been influenced by the following factors.

► *The Distribution of Aboriginal Populations*

The largest pre-Columbian population clusters were found in upland basins in Middle America and Andean America. The Spaniards were attracted to these areas as sources of wealth, labor, agriculture, and converts. The impact of indigenous population centers is reflected in the distribution of population to the present (Figure 5.6). The largest urban area in Latin America—Mexico City—is based upon a major pre-Columbian population concentration at Tenochtitlán. In pre-Columbian South America, the well-peopled Andean upland basins of what is now Bolivia, Peru, Ecuador, and Colombia contrasted with the remainder of the landmass, occupied mostly by shifting cultivators or hunters and gatherers. Exceptions existed in the western deserts where irrigation was practiced and in some eastern coastal regions of what is now Brazil, where more advanced agricultural techniques allowed for higher population densities. But, in general, away from the Andean upland basins, South America was lightly populated. The thinly populated Pampas, which had physical similarities to the Spanish *meseta*, were not settled heavily until after the close of the colonial period.

There is an interesting contrast between the European settlement of the northeastern seaboard of North America by families acquiring farms and the settlement of Middle and South America by men who wanted estates. Whether this distinction would have prevailed if there had been what Robert West, in Chapter 3, calls "a great mass of organized and subservient labor" on the northeastern seaboard of

Traditional dress styles at a rural market in Panajachel, Guatemala.

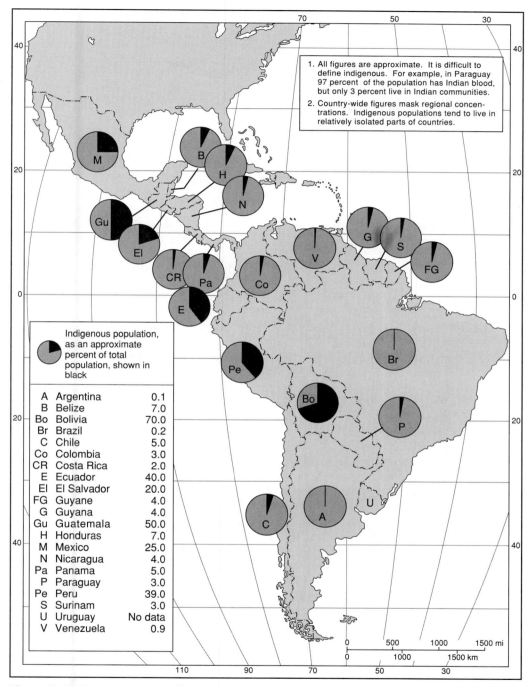

1. All figures are approximate. It is difficult to
   define indigenous. For example, in Paraguay
   97 percent of the population has Indian blood,
   but only 3 percent live in Indian communities.
2. Country-wide figures mask regional concentrations. Indigenous populations tend to live in
   relatively isolated parts of countries.

Indigenous population,
as an approximate
percent of total
population, shown in
black

A	Argentina	0.1
B	Belize	7.0
Bo	Bolivia	70.0
Br	Brazil	0.2
C	Chile	5.0
Co	Colombia	3.0
CR	Costa Rica	2.0
E	Ecuador	40.0
El	El Salvador	20.0
FG	Guyane	4.0
G	Guyana	4.0
Gu	Guatemala	50.0
H	Honduras	7.0
M	Mexico	25.0
N	Nicaragua	4.0
Pa	Panama	5.0
P	Paraguay	3.0
Pe	Peru	39.0
S	Surinam	3.0
U	Uruguay	No data
V	Venezuela	0.9

Figure 5.6 Contemporary indigenous populations.

North America is a subject for debate. Where labor was in short supply, as in the
southeastern seaboard of North America, African slaves were imported to work the
tobacco, cotton, rice, and indigo plantations. The overall effect of the Iberian conquest was to reinforce the existing pattern of population distribution, even if the
immediate result was to lower population numbers.

► *The Location of Major Colonial Settlements*

For strategic and economic reasons it was necessary to attach the New World territories to the imperial structures of Spain and Portugal. New towns and cities were founded at points where they could act as nodes in the administrative and communication network that was constructed to link New World territories with the Iberian peninsula. Many newly founded colonial coastal cities became administrative and trade centers in which population concentrated. Veracruz, Panamá City, Cartagena, Guayaquil, Lima-Callao, Valparaíso, and Rio de Janeiro are among these "points of attachment" that Charles Sargent describes in Chapter 6.

In the nineteenth and twentieth centuries, as Latin America was drawn into the world economy, many of the colonial coastal foundations grew rapidly as centers of commerce, industry, and finance. Cities such as Buenos Aires, Montevideo, and Lima-Callao have become primate cities within their respective countries.

► *The Physical Environment and the Distribution of Mineral Resources*

It is tempting to seek environmental explanations for the distribution of population: deserts are uninhabited because they are deserts; the Amazon Basin is "vast and impenetrable" and therefore sparsely populated. But environmental determinist views do not provide consistent explanations. The west coast desert of South America contains large concentrations of population in association with mining activity. Tropical rain forest regions in Latin America, and elsewhere, can support dense populations. Economic explanations that regard the physical environment and distance from markets as variable cost factors provide a better understanding of population distribution rather than environmental determinism.

Transportation costs make it difficult to exploit the resources found in the interior of South America. In crude terms, the longer the journey into the interior, the greater the costs of transportation and the less competitive any export crops or mineral resources produced there are likely to be. Population densities in the interior of South America remain low. Agricultural colonists in the *Oriente* and Amazonia find it difficult to export low-value crops. Only higher value products, like coca leaves and *aguardiente* (rum distilled from sugarcane), can stand the cost of transportation out of the Peruvian *Oriente*.

The existence of rich deposits of minerals has influenced the distribution of population. The opening of the northern frontier in Mexico was given impetus by the silver deposits at Zacatecas. Mining settlements were also established in the Andes to exploit silver deposits. In the colonial period, the Minas Gerais region of Brazil opened up on the basis of gold and diamond deposits. The town of Ouro Prêto, founded in 1711, became the state capital for a time. Today the historic town (population 65,000) is a national monument. The state of Minas Gerais remains an important center of mining, exploiting iron, manganese, and bauxite deposits. Most mining towns, however, do not become historic monuments. They usually decline as deposits become expensive to mine or the mineral is replaced by another substance. Thus, we cannot point to a good correlation between mineral deposits and high population densities.

In the modern era, governments have taken a regional view of resource development and created urban places to act as economic nodes. For example, the Venezuelan government established the complex of towns that make up Ciudad Guayana on the Orinoco River to smelt bauxite and iron ore from the Guiana Highlands to the south of the river basin. The development is not, however, on the mineral deposits but on the river, close to transportation. Similarly, the Brazilian Volta Redonda steel plant, built during World War II, is located on good transportation lines relatively close to the markets of Rio de Janeiro and São Paulo. The raw materials are brought to the site. Volta Redonda, much like the steel plant at Gary, Indiana, is close to major markets at a point where the raw materials can be economically assembled.

► *Location in Relation to Export Markets*

At the time the Iberians colonized the Caribbean and Central and South America, there were no trade goods leaving the region. Since the sixteenth century, economic development has been bound to the production of agricultural commodities, the mining of minerals, and to a lesser degree, the manufacture of export goods. For exports to be profitable, the cost of transporting products to the market has to be competitive. Some items (for example, silver bullion) are high value and could stand the cost of transportation from northern Mexico or the Andes to the coast and then on to Europe. Most mineral and agricultural products compete with other areas of production and cannot absorb exceptionally high transportation costs.

Location in relation to external markets has played a major part in determining which areas of Latin America produce competitive minerals and agricultural commodities for export. Accessible areas are opened up, attract capital and labor, and have relatively high population densities.

Until well into the nineteenth century, the greatest part of all exports from Latin America and the Caribbean were directed to markets in Europe. In the second half of the nineteenth century, the United States effectively occupied its Pacific shore. From that time it became possible to import into the United States tropical produce from the Pacific lowlands of Central and South America. The opening of the Panama Canal (1914) created access to U.S. markets on the east coast from the Pacific ports of Latin America.

For example, the development of banana exports from Ecuador in the twentieth century would not have been possible without access to North American markets. The bananas are grown by small farmers on the tropical coastal lowlands of Ecuador. Much land was brought into cultivation to grow the export crop, and as a result, population densities increased.

In recent decades, proximity to markets in North America has taken on new dimensions. Some manufacturing operations, in which the labor component is a major cost, can be performed in areas where the price of labor is low. Products are then shipped to markets in economically developed countries. Many parts of Latin America and the Caribbean have abundant, cheap labor that can be trained at reasonable cost to perform assembly work. Puerto Rico has many assembly plants whose products are shipped to the United States and Canada. More recently, plants have located in Central America and the Caribbean to utilize low-cost labor. The most extensive impact of low-cost labor as a locational factor is in the Mexican towns and cities that border the United States. Mexico has important advantages over Taiwan and other low-labor-cost Asian countries. Factories on the border can be easily connected with transportation and distribution networks in the United States. The rapid growth of population in northern Mexico reflects the location of assembly plants seeking cheap labor (Arreola and Curtis, 1993, pp. 22–36).

POPULATION MIGRATION

Latin America has a long history of migratory activity, both voluntary and forced. The Iberian colonizers for the most part were voluntary migrants, coming to the New World to seek social and economic advancement. The indigenous populations, were often forced to move by the Iberian intruders. Labor was forced to move to mines, and many Indian populations were relocated in villages where they could be christianized and used as a labor pool. By the early sixteenth century, enslaved Africans were being forced to migrate across the Atlantic to work on plantations (Table 10.3).

Migratory movement was marked in the colonial period (Robinson, 1990). Evans (1990), by using population data collected by viceroys of Peru in 1575 and 1683–1686, was able to show that there was large-scale migration of population during the period, with people moving into the area to the east of Lake Titicaca, Porco province around Potosí, and into the Yungas region on the eastern flanks of the Andes. Movement was undertaken by Hispanics and Indians, and many of the latter moved of their own decision into new areas of settlement. Swann (1990) found there were high levels of population mobility among free laborers in mining towns in colonial northern Mexico.

In the nineteenth century there was continued migratory movement, both forced and voluntary, as Indian populations were subdued in the southern temperate lands in the process of opening up frontiers for settlement. Immigrants continued to arrive from Europe, particularly in the second half of the century, as Germans settled southern Brazil, and migrants from Italy and Spain entered Brazil, Chile, Uruguay, and Argentina. Frequently, settlers went to farming frontiers, although many stayed in the major cities.

Since the surge in Latin American population numbers commenced in the 1930s, the major movement of population has been from rural to urban areas. By 1940 one third of Latin Americans lived in urban places. Today approximately three quarters of the inhabitants of the region live in towns and cities. In many countries, the movement from the countryside to the city continues, but in others the flow has slowed, for a large majority of people now live in cities. More than 80 percent of Uruguayans, Venezuelans, Argentineans, and Chileans dwell in urban areas. More than 70 percent of Mexicans, Brazilians, Colombians, and Peruvians are town dwellers. By contrast, many Central American and Caribbean countries still have over half their populations living in rural areas.

As a rule, it has been true that the larger the population size of a Latin American city, the faster will be the percentage rate of population growth. Big cities attract more migrants than small towns.

The primate cities of Latin America—places like Lima, Santiago, Buenos Aires, Montevideo, Caracas, and Mexico City—have grown to their present size as a result of immigration. Of course, natural increase has contributed to growth, but when the rate of natural increase in a country was 3 percent, as happened after World War II, the leading cities would often grow at 5 to 7 percent per annum as a result of immigration. Most migrants are young adults, and when they form families, rapid population growth is sustained.

► Theories of Migration

The major cities of Latin America were founded in colonial times. Their present size and importance in the overall distribution of population is a product of recent decades. As Latin American countries have industrialized and modernized, cities have drawn migrants from small towns and rural areas. Sustained rural-to-urban migration has resulted in the rapid growth of cities and a concentration of population in large urban places.

Wilbur Zelinsky (1971) suggested a scheme to describe migratory processes at different stages of economic development. Phase I is represented by a premodern, traditional society in which internal migration is limited. During Phase II, the onset of modernization is experienced, bringing with it "a great shaking loose of migrants from the countryside." In Phase III, the modernization process is well advanced, and although there is still significant movement from the countryside, the rate of rural-to-urban migration slows. Zelinsky saw Phase IV as fitting an advanced society, roughly equivalent to the presently economically developed countries. In this phase, movement from the countryside continues at a much reduced rate and is replaced by inter- and intraurban migration. Phase V outlines possible future characteristics of advanced societies.

Many Latin American countries displayed the characteristics of Phase I in the early decades of this century. For Latin America, rapid migration (Phase II) came in the 1950s, 1960s, and 1970s. Many countries now have an urban majority, and the rate of rural-to-urban migration is declining in them, as in Phase III of Zelinsky's system. Few countries in the region have reached Phase IV.

Two questions concern us about population migration. When the inhabitants of a country begin to migrate, where do they go? And what are their motivations? The initial statement on this problem was made in 1885 by the pioneer geographer E. G. Ravenstein whose "laws of migration" are still the starting point for discussion. The "laws" can be summarized as follows:

1. The majority of migrants move a relatively short distance.
2. The inhabitants of the country immediately surrounding a town of rapid growth flock into it; the gaps left in the rural population are filled by migrants from more remote districts.
3. Each current of migration produces a countercurrent of lesser strength.
4. Migrants proceeding long distances generally go to a major center of commerce or industry.
5. The natives of towns are less migratory than natives of rural areas.
6. Females are more migratory than males over short distances, but males more frequently are involved in international migration.
7. Most migrants are adults. Families rarely emigrate from their country of birth.
8. Large towns grow more by migration than by natural increase.
9. Migration increases in volume as industries and commerce develop and transport improves.
10. The major direction of migration is from agricultural areas to centers of industry and commerce.
11. The major causes of migration are economic.

Ravenstein's analysis was based primarily on the 1871 and 1881 censuses of Britain. Some of the "laws" reflect conditions at that time and may not be applicable to today. For example, in advanced countries most migration is interurban and intraurban, which contradicts Law 5 above, although Ravenstein's statement may have been accurate when he made it. Ravenstein's statement on female migration is related to a phase in economic development. When rural-to-urban migration begins, the male members of society tend to make the initial moves. As they establish themselves in new locations, women from the source communities follow to take jobs in domestic service or light industry. Today in Latin America, more women than men migrate from the countryside to the city, and in many major centers there are more women than men in the population (Gilbert, 1994, pp. 46–46).

Of course, many of Ravenstein's suggestions have been further developed, and frequently the terminology has been revised. We now speak of *direct migration* when a migrant moves from one place to another without settling at an intervening location. The term *stepwise migration* is used to cover situations where a man or woman stops at an intervening point, or points, on the way to a final destination:

for example, moving from a village to a small town and then on to a city. The term *fillin migration* describes patterns in which one group migrates out and another enters to take its place.

We now understand the relative attraction of destinations based on population size. It has been possible to construct gravity models that predict migratory flows on the basis of the relative size of the population centers involved and the distance between them.

However, much of what Ravenstein wrote over a century ago remains true and can be applied to Latin America. The greatest number of migrants within Latin America are young adults in their teens to mid-thirties; most migrants are single; and more women than men move from rural areas to cities. In addition, migrants tend to be better educated and more ambitious than people who choose not to move (Merrick, 1986). Generally speaking, impoverished people do *not* migrate. Migrants going a long distance are usually moving to a large city; big cities have extensive migration fields when compared with small towns. (The migration field of a place is simply the area from which it draws migrants.)

Particular destinations attract migrants with different characteristics. For example, Rio de Janeiro, which has a diverse economic base, is the destination of many female migrants. São Paulo, with a heavy industrial sector, attracts more males than females. Women are attracted to destinations that offer jobs in domestic service and light industry. Young, single adults predominate in rural-to-urban migration.

Families are an important component in migration to farming frontiers like those in Colombia, Venezuela (in the Orinoco region), in Brazilian Amazonia, and the *Oriente* regions of Ecuador, Peru, and Bolivia. Just as families moved to the

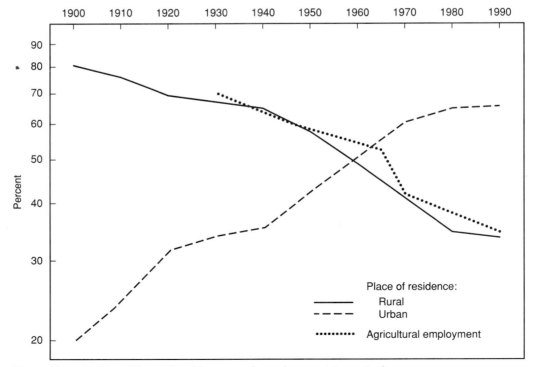

Figure 5.7   Mexico: Place of residence, and employment in agriculture.

TABLE 5.6 Latin American Population Mid-1990s (millions) and Projected Population Numbers for A.D. 2010

	Mid-1990s	A.D. 2010	Natural Increase	Infant Mortality
Mexico and Central America (total)	126.0	163.4		
Belize	0.2	0.3	3.3	34.0
Costa Rica	3.3	4.4	2.2	13.7
El Salvador	5.9	7.6	2.6	41.0
Guatemala	10.6	15.8	3.1	48.0
Honduras	5.5	7.6	2.8	50.0
Mexico	93.7	117.7	2.2	34.0
Nicaragua	4.4	6.7	2.7	49.0
Panama	2.6	3.3	2.1	28.0
Tropical South America (total)	319.0	395.0		
Argentina	34.6	40.8	1.3	23.6
Bolivia	7.4	10.2	2.6	71.0
Brazil	157.8	194.4	1.7	58.0
Chile	14.3	17.3	1.7	14.6
Colombia	37.7	46.1	1.8	37.0
Ecuador	11.5	14.9	2.2	50.0
Guyana	0.8	1.0	1.8	48.0
Paraguay	5.0	7.0	2.8	38.0
Peru	24.0	30.3	2.1	60.0
Suriname	0.4	0.5	2.0	28.0
Uruguay	3.2	3.5	0.7	18.6
Venezuela	21.8	28.7	2.6	20.2
Caribbean Islands (total)	34.0	43.0		
Total	481.0	600.8		

Source: Population Reference Bureau, *1990 and 1995 World Population Data Sheets.*

frontier in nineteenth-century North America, so, too, in the late twentieth century, Latin American families are migrating to new agricultural areas. But Latin American countries lack national policies equivalent to the Homestead Act (1862) that opened up territory and provided secure land titles. As a result, the migratory streams to farming frontiers in Latin America are less than they might be. Return migration is common by those who fail to gain title to land or are frustrated by poor transportation or other problems.

The overall effect of migration is to cause more rapid growth of population in urban areas when compared with the countryside (see for example, Mexico, Figure 5.7). In most rural areas, population numbers grow at rates of less than 1 percent per annum. In large urban areas the rate of growth exceeds national population growth rates. As many Latin American countries now have urban majorities, we can expect migration between cities to become increasingly important.

During the 1950s and 1960s, primate cities such as Caracas and Lima were growing at a faster pace than the average for all urban places. Primate cities often displayed annual population growth rates of 7 to 8 percent per annum, and it seemed that an increasing proportion of the population of many countries would concentrate in fewer and fewer places. In the last two decades, many secondary cities and regional centers have grown economically and now attract migrants. The growth rate of some primate cities is slowing, although it is too soon to predict that the expansion of the second-order places will result in a "rank size" condition. Under the rank-size rule, the largest city is twice as big as the second city, three times the size of the third-ranked place, and has four times the population of the fourth-ranked center. Primate cities, such as Mexico City, Lima, Santiago, and Caracas are many times larger than any other place within their respective countries.

Finally, primate centers remain attractive to entrepreneurs, for they are the major markets, offer a greater range of services than smaller urban places, and are often the seat of government. Primate cities offer jobs and perceived opportunities and attract migrants. In spite of overcrowding, lack of housing, and poor job prospects, Mexico City continues to attract migrants.

► *What Motivates Migration?*

Most authorities, including Ravenstein (1885, 1889) and Zelinsky (1971), assume that economic forces drive the migration process. The work of Ravenstein and Zelinsky is empirical. They examine the evidence, describe the pattern of events, and attempt to ascribe causes for the observed migratory movements. Other researchers adopt a behavioral approach. If done properly, behavioral analysis involves determining the attributes and attitudes of residents of rural communities and, over a period of years, studying the behavior of persons who migrate. Many of the studies done in Latin America have only involved interviewing migrants after they have arrived at their destination. Several problems are inherent in this approach. The questionnaire technique tends to miss migrants who spend a short time at one destination and move on to another place. Studies based on question-naires tend to underreport stepwise migration, for many migrants make a series of moves before reaching a final destination. Questionnaires completed at destinations also underrecord return migration. Finally, human beings are great rationalizers and, when asked about motives for migrating, are capable of manufacturing argu-ments to justify behavior. These arguments may not have been uppermost when the decision to migrate was made. Questionnaires administered to migrants after they have lived at their point of destination for some time tend to confirm that economic motives played the major role in decisions to migrate.

Being concerned solely with economic motivation, however, obscures the social aspects of migration. Migrants do not set off blindly for the nearest large urban center. Migrants move after they have acquired information about conditions at the point of destination. Information is transmitted through a chain of contacts from city to source area. The result of these information flows is frequently to establish migration chains that perpetuate themselves. For instance, once a member of a village has moved to a city successfully, relatives and associates will likely fol-low. Information fields are subject to distance decay factors, so that Ravenstein's points about the attraction of large centers and the length of migratory journeys still apply. But when a migrant arrives in a city, he or she is likely to have knowledge of a neighborhood and may well know some of the inhabitants. The fabric of the migratory process is social as well as economic.

Migratory flows may be a disadvantage to rural areas. Out-migrants tend to be better educated, younger, and more receptive to innovation. The rural loss should be the urban gain, but absorption into the economic life of the city may be slow.

Small provincial towns tend to lose inhabitants to large metropolitan areas. It is not a question primarily of the *number* of people moving but the *type* of people migrating. The lesser urban places lose a significant proportion of their educated population to larger urban centers where opportunities are greater. From the point of view of the professions, metropolitan areas are often the only centers in which the full range of employment opportunities exists. In Latin America, few countries have more than one or two large urban centers. Uruguay, for example, has only one major city, Montevideo, which, with a population of approximately 1.76 million, contains over 50 percent of all Uruguayans. Each of the second-ranked towns by number of inhabitants, Paysandu and Salto, has less than 100,000 inhabitants, and employment opportunities are limited there.

► *International Migration*

International migration has played an important part in the emergence of Latin America as a distinctive region. Western Europeans have settled in the region from the time of Columbus' second voyage. At first, settlers were drawn predominantly from Spain and Portugal, but, later, Italians, Germans, French, and British migrated to the region. The cultural impact of the Iberians was greater than their numbers might suggest, and they became the dominant elements in *criollo* society, even though most of the population was descended from the pre-Columbian inhabitants.

The Europeans were responsible for moving populations into their new world. By 1520 enslaved Africans were imported, through Spain, into Caribbean islands as agricultural laborers. When plantation crops were raised on a large scale for export, Europeans conducted forced migrations of Africans across the Atlantic to the sugar-growing areas of the Caribbean region and northeastern Brazil.

During the nineteenth century, as slavery was suppressed, plantation interests searched for ethnic groups to serve as alternative sources of cheap agricultural labor. For example, in the British West Indies, large numbers of indentured laborers from India and China were persuaded to enter contracts that obliged them to work on plantations at low wages for a number of years. Many indentured laborers never returned home and formed the basis of the Indian and Chinese populations in countries like Trinidad (44 percent Indian) and Guyana (52 percent Indian).

The history of the Caribbean region contains a long record of migrations. When the Europeans entered the region, they came upon a situation in which the Caribs were progressively invading islands occupied by Arawaks. The Europeans dispossessed both groups over time and replaced them with Africans, Indians, Chinese, and those of their own country who would enter indentured labor contracts. Population flows have not ceased in modern times. Jamaicans and other West Indians migrated to Panama to build the canal. Barbadians went to Curaçao to work in the oil industry. Between 1909 and the end of World War II, over 1 million migrants settled in Cuba; the majority of them came from Spain, but other Caribbean islanders were well represented.

The flows of people have not been confined to the circum-Caribbean region. West Indians and Puerto Ricans have moved to New York and other eastern cities of the United States. Many Barbadians and Jamaicans have settled in the central areas of cities in Great Britain. The Caribbean has been described as a region of institutionalized out-migration. The causes of out-migration are straightforward. Most of the islands are small and have a limited economic base. The historical commitment to the production of a few crops, like sugar and coffee, has contributed to the making of narrow economies. Without diversified economies the range of employment opportunities is limited, the chances for advancement are few, and unemployment is high. Not surprisingly, those with ambition, education, and skills are drawn to countries with better job opportunities. The result is that the region is deprived of the trained human resources needed to create viable economies.

Although Europeans were a dominant political and economic force in Latin America, the percentage of population drawn from Europe was small in most countries. There are important exceptions to this statement. Between 1870 and 1914 the population of Argentina increased from 1.7 to 7.9 million persons. The greatest part of the increase was accounted for by migration from Europe. In 1914, 30 percent of the population of the country was foreign born, 40 percent of whom were from Italy and 35 percent from Spain. These figures are comparable in percentage terms to those for the Dakotas and Nebraska, in the 1880s, at the time of their most rapid inflow of Europeans from Scandinavia, Germany, and Central Europe. Other Latin American areas with high European concentrations include southern Brazil (Germans), Costa Rica, and parts of Venezuela.

In the decades before World War I, 30,000 Italian and Spanish workers arrived annually to help with the harvest on the Pampas. These *golondrinas* (swallows) then returned home to work on the harvest in the Northern Hemisphere. In the seasonal and permanent migration flows to the Pampas, males predominated and there was high return migration.

There is a long tradition within Latin America of international migration and settling across borders without legal entry. El Salvadorians have settled on farms in Honduras. Colombians cross the border to farm in Venezuela. Many Peruvians settled in the Ecuadorian *Oriente* in the first decades of this century. Haitians have settled in the Dominican Republic. In the El Salvadorian and Peruvian cases, the illegal settlement was a factor in starting warfare between the countries involved. Although migration from one country to another is more difficult than in the past, selective movement is still permitted when labor and skills are needed. Venezuela encouraged immigration in the oil boom of the 1970s. Many migrants from Colombia and southern Europe moved into the country. Immigration has declined with the slowing of growth in the oil industry.

The largest international migration in the region takes place across the U.S.–Mexico border. The United States absorbed Texas, California, and parts of New Mexico and Arizona at Mexico's expense. The new territories contained Spanish-speaking, Hispanic populations that retained connections with Mexico. The United States acquired potential migration chains as well as new boundaries. Migration was limited in the nineteenth century. In the twentieth century, as economic development gathered pace in the Great Plains, West, Southwest, and California, demand for labor increased, particularly after World War I when European migration to the United States declined. Mexican workers, predominantly young, single males became an important part of the unskilled labor force, especially in agriculture. The entry of temporary Mexican workers (*braceros*) was unrestricted after 1942. The *bracero* program was terminated in 1964; by then there were new populations of Mexicans living in the United States.

Attempts to impose quotas on the entry of Mexican workers after 1965 did not cut the inflow of people. The result was many undocumented migrants living in the United States. Between 1970 and 1990 the Mexican-born population resident in the United States grew from 0.8 million to more than 4 million persons. The 1986 Immigration Reform and Control Act gave amnesty to certain classes of illegal residents and attempted to control further inflow.

The character of the Mexican-American population has altered over time. The *bracero* program brought in temporary workers who were predominantly young, male, uneducated agricultural workers (*Science*, 1991). They usually returned to

Mexico. Later migration has included more females and family members and a higher likelihood that migrants would settle in the United States.

As long as wages are low in Mexico and there is a demand for unskilled labor in the U.S. states that border Mexico, there will be migration from Mexico to the United States. More industry in northern Mexico may slow the rate of movement, but there are now 12 million persons of Mexican origin in the United States, concentrated in metropolitan areas in California, New Mexico, Arizona, and Texas.

SUMMARY

In the first third of the twentieth century, the rate of population increase in Latin America was at a moderate level. During the 1930s, death rates began to fall, birth rates increased, and population numbers started to grow more rapidly. By the 1950s, the age structure of the population had changed, with many people in the age group younger than 25. The rate of family formation increased and with it the birth rate. By the 1960s, the population of the region was growing at around 3 percent per annum. In 1940, the total population of the region was approximately 130 million. Today, the population is approaching 500 million in Latin America (Table 5.6). The percentage rate of natural increase has now slowed to 1.9 percent per annum and continues to fall, but the larger total population produces nearly 10 million new members each year, a gross annual increment of people unthought of in 1940. "Demographic momentum" ensures that the population of the region will grow, even as the rate of increase slows.

As the population has grown, its distribution has changed. In 1930 (Table 5.7), Latin America was a rural region. Today, most countries have urban majorities.

We think of Latin America as a region of primate cities, but in many cases, the emergence of the megacities is relatively recent. When Latin American economies grew in the period 1940 to 1979, the benefits of growth were concentrated in capitals and a few other cities like São Paulo. Economic growth came as countries in the region were adopting nationalistic and statist economic policies. The decision-making process concerning economic development was performed in capitals, and not surprisingly, much new economic activity was located there, too. In addition, royalties on ore and oil exports accrued to national treasuries, as did the revenue from the heavy import taxes imposed to protect and promote manufacturing within national boundaries. The capital cities were the centers of political power, the location for the new consumer goods industries, and home to a growing government bureaucracy that entrepreneurs, national and international, had to deal with. Capital cities attracted migrants of every kind—middle class professionals, skilled workers, agricultural laborers looking for employment in factories, domestics, street vendors, and executives employed by international corporations and agencies. To a degree, the migratory flows perpetuated themselves because the larger the cities grew, the more employment they created, even if the rate of job creation could not keep up with population growth.

Capital cities have grown rapidly. For example, at the beginning of the century, the estimated population of Peru was 3 million; and at 130,000 inhabitants, Lima did not contain even 5 percent of Peruvians. Today, over a quarter of Peru's population lives in the Lima-Callao metropolitan area. The urban area began to expand rapidly in the late 1930s, exceeded a million inhabitants in the 1950s, and today contains over 6 million of Peru's 24 million inhabitants.

At the turn of the century, Caracas, Venezuela, was not quite twice as large as either Valencia or Maracaibo. Today, the capital is many times larger than any other urban place in Venezuela.

TABLE 5.7 Estimates of Urban Population Growth

	Urban Population						Percentage of Urban Population				Number of Cities of Over 500,000 Persons	
	As a Percentage of Total Population				Average Annual Growth Rate (percent)		In Largest City		In Cities of Over 500,000 Persons			
	1930	1960	1990	1994	1965–1980	1980–1988	1960	1980	1960	1980	1960	1980
Haiti		18	29	31	4.2	4.0	42	56	0	56	0	1
Bolivia	14	40	50	58	3.1	4.3	47	44	0	44	0	1
Dominican Republic	7	35	59	60	5.2	4.3	50	54	0	54	0	1
Ecuador	14	37	55	57	4.7	4.7	31	29	0	51	0	2
Colombia	10	54	69	68	3.7	3.0	17	26	28	51	3	4
Paraguay		36	46	51	3.8	4.5	44	44	0	44	0	1
Peru	11	52	69	71	4.3	3.1	38	39	38	44	1	2
Chile	32	72	85	85	2.6	2.3	38	44	38	44	1	1
Mexico	14	55	71	71	4.4	3.1	28	32	36	48	3	7
Brazil	14	50	75	76	4.5	3.6	14	15	35	52	6	14
Uruguay		81	87	89	0.7	0.8	56	52	56	52	1	1
Argentina	38	76	86	86	2.2	1.8	46	45	54	60	3	5
Venezuela	14	70	83	84	4.8	2.6	26	26	26	44	1	4

Source: World Development Report (1990); 1994 World Population Data Sheet, Washington, D.C.: Population Reference Bureau.

In 1940, only 7.5 percent of the population of Nicaragua lived in Managua (Wall, 1993). Today, in spite of being demolished by an earthquake in 1972, the capital has over 1 million inhabitants, and nearly a quarter of all Nicaraguans live there. Of all industrial activity 80 percent is concentrated in the capital, as is the government administrative machine.

The economies of most Latin American countries have enlarged and diversified in the last sixty years. Due to rapid population growth, living standards have not improved rapidly, and increases in income tend to concentrate in the hands of the better-off. The prospects for higher living standards for the majority and an improvement in income distribution will remain poor until the completion of "the demographic transition has eliminated Latin America's labor surplus" (Bulmer-Thomas, 1994, p. 426).

FURTHER READINGS

Chesnais, J. *The Demographic Transition: Stages, Patterns, and Economic Implications: A Longitudinal Study of Sixty-Seven Countries Covering the Period 1720–1984.* Oxford: Clarendon Press, 1992.

Hornby, W. F., and M. Jones. *An Introduction to Population Geography.* 2d ed. Cambridge: Cambridge University Press, 1993.

Robinson, D. J. *Migration in Colonial Spanish America.* New York: Cambridge University Press, 1990.

Skelton, R. *Population Mobility in Developing Countries: A Reinterpretation.* London and New York: Delhaven Press, 1990.

Zelinsky, W. "The Hypothesis of the Mobility Transition." *Geographical Review* 61 (1971): 219–249.

► *Demographic Source Material*

Demographic Yearbook, 1989. New York: United Nations, 1991.

Goyer, D. S., and E. Domschke. *The Handbook of National Population Censuses: Latin America and the Caribbean, North America and Oceania.* Westport, Conn.: Greenwood, 1983. Lists censuses taken by countries noting scope and deficiencies.

Mitchell, B. R. *International Historical Statistics: The Americas 1750–1988.* New York: Stockton Press, 1993.

Statistical Yearbook for Latin America and the Caribbean. New York: United Nations Economic Commission for Latin America and the Caribbean, 1991.

Statistical Yearbook. Paris: United National Educational, Scientific and Cultural Organization, 1994. Annual publication that provides data on educational levels and facilities.

Wilkie, J. W., C. A. Contreras, and C. Komisaruk, eds. *Statistical Abstract of Latin America,* vol. 31, pt. 1. Los Angeles: UCLA Latin American Center Publications, University of California, 1995.

World Population Data Sheet. Washington, D.C.: Population Reference Bureau, 1995.

Chesnais (1992) includes appendices giving crude birth rates, total fertility rates, crude death rates, and infant mortality rates for many countries including Latin American examples.

Many statistics for Latin American countries published before 1950 are suspect. Censuses are irregular. The registration of births and deaths starts late, and registration is often incomplete. Collver reconstructed, recalibrated, and revised many figures relating to age structure, birth and death rates, and total population numbers by extrapolating from data that were known to be relatively accurate. However, most official, published statistics going back to the early decades of the twentieth century have not been revised and are republished in annual statistical handbooks. Mitchell (1993) publishes the "official" birth and death rates which were used to construct Figures 5.2, 5.4, and 5.5. Mitchell also includes Collver's recalculations that show high birth rates in the early part of the twentieth century.

Even in recent years, there are different versions of statistics. Wilkie (1992) reproduces the infant mortality rates for Peru in the 1980s, as published in the U.N. *Demographic Yearbook*. Wilkie also reproduces the Peruvian infant mortality rates as calculated by the Pan-American Health Organization. For 1983, the U.N. *Demographic Yearbook* tells us that out of every thousand live births, 140.6 children did not survive the first twelve months. The Pan-American Health Organization says the figure is 33.8! The statistics do agree that the infant mortality rate was rising in the early 1980s.

Keyfitz, N., and W. Flieger, *World Population Growth and Aging: Demographic Trends in the Late Twentieth Century,* (Chicago: University of Chicago Press, 1990), gives statistics on population numbers, fertility rates, death rates, age structure, urban population, life expectancy, and projections of trends through the year 2025, for every country in the world.

CHAPTER 6

The Latin American City

CHARLES S. SARGENT

Towns and cities have always been at the heart of Latin America's economic, political, social, and cultural development. The Aztec and Incan economic and political empires created city-systems, with their capitals at Tenochtitlán and Cuzco, as noted in Chapter 2. The European conquest of these and other indigenous populations after 1492 led to the establishment of new urban systems, oriented to Europe.

Some of the new cities evolved as ports to link Spain and Portugal, in particular, with the New World. Others became inland urban centers along new transportation routes that focused the harvest of resource wealth toward the ports. Still other cities were the mining centers themselves. New agricultural towns and villages supplied food to the growing ports, trading centers, and mining centers. These new towns and cities were in what were then "colonies" on the periphery of a preexisting European mercantile world and a European political system composed of competing states.

The colonial city-systems were quickly in place, for towns provided the foothold for the political occupance of the land and the development and extraction of natural wealth. Towns were stepping-stones toward the political and economic control of the land; they were both the symbol and the mechanism for the colonization process. The larger Spanish and Portuguese colonial cities were centers of conquest and administration as well as centers of trade. The new cities promoted the settlement and development of rural lands and many small rural places were linked to a wider world by colonial cities that are still the major urban centers of Latin America.

With the national independence movement in the early 1800s, the city-systems entered a new era. Colonial capitals, by and large, became national capitals, and most regional centers retained roles as centers of growth and change aided by introduced European technologies, especially the railroad. Some regional centers, however, found that economic and political change worked to their disadvantage, and they declined in importance.

Particularly since the 1930s, large numbers of people have migrated from villages and towns to cities, some of which became rapidly growing major metropolitan areas. The harshness of rural life, the exploitation of both landed and landless peasants, the economically marginal nature of much agriculture, and the limited amenities of the smaller towns enhanced the perception of the larger city as an attractive alternative. Unfortunately, the inward flow of migrants was not matched by job creation, adequate housing, or urban services; urban life, for many, is difficult and marginal. The result is a growing social, economic, and political urban crisis in many parts of Latin America.

MAJOR PERIODS OF URBAN
DEVELOPMENT AND TRANSFORMATION

Four major periods of town foundings, urban growth, and transformation have attended the 500 years since the first Europeans arrived in what became known as Latin America. Two of the four phases took place in the 300-year colonial period between 1492 and 1820.

1. Creation of an urban framework related to the economic and political interests of the Spanish and, in Brazil, of the Portuguese. This period of rapid colonial town foundings lasted until about 1600 in both Spanish America and Brazil.

2. Adjustment of the early colonial urban network. In Spanish America, the growth of some cities was favored, but (except in central Chile and northern Mexico) relatively few new towns were founded between 1600 and the early 1800s, when most countries became independent of Spain. In Brazil, on the other hand, the 1700s saw the settlement of major new regions and the creation of numerous important settlements.

3. The early "modern" period (1820–1930), when newly independent Latin American countries established wider ties to the markets and industrial nations of Europe and the United States. In Argentina and Brazil, with large areas not effectively settled in 1820, the period saw many new town foundings. In Brazil, the process continues to the present. Throughout Latin America, this period led to the emergence of dominant cities, usually national capitals and/or the major port of the country.

4. The post 1930–1940 period, when large-scale rural-to-urban migration and selective urban growth resulted in the present-day urban hierarchy, or ranking of cities by importance and size. As before, the national capitals—which are the focus of influence and prestige—and the port cities, which serve as the link between Latin America and the markets and industries of the world—have been the center of this growth. The result is an increasing number of very large metropolitan areas, comprised of a major central city and nearby surrounding smaller cities. In the near future, Mexico City—with some 25 million residents—may become the largest metropolitan area in the world.

► *The First Century of Colonial Settlement: 1492–1600*

Much of today's urban network in Latin America was put in place in the century that followed the European discovery of the New World (Figure 6.1). Spanish towns were laid out in areas of relatively dense native populations and in mineralized districts, while Portuguese ports evolved along the coast of Brazil. The presence of a native workforce was a key to development, given the small number of Spanish and Portuguese on the continent. When native populations died of overwork and introduced diseases, enslaved Africans were utilized. In Brazil, where sugar cultivation became the foundation of the colonial economy, slaves were of particular importance.

Throughout Spanish America, towns were founded with three major objectives in mind: to expand empire, to exploit the mineral and agricultural wealth of the land, and to bring the Holy Faith, the *Santa Fe*, to the Indians. An exploitative and administrative character was common to all the early Spanish cities. In Spanish America, the colonizers—like the Indians before them—were drawn to the highland regions whose climate, terrain, and mineral potential were to their liking. In coastal Brazil, with its emphasis on agriculture, the urban focus was less strong.

Because of the physical contrasts between regions, the size of Latin America itself, and the initial zones of European contact with the unknown landmass,

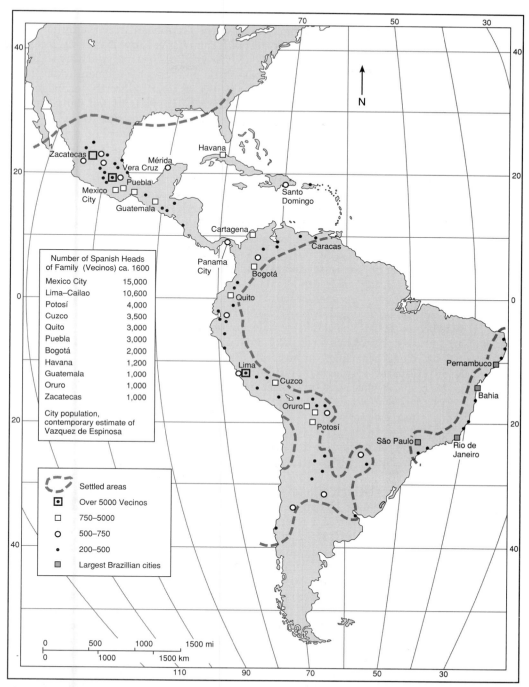

Figure 6.1   Major Latin American settlements ca. 1600.

significant regional distinctions arose in the timing, pace, and nature of town foundings. First came settlement of the Caribbean area, followed by conquest of Mexico
and Central America, and concurrent movement into the Andean region of South
America. The Paraná River basin and the coast of Brazil represented settlement
zones farther south and east.

The Caribbean Threshold

The islands and the continental shores of the Caribbean were the initial areas of Spanish contact with the New World, with many small towns quickly founded throughout the area as the Spanish searched for gold. Where no gold was found—a geological reality for many of the islands—little permanent settlement took place, leaving a vacuum that was filled more than 100 years later by North European colonizers. Where gold was found, as on the islands of Hispaniola and Cuba, towns developed and the local Indians were exploited both for mining and for agricultural production. When no local Indians were available, distant populations were forcibly brought to the mining sites. On some islands hostile Caribs made settlement difficult.

Santo Domingo, founded in 1496, was a major early town. While not the first town, or *villa*, on Hispaniola, it is the oldest to survive and, consequently, the oldest European city of the Americas. Fifteen *villas* were established on Hispaniola by 1505 and another seven in Cuba by 1515, including Trinidad (1512), Havana (1514), and Santiago (1515). On Jamaica, the *villa* of Seville La Nueva was established in 1509, but no longer exists.

The resiting and abandonment of early towns was common as disasters such as hurricanes, volcanic eruptions, earthquakes, and fires took their toll. Some towns were rebuilt; others were abandoned when anticipated riches did not materialize or exploration routes changed. Many towns disappeared as the native labor force was decimated by disease. The Indian population of the island of Hispaniola, for example, was greatly reduced as a result of contact with Europeans. In addition, mining settlements based on placer, or streambed, deposits were particularly short lived, while those related to the exploitation of veins of ore continued longer. By and large, the major towns were ports that tied the new lands to Spain. A number survived because of this role and investment in the port defenses, churches, public buildings, and residences.

The Early Cities of Mexico

When the Spaniards entered the Aztec state of central Mexico in 1519, they encountered gold artifacts, millions of sedentary Indians, and an established network of Indian towns. The Aztecs had established ascendancy in the Valley of Mexico after 1300, and about 1325 had begun to develop the capital city of Tenochtitlán on islands in Lake Texcoco. For a century or more the city was a rudimentary center of trade and tribute, but by 1519 had a population in excess of 100,000, with many temples and an impressive separation of land uses. Hernán Cortés described it as a magnificent city, divided into four districts by major streets, with a great plaza, temples, pyramids, and mansions. The city of Texcoco, across the lake, possibly had a population of over 100,000. Yet another large city, Tlaxcala, whose residents came to support Cortés against the Aztecs, lay between Tenochtitlán and Veracruz, the Spanish base established in 1519 on the Gulf coast. In addition, there were numerous smaller cities. Not surprisingly, the Spaniards were impressed by Tenochtitlán, as only a few European cities—Paris, Naples, Venice, and Milan, in that order—could claim more than 100,000 population at that time. Seville, just prior to the departure of the *conquistadors*, was a city of perhaps 50,000. Impressive also were the remains of Teotihuacán, a religious, market, and trade center 8 square miles in extent that lay 30 miles northeast of Tenochtitlán. Active between about 100 B.C. and A.D. 800, Teotihuacán had a peak population in excess of 125,000 who left behind impressive pyramids, temples, and extensive ruins.

The defensive position of Tenochtitlán, which was accessible only by causeways or boat, and its dominance as a trade and tribute center largely explain why Cortés founded Mexico City on the embattled site of Tenochtitlán in 1521.

Moreover, the temples and other buildings provided a ready supply of construction materials for the new city. What is today the largest cathedral in Latin America rose on the site of an Aztec temple.

Other town foundings took place after the conquest, but all focused on Mexico City. Among the more important were silver centers such as the Indian towns of Taxco and Pachuca. Others, such as Guadalajara (1531), Puebla (1532), and Oaxaca (1536), were agricultural towns established in already densely settled and fertile areas. These and many of the smaller new towns were administrative centers and religious *congregaciones* into which Indians were herded for religious proselytization and instruction in Spanish trades, and became pools of agricultural or mine workers. The result was a rapid development of an extensive yet integrated nuclear pattern of towns and cities that focused on Mexico City (Figure 6.1).

Beyond the core, focused on Mexico City, towns such as Tepic (1531) and Culiacán (1532) were established as way stations along lowland Pacific exploration and slaving routes. Acapulco was sited as a port for silver shipments to Spanish-held Manila in the Philippines, and as a receiving port for Asian silks and other goods that made their way to Spain through Mexico City. Manila came under the administrative control of Mexico City in 1583. To the north of the core, additional towns were founded in the early 1540s after the discovery of silver at Zacatecas in 1546. In time Zacatecas, which was central to numerous rich silver veins, became the second largest city of New Spain, or Mexico. A number of other silver camps, or *reales de minas*, such as San Miguel (1542), Guanajuato (1548), and San Luis Potosí (1592) also became important towns. New agricultural settlements such as León and Celaya supplied food and animals to the mines. In addition, the continued presence of hostile Indians meant that a number of fortified towns had to be established on the route between Zacatecas and Mexico City. To the northwest of Zacatecas, mining towns such as Fresnillo, Sombrerete, and Durango (1563) developed along the silver-rich foothills of the Sierra Madre Occidental as far north as Santa Barbara (1570). Town development in the north, beyond the Silver Belt, was delayed until about 1600, but the silver towns, agricultural settlements, and missions represented the foundation of the linear pattern of cities characteristic of northern Mexico. The pattern contrasted with the nuclear grouping of towns in central Mexico.

Central American Urban Corridor

The Spaniards converged from Mexico and Panama on Central America. The overland entry from the north was through Indian settlements such as Huehuetenango, Quetzaltenango, and Chichicastenango in the Guatemalan highlands, which still exist today. New towns were created at San Salvador (1524) and Antigua Guatemala (1526), and ports established at Trujillo and Puerto Caballos (Puerto Cortez). Inland, along the more densely settled and fertile river valleys, towns such as San Pedro Sula (1536) were founded. The Yucatán Peninsula, the rain forests of Petén in lowland Guatemala, and the Caribbean coast in the area of modern Belize were recognized as physically difficult areas without gold and only sparsely settled since the disappearance of the Mayans. Mérida and Valladolid, in Yucatán, were not founded until the early 1540s, and the area was relatively deficient in town development.

Explorers from Panamá City, a base for explorations to both the north and the south, sailed to Nicaragua and established the agricultural colony and slaving base of Granada (1524) in fertile, densely settled lowlands. At the same time, another administrative and slave center was founded at León. The Nicaraguan highlands of Nueva Segovia became a major gold area after 1527, to which were later added the Honduran gold and silver districts around Comayagua (1537) and Tegucigalpa

(1578), the latter becoming a mining and smelting center for a number of smaller mining camps.

Movement south from Nicaragua led to the Costa Rican agricultural colony at Cartago (1562). But here there were few Indians because of the earlier arrival of European diseases, and Costa Rica quickly became an almost totally European district following planned immigration of European settlers. It remains so today. By 1650, in fact, the Indian population of Central America had declined in numbers. Depopulation was especially widespread in the warmer, more disease-prone lowlands, a fact that helps explain why most of the major cities in Mexico and Central America today are highland cities.

After the 1521 conquest of Mexico and the exploration of Central and South America, most of the Caribbean quickly became a support area, and a number of towns turned to port, supply, and defense roles. Others disappeared. Among the most important to develop were Havana, Cartagena, Nombre de Dios, and Veracruz. Havana, founded in 1514, became the chief fortress of the Spanish Main, guarding the approaches to the Gulf of Mexico. Veracruz, Nombre de Dios [(1510), later replaced by Portobello], and Cartagena (1533) were the only continental ports initially authorized to trade with Spain. They were developed as the keystones in the Caribbean defense and trade system. Panamá City (1519) became the corresponding port of the Pacific side of the isthmus and the most politically and economically influential city between Mexico City and Lima, to which it developed close trading ties. Veracruz, Nombre de Dios, and Cartagena were the treasure terminals of the convoys that linked the New World to Spain between 1561 and 1748. The post-1520 fortification of these and other cities such as San Juan (1521), Santa Marta (1525), and St. Augustine (1565) was designed to fend off pirates and the navies of foreign powers that sailed this transportation corridor, the famous Spanish Main. Cartagena, for instance, was fortified between 1558 and 1735 with a 40-foot-wall and forts that protected the approach from the sea.

Farther east, along the Venezuelan coast, early town development was primarily the result of the interest of a German commercial house, the Welsers of Augsburg. Some towns, including Cumaná (1516), Coro (1527), and Maracaibo (1529), grew up along the coast, but they were impeded by native hostility earlier engendered by Spanish slaving expeditions sent out to procure labor for the Caribbean islands.

Cities and Towns of the Northern Andes and Upper Peru

In their search for mineral wealth and imperial expansion, the Spaniards penetrated the interior of South America from bases established along the Caribbean coast and from newer settlements on the Pacific coast (Figure 3.2). As in Mexico and Central America, permanent and large-scale settlement was dependent on finding minerals or fertile land and on the availability of indigenous labor for mines and fields.

In the Andean region the Spaniards superimposed their economic system over the earlier Indian urban network and reoriented it to suit their interests. Unlike central Mexico, however, where large basins and high plains facilitated movement and a major clustering of important towns and cities, the isolation of the high Andean mountain basins from one another and the mountain barrier between the Pacific lowlands and upland zones precluded the growth of a closely knit urban network. Instead, a strongly linear pattern of towns evolved. But the north–south linkages that had been established in the Inca period with the construction of a well-developed pedestrian road system were now balanced by shorter, east–west trade routes between inland towns and Pacific ports (Figure 3.5). There was relatively little economic interaction between highland towns, except as way stations on the bullion routes or centers of agricultural surplus.

There was little mineral wealth in the eastern segment of the Andes that sweep northward through Colombia into Venezuela, and urban growth was slow. By 1600, what is today the small town of El Tocuyo (1545) was the largest place in Venezuela, while Barquisimeto (1552), Valencia (1556), Trujillo (1557), Mérida (1558), and Caracas (1567) were small farm towns in isolated Andean valleys. The towns of eastern Colombia had similar agricultural functions. The small village of Bogotá (1538) was in the midst of a densely settled and fertile area, the *sabana*, but was initially of little importance to the Spanish. Nearby was the Indian town of Zipaquira, already noted for its salt mines, and Tunja (1538), which, like Bogotá, gradually became more important.

In contrast to these marginal areas, other portions of Colombia were rich in gold and major mining areas developed by the middle 1530s in the Cauca Valley, on the Antioquian highlands, and along the Pacific coast. Central Colombia was soon a landscape of impermanent gold camps, although a few, as exemplified by the town of Antioquia (1546), became important mining and trade centers. Other towns, including Popayán (1536), Cali (1536), and Pasto (1539), were founded in the same period as the Spaniards moved north into the area from the Pacific coast. What is now Ecuador had little mineral wealth and was a land of isolated mountain basins. The major settlement was the northern Inca subcapital of Quito, just to the south of which a Spanish administrative and exploration center was laid out in 1534. Far from the sea and the mines of Colombia, Quito grew at a slow pace, tied to the outside by the port of Guayaquil. A string of towns including Cuenca (1557) and Ambato was founded in the populated basins to the north and south of Quito to serve as way stations.

The highlands of Peru and Bolivia (Upper Peru) became the major focus of Spanish interests. Like Quito, the Inca capital of Cuzco was occupied and made into a Spanish town after 1534. Spanish Cuzco was built directly on the earthquake-proof Inca walls and foundations, so that today the heart of the city still features narrow streets faced with the famous Incan masonry. There was also the occupance of other Indian towns, including Arequipa (1540) and Arica. Farther south, on the Altiplano and eastern lowlands of Bolivia, a number of towns were founded, including Sucre (1538).

Lima served as the capital of the Andean region. A coastal base for the conquest of the interior, Lima (1535) soon became the administrative capital of the viceroyalty of Peru, with districts as distant as Bogotá and Buenos Aires nominally under its control. Lima was a religious, educational, and commercial center for trade with Spain, located on the coastal desert, and watered by the Rimac River that descended from the Andes. By 1600, the city's population was about 14,000, with the small, recently founded port at El Callao on the coast 8 miles to the west. Elsewhere, the coastal lowlands—except where crossed by streams like the Piura, Rimac, Pisco, Ica, and Liuta—was an inhospitable desert.

In 1545, the attention of the Western world was drawn to a rich hill of silver, the *Cerro Rico*, at the base of which the mining center of Potosí was quickly established. Postosí's location was difficult and its climate harsh because of its elevation of nearly 14,000 feet that limited local agricultural production and meant most food had to be brought in from far away. Nevertheless, Potosí had a population of about 120,000 in 1570. The population peaked at some 160,000 inhabitants in 1600, and numerically it was the largest settlement in Latin America and one of the largest in the Western world. The Spanish population probably did not exceed 20,000; most of the remainder were Indians forced to labor in the mines, mills, and other activities tied to the rich Cerro. The population figures are deceptive. Despite its size, fine churches, theater, convents, and expensive homes, Potosí was functionally more an enormous mining camp than a city of trade and manufacturing and should not be strictly compared to Lima, Mexico City, or other large Latin American centers.

Silver was the principal commodity that sustained Lima as an administrative and trade center. Potosí was an enormous market for agricultural produce and mine supplies from a vast area. A number of towns, including La Paz (1548), now Bolivia's largest city, developed as way stations on the bullion route from Potosí to Lima; lowland settlements such as Ica became food sources. Huancavelica was quickly settled after mercury was discovered there in 1563, obviating shipments from Spain for use in the processing of silver ore. Arica was selected as the principal port serving Potosí, and the mercury from Huancavelica and much merchandise for the mining town passed through, as did the silver en route to Lima.

Settlement Fringe at the Southern Andes

The initial settlement pattern in Chile was strongly linear and, as elsewhere, related to the search for gold and silver and the extension of the empire. Mineral exploration proved less successful than in the Mexican Silver Belt. After 1600, Chile became more an agricultural area than a linear frontier of military outposts. Central Chile, between the arid north and the hostile Indian territories in the south, was the focus of Spanish development. Its long valley was fertile and the Mediterranean-type climate familiar to the Spanish. Santiago (1541) was eventually selected as the administrative center while a port developed at Valparaíso (Paradise Valley, 1544) as the link to Lima and Panamá.

As the Spaniards advanced south from Santiago toward the hostile Araucanian Indians, they built fortified outposts, one of which, Concepción (1552), was designated the early administrative capital of the Chilean district. Farther south, walled towns were constructed at Angol, Imperial, Temuco, Villa Rica, Valdivia (1552), and Osorno (1558). Finally, on Chiloé Island, the town of Castro was established 26 years after the founding of Santiago. But the Indians proved resistant, especially south of Concepción, where forts were abandoned by 1598, to be reestablished off and on during the colonial period. Araucanian resistance in the southern forests ended only in the 1860s.

Until 1776 the Cuyo district of western Argentina was administered from Chile, and towns were founded there by expeditions from Santiago that crossed the fertile Vale of Chile and went over the Andes via Uspallata Pass to the piedmont oasis of Mendoza (1562), which was an important trade and agricultural station on the east side of the mountains. San Luis (1596) was also on the trade route. But a number of other oasis towns such as San Juan (1562) proved to be off the trade routes to Chile or Bolivia and either grew little or disappeared. As late as the middle of the nineteenth century there were no towns south of Mendoza because of hostile Indians.

Potosí was an important influence on the founding and growth of a number of towns in lowland Bolivia and northeastern Argentina. In Bolivia these included Santa Cruz (1557) and Tarija (1574). In Argentina, Santiago del Estero (1553) was the oldest of these towns but more important were Tucumán (1565) and Salta (1582), a trading center that provided the Potosí mining operation with mules and horses from eastern Argentina: from 30,000 to 60,000 mules a year moved through the annual Salta animal fair. Córdoba (1573) was the economic link between the mining districts and the mule and cattle grazing lands to the east.

Several towns founded from Santiago del Estero no longer exist. One was Londres (London, 1558), established in a region known briefly as New England (Nueva Inglaterra) to commemorate the marriage of Philip II of Spain to Mary Tudor. But Londres, like many other settlements, was destroyed by Indians. As elsewhere in the fertile intermontane valleys at the edge of the Andes, a number of small isolated Indian settlements, effectively beyond the Spanish commercial orbit, were spread throughout the Argentine northwest. Ultimately, the effect of Potosí's silver was felt as far south as Buenos Aires (1580), which remained, by law, a minor port in this period with no legal export trade in silver from the mining districts.

Cities and Towns of the Río de la Plata and the River Paraguay

The small port of Nuestra Señora Santa María del Buen Aire—Buenos Aires— was first founded in 1536. Buenos Aires was sited as far north on the Atlantic coast and as far inland as possible, in order to establish an Atlantic route between Spain and Peru and to serve as a barrier to southern advances by the Portuguese. As a consequence of Indian raids and food shortages, most of the colony moved upstream to found Asunción in 1537. The site of Buenos Aires was abandoned by 1541 but resettled in 1580, when colonists moved south from Asunción, founding Santa Fe (1573) en route.

Until late in the colonial period, Buenos Aires had little economic importance because Lima–Callao was the official port for silver exports to Spain. The merchants of Seville, Lima, and Panamá objected to Buenos Aires as a port at all and restricted its use so that as late as 1753 the city had a population of only 15,000 and dealt heavily in contraband. Asunción, a minor and isolated point of departure for expeditions to the north and west, was smaller. As in Chile, the towns asserted Spanish claims to southern lands and were not fully part of the economic realm that centered on Potosí and Lima.

Settling the Brazilian Shore

The Portuguese found no mineral riches along the northeast coast of Brazil, the nearest landfall from Portugal. As late as the 1530s, there were only small trading posts along the coast, where the Portuguese bartered goods with Indians and exported *pau-brasil*, a red dyewood after which the country was named.

The Portuguese crown granted large estates, or *captaincies*, that stretched from the coast far inland. Each grantee was responsible for the colonization, development, and defense of his captaincy, a system in force until royal control was later reasserted. The coastal towns of the different captaincies functioned as administrative centers, entrepôts for European products, and collecting points for exports. As lands were opened to agriculture, new towns were often founded inland. Most ports were small, serving a thinly settled agricultural hinterland to which they were economically subservient. Unlike Spanish America, the focus of political power was not in the town, but in the hands of the rural landowners.

Both a port and agricultural center, São Vicente (1532) was the first town in Brazil, soon followed by others that stretched along the coast as far north as the hump of South America (Figure 6.1). By 1550 there were 15 small towns along the coast, among them the short-lived Ilheus (1532), Salvador (1534), Recife (Pernambuco, 1536), and nearby Olinda (1537). A number of other towns were founded by plantation owners, ostensibly as acts of piety, but equally as acts of land speculation and labor exploitation. In these instances, land was given to the church, a plan drawn up, and the town established in order to enhance surrounding land values and assure a labor supply for agriculture. One of the first of these so-called *patrimonios* was the coastal town of Santos (1545), near São Vincente. After 1550, Jesuit missions were established on lowlands adjacent to the towns. A few were in the interior, the most notable being what became São Paulo (1554). The fortified military settlement of São Sebastião do Rio de Janeiro was laid out on its present site in 1567, supplanting the destroyed French colony of Antarctic France (1555); the site had been occupied and fought over since 1531. With an excellent harbor and strategic location, Rio became an important port and administrative center for southern Brazil.

By 1600 the major towns were the capital at Salvador, Recife–Olinda, Rio de Janeiro, and São Paulo. With few exceptions the other towns were small ports or collections of dwellings near a church or fort. Nonetheless, the littoral was occupied and an urban network established that was to remain unchanged until the discovery of gold late in the seventeenth century led to new town foundings and restructured the Brazilian urban hierarchy.

▶ Fine-Tuning the Colonial Urban World: 1600–1800

By 1600, the Spaniards had delineated the major regions of mineral wealth, high-yield agriculture, and tractable Indian populations. Urban development after 1600 in Spanish America was a process of gradual urban "in-filling" and adjustment of the early settlement pattern. The rise and decline of towns and cities was common as mines opened and closed, agricultural areas were developed, and political decisions led to a reorganization of administrative centers. Places that were distant from the already-productive areas and were perceived to have little development potential in terms of the economics, technology, and transportation of the time remained unsettled. The Spaniards' modest population level (estimated at only 150,000 in the Americas in 1574) precluded major extensions into new areas.

If the opening of new lands was relatively insignificant in Spanish America, just the opposite was true of Brazil, where the settlement of the interior highlands led to a restructuring of the urban hierarchy and the creation of many new towns (Figure 6.2). The impetus for change was the discovery of gold and diamonds in the interior. To a lesser extent, the occupation of northeastern Brazil by the Dutch played a part in the urban transformation. Along the southern coast of Brazil, the Portuguese extended their claims and founded a series of ports as far as the Río de la Plata, where they interfaced with the Spanish sphere.

Throughout South America, most towns were slow to grow and small in size. Limits were imposed primarily by poor transportation, restrictions on trade both within and beyond the colonies, a declining or only slowly increasing Indian population base, a relatively small number of Europeans, and the dominance of a local agricultural economic base. Few towns had a population in excess of 5,000; most were considerably smaller.

The exceptions were the ports, political centers, and mining towns. Over time, cities such as Buenos Aires, Bogotá, and Caracas came to share the political and economic roles jealously guarded earlier by Lima. In Brazil, Rio de Janeiro came to dominate as both a port and a capital.

The largest cities of colonial Latin America were small in comparison with today's urban areas. At the beginning of the nineteenth century, the largest city of Spanish America was Mexico City, but its population of 128,000 was no more than that of its predecessor, Tenochtitlán, at the time of Cortés's arrival in 1519. In 1800, Lima was a city of only 64,000, Buenos Aires, 45,000, and Caracas, 38,000. Rio de Janeiro was Brazil's largest city, with 100,000 inhabitants, followed by Salvador, 50,000, and Recife, 25,000. Most of the major inland regional centers and entrepôts throughout Latin America in 1800 ranged from 10,000 to 30,000, depending on the wealth of the region.

The population levels of mining towns were volatile, and major centers like Potosí experienced declining numbers by 1800. Important ports such as Veracruz, Santos, and Callao were small towns of 11,000, 7,000, and 2,000, respectively, in 1800; even smaller were the ports of the Caribbean. Most of the towns in the interior of Latin America were mere agricultural villages of modest dwellings and few commercial activities. In short, while many towns had been founded in the colonial era, few were very large.

Figure 6.2   Major settlement areas, 1600–1800.

Mexico

Mexico, after 1600, remained a key Spanish realm and, by 1790, was still producing over half the world's silver. Mexico and Brazil together accounted for
approximately 90 percent of the world's output of precious metals around 1800. In

both countries, the mining base created demand for goods and services. The manufacture or trading in mining equipment, leather products, pottery, woolen and cotton goods, furniture, silver artisanry, and the like were urban pursuits, as was the exchange of foodstuffs and livestock. The ports of Veracruz, Acapulco, Tampico, and Campeche serviced an increasing number of ships trading in the Caribbean and the Pacific. The inland centers established in the earlier period became more important regional, political, and commercial centers.

The Mexican frontier after 1600 lay beyond the silver mining districts of Durango and Santa Barbara. In an extensive area that included present-day California, Arizona, New Mexico, and Texas, a linear network of forts (*presidios*) and missions developed that were outposts of Spanish political interests and centers for the conversion of the Indians to Catholicism. Both *presidio* and mission represented the interests of the state and acted as buffer to the expansive policies of other nations. Other settlements were created as Spanish *pueblos*, or legal towns.

The earliest move to the north took place after 1598 into what is now New Mexico, 700 miles north of the Santa Barbara mining district. An administrative center was founded in Santa Fé (1608) that encompassed the Indian *pueblos* that stretched 200 miles along the Rio Grande, both to the north of Santa Fé (*río arriba*) and to the south (*río abajo*). Initially, there had probably been 40,000 Indians in 50 or 60 villages, but this total was reduced to perhaps 8,000 by the beginning of the nineteenth century. The settlement void between the Rio Grande and Santa Barbara was later bridged by missions, including the one at El Paso del Norte (Juárez, 1659) and by the mining town of Chihuahua (1703). San Antonio, Texas (one of its missions now known as the Alamo), was founded in 1718—the same year that the French established New Orleans. The Spanish held New Orleans from 1762 until 1800. The mission at Laredo dates from 1769.

To the west of the Rio Grande Valley, the Spaniards moved steadily northward with both *presidios* and missions into present-day California and Arizona, founding *presidios* at places such as San Diego (1769), Tucson (1776), and San Francisco (1776). Northernmost of a long chain of missions was that at Sonoma, California, where the Spanish sphere of influence met the Russian fur-trading realm operating out of Fort Ross (1812). Numerous missions were the core of subsequent urbanization, as the string of towns along the *camino real* (royal road) in California indicates. Early *pueblos* included San Jose (1777) and Los Angeles (1781).

Central America and a Refocused Caribbean

Relatively few Spaniards were attracted to Central America and towns like Antigua, León, Granada, and Cartago failed to become important centers. In 1773, Antigua was devastated by a severe earthquake that led to its decline and the creation of Guatemala City (1776) as the new capital. The English, French, and Dutch began to occupy Caribbean islands and the marshy coastal lowlands of the Guianas in the early decades of the seventeenth century. With few exceptions, these areas were viewed as marginal by the Spanish and Portuguese, given the lack of mineral wealth and Indian labor.

Although a handful of early settlements, including Belize (1630) and Greytown (San Juan del Norte), functioned as pirate bases, most evolved as sugar producers, following the earlier successes of the Dutch in northeastern Brazil. The French colony of Sainte Domingue (Haiti), acquired in 1697, became the most productive of the sugar islands, leading to the founding of a number of port towns, including Cap Haitien and Port-au-Prince (1729) and the growth of some older places such as Jacmel.

Jamaica was taken by England from Spain in 1655, and Kingston (1692) replaced Port Royal as the principal settlement when the latter, a notorious pirate base, was destroyed by an earthquake and flood. Holetown (1627) was the first English settlement on Barbados, but lacked a good natural harbor, and Bridgetown rose to prominence as a sugar-exporting port in the 1640s. Basseterre, on Nevis, became important after 1675. The northern edge of the Spanish Caribbean was increasingly threatened by the French and English, leading to the fort at Pensacola in 1698. New Orleans, Savannah, and Charleston were all to impinge on the Spanish realm.

In what became the Netherlands Antilles, the port of Willemstad evolved after 1634. On the Guianas shoreline, encroaching upon the Portuguese, the Dutch established Paramaribo (1640) and the French created Cayenne (1643). The Caribbean area, whether Spanish, English, French, or Dutch, was essentially plantation-driven, with a single major port on each island.

Andean Urban Adjustments

In Colombia, as late as 1700, settlement was still largely restricted to the Caribbean coast and the Magdalena and Cauca valleys. Cartagena, storehouse of the wealth awaiting export to Spain, became the most frequently besieged city on the Spanish Main, a reality that, combined with hostile Indians and the difficult terrain of Colombia, diverted official attention from interior development. The exceptional case was the highland region of Antioquia, where gold was mined and where Medellín (1650) developed as a regional service center.

After 1700, however, the accession of the French Bourbons to the Spanish throne brought changes in colonial policy which affected the relative importance of cities. Bogotá was made the seat of a viceroyalty in 1739, an act that focused attention on an area that had been ignored by the authorities and merchants of Lima. Later, Caracas was designated a political center, favoring urban development there. Venezuela began to develop its agricultural potential, exporting coffee, cotton, and indigo as well as expanding cattle production on the Llanos of the Orinoco.

At the other end of the continent, the viceroyalty of the Río de la Plata, with Buenos Aires as its capital, was created in 1776. One motive for its political rise was to offset Portuguese expansionist tendencies reflected in the establishment of the fortified port of Colonia across the estuary. This and the "free trade" regulations of 1777 and 1778 enhanced the port town, earlier suppressed by interests in Lima. Despite restrictions, Buenos Aires was already the largest town on the Río de la Plata/Paraná–Paraguay navigation system. With a population of 20,000 in 1776, the new administrative and military role simply reinforced its dominance in the region. On the north side of the La Plata estuary, the fortress settlement of Montevideo (1724) had been established to stop Portuguese advances to the south. Development was favored by its designation as an official port on the Cape Horn route. In time, the cape route led to a decline in the importance of towns in Panamá and a rise of Pacific ports, particularly Valparaíso.

There was a resurgence of town foundings in central and southern Chile in the eighteenth century. Ancud (1768) and Constitución (1789) became ports, while Rancagua (1743), Curicó (1743), Linares (1789), Parral (1789), and others were new agricultural centers. Copiapó (1745) was one of several new mining centers. In all, over 30 cities were reestablished or created.

As a consequence of these political, transportation, and commercial realignments, Lima gradually lost its preeminence. Although Peruvian and Bolivian mineral products remained important as a result of new finds at Cerro del Pasco (1630) and elsewhere, Lima's boom days were over. The city's decline was compounded in 1746 by a severe earthquake that took an estimated 16,000 lives; the port of Callao was destroyed by the resultant seismic wave. If a more architecturally attractive Lima was subsequently created, it was nonetheless downgraded commercially and politically.

Portuguese America

In contrast to Spanish America and the Caribbean area, there had been little urban development in the interior of Brazil as late as 1650. There were few towns of note even within the initial coastal zone, because that region's development was rural and plantation oriented. Brazil's growth was adversely affected by Spanish control of Portugal, after 1580, and by the occupance of northeastern Brazil, including Recife and Bahia (Salvador), by the Dutch West India Company from the 1630s to 1654. The Dutch left behind a strong architectural heritage, particularly at Recife (Pernambuco).

Gold Towns. The development of towns in the interior came in the eighteenth century following gold finds by *bandeirantes*, the quasi-military expeditions that had searched the interior for minerals to mine and Indians to enslave. In 1693, gold was found in stream gravels throughout Minas Gerais, but notably in the Serra do Espinhaço; 20 years later deposits were located to the west in Mato Grosso. In both areas, the mining camps evolved into settlements, particularly after about 1720 when vein mining replaced the peripatetic panning of streams.

The most famous of the gold towns came to be Vila Rica do Ouro Prêto, the Rich Town of Black Gold, so called because the gold matrix was black quartz. Founded in 1712 at the center of the mineralized districts, the gold mined there financed fine churches and town houses that make the town, now a national monument, a gem of Latin American baroque architecture. At the height of the gold rush, between 1725 and 1750, the town had a population of over 60,000, but by 1800, with the rush terminated, it had declined to 8,000; scores of mining camps disappeared. Small gold towns such as Sabará, Marianna, São João del Rei, and Congonhas do Campo became villages.

In the Mato Grosso, deposits of gold were found at Cuiabá in 1718 and Goiás in 1725. By the mid-1720s, Cuiabá had a population of 7,000, but it, Goiás, and other towns declined well before 1800, becoming small service centers in an area of quasi-subsistence agriculture. Elsewhere, diamond fields found after 1729 led to the development of mining towns in the modern states of Minas Gerais and Bahia. One of the largest towns, Diamantina, was founded in 1730. Unlike gold, however, diamonds were declared a crown monopoly, thereby restricting their trading and the number of mining towns that came into existence.

Not surprisingly, thousands arrived from Portugal for the mining fields and thousands more left the settled coastal zone, contributing to the decline of the coast in population and economic significance. The urban hierarchy (the ranking of towns by size and importance) was restructured in favor of the interior and in favor of the port of Rio de Janeiro, then the sole official port for the export of gold bullion. Because of its central location vis-à-vis the mining interior and its expanded trading role, Rio replaced now poorly located Salvador as the capital in 1763. Decades later, in 1808, Rio de Janeiro became the residence of the Portuguese royal family, who fled Napoleon and set up court there until 1822, when Brazil declared its independence and was ruled by an emperor until 1889, when the republic was established.

As with the silver district of Potosí in Upper Peru, the Brazilian mining districts could not procure enough food or pack animals in surrounding areas. This encouraged the expansion of agriculture elsewhere in Brazil and led to the formation of new towns. Many overnight stopping places, or *pouso*, and stations for collecting taxes on gold shipments came into existence and became small towns whose names, such as Pouso Alegre and Registro, reflect their initial function. Most of the mules destined for mine work and for goods transport came from the south, and the town of Sorocaba played the same role that Salta had performed for the mines of Upper Peru. Still other towns and overnight stops developed on the trail northward from Sorocaba to the mining districts. Cattle to feed the mining districts of Minas Gerais came from the *sertão* of the Northeast, where the new markets in the mines to the south encouraged the expansion of ranching.

By the late seventeenth century a line of military posts adjacent to the São Francisco River had been established to fight banditry along the trade routes; some of these posts became small highland towns. In short, many preexisting settlements in the interior flourished, and new towns were founded on the basis of the mining and export of gold.

In the highlands south of São Paulo, gold was discovered near Curitiba in 1654, with that town evolving as a mining center. A second zone of exploration was established along the coast south from São Vicente. Farther to the south, forts and garrison towns were founded at São Francisco do Sul (1660), Florianopolis (1726), and Porto Alegre (1743) to protect against potential Spanish and French encroachments. The southernmost extension of Portuguese influence had already been established at Colonia do Sacramento (1680) on the estuary of the Río de la Plata, opposite Buenos Aires.

To the north, political considerations led to the siting of towns on the coast and in the Amazon Basin. The intent of the settlement at Fortaleza (1609) is evident in its name, but other towns, such as São Luis (1594, the French town of St. Louis taken in 1615), had the same role—to secure the north from Dutch, English, and further French occupation. A fort and, soon after, a mission were founded in 1616 at Belém (Bethlehem), followed by similar foundings at Santarém, Óbidos, Manaus (1674), and elsewhere along the Amazon.

A Model of Urban Network Development

The history of Latin American urbanism shows that many of the earliest towns were ports meant to *attach* the New World with the Iberian peninsula. In the interior, mining towns developed that supplied gold, silver, and precious stones exported to Spain and Portugal through the ports. Other towns were established as agricultural settlements to supply food to the ports and the mining areas or to provide an export crop such as sugar. A significant number of towns developed as the way stations to move local agricultural produce and European manufactured goods to the mining districts and to transport New World products to the ports, the *points of attachment* to Europe. In time, several interior towns became more important than some of the early ports. They came to combine political roles with growing importance as internal production and trade centers, while benefiting from their position as collection points in the chain of towns engaged in long-distance overseas trade.

A simple four-stage model of (1) exploration, (2) initial settlement, (3) expansion of the network, and (4) in-filling of the network sets out the "mercantile" or trading motivation that generated the urban network (Figure 6.3). The mercantile model indicates that long distance, *exogenic* international trading and political ties, not the development of local or regional markets are the major impulses in the growth of an urban network. The urban system of Latin America was tied to Iberian and broader European mercantile policies more than to local factors. The New World city-system, in short, was an element of Europe's political, religious, and economic policies.

Cities were important to their developing hinterlands and the growth of local markets. In time, these ties became more important, and *endogenic* relationships were of increasing significance as towns became "central places" for local and regional trade. As the mercantile model suggests (Stage 4), this central place aspect of a town's development is an addition to the outward-looking, *exogenic* ties and

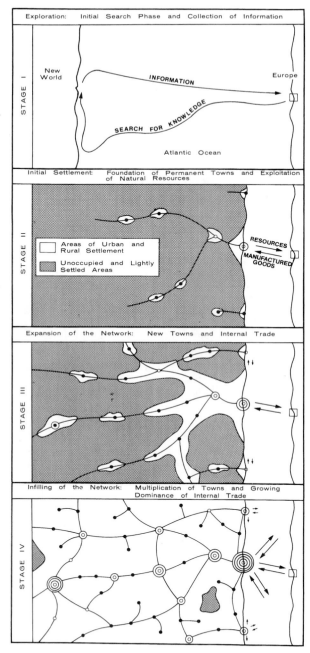

Figure 6.3 Mercantile model of urban settlement
(after Vance).

the development of economies within Latin American countries. The mercantile model accommodates the early trading networks with the nineteenth and twentieth century "in-filling" of a colonial frontier.

Because of Latin America's size, physical and cultural diversity, and different political regimes, the four stages of the mercantile model did not occur everywhere at the same time. The timing of stages of urban development, for instance, varied greatly between the Andean highlands, the forests of Amazonia, the Pampas of Argentina, and the islands of the Caribbean. The mercantile model should be viewed in a regional context that takes the chronology of development into account. For instance, the initial search phase, Stage 1, encompassed the four voyages of Columbus (1492–1504) and the exploration of inland Brazil in the 1600s and of California in the 1700s. The second stage includes the harvesting of brasilwood after 1500, the early mining of silver and gold in Spanish America, Caribbean mahogany production in the 1700s, and the gathering of latex for rubber in the Amazon in the latter nineteenth century. Yet another example of the time variation is the development of towns and the sugar economy of coastal Brazil after 1530 (Stage 2) in contrast to the development of agriculture and towns in interior Brazil after 1700 (Stage 3). The settling of southern Brazil's interior took place only in the late nineteenth century.

► *The Latin American City System in Transition: 1800–1930*

For most of Latin America, the early decades of the nineteenth century brought political emancipation. Later decades saw broader entry into world markets and the diffusion into Latin America of Europe's ongoing agricultural and industrial revolutions. With these changes came a reshaping of urban systems. The timing of changes varied, but all cities were influenced by foreign investment, modern industrial technology, new modes of transportation, population increase, agricultural change, and other factors.

The exploitation of mineral resources was broadened from precious metals to the large-scale extraction of industrial minerals such as copper, tin, and iron ore. Agricultural areas were opened or transformed from pasture to wheat field as newly introduced railroads tied them to European markets by extending the hinterlands of the ports. Technological advances such as the marine steam engine, steel hull, screw propeller, and refrigeration linked the products of Latin American pastures and fields to world markets.

Colonial ports and political capitals continued to be favored in this new commercial era. Existing cities were the source of a relatively skilled labor supply, especially as European immigrants began to arrive in the port cities. Cities were the nodes of transportation networks that collected raw materials and shipped finished goods. Cities were sources of investment capital and major markets in their own right. Urban growth became more intense than in the colonial period and tended to focus on cities already established as administrative centers, transportation nodes, manufacturing centers, and sources of capital and information.

The new economic system imposed by dependence on world markets meant that hundreds of settlements became more peripheral to the commercial-industrial order that was evolving. The urban stagnation that followed for such towns was acute in parts of Central America and along the Andes where physical limitations, poor access, economic restraints, and the attraction of larger cities precluded growth. Many towns stagnated. They were bypassed by modern technology and lost their traditional economic base as imported manufactured goods and agricultural products worked to their disadvantage. By 1930, the stage was set for the large-scale, rural-to-urban migration that transformed the Latin American city and created the contemporary city hierarchy.

Latin America made its political and economic transition in the nineteenth and early twentieth centuries from a set of colonial, urban systems tied to European powers, to an adjusted set of mercantile and influential cities increasingly linked to broader world markets. The change was particularly notable in the major cities, but change also created new towns in some formerly marginal areas. The new urban zones (at the edge of the older urban systems) were widely spread across Latin America:

1. Mexican, Central American, and Caribbean fringe areas
2. Chilean Urban Frontiers—South and North
3. The Pampas Grasslands of Argentina and Uruguay
4. Towns of Patagonia and Tierra del Fuego
5. New and Expanded Towns of Southern Brazil
6. Amazonia

Mexico, Central America, and the Caribbean Coasts

After 1854, the establishment of the present U.S.–Mexican border slowly created a linear pattern of towns along the frontier. Some were twins of U.S. towns; others were not. Without exception, the towns remained small until after the 1930s.

Farther from the frontier, the colonial city of Monterrey gained a new strategic importance in addition to its long-term administrative and trading function. Monterrey became a point of commercial connection linking Mexico City and Texas. During the U.S. Civil War, Monterrey was a center for the distribution of products going to and from the Confederacy. Mid-century agriculture and commerce generated capital that was later invested in industrial activity. Improved transportation helped the city's development as mule-drawn vehicles were gradually replaced by the railroads after 1882, when the first rail line arrived. By 1887, Monterrey had rail connections with Laredo, Texas; Tampico on the Gulf coast; and Mexico City, the center of Mexico's rail system. In addition to its commercial importance, the railroad spurred industrial development at Monterrey and on the Gulf coast, where early in the twentieth century, the old towns of Tampico and Tuxpan grew as oil fields were developed.

At Mexico City, a Federal District was created in 1824 outlined by a circular area with a radius of 6 kilometers from the Plaza Mayor. Until the early twentieth century, Mexico City was framed by these Federal District limits.

The expansion of irrigated agriculture impelled the creation of new towns in northern Mexico, including a number of small towns along the U.S.–Mexican border once it was defined by 1860. Along the Gulf of California, agricultural centers such as Cuidad Obregón, Los Mochis, and colonial Culiacán have been favored by an expansion of irrigated agriculture as dams have been constructed to hold the waters from the Sierra Madre Occidental.

Unlike much of Latin America, the present hierarchy of cities in Central America does not represent a simple continuation of colonial patterns. Political centers have shifted since independence from Spain in the nineteenth century. San José became the capital of Costa Rica in 1823, replacing nearby Cartago. The independence of Nicaragua sharpened a rivalry between liberals based in León and conservatives based in Granada; this was resolved in 1856 by making the old Indian town of Managua a compromise capital. The capital of Honduras moved from Comayagua to Tegucigalpa in 1880.

The creation of large banana plantations in the lowlands of Central America after the 1870s and 1880s led to the appearance of company towns, some new ports, and the growth of existing places. Trujillo and Puerto Castilla grew on the north coast of Honduras. In Costa Rica, the Pacific port of Puntarenas (1814) and the Atlantic port of Puerto Limón (1867) were established. Coffee and banana production provided exports for these and some other Central American ports. The cutting of hardwoods led to the growth of small towns and ports. The construction of a railroad across the then-Colombian Isthmus of Panama created the railroad town and port of Colón on the Caribbean side.

Caribbean

In the Caribbean, the major towns of the colonial period remained the magnets for wealth and population. In common with the other islands, virtually all of Cuba's major colonial towns were sited on the coast, with contact between them principally by water. The network expanded with the concurrent development of sugar plantations and the first railroad in Latin America (1836) that took sugar to coastal ports. As plantations developed, so did small towns. Havana nevertheless maintained its position as the principal metropolitan area.

Venezuelan town development after independence was strongly limited between 1830 and 1900 by political anarchy: there were 39 major revolts and over 120 lesser uprisings. All of them contributed to the absence of economic and urban integration. It took the discovery of oil in 1917 in Lake Maracaibo to produce major changes in Venezuela's economy and urban system. The oil industry transformed the city of Maracaibo and the surrounding area. Oil royalties accrued in Caracas, the capital, and that city enhanced its status and became strongly primate at the top of the Venezuelan hierarchy of urban places. At the beginning of the twentieth century, Caracas had a population of approximately 75,000 people. The second ranked place, Valencia, had some 38,000 inhabitants, Maracaibo had a population of 35,000, and Barquisimeto around 32,000. The oil revenues increased the political power and wealth of the capital, which became many times larger than any other city in Venezuela.

Chilean Urban Frontiers—South and North

From the mid-1800s, Santiago and Valparaíso thrived as Chile increased production of wheat, copper, and nitrates for sale in the world market. Concepción and Valparaíso were the first ports after the difficult passage around Cape Horn. By the beginning of the twentieth century, Santiago had a population of approximately 300,000 inhabitants, Valparaíso had about 40,000 people, and Concepción around 50,000 inhabitants. No other place exceeded 40,000 residents. Santiago and Valparaíso were linked by rail in 1863, and began to establish the dominance in the urban hierarchy that they retain to the present. Provincial towns remained small and unattractive.

In the second half of the nineteenth century, southern Chile was colonized, and agricultural, lumbering, and fishing resources were developed. As the Indian "problem" disappeared with wars on the frontier, German colonization in the south signaled a new period in Chilean settlement. German immigrants settled in lands south of Concepción, especially around Osorno, Valdivia, and a number of new towns, including Puerto Montt (1853) and Puerto Varas (1854). The Germans found the climate and topography similar to their homeland. A "longitudinal" railroad down the Central Valley opened the south of Chile to additional development early in the twentieth century.

Until the War of the Pacific in 1879, Chile's new towns had been appearing only in the south. With the defeat of Bolivian and Peruvian forces in 1883, Chile focused attention on lands to the north. The victors claimed 600 miles of territory stretching from Copiapó to the port of Arica (at the present Chilean–Peruvian border). Victory in the "Fertilizer War" gained Chile valuable nitrate territories in the Atacama Desert and an in-place mining workforce. Settlements evolved as mining centers, small ports, and scattered oases. There was little rural population, for the environment would not support agriculture. The mining economy did not produce large cities because the wealth from nitrates was invested by the elite in central Chile.

The Pampas Grasslands of Argentina and Uruguay

Argentina is one of the world's outstanding examples of the impact of industrial technology, agricultural advances, improved access to markets, and large-scale immigration upon regional and urban development. The transformation of the sparsely settled and unfarmed Pampas into an intensive network of towns and fields was initiated by a 100-year campaign of frontier expansion and Indian extermination that began in 1779 and ended with the so-called "Conquest of the Desert" between 1879 and 1883. The expansion of the frontier resulted in a network of forts and adjacent small settlements. The earliest of these garrison towns were less than 100 miles south of Buenos Aires, along a line that lay to the north of the Río Salado. In 1815, it was estimated that 600 persons of European stock lived south of the Salado. By 1870, there were 100,000 Europeans south of the river, most of them in the expanded network of fort towns. By 1895, some 360,000 lived south of the Salado. In all, about 50 settlements were established on the Pampas in the 100 years after 1779.

Railroad construction, agricultural development, and European immigration onto the Pampas turned many of these small, crude garrison settlements into farm towns late in the nineteenth century. Political decisions made some of them county seats. A number of new settlements were laid out along the post-1860 railroads, as the intent was to establish stations every 12 miles along the rail lines. All the towns

Avenida 9 de Julio, Buenos Aires.

Montevideo, Uruguay.

depended on the railroad to move wheat and cattle from the Pampas to the ports of Buenos Aires, Rosario, and Bahía Blanca.

The focus of the rail network was Buenos Aires, which grew from 178,000 in 1869 to 1,561,000 in 1914. The population of its suburbs, reached by suburban rail service over the same lines that went to the Pampas, totaled 450,000 in 1914, giving the metropolitan area a population of 2 million. Not far away, the city of La Plata was laid out after 1882 as the capital of the province of Buenos Aires. Buenos Aires city, the earlier provincial capital, became the national capital.

Across the broad, shallow estuary of the Paraná River—the Río de la Plata—the city of Montevideo functioned like Buenos Aires on a smaller scale. As Uruguay's major port, capital, and business and industrial center, Montevideo dominated the urban network. Towns such as Fray Bentos (1859), Paysandu (1823), Salto (1817) along the Uruguay River, and Durazno (1931) on the plains developed as outlying centers in a settlement pattern that has always focused on Montevideo. Along the Atlantic coast some small settlements, notably Piriapolis (1893) and Punta del Este (1880), evolved into beach resorts.

Towns of Patagonia and Tierra del Fuego

The Gobernación de la Patagonia was established in 1878, and the southern campaigns of the Conquest of the Desert followed. The Patagonian Indians were less hostile than the Araucanians on the Pampas and infrequently hindered European settlement. Among the early settlers in the area were Welsh farmers who began the small port of Puerto Madryn (1865) and villages in the Cwm Hyfrwd (Gorgeous Valley) of the lower Río Chubut——Trelew (1871), Rawson (1874), and Gaiman (1874). Farther up the valley, outposts were created at Trevelin (1888) and at nearby Esquel (1906) in the Andean foothills. A narrow-gauge railroad ran up the Chubut valley after 1889. Welsh immigration effectively ended in 1911, and by 1915 Spanish and Italian immigrants outnumbered the 3,000 Welsh-Argentines.

At the edge of the Andes, between Mendoza and San Carlos de Bariloche, small towns developed parallel to the new Argentine–Chilean border, created as a result of the 1881 treaty between the states. Earliest to develop was the territorial capital at Chos-Malal (1884) along the Río Negro. Within a few years, the 1881 border was in dispute, and the Argentinian villages of San Martín de los Andes, Junín de los Andes, Norquin, and Las Lajas were all started in the 1880s as forts along the border with Chile. War was imminent before boundary adjudication in 1902. For Bariloche, the growth of tourism became important with the establishment in 1909 of Nahuel Haupi National Park, its inauguration in 1922, and the development of tourist excursions after 1924.

Downstream on the Río Negro, the earlier Conquest of the Desert led to the 1879 creation of Fuerte General Roca. Nearby Neuquén (1899) became the major settlement only decades after the territorial capital was transferred there in 1904 from Chos-Malal. Rail service entered the valley at the turn of the century and led to additional small centers, notably Cipolletti (1903) and Allen (1907) as irrigated agriculture—fruit and alfalfa—expanded. Oil development began about 1918, leading to other small towns, and an Italian colony was established at Villa Regina in 1924.

Along the southern coast of Patagonia, the territorial capital at Puerto Santa Cruz (1894), San Antonio Oeste (1895), and Río Gallegos (c. 1897) were all centers of small agricultural colonies; in 1905 the small bank at Río Gallegos was robbed by Butch Cassidy and the Sundance Kid. Drilling for water led to the accidental 1906 discovery of oil at Comodoro Rivadavia, created in 1901 as the port for Colonia Sarmiento (1897), located 100 miles inland. San Julián also dates from 1901, while Puerto Deseado, previously settled in 1780, 1792, and 1884, was resettled in 1909.

The southernmost town in the world, Ushuaia, began in 1870 as an Anglican mission, whaling station, and shipwreck refuge on the Beagle Channel. Contact with European diseases exterminated the Indians and closed the mission, but the 1886 introduction of sheep into Tierra del Fuego expanded the economic base of Ushuaia, as did the establishment of a national prison and *presidio*. In 1897, another church mission was the nucleus of Río Grande on the north side of "the island." Especially before the opening of the Panama Canal early in the twentieth century, Ushuaia, Port Stanley (first settled on the Islas Malvinas–Falkland Islands by Argentina in 1789 after brief occupation by the French and English), and Punta Arenas (1849) were important points of ship repair and refuge from storms and shipwreck in the South Atlantic.

New and Expanded Towns of Southern Brazil

The sugar and gold booms of colonial Brazil were followed in the late nineteenth century by a coffee boom. First grown in quantity in the Paraíba Valley near Rio de Janeiro, coffee expanded into the highlands north and west of São Paulo in the 1880s, especially into zones of *terra roxa* soils. With the expanding coffee frontier came new towns that filled with European immigrants. As in Argentina, the expansion of the railroad went hand in hand with this movement. Lines radiated from São Paulo, a transportation node itself connected with the port of Santos by the single rail route opened in 1867 over the Great Escarpment. The rail network in time became second only to that of the Argentine Pampas, and many coffee towns of all sizes, such as Ribeirão Prêto, Rio Claro, São Carlos, and Botucatu, prospered. Campinas, as a transportation node, grew rapidly. Santos became the world's principal coffee port.

As in Argentina, dissatisfaction with poor rural conditions and low wages encouraged many European immigrants to move into the cities. They moved especially into São Paulo, which grew from about 30,000 in 1870 to 240,000 in 1900 and 1 million in 1930. By 1900, as coffee profits were locally invested in industry and commerce, São Paulo was becoming an industrial and financial center. It remains so today and ranks ahead of Rio do Janeiro despite economic decentralization to nearby cities such as Sorocaba, Jundiaí, and Campinas.

Farther south, Brazil encouraged settlement in the zone of contact with the Spanish. The German settlement at São Leopoldo (1824) was the first of the government-sponsored agricultural colonies in the state of Rio Grande do Sul and the nucleus of a settlement region into which more than 20,000 Germans moved by 1869. The growth of Porto Alegre was favored by this hinterland development. A second, more northerly concentration of German colonies occurred in Joinville (1849) and Blumenau (c.1850). Small Italian settlements in the south developed in the 1870s and 1880s concurrent with the large-scale Italian immigration into the more northerly states of Minas Gerais and São Paulo.

Several new state capitals were created in this period. Teresina was founded in 1852 as an inland administrative and commercial center for northeastern Brazil, and Belo Horizonte was laid out in 1896 as a new capital for the state of Minas Gerais, replacing the colonial mining town of Ouro Prêto. The site for a new national capital (Brasília) was pinpointed at this time, but that city was inaugurated only in 1960.

Amazonia

Belém had been founded in 1616 in an effort to secure the Amazon area from intrusions by the Dutch, French, and British. Defense, exploration, the search for gold, and the subjugation of indigenous peoples were its colonial functions. The earliest substantial structures were churches intended to help convert the region's native peoples. By 1800, it had perhaps 10,000 people, reflecting the "economic backwater" status of the Amazon. In the heart of the Amazon, Manaus grew around the fort of São José do Rio Negro (1699). The first major impulse for development came in 1867, when the Amazon was opened to the international shipping of rubber. The Amazon rubber boom ran from the 1870s and peaked in 1910; the population of Manaus reached 236,000 in 1920.

Belém was the major rubber port. Its adjacent forests of Pará produced large amounts of rubber, but even the rubber in the neighboring province of Amazonas was largely directed to Belém from Manaus. Manaus, 10 days from Belém by steamer, even received most of its supplies through Belém. The rubber boom was ended by competing plantation-grown Asian rubber and Belém, Manaus, and other Amazonian towns grew at a greatly reduced rate.

THE CONTEMPORARY LATIN AMERICAN CITY

The typical Latin American city was small at the close of the colonial period and remained so for almost 100 years. The median population of the 40 largest cities in 1880 was about 35,000, and there were only 8 cities in Latin America with a population in excess of 100,000. By 1970, there were about 150 cities with over 100,000 inhabitants (Table 6.1). In 1880 the largest city was Rio de Janeiro, with a population of about 350,000; by 1990 the largest metropolitan area was Mexico City, with an estimated population of about 20 million.

TABLE 6.1 Population Estimates of Selected Latin American Cities (in 000's)

City	1880	1905	1930	1950	1960	1970	1980	1990
Mexico City	250	450	900	3,419	5,000	8,605	13,625	20,000
Guadalajara	70	110	150	461	735	1,445	2,331	3,600
Monterrey	15	70	100	411	600	1,167	1,883	2,970
Guatemala City	50	90	120	337	380	1,067	1,430	1,500
Managua	10	30	35	109	190	385	662	850
San Salvador	30	50	110	213	230	565	858	1,300
Tegucigalpa	10	20	30	72	105	295	406	800
San José	30	30	60	146	220	401	508	1,000
Panamá City	15	35	70	217	255	520	794	1,200
Havana	200	300	600		1,220	1,681		2,100
Kingston–Spanish Town	40	60	120		380	505	650	750
Port-au-Prince	25	100	120	143		385		1,000
Santo Domingo		20	35	182	365	671		2,200
San Juan	20	40	80		590	700		1,000
Caracas	55	100	200	686	1,355	2,058	3,208	4,100
Maracaibo	20		100	252	455	681	1,037	1,400
Bogotá	40	110	235	645	1,125	2,540		4,850
Medellín	35	55	120	359	800	1,400		1,700
Guayaquil	20	80	110	260	450	855		1,520
Quito	75	70	120		315	550		1,110
Lima-Callao	100	150	250	1,229	1,965	3,318	4,679	6,300
Santiago	150	300	600	1,350	1,700	2,850	3,902	5,200
Valparaíso-Viña	100	195	240		380	480		593
Buenos Aires	290	1,200	2,000	4,722	6,765	8,353	10,240	10,900

TABLE 6.1 *Continued*

City	1880	1905	1930	1950	1960	1970	1980	1990
Rosario	30	160	400	500	670	795	955	1,100
Montevideo	90	300	450	609	860	1,300	1,300	1,760
Asuncion	25	80	110	207	400	550	605	1,000
Rio de Janeiro	350	800	1,500	3,044	4,675	6,847	9,619	11,206
São Paulo	25	400	1,000	2,336	4,430	7,838	12,273	17,113
Belo Horizonte	130	250	350	406	655	1,505	2,279	3,615
Salvador	130	250	350	396	655	1,067	1,563	2,425
Brasília	0	0	0	0	130	538	1,082	1,804
Manaus	30	60	100	140	174	312	633	1,114
Belem	60	100	240	255	400	633	933	1,418
Recife	115	120	300	525	788	1,060	1,200	2,815
Porto Alegre	45	75	220	395	635	885	1,125	2,907

Sources: J. P. Cole, *Latin America*, London: Butterworths, 1975. R.W. Fox, Urban Population Growth Trends in Latin America, Washington, D.C.: I.A.D.B., 1975. *Statistical Abstract of Latin America*, 1978. *Mexico and Central American Handbook*. Bath, England: Trade and Travel Publications, 1990. *Caribbean Islands Handbook*. Bath, England: Trade and Travel Publications, 1991. *South American Handbook*. Bath, England: Trade and Travel Publications, 1990. Alan Gilbert, 1993. Brian Godfrey, 1991.

Notes: Particularly from 1950 onward, the figures reflect metropolitan area population.

At the beginning of the 19th century, the largest city was Mexico City (128,000), followed by Rio de Janeiro (100,000). Other large places were Lima (64,000), Salvador (50,000), Buenos Aires (45,000), and Caracas (38,000). Other major regional centers ranged from 10,000 to 30,000.

The transition to the large metropolis, or "megacity," has not occurred at an even pace throughout Latin America. Some of today's megacities began to appear in the late nineteenth century with the expansion of commercial agriculture tied to increasing federal expenditures, the development of industry, and large-scale immigration. This pattern is exemplified by Buenos Aires, Rio de Janeiro, and São Paulo. Most of today's large cities grew rapidly between 1900 and 1950. But this increase was modest compared with the urban "explosion" of the last 50 years, linked to a rapid rate of natural population increase and a steady influx of people from the rural areas and smaller cities and towns. Foreign—and especially European—immigration is now a minor factor.

Until the 1930s, the large cities were predominantly commercial and administrative centers with limited industrial production. The bulk of the major manufactured items consumed came from Europe and the United States through the ports that exported agricultural and mineral products. Local industry typically produced items that were protected from larger-scale foreign manufacturers by distance, transportation costs, and duties. Preeminent among these industries were food processing, textiles, and household goods, such as mattresses and other bulky, low-value products.

Manufacturing in the Latin American city changed with the economic depression of the 1930s and World War II. Both events stimulated policies of import-substitution, which sought to replace imported products with domestic manufactures. The rise of a

Resort town of Mar del Plata, Argentina.

broader spectrum of large-scale manufacturing throughout Latin America can generally be dated from this period. Consequently, the major metropolitan areas are industrial as well as commercial centers today.

There are sound reasons to locate industries in larger cities, including the *external economies* created by direct linkages between industries, so that the finished output of one factory becomes the "raw material" of another. There are also *urbanization economies* such as the existence of local capital, a large, skilled labor pool, and an existing supply of workers' housing, however inadequate it might be. In addition, the cities are the centers of innovation, invention, and information—valuable prerequisites for industrial expansion. Finally, large cities are major markets because of large populations and the higher purchasing power of urban residents compared to rural populations.

The number of cities sharing these advantages has expanded in the last 50 years so that there has been commercial and industrial growth in regional centers outside the major metropolitan areas. Cities such as Córdoba, Concepción, Chimbote, Cali, Medellín, Ciudad Guayana, Guadalajara, Monterrey, and a number of cities in the Rio de Janeiro–São Paulo regions have become more important industrial centers. Some regional "growth poles" are protected by distance from the competition of major cities. Others are important centers for the production of goods for national and international markets that benefit from planned transportation improvements that draw regional industrial cities closer together and facilitate the interchange of products and ideas. In the heart of the Amazon, rapidly growing Manaus received a major boost in 1967 when the city became a "free trade zone" and Amazonian "growth pole."

Several planned capital cities have been conceived across Latin America. The best-known is Brasília. The search for an internal capital location to replace Rio de Janeiro dates back to the creation of a republic in 1889, but the project was initiated only in 1956. The new city was meant to serve two principal purposes: first, to remove political activities to a neutral location away from the dominant Rio de Janeiro–São Paulo axis; and second, to provide a "growth pole" for interior economic development. Inaugurated in 1960, Brasília already had a population of about 130,000; by 1970 the population was 550,000, well above the population limit of 500,000 initially planned. By 1980, it exceeded 1 million.

The miscalculation of population aside, complaints about Brasília range from distress over its distance from Rio de Janeiro (about 600 miles) to problems of inadequate water supply, a housing shortage reflected in suburbs of squatter settlements, and a sterile urban environment. On the positive side, Brasília has drawn investment capital to the undeveloped interior, whatever the long-term environmental impact of interior development.

In Venezuela the exploitation of iron ore deposits south of the Orinoco promoted urban growth at El Pao and Ciudad Bolívar and encouraged a planned center of heavy industry. In addition to iron and steel production, Ciudad Guayana has metal fabricating industries, an aluminum plant, and other enterprises in order to promote national industrial diversification and provide a viable alternative to the major cities as migration targets and growth poles.

► Border Towns

One of the most spectacular examples of rapid urban growth in the twentieth century is along the U.S.–Mexican border where, from Tijuana ("bright place by the sea") on the Pacific to Matamoros on the Gulf of Mexico, formerly insignificant towns have become sizable cities (Table 6.2). The origins of these and smaller border towns is uncommonly diverse: El Paso del Norte (Ciudad Juárez after 1888)

TABLE 6.2 Mexican Border-City Populations, 1910–1990

City	1910	1930	1950	1960	1970	1980	1990
Tijuana	733	8,384	59,952	152,473	277,306	461,257	742,686
Mexicali	462	14,842	65,749	179,539	263,498	510,664	602,390
San Luis	—	910	4,079	28,545	49,990	92,790	111,508
Nogales	3,177	14,061	24,478	37,657	52,108	68,076	107,119
Ciudad Juárez	10,621	39,669	122,566	262,119	407,370	567,365	797,679
Nuevo Laredo	8,143	21,636	57,668	92,627	148,867	203,286	217,912
Reynosa	1,475	4,840	34,087	74,140	137,383	211,412	281,618
Matamoros	7,390	9,733	45,846	92,327	137,749	238,840	303,392
Total: 8 Border Cities			1,031,598		1,619,213	2,607,324	3,501,217

Source: Arreola and Curtis, 1993.
All cities over 100,000 population in 1990 are included.

began as a mission and presidio town in 1659, while Matamoros developed in 1700 as a mission and after 1824 as a refugee community for blacks escaping Texas. Nuevo Laredo was founded around 1850 by Mexican nationals who moved south after lands north of the river became part of the United States. Nogales evolved as a railroad and market town on an 1880s cattle trail, while Mexicali was founded by U.S. interests in 1902 as a consequence of irrigation development. Tijuana, a ranch settlement as early as 1840, was laid out as a town only in 1889.

All the Mexican border towns and cities share a strong symbiotic relationship with the United States, which is a function of dependence on U.S. markets, tourism, and employment on the U.S. side for many Mexican nationals. The distance of these towns from the economic heart of Mexico and the arid character of the intervening space reinforce the close relationship with the United States.

In the last three decades, duty-free *"maquiladora"* zones have been established as part of the "border industries program" to attract U.S. industry to the Mexican side of the border and provide employment after termination of the *"bracero"* farm worker program. Electronics firms, garment manufacture, and furniture making are among the most common enterprises that have moved into the duty-free *"maquiladora"* zones since the first one opened in 1965. *"Maquila"* is literally the grain taken by a miller in payment for grinding flour.

More than 90 percent of the *maquiladoras* are in the Mexican states bordering the United States. Tijuana, Cuidad Juárez, Reynosa, and Matamoros account for almost half the plants and about 60 percent of the workers. The Mexican government originally controlled the locations, hoping to reduce unemployment along the border. Initially, the *maquiladoras* had to be sited within 20 kilometers (12½ miles) of the border. In 1972, new rules permitted *maquiladoras* to locate anywhere in Mexico except in metropolitan Mexico City. A major negative impact has been the degradation of the physical environment owing to chemical spills, weak infrastructure, and other burdens on the physical environment. Much of the labor force is low-paid, female, and vulnerable.

Once resident in a border town, some do not return to the interior, and many who do return are replaced by still others. The turnover rate is high. A combination of high immigration and natural population increase is reflected in the population growth of these border towns. Rapid growth came especially after about 1940 as a result of expanded irrigation projects, the legal employment of Mexican *braceros* in U.S. agriculture between World War II and 1964, increased American tourism, and the development of local manufacturing. In 1900, the six major towns (Tijuana, Mexicali, Nogales, Cuidad Juárez, Nuevo Laredo, and Matamoros) had a combined population of approximately 30,000 persons; by 1990 the same cities had a combined population of almost 3 million.

► Urban Problems

As late as 1930, most Latin American cities were still relatively small in population and compact in form. Intraurban transportation was poorly developed. There was limited modern industrial production and few of the prerequisites for modern metropolitan life. It was only after 1930 that the rural-to-urban movements, now so characteristic of Latin America and other parts of the developing world, began on a large scale and that a number of cities grew rapidly. Today, large Latin American cities share the common problems of urban expansion, including inadequate housing and transportation, environmental deterioration, and limited employment opportunities for the growing population. The contemporary housing crisis is tied to the rapid urbanization of a growing population, the gap between the income of most workers and the cost of dwellings, the rising cost of land, the limitations of national financing agencies, the absence of housing policy in most countries, bureaucratic obstacles to construction, and the backwardness of the building industry itself. Continued housing deficits are likely.

The major Latin American cities are visually striking because of the modern high-rise office buildings and apartment houses. These buildings, however, serve what are increasingly minority groups in the city: the stronger commercial elements and the more affluent city dwellers.

More and more common to the major cities of Latin America are the related problems of traffic congestion and environmental deterioration. The narrow colonial streets of the center city, highly unsuited to modern cars and buses, are greatly overburdened, given the continued vitality of the downtown business district in the Latin American metropolis. In some cities continued observance of the long midday *siesta* creates an extra set of daily traffic peaks as people go home for lunch, but this custom declines as cities grow larger and make the two extra journeys too time-consuming and costly.

In response to the traffic problem, a few cities have turned to mass transit. Buenos Aires was the first city to employ the subway, or "metro," in 1913. Today Mexico City has the longest system (8 lines, 141 km), but did not inaugurate the first line until 1969, followed by São Paulo and Santiago in 1974, Rio in 1979, and Caracas in 1983. Systems are being planned and built for Lima, Medellín, and Brasília. A handful of other cities have adopted surface light-rail systems. Most cities, including those with metro systems, rely on the overburdened, slow, and accident-prone bus. Only a few trolley lines are still in operation, and significant suburban train service is limited to the larger cities.

Traffic volumes, fires for cooking and waste disposal, and industry are contributors to increasing air pollution. The dissipation of pollutants occurs in most cities through normal air circulation; all are becoming larger *heat islands*, and some are subject to serious smog conditions, including Santiago de Chile, Mexico City, São Paulo, and Río de Janeiro. The absence of sewers in the rapidly growing squatter settlements at the periphery of the major cities worsens the air pollution as powdered human fecal matter is blown into the air.

The Latin American city is being engulfed by the unrestrained growth of the peripheral residential *barrio*, the squatter settlement, and the industrial tract. There is an inability to finance adequate public services, and there are growing traffic congestion problems and a housing crisis of proportions that even the best of public housing policies could not handle effectively. Inadequate parks and a shortage of open space for new parks and playgrounds give the gridiron pattern an even glummer appearance. Indications are that conditions are worsening, if at a slower pace since the 1980s, and that the Latin American city is in crisis. Poverty, disease, crime, corruption, and political instability are all exacerbated by the unceasing growth of cities.

► *Squatter Settlements*

The lure of the major cities in terms of both employment and the *demonstration effect*, a heightened awareness that urban dwellers generally live better than their rural counterparts, has attracted vast numbers of rural people to the city. Urban attraction is reinforced by negative factors such as the harshness, poverty, and hopelessness of much rural life. Together, these push and pull forces have led millions to the cities, whether or not employment and housing existed. As noted in Chapter 5, this migration often occurs as a *step migration*, the first step being a move from the countryside into a smaller city and subsequently the move from the smaller to the larger city. Others come directly from rural areas into the larger cities. Either way, this rural-to-urban flow accounts for a large portion of the rapid

All major Latin American cities possess squatter settlements. In the picture we see part of a *favela* in Rio de Janeiro, Brazil. Squatter settlements frequently grow in difficult physical conditions. In this case, the slope is steep and in heavy storms houses are likely to be washed downslope onto the neighbors below. Squatter settlements across Latin America are overcrowded, often lack space for a garden, have narrow, unplanned streets, and lack some utilities. The *favela* above is moving toward maturity, as a number of houses are beginning to display recognizable features of domestic architecture, although many homes are still crudely built shelter boxes.

population growth of Latin American cities since 1940 (Table 6.1). High birth rates in the city have been a factor in growth.

In few cities, if any, does a housing shortage affect the wealthy. The problem is housing the growing lower-income masses that flow in from smaller cities and the countryside. The result has been proliferation of the central-city slum with its high population densities and poor living conditions and, in the last 40 years, the development of numerous squatter settlements in and on the periphery of all the larger cities of Latin America. The squatters are either new in-migrants to the city or they come from inner-city slums that are more crowded and more expensive than the squatter settlements. Major cities have 25 to 40 percent of their population living in squatter settlements.

Squatter settlements are usually in areas that home builders generally avoid. Here, wooden platforms on stilts comprise a building site. The picture is full of contrasts: the tile-roofed shelters lack connection with water or sewage pipes, but electricity seems to be available.

The generic name of these squatter settlements varies by country, being known as *favelas* in parts of Brazil, *poblaciones callampas* (mushroom towns) in Chile, *villas miseria* in Argentina, *ranchos* in Venezuela, *barriadas* in Peru and Panama, and *poblaciones tugurios* in Mexico. By whatever name, a common characteristic of these *barrios populares* is their location on marginal lands, either on the periphery of the built-up zone, along railroad tracks, on vacant pieces of public property, or on lands unsuitable for conventional uses. Some of the *favelas* of Rio de Janeiro are good examples of the latter; they are in an inner-city location, but on steep slopes unsuitable for conventional commercial or residential usage. In a storm, *favelas* can be washed from steep hillsides.

Squatter settlements are most common on publicly owned land, but some are on private land, and rent is paid. There are many cases of planned "invasions" of vacant land. In such an instance makeshift homes are constructed, often overnight, with materials collected in advance and in accordance with a general plan for land occupance. The intent is to make it morally and physically difficult to be dislocated by authorities.

Although some dwellings in squatter settlements are of cinder block, many others are constructed of whatever materials are available—scrap wood, cardboard, and flattened tin cans. Public services are scarce, with few sources of water, little or no paving, no sewer lines, no garbage collection, and only limited provisions of electricity. Sanitation conditions in shanty towns are marginal, and rats abound. Over time, however, the tendency is for the dwellings to become more substantial as the residents acquire better building materials and basic services are provided.

Mexico City, the San Angel dump. Many materials for the hovels in the squatter settlement have been retrieved from the garbage.

The occupants of the squatter settlements are generally employed or underemployed low-income families, since the creation of squatter settlements is more a consequence of the housing shortage than it is a problem of unemployment. Up to 90 percent of the potential labor force in such settlements is employed as artisans, street peddlers, shopkeepers, construction laborers, mechanics, and general laborers. For many, squatter conditions offer hope rather than despair and are viewed as part of the long process of material progress.

► *Too Many Big Cities*

Most countries of Latin America are characterized by an urban hierarchy where one city dominates demographically, economically, socially, and politically. Such a condition is termed *primacy*, and *primate* cities are many times larger than any other city in the country. The relationship between primacy and the level of economic development is unclear. Primacy occurs in relatively developed countries such as Chile and Argentina and in less developed countries like Peru, Guatemala, and Haiti. A dominant city has also been important in the history of major European kingdoms, including England (London), France (Paris), Denmark (Copenhagen), and Sweden (Stockholm).

Virtually all the Caribbean islands have only one major city and, except in Colombia, Brazil, and Ecuador, the condition is characteristic of the mainland. In Colombia rugged terrain and varied economic bases have historically made regional interaction difficult and fostered "twin primacy": Bogotá and Medellín. In Brazil, coffee gave impetus to São Paulo as an important competitor to Rio de Janeiro, and the entry of Ecuador into world commerce has brought the port of Guayaquil to the fore at the expense of Quito. São Paulo surpassed Rio in population in the 1960s (Table 6.1). The dominance of a few cities was a characteristic of Latin America from the beginning of the colonial period and a feature of its mercantile system. The colonial administrative centers such as Mexico City and Lima, and the major ports, clearly dominated the urban network, just as the commercial-industrial-political centers do today.

At what point does a city become too large? Some feel that primate cities hinder the growth of smaller regional centers by drawing off capital, skilled labor, and the educated. In addition to this "parasitism" and "brain drain," it is argued that the largest cities cannot provide urban services as economically as smaller places. Large cities compound environmental problems more than smaller cities do.

On the other side, primacy offers some advantages, including the availability of goods and services that smaller cities cannot offer because of an inadequate market size. The larger the city, the greater and broader the variety of goods and services. The larger city is given credit for helping to modernize society, enhance the social integration of the masses, and provide better educational opportunities than smaller cities or rural areas. It is maintained that they provide better health services and represent a viable concentration of scarce resources. It is only when the largest, or *primate*, city reaches a size where "diseconomies of scale" in the form of traffic congestion, long journeys to work, extremely high land values, environmental pollution, and so forth outbalance the advantages that the value of directing growth to smaller centers becomes apparent.

There is no effective way to determine an optimum size for cities. The optimum will vary from region to region and over time. It will be related to changing technology and its application and variations in consumer real income.

Large cities such as Mexico City, São Paulo, Rio de Janeiro, and Buenos Aires may have reached a point where more is lost than gained by increased centralization of commerce, industry, and population. Elsewhere, it is not so clear that the optimum size has been attained: Who can argue that the primate cities of the various Caribbean islands, much of Central America, and Paraguay, Bolivia, or Uruguay are too large? Without at least one large city in each of these countries, many higher-order goods and services could not be economically provided. These dominant centers are industrial growth poles and the only viable site for many cultural, educational, and health facilities. On balance, there is arguably a strong case in favor of the primate city in most Latin American countries.

THE MORPHOLOGY AND CHARACTER OF THE CITY

The morphology, shape, or form of a city reflects its past, and the clearest reflections are found in the street patterns, the size of city blocks, the dimensions of urban lots, and the surviving colonial architecture. In the typical Spanish American city, the colonial past is on view in what is now the city core, which is commonly laid out in a gridiron of square or rectangular blocks subdivided into long, narrow lots along equally narrow streets.

▶ *Nature of the Colonial City*

In Spanish America, the grid pattern was a response to royal instructions, which included guidelines on the location and proportions of the main plaza, the distribution of public buildings around it, the cardinal directions of the streets, and other urban elements (Figure 6.4). Initially, royal instructions on town foundings were so broad and undefined as to preclude a uniform application. By 1514, however, royal orders were coherently formulated. The postulates were more formally structured in 1523, but it was 1573 before the so-called Laws of the Indies definitively outlined 28 rules and regulations for town siting and layout. By that time, of course, many colonial towns and cities were already in existence, sited and platted according to

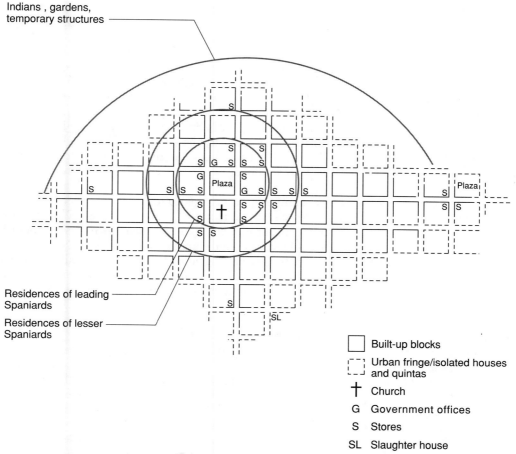

Indians , gardens,
temporary structures

Residences of leading
Spaniards

Residences of lesser
Spaniards

□ Built-up blocks

⌐ ⌐ Urban fringe/isolated houses
∟ ⌐ and quintas

✝ Church

G Government offices

S Stores

SL Slaughter house

Figure 6.4   Colonial town plan.

the more informal orders and the realities of the physical setting. Some towns proved to have unsuitable locations and moved several times before trial and error led to a good site. Generally, however, they fell within the constraints of the settlement regulations.

Most mining towns were exceptions to the royal mandates because waterpower was essential to the operation of milling equipment and because most mineral finds were in mountainous locations. For these reasons the town was usually located at the bottom of a long, narrow valley or on a hillside that offered little level land for large plazas and a gridiron pattern of streets. Instead, there was an elongated and irregular pattern of winding streets, irregularly sized and shaped blocks and plazas, and other informal features that today give towns such as Guanajuato and Taxco much of their charm.

Dating back to Vitrivius, a Roman planner of Greek background, the gridiron as it was imposed in Latin America reflected multiple influences—those of the Italian Renaissance, the planned thirteenth-century *bastide* towns of southern France, and the military towns established during the *Reconquista*, when Spain was slowly recaptured from the Moors by Christian forces in the centuries prior to the discovery of the Americas. But the pre-Columbian Indian cities in America also exhibited an essentially gridiron layout. In Mexico, both Teotihuacán and Tenochtitlán had essentially gridiron layouts, as did the pre-Incan cities along the Peruvian coast. The largest of these cities, Chan-Chan, was rectangular, with a symmetrical pattern of streets and irrigation canals.

The Incan town was characterized by two principal axes that met at a central square that was a commercial and religious center. The placement of religious edifices and the homes of the noble class was similar to the Spanish practice. Thus, the later layout of a number of Andean colonial towns easily fit into the general lines of the Incan city, utilizing existing street patterns and the firm Incan stone foundations for the construction of new buildings. The Incan capital at Cuzco is a notable example of this blending. In the final analysis, the simple grid of blocks seems to have been independently "invented" in various parts of the world, including pre-Columbian America, as a way to subdivide land and provide for extensions as the town grew.

The towns of Brazil were laid out with reference to function rather than royal orders. In some places, houses clustered around forts, forming a tight nucleus. Elsewhere, linear towns stretched out along a road. Mining towns, as in Spanish America, were more influenced by terrain and mining considerations. Nonetheless, most towns did have some sort of "natural" gridiron pattern, especially the early mission settlements that centered on the church, which itself faced a large rectangular *praça*, or plaza. Other towns informally evolved around the *rossio*, a communally held piece of land.

► *The Pattern of Land Use Within the Colonial City*

In any city, both a segregation and congregation dynamic is in operation. Some land uses, such as slaughterhouses and fine residences, are incompatible and are therefore separated. On the other hand, it is often advantageous to cluster similar activities together. Public buildings or people of like ethnic backgrounds or income congregate. These dynamics are expressed in downtown business districts, residential areas, and other types of land use and are forms of land use clustering as old as the city itself. In Teotihuacán, merchandise was sold in areas exclusively assigned to those goods, and the Incan capital of Cuzco had segregation and congregation of functions.

There was clear separation of land uses in both the Spanish and Portuguese colonial city. The heart of the city was its principal plaza, along whose sides were located a church, government buildings, and some businesses. Nearby were the homes of many of the wealthy who sought a central location in order to enjoy easy access to the commercial outlets, government, and religious offices. Some of the poor were intermixed with the more affluent, but most of the middle- and lower-income group lived in *barrios* farther from the center. The edge of town was largely reserved for the poor, the Indians, the municipal cemetery, slaughterhouse, and other undesirable uses. The multitude of Indian "barrios" were particularly grim in appearance. The formal laws were explicit about the center versus the periphery, noting, among other things, that leprosy houses, gambling establishments, brothels, and sewer outlets shall be located on the downwind side while the interior of the city should be attractive and pleasant. Those with wealth and prestige lived adjacent to the central plaza, and regulations precluded Indians from the central city.

The "better houses" toward the city center were typically two-story structures with a decorated doorway, an interior patio, and few windows facing the street. They were protected by wrought-iron grilles, or *rejas*, while the second story might have an ornate balcony for observing the street scene. Architecturally, elements of the floor plan can be traced back to houses of imperial Rome. Sharp economic, social, and political class distinctions were at the heart of colonial society. At the apex of the social pyramid was the *peninsular*, or immigrant from Spain; next came the *criollo*, born in Latin America of Spanish parents; the next level encompassed

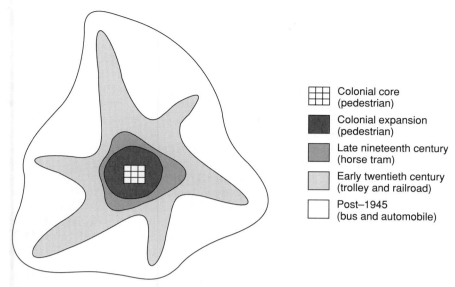

Legend:
- Colonial core (pedestrian)
- Colonial expansion (pedestrian)
- Late nineteenth century (horse tram)
- Early twentieth century (trolley and railroad)
- Post–1945 (bus and automobile)

Figure 6.5 Morphological change.

the *mulattoes, zambos* (Indian and black), and blacks. Income closely followed social rank. The Indians were a separate class altogether, under the system of the *encomienda* that distributed both Indian lands and the Indians to Spanish aristocrats. Brazilian towns were quite different; many of the elite lived not in town but on the coastal plantations, in a rural society.

The colonial town was pedestrian. It was a short walk from the fields, gardens, and hovels of the periphery to the heart of the town with its plaza, shops, and amenities (Figure 6.5). The center of the city was the location of whatever public services—paved streets, municipal water, garbage collection, street lighting—did exist, and this enhanced its appeal. There were few services before the end of the eighteenth century, however; for the colonial period all but the largest towns of Latin America have been characterized as "miserable" settlements with few buildings beyond the core with an appearance of elegance or permanence. Most houses were crude thatch huts or adobe shacks, and streets were unpaved, filled with swirling dust in dry periods or mud in rainy times. Hogs scavenged garbage, and dead animals lay in the streets to rot. Outdoor toilets and private wells were close together, a source of infection and disease.

Towns grew slowly despite periodic epidemics, as new, marginal *barrios* formed on the edge of town, as did new plazas, which often became business subcenters. Significant commercial, political, religious, and industrial activities remained in the central city. The cities of the colonial period can be described as mononuclear, with slow growth taking place in a more or less concentric pattern out from that center.

► Internal Structure of the Modern City

Even today many of Latin America's towns and small cities are not significantly different in *form* or *function* from what they were in the colonial period. They do not look the same because they now have electric streetlights, automobiles, high-rise apartments, and other products of modern technology. The picture is starkly different for the regional and national capitals, however, where nineteenth- and twentieth-century population increases are combined with new forms of transportation, urban sprawl, and the development of speculation in urban land on a large scale. Here the overall *form* and the *internal structure* have been strongly affected.

The colonial city was mononuclear, with the focus on the commercial and governmental areas of the central city. Residential *barrios* lay at the edge of the center and access between the two was by foot. As the larger cities grew, other means of access were devised. The horse-drawn tram on rails was introduced around 1870, but its inherent limitations precluded any real effect on the nuclear shape and pedestrian orientation of cities. Service was usually too infrequent, slow, and expensive to allow workers living in the central city to move out to newer suburbs. In any case, most workers could walk to work from their crowded inner-city apartments or lodgings. None could afford land on the periphery.

The electric trolley had a greater impact. Introduced into Latin American cities around 1900 and widespread by the 1920s, it was faster and more dependable than the horse tram and allowed middle-income city dwellers to live farther from the city center. In Lima, for instance, outlying beach towns became upper- and middle-income suburbs after the adoption of the trolley shortly after 1900. In Rio de Janeiro the development of the beach areas at Copacabana and Ipanema was tied to trolley access. The extensive trolley network of Buenos Aires, begun in 1898, opened large areas of the adjacent Pampas to settlement. Similarly, Mexico City and Caracas saw suburban development with the trolley, and even smaller cities, such as Quito, were markedly affected by the trolley.

Suburban rail service became significant only in a few of the largest cities. The bus was widely adopted in the Latin American city in the middle 1920s. Its widespread use, especially in the past four decades, has helped in the large-scale development of residential areas that formerly were too far from the city core or too distant from trolley lines to be settled. Increasing use of the automobile, still not widely used because of its high cost for most urban dwellers, also favors suburban development.

All these transportation modes usually concentrated on the central business district (CBD), making it the point of maximum accessibility for the entire urban area, just as it was in the colonial period. Today, the CBD of the typical Latin American city is strong and growing, although a number of outlying shopping streets or districts siphon off part of the trade. In some of the large cities the partial *replication* of the CBD in terms of shops supplying elegant goods is taking place in prosperous neighborhoods but does not challenge most central-city stores.

As industry developed in the city, certain districts came to have a strong industrial character. Noxious industries and those requiring much space, such as stockyards, were located on the edge of the city, even in the colonial period, and had to be relocated as the city expanded. Recently, planned industrial areas have been developed on the periphery of some cities. Ports evolved as industrial zones as imported raw materials and coal for power generation were often essential to large-scale production. Rio de Janeiro, Lima–Callao, and Buenos Aires, for example, all have industrial districts near navigable water. Except for large-scale, noxious, or port industries, however, there is still little segregation of small-scale industry in the city. It is common today to find numerous manufacturing operations in middle- and upper-income residential areas as well as in working class districts. City planning and restrictive measures such as zoning are recent and have had limited effectiveness in controlling land use. As elsewhere, including the United States, private interests hold the upper hand in influencing the direction and form of urban growth.

► A Model of the Modern City

Just as the colonial city exhibited common characteristics, so the contemporary large Latin American city can be summarized by a generalized model of city structure. One such model (Griffin and Ford, 1980), indicating the major structural

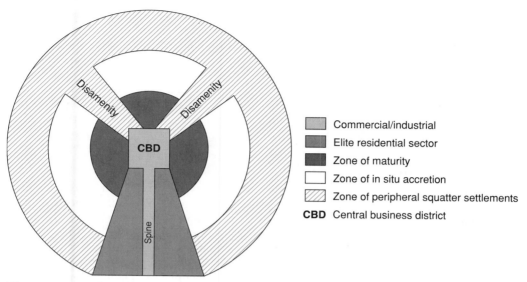

Figure 6.6 Model of city structure.

elements of the large city, is shown in Figure 6.6. It is primarily based on Bogotá, Colombia, and is characterized by concentric rings and racial sectors.

1. The central business district (CBD) remains the economic and administrative core of the city, with expansion of those activities outside the core being rare before the 1930s. Such expansion is now commonplace as downtown activities are "replicated" elsewhere in the metro area. Residual and crowded slum housing is not uncommon.

2. One major outlying focus is the affluent commercial spine extending out from the center city and surrounded by an elite residential sector. An extension of the CBD, the "spine/sector" contains some office buildings, fine stores, the professionally constructed upper-class and upper-middle-class housing stock and the most important urban amenities, including boulevards, major parks, museums, and many of the best theaters. It contains only a small percentage of the metro population, usually less than 5 percent.

3. Another attractive residential district is the "zone of maturity," an area of "better residences" that have filtered down to middle-income residents as the elite move into the newer spine/sector. The zone is fully serviced with urban amenities, but the sector as a whole shows signs of both deterioration and improvements and its population is stabilized.

4. The zone of *in situ* accretion is characterized by a wide variety of housing, with deteriorated dwellings or hovels adjacent to good housing units. Public services are slowly being added to the area. As such, it is transitional between the older, better residential areas of the center city and the newer, poor housing at the fringe.

5. The zone of peripheral squatter settlements is the worst section of the city in terms of housing and public service provision, although isolated areas elsewhere can match its characteristics, given the ubiquitous character of the housing problem. Equally, older villages that have been incorporated into the city by urban sprawl have nodes of "better housing" and commercial activities. Not infrequently, the residents of the squatter settlements first lived in a center-city slum upon migrating to the city and have now moved to the periphery where they begin the process of "self-help" housing. The houses are small and fragile with few public services. The "disamenity" sectors are marginal environments—such as river beds and railroad corridors—that are nevertheless zones of high density, poverty, and persistent slums. The quality of life is as marginal as the location, and long journeys to work are common.

Although this model is a reasonable approximation of the Latin American city, it does not differentiate commercial, industrial, and residential land uses within the sectors nor does it suggest the important role of small-scale neighborhood retailing and services.

► A Model of the Border City

The model of the border city (Figure 6.7) is based on eighteen cities along the United States–Mexico border. These cities are nonetheless morphologically similar to urban places elsewhere in Mexico and throughout Latin America. It is primarily the border itself, tourist districts, and zones of *maquiladora* factories that differentiate the form of border cities and landscapes from the other cities in Latin America. This dynamic model tells us much about Latin American urban form and function.

The model depicts two stages of urban development. The "newer" areas are particularly those that developed after about 1950; the "older" areas are more traditional in form, including the central business district, with a well-developed foreign tourist district. As is characteristic of Latin America, the quality of housing declines toward the periphery, although there are major areas of inner-city slums. New industries, including the *maquiladoras* in this instance, tend to be at the edge and near the border. Newer elite and middle-income residential areas are increasingly toward the suburban periphery.

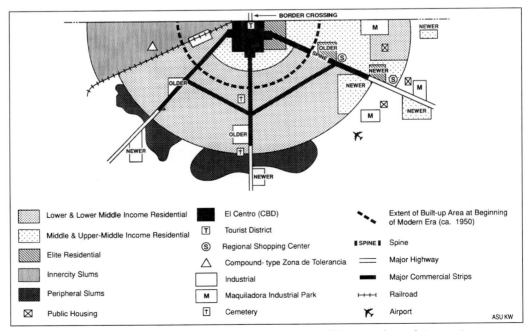

Figure 6.7 Model of a Mexican border city. From Daniel D. Arreola and James R. Curtis, *The Mexican Border Cities: Landscape Anatomy and Place Personality* (1993) with the permission of the authors and The University of Arizona Press.

PROSPECTS FOR THE LATIN AMERICAN CITY

The problems of the Latin American city will not be significantly ameliorated in the future. A careful reading of past processes and future trends indicates that the urban landscape—an expression of the broader economic and social framework of a country—will continue to be severely challenged.

There is reason to forecast increasing *urbanization*. The effect of urbanization will be to place pressures on urban housing and to overtax public utilities such as water, sewers, and streets. Underemployment and unemployment will remain high; street vendors hawking shoestrings, lottery tickets, or shoe shines may strike the visitor as picturesque, but they are in reality symbols of *underemployment*, where a worker is engaged in activities well below his or her potential, and the only alternative is unemployment or illegal activities.

As the cities get larger, traffic congestion and air pollution in the narrow streets of the colonial central business district will worsen as automobile ownership rises and bus systems expand their fleets. No alternative to the bus exists in most cities; while this results in an admirable network of routes, travel times slowly creep upward as congestion increases. The high cost of land in the densely built-up central portion of the city reduces the possibility of buying land to create an adequate network of boulevards and freeways.

The housing crisis promises to get worse. The Romans were said to have vented their discontent with crowded tenements by turning thumbs down on the gladiators. It is likely that a share of the crime in Latin America is traceable to the frustrations of modern urban life as well as the nationwide political and social inequities that are so apparent in cities where all socioeconomic spectrums can be easily observed and compared. But most Latin Americans compensate in more passive forms. Neighborhood bars, plazas, and streets are important social outlets for both tenement and shanty-town dwellers, and professional soccer replaces the gladiator. Bullrings are found in a few countries of Latin America, notably Mexico, Venezuela, Colombia, and Peru.

Many city residents who are not apartment dwellers have a private patio or backyard as a compensation for public parks that tend to be inadequate in number, size, facilities, and maintenance. In this aspect of urban life, like so many others, there is little prospect for improvement because of tight city budgets. Indeed, entire new suburban *barrios* are virtually without basic services such as water and sewers, and the streets are commonly unpaved, unpatrolled, and uncleaned by the city.

To this must be added the prospect of long-lasting two-digit or three-digit inflation that eats up low salaries, reduces the incentive to save, increases speculative investments in apartments and land by higher-income groups, and raises the price of raw land on the periphery, putting it out of the reach of most city dwellers. Fragile economies and corrupt governments often add to the instability and precariousness of urban life in much of Latin America.

A number of Latin American cities have instituted zoning and practiced urban planning since the 1930s. The agencies are inefficient and are unable to *implement* effective long-range plans. Additionally, planning is often devoted to aesthetics more than to urban problems, and usually in the hands of architects who lack a background in urban studies.

While the continued rapid population growth and physical expansion of the major cities offer little hope for improved conditions, guarded optimism can be felt in anticipating the growth of middle-sized cities and the changing hierarchy of cities within regions. New areas, such as Patagonia in southern Argentina and the Amazon Basin, will hopefully be a positive element in the agricultural and industrial growth of the countries. In the process, they could become areas in which to decant some of the population from overcrowded cities and poorer regions. There are few such large virgin areas left, however, and the potential of Amazonia and other similar, if smaller, regions should not be overrated. Particularly in Amazonia, environmental and agricultural disasters loom on the horizon.

As a consequence of population growth and increasing urbanization of the population, the growth of towns can be anticipated at all levels of the existing and urban hierarchy. Small towns in all but the most marginal areas will become small cities, providing new services and enhancing life-styles in regions removed from the larger cities. The growth of the primate city is probably inevitable and sometimes desirable. The development of many planned new towns is unlikely, since the capital and economic infrastructure for the development of new towns is unavailable to most countries. In 1987, the Argentine national government announced plans for a new capital city to replace Buenos Aires, but a subsequent change in government and financial realities preclude the conversion of the small littoral town of Viedma into the national capital. The old colonial interior of Argentina would arguably be a better location for the new capital, helping to restore those districts to an economic health they earlier enjoyed and drawing to them some of the migrants who now go to Buenos Aires. Brasília is probably not the appropriate model for Argentina, and even Brasília has not measured up to expectations. Sterile in plan and architecture, it is surrounded by a sea of squatter settlements and is far from its hoped potential.

Cities have been a reality in Latin America since before the *conquistadors* arrived. Another reality, however, has been unbalanced resource distribution, debilitating climate in some areas, recurrent natural disasters, a strong dependence on foreign markets and capital, increasingly unfavorable terms of trade, and the curse of the *kakistocracy*—government by the petty and the corrupt. There is little reason to anticipate any change in these realities. Population growth will continue to exacerbate the urban crisis.

FURTHER READINGS

Arreola, D. D., and J. R. Curtis. *The Mexican Border Cities: Landscape Anatomy and Place Personality*. Tucson: University of Arizona Press, 1993.

Augustin, B., and M. Kohout. "The Border Industries Program: A Geography of *Maquiladoras.*" In *A Pathway in Geography Resource Publication: A Geographic Glimpse of Central Texas and the Borderlands: Images and Encounters*, No. 12. Indiana, Pa. National Council for Geographic Education, 1995, pp. 75–83.

Bromley, R. D. F. and G. A. Jones. "Identifying the Inner City in Latin America." *Geographical Journal* 162 (1996): 179–190.

Gilbert, A. *The Latin American City*. London: Latin American Bureau, 1994.

Godfrey, B. J. "Modernizing the Brazilian City." *Geographical Review* 81 (1991): 18–34.

CHAPTER 7

Mining, Manufacturing,
and Services

ALAN GILBERT

INTRODUCTION

When the Spanish and Portuguese conquered Latin America, their aims were clear.
They wished to exploit the resources of the region, particularly precious metals such
as gold and silver. Although they never discovered the mythical El Dorado (a country
full of gold), they did occupy areas with rich mineral deposits. Mining was important
to many pre-Columbian peoples and has been a significant element in the regional
economy ever since. In several countries, mineral products are still the largest single
generator of export earnings: oil in Trinidad and Venezuela, and copper in Chile.
Elsewhere, minerals contribute substantially to export earnings: oil in Colombia,
Ecuador, and Mexico; bauxite in Guyana, Jamaica, and Suriname; copper in Mexico
and Peru, natural gas in Bolivia, nickel in Cuba and Colombia, iron ore in Brazil and
Venezuela, and lead and zinc in Bolivia and Peru.

Gradually, however, manufacturing developed and by the 1940s generated a
higher proportion of gross domestic product (GDP) in most countries than mining.
Industrial development was encouraged throughout the region by deliberate government action. Planning was integral to the strategy known as import-substituting
industrialization (ISI)—the replacement of manufactured imports with national production. Between 1950 and 1975, ISI was the heart of development policy throughout the region. Later, when this strategy of development failed to produce the
affluence the region had hoped for, a new strategy was adopted. This was the phase
known as export-oriented industrialization (EOI)—an attempt to produce manufactured goods that could be sold abroad. EOI is the development strategy operating currently in most Latin American and Caribbean countries. It fits into the new
world economic order, which recommends competition and condemns any protection of domestic industry behind high tariff walls. The ability of many countries in
the region to increase manufactured exports in recent years has been impressive.

If manufacturing and mining have been of immense importance to Latin
America's development, neither has ever been as significant a generator of employment as the service sector, which now contributes more than half of GDP in most
countries of the region (Table 7.1). In urban areas, services are the major source of

TABLE 7.1   **Structure of Production for Selected Latin American and Caribbean Countries, 1993 (% of gross national product)**

Country	Agriculture	Industry	Services
Argentina	6	31	63
Brazil	11	37	52
Colombia	16	35	50
Guatemala	25	19	55
Jamaica	8	41	51
Mexico	8	28	63
Paraguay	26	21	53
Peru	11	43	46
Puerto Rico	1	42	57
Uruguay	9	27	64
Venezuela	5	42	53
Total*	9	34	57

* For 14 countries: Argentina, Brazil, Colombia, Costa Rica, Ecuador, Jamaica, Mexico, Panama, Paraguay, Peru, Puerto Rico, Trinidad, Uruguay, and Venezuela; author's calculations.
Source: World Bank, *World Development Report*, 1995, Table 3.

both income and work. The debt crisis of the 1980s increased the contribution of services even further because it badly hit the manufacturing sector in most countries.

This chapter will discuss mining, manufacturing, and services in turn. The section on mining will illustrate the contribution that the sector can make to development by examining the contrasting experiences of Venezuela and Bolivia. The section on manufacturing will look at ISI and EOI, and provide a brief description of two particularly important industries, automobiles and steel. It will then discuss the impact that changing development strategies have had on industrial location. Finally, the section on services will consider the different functions that the sector plays in the national economy and will look at the growth of tourism.

MINING

The civilizations of precolonial America made extensive use of silver and gold. After the conquest, the silver mines of Mexico and Peru helped to sustain the weak Spanish economy and the gold of Minas Gerais that of Portugal. Indirectly, Latin American mining aided the development of European industry.

Mining has had a controversial history in Latin America, and the reasons are not difficult to establish. First, mining can produce wealth quickly and transform a national economy in a few years. In Peru, for example, the guano boom in the 1850s and 1860s and the development of nitrates before 1879 totally altered Lima, the size of the government bureaucracy, and the relationship between national and provincial economies. Similarly, in Chile the growth of the nitrate industry after the War of the Pacific (1879–1884), and especially the second copper boom after 1910, brought relative prosperity and allowed the government's budget to expand rapidly (Blakemore and Smith, 1983). Unfortunately, while mining revenues can soar, they can decline equally quickly. The discovery of alternative supplies, a change in manufacturing technology, or the development of a substitute can devastate a national economy.

Second, until recently, the bulk of mineral exploitation was in the hands of foreign companies. Large, multinational companies such as Kennecott and Anaconda were involved in the Chilean copper industry and Shell and Standard Oil in the exploitation of petroleum in Mexico and Venezuela (Odell, 1974). Such foreign involvement in the main source of export earnings was tailored for controversy, especially since many foreign companies established highly profitable deals with unsophisticated and sometimes corrupt national governments. The history of petroleum deals in Venezuela before 1958 is scandalous and the experiences of Bolivia up to 1952 little better (Klein, 1965; Lieuwen, 1965; Osbourne, 1964; Randall, 1987). Gradually, control over foreign companies tightened and in many cases the state took over mining activities. The Mexicans nationalized their petroleum industry in 1938, and after World War II, the pace of nationalization quickened. The Bolivian tin industry was nationalized in 1952, the Peruvian petroleum industry in 1968, Chilean copper in 1973, and Venezuelan oil in 1976. During the 1980s and 1990s however, some of these companies were sold back to the private sector: the Argentine petroleum company, the Mexican copper company, and some major Bolivian tin and silver mines.

Third, the nature of mining has complicated the relationship between companies (whether foreign or state-owned) and society. Mining activities tend to be highly organized, large-scale, and capital-intensive, and in Latin America they are superimposed on poor economies. In the past they were run by foreign companies almost as separate states. Even today they are distinctive insofar as they create huge revenues but employ few workers and generate few economic linkages. This "isolation" from the rest of the national economy has been magnified by the physical location of many mines. Most of the copper of Chile and Peru and the petroleum of Venezuela, Ecuador, Colombia, Mexico, and Peru come from areas far away from the main cities. In addition, miners are better paid and politically more active than other groups. Mining regions often form enclaves largely unconnected to the rest of the national economy.

Despite the inherent problems, few Latin American governments have been willing to forgo the opportunity to develop mineral resources. Since the mid-1960s, large deposits of iron ore have been exploited in Peru and Brazil; copper in Peru, Chile, and Mexico; nickel in Colombia; and petroleum and coal in Bolivia, Colombia, Ecuador, and Mexico. Throughout the continent new reserves of minerals are being discovered. Even steam coal, previously scarce in Latin America, is now being exported in large quantities from the Guajira Peninsula of Colombia.

Cerro Bolívar iron ore mine south of Ciudad Guayana.

Mining activity will continue to influence the economic geography of the region albeit in widely varying ways. Mineral development has sometimes had a beneficial influence on a country's development, sometimes a terrible impact. This paradox is clearly demonstrated by the experiences of Venezuela with oil and Bolivia with tin.

► *Venezuelan Oil*

From the 1920s, when oil began to be exploited in large quantities, until the 1990s, oil consistently contributed more than 90 percent of Venezuela's foreign exchange. It also generated the bulk of government revenues and was almost singly responsible for Venezuela's high per capita income (Table 7.2).

Although Venezuela has been favored by petroleum wealth, reading its modern history creates the impression that the oil industry has been a major liability (Ewell, 1984; Lieuwen, 1965). This paradox is explained by the controversial relationship that has existed between government and foreign companies since the 1920s and the failure of most Venezuelans to benefit from the oil revenues.

Venezuela became a major producer in 10 years. Overgenerous concessions encouraged the major oil companies to tap the huge reserves under Lake Maracaibo and turn the country into the world's second oil producer and leading exporter by 1928. Under the Gómez dictatorship (1910–1935), Venezuela became rich while most of its people remained poor. Petroleum changed the economy radically without adding to its productive potential. The country became dependent on foreign supplies of manufactured goods and even food.

Changes were made after Juan Vicente Gómez's death in 1935. Oil laws less favorable to the foreign companies were introduced, and plans were devised to diversify the economy. Oil revenues were to be used to increase agricultural production, accelerate industrial growth, and build social and economic infrastructure. The new oil law, signed in 1943, set the ground rules for the next 30 years. After the fall of the Medina dictatorship in 1945, demands for a fair deal from the companies and greater investment in other economic sectors underlay government policy. This period of enlightenment faded with the rise of the Pérez Jiménez dictatorship. The favorable treatment of Venezuelan labor was forsaken, new concessions were granted to the major oil companies, and foreign earnings poured in. Admittedly, most of this money no longer went to the president and his associates, and large investments were made in developing the iron industry and in building roads. But the most notable "development" goal was to turn Caracas into a national show-

TABLE 7.2 Venezuelan Oil Production, 1920–1994 (average annual production in million barrels)

Year	Production
1920–1929	37.9
1930–1939	150.4
1940–1949	310.4
1950–1959	782.4
1960–1969	1,211.5
1970–1979	1,030.0
1980–1989	648.9
1990–1994	846.8

Source: Wilkie, Contreras, and Komisaruk, (1994).

Oil refinery, Curaçao, Netherlands Antilles.

piece. Half of the government's revenues were spent in the city; the government bureaucracy increased, skyscrapers sprouted, urban motorways were built, and the city's population grew from 695,000 in 1950 to 1,248,000 in 1961.

The return to democratic government in 1958 changed the role of oil in the national economy. Deals between the foreign companies and the government favored the government more, and in 1976 the oil industry was nationalized. More government funds were invested in transport, industry, and agriculture. However, the nature of relations between the foreign oil companies, the dependent middle class, the government bureaucracy, and the poor remained unaltered.

The very nature of the oil industry caused problems. Oil has always been an enclave in Venezuela, and although direct control by foreign companies has ended, it is likely to remain so. First, oil extraction employs few people. In 1990, oil generated 80 percent of national exports and 23 percent of the gross national product but employed fewer than 1 percent of the labor force. Second, petroleum creates little related manufacturing employment. Venezuelan industry uses little crude oil or petroleum derivatives except as fuel, and most petroleum industry equipment is imported. Third, the main oil-producing regions—Zulia to the west and Anzoátegui and Monagas to the east (Figure 7.1)—are isolated from the dynamic economic

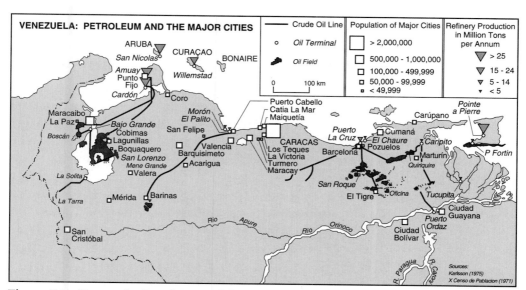

Figure 7.1 Venezuela: Petroleum and the major cities.

TABLE 7.3   Population Growth of Caracas and Maracaibo, 1926–1990 (1000s)

	1926	1936	1941	1950	1961	1971	1981	1990
Maracaibo	75	110	121	270	461	682	962	1,358
Annual growth (%)		3.9	2.0	7.7	5.0	4.0	3.5	4.0
Caracas	182	273	369	684	1,347	2,175	2,642	2,990
Annual growth (%)		4.2	6.2	7.6	6.6	4.5	2.0	1.4
Venezuela	3,027	3,364	3,850	5,035	7,524	10,722	14,517	19,405
Annual growth (%)		1.0	2.7	3.0	3.7	3.6	3.1	3.3

Source: Villa and Rodríguez (1996) and national census figures.

186

regions of the country. During the oil boom, Maracaibo grew from a small town to the country's second city (Table 7.3), but the western oil capital has been gradually losing importance relative to the central area. Most industrial and commercial activities are concentrated along the Caracas–Valencia corridor and are continuing to locate there.

The geography of Venezuela's development is paradoxical. Apart from the initial expansion of the petroleum economy in Maracaibo and later in the east, the principal regional beneficiaries from oil have been the cities of Caracas, Maracay, and Valencia. All three cities benefited in the Gómez period as funds were directed toward them. Between 1948 and 1957 Pérez Jiménez boosted government spending in the capital and initiated the policy to make Caracas an urban showpiece. Distant from the oil fields, Caracas has gained most from the oil revenues; its bureaucracy, industrial sector, and middle-class population have flourished.

For years, Venezuela was the region's most affluent country. Oil made it rich but also created the conditions that have brought about its recent fall from economic grace. During the 1970s, the government accumulated a huge foreign debt, which led to severe recession in the 1980s. Because the loans were used poorly, Venezuela gained little from accruing the debt. Between 1981 and 1989, per capita income fell by 25 percent. In the process the country fell behind Trinidad, Brazil, and Uruguay in the Latin American prosperity league. Many blame oil for the country's current predicament. More accurately, it is Venezuelan government and society that should be blamed for their misuse of oil revenues. Without oil, Venezuela would probably have remained the backwater that it was in the 1920s.

► Bolivian Tin

While petroleum created wealth in Venezuela, tin never had the same success in stimulating the Bolivian economy. Today, the Bolivian tin industry has almost disappeared, an outcome of increasing world competition, dwindling reserves, and the falling price of tin. Bolivia remains one of Latin America's poorest countries.

Exploitation of tin began at the end of the nineteenth century and exports rose to a peak of 47,000 tons in 1929. Mining was controlled by the Patiño, Hochschild, and Aramayo families. The families created international enterprises, smelting tin ores in Britain and the United States. Their influence over the Bolivian government reached extreme levels with the Patiño enterprise at one stage providing half of the government's revenues and four fifths of the country's export revenues (Klein, 1965). The political influence of the companies was so resented that they became prime targets in the 1952 revolution. The incoming government nationalized the tin mines and created the state mining corporation, COMIBOL.

At nationalization, Bolivia's best ores had been depleted and the tin that remained was expensive to extract and difficult to refine. Production was complicated by the fact that COMIBOL inherited a difficult industrial relations problem. The tin workers had played an important role in the revolution and were politically united behind a powerful union. They were able to demand better conditions and became a relatively privileged group in Bolivian society. Production costs rose as the labor force grew. The industry was in a bad way when the central government was required to subsidize tin by up to 30 percent.

Bolivian tin was also poorly located to compete internationally, the concentrate moving by rail from the Altiplano to the foreign ports of Antofagasta, Arica, and Matarani (Figure 7.2).

Despite these problems, the tin industry remained vital to the Bolivian economy. In 1965, it generated four fifths of the country's exports, and as late as 1978 it was producing almost three fifths of the foreign earnings. When the tin price collapsed in October 1985, output fell dramatically. From almost 27,000 tons in 1983,

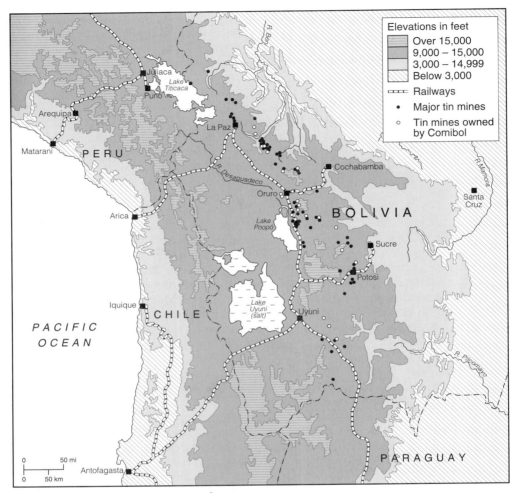

Figure 7.2 Bolivian tin: Mines and export routes.

tin production plummeted to less than 6,000 tons four years later; exports worth $378 million in 1980 fell to less than $50 million in 1987 (Dunkerley, 1990). In two years, COMIBOL's labor force declined from 28,000 to 5,000.

The experiences of Bolivia and Venezuela show that mining offers a problematic route to sustained economic and social progress. The revenues from mining are concentrated in few hands, even when the industry is nationalized. The poor gain some jobs, but most benefits accrue to privileged groups in society. Few if any Latin American countries have avoided this outcome.

Despite the problems, the economic advantages of mineral exploitation continue to tempt Latin American governments to look for new mineral deposits. During the 1980s governments again began to welcome foreign companies, granting them exploration rights and sometimes selling them previously nationalized companies. As a result, the level of mining activity has been rising during the 1990s. However problematic the relations with foreign companies, however limited the new employment, however great the concentration of the benefits of mineral development, Latin American governments continue to encourage the mining sector.

MANUFACTURING

Industrialization began in Latin America as a spontaneous reaction to a boom in exports of primary products. As the expansion of meat, wool, and cereal exports brought prosperity to Argentina and Uruguay and coffee production stimulated the economy of southern Brazil, the development of manufacturing was a natural outcome. Local profits were invested in industries such as food processing and textiles. By 1914, Argentina, Brazil, and Uruguay had developed manufacturing sectors. Progress was slower during the subsequent years but accelerated again during World War II, when the markets of urban Latin America were isolated from the developed countries and their manufactured exports. At that time more countries began to industrialize, and by the 1950s all had adopted the new development strategy propounded by Raúl Prebisch in 1949.

► *Import-Substituting Industrialization*

Prebisch diagnosed the economic problems of Latin America in terms of dependence on primary export products and deteriorating terms of trade. Whereas developed countries sold a range of manufactured goods, Latin America continued to export only mineral and agricultural products. Over time, the prices of manufactured goods had risen relative to those of primary products. The result was that Latin American countries received less in the way of manufactured imports from the developed countries for each ton of primary products exported. Prebisch suggested that Latin America should remedy the problem by building its

The Guri Dam on the Caroni River, a tributary of the Orinoco River, provides hydro-electric power to Ciudad Guayana and other parts of Venezuela.

TABLE 7.4   Structure of Manufacturing for Selected Countries, 1992 (% of gross domestic product)

Country	Foods	Textiles	Machinery	Chemicals	Other
Argentina	21	10	13	12	44
Brazil	15	11	22	14	38
Colombia	29	14	10	16	31
Guatemala	42	9	3	16	28
Jamaica	44	6	9	7	34
Mexico	24	5	25	17	30
Paraguay	55	16	0	12	17
Venezuela	22	5	9	12	52

Note: Foods also include beverages and tobacco; textiles include clothing; machinery includes transport equipment.
Source: World Bank, *World Development Report*, 1995, Table 6.

own manufacturing base. This could be achieved if governments protected domestic markets from imports, replacing them with locally produced substitutes. The 1950s and 1960s were characterized by the widespread adoption of import-substitution policies. Quotas and import restrictions were established, and several governments set up industrial institutes to encourage manufacturing development.

In the largest countries, the new strategy led to the development of a diversified industrial sector; by the 1960s Mexico, Brazil, and Argentina contained most kinds of manufacturing activity. In the medium-sized countries—Colombia, Venezuela, Peru, and Chile—intermediate-goods industries and even plants producing steel and automobiles were established. The new approach was less effective in the rest of the continent. In the smallest and poorest nations the combination of small populations, low per capita incomes, and acutely concentrated distributions of income meant that markets were too limited for most kinds of industry to develop. Industrial development seldom went beyond the production of consumer goods (Table 7.4). The strategy of import substitution was soon exhausted in all except the six largest nations. A further difficulty affected every country: While production often increased rapidly, employment did not. The reason was simple: Modern capital-intensive technology raised productivity and used relatively little labor. For example, between 1958 and 1967 industrial value added in Colombia doubled while employment increased by 25 percent. This was a particular problem in those countries where the population was expanding at an annual rate of 3 percent. Admittedly, industrial employment usually grew more quickly than national or even urban populations but, because of the small base on which expansion was occurring, too few jobs were created. Unemployment and underemployment in the cities remained a major problem.

Another dilemma was that instead of improving the balance of payments, import-substituting industrialization often led to a deterioration. Local manufacturing created a demand for capital goods with which to make the substitutes. The structure of imports changed, but the total import bill rarely declined. Many of the new products used foreign patents and technology on which royalties had to be paid in hard currency. Many of the new industries were subsidiaries of foreign or multinational companies. The companies required a return on their investments and wanted to repatriate profits. A further drain on foreign exchange was the repayment of loans; industrialists borrowed money to establish plants and governments borrowed to finance economic infrastructure.

The result of import substitution was an industrial sector that solved neither unemployment nor the trade deficit of most Latin American countries. Worse, many companies were inefficient, especially where the market was small or where competing plants produced similar products. When national governments encouraged foreign companies to reinvest their profits in the local economy, further difficulties arose. With ready access to capital and technology, foreign subsidiaries tended to be more competitive and bought out local companies. Gradually, their share of the industrial sector grew, and governments feared that control of the national economy was passing into foreign hands.

By the 1960s most of the larger countries realized that there was little future in import-substituting industrialization. They now had high-cost manufacturing sectors with little potential for growth. An alternative path was required.

► *Export-Oriented Industrialization*

For many years it was thought that Latin American countries could not export manufactures to the developed world. Their industry simply could not compete with German, Japanese, or U.S. producers. However, the need to increase foreign earnings remained, and increasingly Latin Americans turned their attention to the idea of exporting more manufactures to one another.

From the late 1950s on, national politicians had begun to encourage inter-American trade (Dell, 1966). In 1965 six Central American countries established a common market, which promised to create one decent-sized market out of their tiny economies. The Latin America Free Trade Area was established in 1960, and in 1969, the Andean Group was formed by Colombia, Ecuador, Bolivia, Peru, and Chile. Unfortunately, none of these schemes was really successful. National rivalries and different approaches to development made the difficult task of integrating a series of less-developed economies into one that was well nigh impossible.

V.W. Factory in São Bernardo Campo, Brazil.

However, the world was changing. Transnational corporations were modifying their trading and manufacturing strategies, dispersing their activities more widely than ever before. Faced by higher labor costs and tougher environmental controls at home, they found the case for producing in less-developed countries attractive. New forms of technology and improved communications allowed them to produce abroad. Use of the silicon chip reduced the weight of electronic goods and allowed companies to use cheap labor to produce radios and televisions before shipping them to the United States or Europe. Faster and cheaper transport increased the attractions of off-shore production. Transnational corporations operating in Latin America were not blind to this development. Companies such as Volkswagen envisaged the day when they would make cars in Brazil for sale in Germany.

With prompting from the World Bank and intellectual support from neoclassical economists, a number of Latin American governments began to modify their development policies. The military government that took power in Brazil in 1964 was one of the first to adopt a strategy of export-oriented industrialization. Transnational corporations reacted positively to Brazil's combination of export incentives, acceptance of foreign investment, lax attitude toward environmental pollution, and strict controls on union activity. They increased investment steadily and expanded production for export. The share of manufactures in total exports rose steadily, rising from 5 percent in 1960 to 40 percent in 1982. Motor vehicle manufacturers were especially successful with vehicle exports increasing from 14,000 units in 1972 to 213,000 in 1981 (Gwynne, 1990). At first, this strategy promised to transform the Brazilian economy, and between 1967 and 1974, the gross domestic product grew annually by 10 percent. The economic miracle was terminated, however, by the world oil-price rises of 1973. Dependent for most of its oil on foreign suppliers, the rise in prices caused balance-of-payment problems. An attempt to borrow its way out of recession ended disastrously. By 1979, the growth rate had slowed, and the country's foreign debt was approaching mammoth proportions; by 1982 debt service accounted for 85 percent of export earnings. Brazil had become the largest debtor in the Third World.

During the 1970s and early 1980s, most Latin American countries accumulated massive foreign debts as a consequence of oil price rises, internal mismanagement, the

TABLE 7.5 External Debt of Selected Latin American Countries, 1987

Country	Total Debt (U.S. $ billions)	Debt per Capita (U.S. $)	Interest Due as % of Export Earnings
Latin America	410.5	1,006	29.7
Argentina	56.8	1,818	51.0
Bolivia	3.9	578	43.9
Brazil	114.6	808	33.1
Chile	19.1	1,526	26.4
Colombia	15.9	537	20.7
Dominican Republic	3.8	566	14.7
Ecuador	10.5	1,051	32.8
Guatemala	2.8	331	13.6
Mexico	96.7	1,159	29.8
Peru	16.2	824	21.9
Venezuela	31.9	1,728	23.7

Source: UNECLA, *Anuario Estadístico para América Latina y el Caribe*, pp. 24–25, New York: United Nations, 1988.

TABLE 7.6   Latin America: Export Performance of Selected Countries, 1965–1993

Country	Exports as % of GDP			Manufactured Exports as % of Total Merchandise Exports		
	1965	1983	1993	1965	1982	1993
Argentina	8	13	6	6	24	32
Brazil	8	8	8	8	39	60
Chile	14	24	28	4	8	19
Colombia	11	10	17	7	25	40
Jamaica	33	40	60	31	60	65
Mexico	8	20	13	16	12	52
Peru	16	21	10	1	14	17
Venezuela	26	26	26	2	3	14

Source: World Bank, *World Development Reports*, 1985, 1992, and 1995.

rise in interest rates on existing loans, and the conservative economic policies introduced by developed countries in the early 1980s (Table 7.5). When Mexico declared in 1982 that it could not meet its debt obligations, the debt crisis had arrived. Few Latin American countries could continue as before. They were forced to reschedule their external debts, and many were obliged by the International Monetary Fund to alter their development strategies. Structural adjustment required cuts in government budgets and in consumer demand, opening up the economy to external competition, and giving greater encouragement to exporters. If Latin American countries would produce more and consume less, they could increase exports and pay back their loans.

Table 7.6 shows that the new strategy had an impressive effect, increasing the share of manufactures in many countries. Manufactures now account for over half of Brazil's and Mexico's exports. Even countries that had previously exported few manufactured products now export a great deal. In some countries, such as Chile and the Dominican Republic, exports as a proportion of gross domestic product have increased substantially.

Nowhere is the effect of export orientation shown more clearly than along the U.S.–Mexican border. Of course, Mexico has the advantage, at least with respect to foreign trade, of bordering on the world's largest consumer market. This potential advantage was first realized in 1933 when Mexico created free-trade areas in Tijuana and Ensenada and later extended this incentive to other cities along the border. The incentives were increased in 1965 when the Frontier Industrialization Program was initiated, permitting companies to import materials duty-free, providing that the final product was exported. In this way foreign firms could take advantage of low Mexican wages; in the middle 1960s minimum wages in the frontier area were between one fifth and one sixth the average wages in the United States. By 1966, 57 plants were operating in the border area employing more than 4,000 workers; 30 years later there were over 2,000 plants with more than 500,000 employees (Table 7.7).

With the debt crisis affecting Mexico badly and more and more transnational companies adopting "flexible production" methods, the attractions of the border have increased. Costs of production in Mexico remain low due to a spectacular decline in the value of the peso in relation to the dollar. In 1981, one dollar bought 24 Mexican pesos, in 1983 a dollar bought 119 pesos, and by March 1995 a dollar was worth the equivalent of 7,450 pesos. As a result, North American companies, together with a number from Japan and Europe, continued to establish plants in

TABLE 7.7 Mexico's *Maquiladoras*, 1966–1994

Year	Companies	Employees	Foreign Exchange Earnings (U.S. $ million)
1966	57	4,257	na
1975	454	67,214	454
1980	620	119,546	773
1985	760	211,964	1,450
1987	1,125	305,253	1,598
1991	1,925	467,352	4,134
1994	2,121	552,239	5,803

Sources: Wilkie, Contreras, and Komisaruk (1994, p. 518); Maldonado (1995, p. 494); Rodríguez and Alvarez (1995, p. 531–532)

Mexico. By 1994, there were 2,121 *maquiladoras*, some located well beyond the frontier cities, employing more than 550,000 people. The entry of Mexico into NAFTA should strengthen this trend especially after the huge devaluation of the peso in December 1994.

EOI has also been a significant factor in the development strategies of most Caribbean countries. The vast majority of the islands have established Export Processing Zones (EPZs). Most are located close to ports or international airports and offer foreign companies ready access to cheap labor. Companies are permitted to import materials free of tax, providing that the majority of the final assembled product is re-exported. Exemption from income taxes is available, sometimes for up to 15 years, as well as subsidized factories and infrastructure (Klak and Rulli, 1993).

These zones have helped manufactured exports from the Caribbean increase dramatically during the 1980s; between 1983 and 1988, exports to the United States increased two and a half times (Deere et al., 1990, p. 158). In the process, a new range of industrial products has been assembled in the Caribbean: from Haitian baseballs to Dominican soap; from Barbadian medical products to Trinidadian electronic aircraft equipment (Klak and Rulli, 1993). The main problem with EOI is that it has proved to be rather unstable. Foreign firms are highly mobile and are likely to leave one island the moment local conditions become unsuitable. If labor rates rise or political conditions "deteriorate," the plant will be moved. Islands are in competition with each other and with other parts of the Third World. As Klak and Rulli (1993, p. 144) put it, "Foreign firms have demonstrated in recent years how they can invest, move profits, and withdraw operations from particular sites with remarkable ease. Thus foreign investment for industrial exports has taken most Caribbean countries for a very uneven ride."

▶ The Automobile Industry

Major automobile industries have developed in several countries of the region, notably in Argentina, Brazil, and Mexico. At times, the combined annual production of these three countries has reached 2 million vehicles (Table 7.8). In the past, the big three U.S. companies dominated production in most parts of the region, but recently Japanese and European producers have become increasingly important.

Argentina and Brazil have the only integrated vehicle industries. Until recently, Brazil produced all the parts for its own vehicles; removal of protection, however, is rapidly cutting the country's level of self-sufficiency.

Brazil has long sold vehicles to other parts of the region and in 1994 exported almost 400,000 units. Argentina also exports cars to Brazil, Venezuela, and Colombia. However, Mexico is now the leading exporter. Since 1985, several major U.S.

companies have established state-of-the-art factories in northern Mexico. Ford's plant at Hermosillo is capable of producing 400,000 cars, all for export to the United States. The success of these plants is illustrated by recent sales figures: in 1993, 593,000 vehicles were sold in Mexico and 472,000 abroad, the vast bulk in Canada and the United States (Maldonado, 1995, p. 493). It was estimated that Mexico would export 800,000 units in 1995 (*The Economist,* April 15, 1995).

► Steel

Most of the larger countries have had steel plants since the 1940s. National governments established most of these plants, believing that steel production was an essential element in the creation of a viable manufacturing sector. SOMISA in Argentina, Volta Redonda in Brazil, Paz del Río in Colombia, SIDERPERU in Peru, SICARTSA and SIDERMEX in Mexico, and SIDOR in Venezuela were all established by state companies. Now there is a strong movement to privatize production.

At the beginning of the 1960s, most producers used the Siemens-Martin open hearth method. Today, most production comes either from electric ovens or oxygen converters; Bessemer converters and Siemens-Martin open hearth ovens have virtually disappeared.

Regional production was less than 5 million tons in 1960 but had trebled by 1976. After that, it rose dramatically as countries tried to expand capacity for export. Production reached 30 million tons in 1980 and passed 40 million tons in 1989 (Table 7.8).

► Industrial Location

Before the turn of the century and in the smaller countries even up to 1950, most industrial production was in the hands of artisan producers. Small companies operated in the food, drink, and clothing sectors, supplying local markets and using local materials. Manufacturing concentrations existed in the larger cities, especially in rich resource areas and port centers, but industrial production generally was more widespread than today (Gilbert, 1974).

TABLE 7.8 Latin American Automobile, Oil, and Steel Output, 1990s

Country	Vehicles (000s)[1]	Steel (million tons)[2]	Oil (000 bpd)[3]
Brazil	893	25.7	700
Mexico	989	10.2	3,158
Argentina	140	3.3	658
Venezuela	50	3.9	2,506
Colombia	50	0.7[1]	440
Chile	10	0.8[1]	—
Others	70	0.8[1]	476
Total	2,202	38.4[1]	7,938

[1]1991
[2]1994
[3]1996; bpd-barrels per day.

Sources: Latin American Weekly Report, November 19, 1992.
      Latin American Economic Bulletin, September 1995.
      Siderurgía latinoamericana, May 1991.

Hispanic, North American, third-world landscape in Morelia, Mexico. The street vending of foodstuffs, shoelaces, and lottery tickets is part of the urban scene in Latin American cities.

The rise of manufacturing industry in the late nineteenth century in Argentina and southern Brazil and in the early twentieth century in Mexico, Chile, Peru, and Colombia altered the pattern. Most of the new textile, tobacco, and cement factories were established in the largest cities. Industries processing exports were located in port cities. The trend toward spatial concentration had begun.

Industrial development, however, did not always favor the capital cities. By 1920, Monterrey, with its iron and coal reserves, had developed into Mexico's principal industrial city and remained so until the 1940s. In Colombia, entrepreneurs in Medellín established large, efficient textile plants before industrial development had really commenced in Bogotá. In Brazil, manufacturing developed more rapidly in São Paulo than in Rio de Janeiro because of the availability of capital generated by the booming coffee economy. Nevertheless, Monterrey, Medellín, and São Paulo were all major cities, and their development reflected the basic pattern whereby industrialization developed principally in major urban areas. After 1940, this tendency was to become even more marked.

Import substitution emphasized this trend. New companies producing foodstuffs, textiles, and clothing catered mainly to an urban market. Most consumers lived in the largest cities, and those cities lay at the center of national transport networks. Buenos Aires, for example, was the best point from which to supply the rest of Argentina, as the city was at the core of the rail and road systems. The consumer goods companies, which were characteristic of the early stage of import-substituting industrialization, located mainly in the largest cities.

Other factors also encouraged spatial concentration. Services and infrastructure were only available in the larger centers because few small cities had adequate electricity, water, or telephone systems. Supplies of skilled labor were another factor. Few professional or managerial people were prepared to live outside the main urban centers. Imports could also be obtained most readily in the larger cities because many were located near the coast.

As import substitution advanced, national industries began to supply more of the inputs required by other companies. As industrial linkages developed, there was an advantage for new companies to locate close to their suppliers or clients. Agglomeration economies favored the further expansion of the major cities. Of course, some kinds of industry were still tied to particular locations by their demand for natural resources—most large steel works still located close to supplies of iron ore. But increasingly, new industries were not of this kind and did not depend on natural resources. Supposedly, such industry was "footloose"; it could locate anywhere. However, one of the ironies of Latin American geography is that most footloose industries were established in the largest cities.

Industrial concentration was reinforced by government policies. Import-substitution was directed by government agencies. Systems of quotas and import licenses regulated which materials would be imported and which companies would be protected from manufactured imports. Governments made finance and capital available for industrial development and established agencies to help small companies and to encourage export production. As the industrial sector became larger and trade unions more powerful, governments became involved in the control of prices, wages, monopolies, transport rates, social security, and related matters. When superimposed on traditional Latin American political and business practice, the increasing links between business and government constituted a strong centralizing force. Managers found themselves in negotiations with one branch of government or another several times a week. The traditional failure in Latin America to delegate authority meant that minor matters were determined in meetings at the executive level. Many managing directors whose companies might have located in provincial cities decided that their meetings with top government representatives were of such importance that the plants and/or the head offices had to be established in the capital city. Only where provincial cities were sufficiently large that ministers and high government officials visited frequently was capital city bias avoided.

► *Industrial Deconcentration*

Most other economic activities exhibited a similar trend toward geographical concentration. As economic development proceeded, commercial, business, and government activities all centered in the larger cities. The growth in employment opportunities encouraged migration, which, together with high rates of natural increase in these cities, led to dramatic rises in large city populations. Between 1950 and 1970 Mexico City grew from 3,419,000 to 8,605,000 inhabitants and São Paulo from 2,336,000 to 7,838,000. Growth on this scale began to worry governments. Everywhere urban expansion was associated with larger numbers of people living in slums and squatter settlements. Local governments became alarmed by their inability to service the growing number of people; the youth of the urban population caused schooling to become a major problem; traffic congestion worsened; and air pollution began to be noticed. In certain cities, local factors emphasized particular problems and focused attention on the question of urban diseconomies. In Mexico City air pollution and the provision of water were more serious problems because of the topography and climate. The location of Caracas in a narrow valley increased the value of land, magnified the problems of physical growth, and encouraged the growth of squatter settlements on hillsides subject to landslides.

National governments were aware that the concentration of economic activity in the major cities had failed to bring benefits to the poorest parts of countries. The stagnation and poverty of areas such as the Peruvian *sierra* or the Brazilian Northeast contrasted with the expansion of Lima, São Paulo, and Rio de Janeiro. Sometimes the contrast was underlined by political controversy in the poor regions. In northeastern Brazil recurrent droughts and consequent starvation demanded a political response (Hirschman, 1965). One possible solution for poor regions (and a means for slowing migration to the larger cities) was a policy of economic decentralization.

Parallel to these developments came a change in thinking about planning. During the 1950s and 1960s, international organizations such as the United Nations Development Program (UNDP), the United Nations Economic Commission for Latin America (UNECLA), the Organization of American States (OAS), and the World Bank were emphasizing the need for planning, initially at a national scale but later at a regional level. The establishment of national planning offices in the late 1950s and the 1960s was followed by a realization of the need for regional development programs. The programs differed from government to government, but by the late 1960s, most countries had adopted some form of regional development or economic decentralization policy. A common element was the dispersal of industrial activity away from large, primate cities.

Latin American governments established a variety of schemes to disperse industry. Sometimes they acted directly and set up state firms in poorer regions. This approach often embraced related national development motives such as the wish to exploit natural resources, as in the case of Ciudad Guayana—a steel and power complex in Venezuela (Figure 7.3). Direct government action usually guaranteed that industrial development would take place. Schemes that relied on the private sector were less predictable. But direct action was an exceptional response in most countries; more common were efforts to provide provincial cities with basic infrastructure in the hope that private industry would locate in the area. Industrial estates were established in practically every Latin American country.

Companies locating in poorer areas were often given tax incentives. The governments of Argentina, Mexico, Puerto Rico, Peru, and Venezuela attracted companies to poorer regions with the promise that they would not have to pay income tax for a certain number of years. An interesting variation was the Brazilian 34/18 mechanism, which offered existing companies tax relief if they invested the unpaid taxes along with new funds in poorer areas of the country. At first, the Brazilian scheme embraced only the Northeast but was extended to include the Amazon region. Chile, Colombia, and Venezuela experimented with industrial licensing and banned specific industries from the largest cities.

Many governments were successful in building plants in backward regions or in persuading private companies to move there. The Brazilian government succeeded in stimulating industrial development in the Northeast and Amazon regions. Recife, Fortaleza, Salvador, Belém, and Manaus were all major beneficiaries of the regional program. Large petrochemical plants were built near Salvador, a steel mill in Manaus, and major jeep, paint, and refrigerator manufacturers in Recife. The Mexican government persuaded many North American companies to take advantage of cheap labor and free-trade facilities on the Mexican side of the border. More surprising was the ability of the Argentine government to attract electronics companies to Tierra del Fuego. Clearly, where there was a strong political or economic motive, governments were able to disperse industry.

Whether industrial deconcentration made economic sense and whether it helped the poorer regions are different questions. Undoubtedly, problems arose from industrial dispersal. It was sometimes very inefficient. In 1958, the Chilean government decreed that all automobile plants be established in the northern port of Arica.

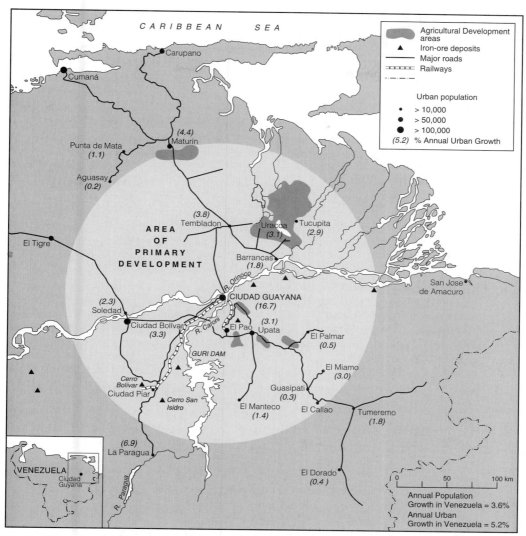

Figure 7.3 Ciudad Guayana and its regional impact, 1970s.

Despite protests, a number of producers established factories, importing components from abroad and shipping the finished vehicles to the main markets in the south. Later, when increasing numbers of parts were produced in Chile, the problems became apparent. Since most of the engineering companies that made car parts were located in or near Santiago, a double transport haul was required. Components were shipped 1,300 miles from Santiago to Arica, assembled, and hauled back to the central region. Such inefficiency was expensive, and when the decree was lifted, four car manufacturers immediately moved their plants to the central region.

Industrial dispersal was also problematic in that it generated few jobs in the poorest regions. In northeastern Brazil, a great deal of investment created little employment. In the Amazon, the declaration of Manaus as a free-trade zone in 1967 attracted 30 new companies, including an oil refinery and a steel mill, but generated few jobs (Kleinpenning, 1975). Nor did the new industries create much related employment, mainly because their linkages were with foreign companies or to plants in the major cities. A study of Arequipa, for example, showed that one third of industrial inputs were imported and a further quarter came from outside southern Peru (Waller, 1974). The local multiplier effects were limited. As a result, such programs did not reduce poverty. This was demonstrated by the experiences of Recife, a major city of northeastern Brazil and one of the principal beneficiaries of the Northeast program. Although by the early 1970s, Recife had received around one quarter of the total investment from the 34/18 mechanism, the impact on incomes was minimal. Family household surveys showed that per capita incomes had scarcely risen and that among the poorest two thirds of the population they had marginally declined (Gilbert and Goodman, 1976).

These issues are well illustrated in Venezuela's Ciudad Guayana program. The project was intended to incorporate the region into the national economy by exploiting mineral and power resources and accelerating economic growth (Friedmann, 1966; Rodwin, 1967). The physical development of the city and the manufacturing sector was impressive. Major plants were erected, capable of producing 4.8 millon tons of steel and 1.1 million tons of alumina. The Guri Dam and its related hydroelectric complex today produce almost half of Venezuela's electricity. Associated with these industrial and infrastructural projects has been rapid urban growth. Santo Tomé de Guayana, a town of 3,803 inhabitants in 1950, had over 400,000 inhabitants 40 years later.

However, the growth of the city was not free from problems. Industry did not create as many jobs as had been anticipated. According to one early estimate, the city was expected to have around 30,000 manufacturing workers in 1971, but it actually had 8,825. By comparison, rates of population growth were nearer to planning targets (61 percent of the target compared to 29 percent for employment). High rates of migration maintained unemployment in Ciudad Guayana at a level similar to that in the rest of the country. The housing program was unable to keep pace with the growth of the city. It had been hoped that most workers would be housed in the new city. In fact, three quarters still live on the other side of the Caroní River in the unplanned city of San Félix.

Planners hoped that Ciudad Guayana would act as a growth center for the surrounding rural areas and stimulate agriculture, forestry, ranching, and mining activities. As Figure 7.3 shows, this did not occur. Most of the surrounding towns grew slowly during the 1960s, and several did not even match the national population growth rate. Only La Paragua, close to the main iron-ore deposits, and Santo Tomé de Guayana itself grew at an annual rate of more than 4 percent. Few industrial linkages benefited the local area. Most were felt outside the Guayana region, some in Caracas and in the Maracay–Valencia area, and others outside Venezuela altogether.

The experience of Ciudad Guayana shows how hard it is for regional planners to create jobs in less developed regions. The establishment of dependent, capital-intensive industry in a poor area cannot resolve the problem of poverty and backwardness.

By the 1980s, few governments were attempting to modify the distribution of industry. Apart from the intellectual doubts about the policy, there were additional reasons why directed decentralization became unfashionable. Economic recession and a new philosophy of development meant that few governments were able or willing to intervene. They were less able to intervene because recession and the need to repay the foreign debt cut government revenues; they had less money with which to disperse industry. They were less willing to intervene because development policy, as laid down by the International Monetary Fund, required that national economies rely more on market forces. This model has required cutbacks in government spending, a program of privatization, reduced government intervention, the liberalization of the trade regime, and encouragement for exporters. Governments were instructed to intervene less.

Second, deliberate policies to disperse industry were less necessary. The Latin American industrial world had changed in a vital way; the economies of most of the major cities had stopped expanding. Economic recession and the opening up of Latin American economies to external competition hit industry in the major cities hard. The cities that benefited most from the previous development model lost a major ingredient in their growth. The shift from import-substituting industrialization to export-oriented industrialization had a profound effect. Many of the advantages of metropolitan location have been reduced, and the new approach has encouraged competition in the domestic market. The flood of manufactured imports unleashed by the lowering of import tariffs in many parts of Latin America has sometimes devastated local industry. Chile suffered particularly badly in the early 1980s, and Mexico experienced something similar beginning in 1982. Worse still, the domestic market has shrunk as the twin forces of the recession and structural adjustment policies have taken hold. The largest cities have borne the brunt of the shift. Between 1980 and 1988, industrial employment in Mexico City declined by 25 percent and Monterrey also suffered severely from the crisis (Cordera and González, 1991; Ruiz, 1990).

Even before the recession, expansion of the largest cities had been slowing because many companies had decided to move to other locations within the metropolitan region. Faced with increasing operating problems within the major cities, companies have begun to establish new plants outside. On the whole they have not moved far; new factories being set up in smaller cities within the metropolitan region. In Mexico, the problems of operating in Mexico City have convinced many companies to produce nearby. Cities such as Puebla, Toluca, and Cuernavaca have been beneficiaries. Similarly industrialists around São Paulo have been moving to small cities up to 200 kilometers from the regional center. Volkswagen has established a plant in Taubate and General Motors one in São José dos Campos. Other firms have established factories to the northwest of São Paulo in towns such as Campinas located along the Paulista railway line (Townroe and Keen, 1984; Storper, 1984). Similar tendencies have been apparent around other major Latin American cities, with industry locating in towns outside, but near, the metropolitan centers of Bogotá, Buenos Aires, Caracas, Guadalajara, and Monterrey.

In addition, growth in the major cities has sometimes been slowed by the preference of export-oriented plants for other regions. The most obvious instance is the development of *maquila* plants in northern Mexico. Figure 7.4 shows that the *maquiladoras* have generated large numbers of jobs in cities such as Ciudad Juárez, Mexicali, and Tijuana. Admittedly there are distinct problems with this form of development (Sklair, 1989; South, 1990). Wages are low because most of the jobs are intended for unskilled female workers. The spin-off for Mexican manufacturing industry is also limited. As Shaiken (1990, p. 13) points out, "Despite strong growth, the *maquila* industry has developed few backward linkages to Mexican suppliers for parts and materials, drawing most of its inputs from the United States and the Far East." Most *maquila* plants are more closely linked to the United States economy than to Mexico's (Herzog, 1991).

Nor are the local effects of *maquila* development always beneficial. Certainly, few of the cities that have grown so spectacularly along the U.S.–Mexican border strike visitors as desirable places to live. Compared to most Mexican cities farther south, Tijuana looks and feels awful. Terrible problems are being caused by the disposal of industrial waste and untreated sewerage. The pollution has now spread across the international border, and the Rio Grande is badly contaminated.

To this catalog of criticism a few mitigating points should be added. First, although wages are approximately one sixth of those paid in the United States, they are higher than in other Mexican industrial plants (Escobar and Roberts, 1991, p. 104). Second, whatever the deficiencies, the *maquila* is less susceptible to ups and downs in the Mexican economy (South, 1990, p. 566). During the 1980s, the

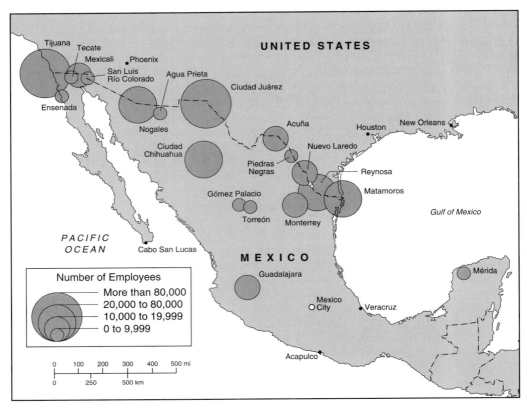

Figure 7.4 *Maquiladora* employees 1994 (After Augustin and Kohout).

maquilas created work when the rest of Mexico's manufacturing was contracting and shedding labor. Finally, there are signs that new export plants are being set up in areas farther away from the border. One vehicle assembly plant has been built in Aguascalientes. Indeed, with the establishment of NAFTA, *maquilas* are more likely to move south into the "real" Mexico.

The result of these new forces is to reduce the tendency for spatial concentration in Latin America. Table 7.9 shows the impact of recent developments on the distribution of industrial employment in Mexico. Between 1980 and 1988, the number

TABLE 7.9 Concentration of Industrial Employment in Mexico City, 1940–1988

Year	Mexico (000s)	Mexico City (000s)	Mexico City % of National Total
1930	285	54	19.0
1940	363	89	24.6
1950	626	157	25.0
1960	885	407	46.0
1970	1,596	658	41.2
1980	2,259	1,059	46.9
1988	2,587	756	28.8

Source: Garza (1987, p. 100); Rowland and Gordon (1996).

TABLE 7.10 Industrial Production by Region in Brazil, 1950–1990 (%)

Region	1950	1970	1990
North	0.6	0.7	3.1
Northeast	9.3	5.7	8.4
East	28.9	22.7	20.0
São Paulo	46.6	58.1	49.3
South	14.0	12.0	17.4
Center-West	0.6	0.8	1.8

Source: Diniz (1994, p. 302).

of industrial workers in Mexico City plummeted, even though employment in the country as a whole increased. If we remember the figures presented in Table 7.7, it is clear that much of the expansion in manufacturing has been concentrated in the north of the country. In Brazil, a similar if less dramatic trend toward geographical deconcentration is apparent. Table 7.10 shows that the state of São Paulo's grip over the Brazilian industrial economy was greatly reduced after 1970. Fewer companies were establishing plants in the state; although the new factories were still in the south of the country, many were located in neighboring states (Diniz, 1994). Clearly, the Latin American industrial world has changed profoundly.

THE SERVICE SECTOR

In 1940, most people in Latin America and the Caribbean lived in the countryside, and the majority of workers were employed in agriculture. Today, most people live in cities, and only in the poorer countries of the region does more than one third of the labor force work in agriculture. We have already seen that industrialization was only a partial success in terms of job creation. While it created jobs, the employment it generated rarely kept pace with the rapidly growing labor force. Nor was the industrial sector sufficiently dynamic to maintain a fast rate of economic growth. Throughout the region the tertiary sector has grown rapidly; services generate at least half of the gross domestic product in every country listed in Table 7.1. In terms of jobs, the sector is vital. Table 7.11 shows how important services are as a source of work in a number of major cities for which there are data. In Latin America, there are few towns or cities where mining and manufacturing employs more than one third of the labor force, and often the two sectors employ less than 10 percent.

TABLE 7.11 Structure of Labor Force in Selected Major Cities

City	Primary	Secondary	Tertiary	Year
Bogotá	1.5	28.3	70.2	1991
Buenos Aires	0.7	29.3	70.0	1992
Guadalajara*	na	27.1	64.8	1988
Monterrey*	na	28.6	62.6	1988
Rio de Janeiro	2.2	22.8	75.0	1988

*A considerable proportion of activities were undeclared in the survey of these cities, which is why the total does not add up to 100%.
Source: Official sources.

Why has the service sector provided so many of the new jobs? At one level the reason is the same as in the developed countries, where services have long dominated the employment structure. As national economies grow, they create more jobs in the service sector. Some of these are in production services: accountancy, consultancy, advertising, and the like. Some are in public services like education, health, and garbage collection. Some are in consumer services: retailing, restaurants, and recreation. Every major city in Latin America has large numbers of workers occupied in services (Table 7.11). While many of these workers are paid little, there has also been a major expansion in the number of "middle class," "respectable" jobs. In 1988, 48 percent of the labor force of Guadalajara worked in white-collar occupations; in Caracas, the figure was 44 percent.

In some countries, the service sector has grown because it generates foreign exchange. Tourism, for example, is a major earner even if it does not always manage to offset the expenditure of Latin American tourists to London, New York, and Disneyworld. In 1992, 41 million tourists visited Latin America and the Caribbean, earning some U.S. $29 billion in tourist receipts (U.N., 1994: Table 96).[1] Caribbean beaches have attracted tourists for generations as have the archaeological sites of Mexico, Guatemala, and Peru, and the animal life of the Galápagos. The beauty of the Iguaçu Falls, the allure of Rio de Janeiro, and the colonial architecture of Ouro Prêto bring large numbers of tourists to Brazil. In 1992, the principal destinations for international tourists were Mexico, a number of the Caribbean islands, and Argentina, Brazil, and Uruguay in the southern cone (Table 7.12). Peru had ceased to be a major destination in the early 1990s, as a result of the activities of the Shining Path guerrillas, while Cuba was actively developing its tourist business.

Tourism not only attracts people from outside Latin America, it also caters to the increasing taste for travel among the region's middle class. Fashionable coastal resorts such as Cartagena, Viña del Mar, Punta del Este, and Margarita Island attract large numbers of local people. In Mexico, new resort complexes such as Cancún and Ixtapa–Zihuatanejo rival established resorts such as Acapulco for both national and foreign tourists.

Unfortunately, beyond the glamour of tourist resorts and the rewarding life offered by professional jobs, there is another side to the service sector. Many of its workers are doing little more than eking out an existence. Many of those employed in the so-called "informal sector" have jobs that pay very badly. Such occupations include archetypal "Third World" jobs such as shoe cleaning, garbage picking, begging, and prostitution. Few of these workers are well paid and most work very long hours.

For years, concern has been expressed about the numbers of people involved in services in Latin America's cities. According to Gino Germani (1973, p. 32) the problem was that "a considerable proportion of the working-age population residing in cities were not absorbed in services of 'modern' type and remained marginal or relatively marginal to the modern forms of the economy." It is true that too many of the poor are forced into unsavory kinds of work. Because of their poverty, they are forced to build their own homes, to live without adequate services, and to eat the wrong kinds of food.

Much tertiary activity is only undertaken as a way of avoiding unemployment. Since there are no unemployment benefits, the poor are forced to find work of some kind. Adults in the poorest families in Latin America are rarely unemployed. This fact is illustrated by the poor's behavior during the great recession of the 1980s. According to González (1990, p. 118), "To avoid a drastic reduction in food

---

[1]Some 10 million of the visitors to Mexico are actually day visitors from the United States.

TABLE 7.12   Tourism in Selected Countries of Latin America and the Caribbean, 1992

Country	Visitors (000s)	Visitors per 1,000 Inhabitants	Tourist Receipts (U.S. $ million)
Argentina	3,031	92	3,090
Aruba	542	8,090	443
Bahamas	1,399	5,339	1,244
Barbados	385	1,486	463
Brazil	1,475	10	1,307
Chile	1,283	94	706
Colombia	1,076	32	705
Costa Rica	505	158	431
Cuba	410	38	382
Ecuador	403	37	192
Guatemala	541	56	243
Jamaica	909	379	858
Mexico	17,271	203	5,997
Netherlands Antilles	883	4,551	534
Puerto Rico	2,640	733	1,511
Uruguay	1,802	581	381
Venezuela	434	21	432
Virgin Islands	487	4,919	792

Note: Only countries with more than 350,000 visitors have been included.
Source: U.N. (1994, Table 96).

consumption, households have sent more members to the job market; more youths, women, and children have entered the work force in order to earn the income needed for the survival of the group, the domestic unit." The contribution of the informal sector to non-agricultural employment in Latin America rose from 40.2 percent in 1980 to 54.4 percent in 1990, rising especially quickly in the economies most affected by recession (Thomas, 1995, p. 46).

The service sector, therefore, is very mixed. It contains well-paid professional workers as well as the desperately poor. The fact that the sector has expanded is a sign both of the progress that has occurred in the region and of its continuing poverty.

SUMMARY

Mining was important to many pre-Columbian peoples and has been an important element in the regional economy ever since. Mineral production still contributes significantly to exports in most parts of the region. In general, however, mining does not create many jobs. By the 1940s, manufacturing contributed more than mining to both employment and GDP. Government action encouraged manufacturing development after 1940 through a strategy known as import-substituting industrialization (ISI). Between 1950 and 1975, ISI lay at the heart of development policy throughout the region. Later, export-oriented industrialization was adopted, and most Latin American and Caribbean countries are currently following this development strategy. Most have managed to increase their manufactured exports.

ISI encouraged industry to locate in the region's major cities. The concentration of manufacturing helped to explain the rapid expansion of Mexico City and

São Paulo. The debt crisis and EOI have reduced the advantages of established industrial regions. Government intervention has been reduced and market forces have tended to encourage the deconcentration of manufacturing production. This tendency has been marked in Mexico with industrial employment falling dramatically in Mexico City but increasing along the U.S.–Mexico border.

Today, neither manufacturing nor mining is as important to the region's economy as the service sector. Most people now live in cities, and services are now the major source of both urban income and employment. The debt crisis of the 1980s increased the contribution of services even further because recession hit the manufacturing sector particularly hard. Currently, the service sector is the largest contributor to GDP and the biggest generator of jobs. Services contain many of the best-remunerated jobs and most of the worst-paid. The relatively low rates of unemployment in the region are largely explained by the numbers of people forced to work in the service sector.

In the future, the expansion of services is likely to continue. Manufacturing activity will not increase employment dramatically because the sector is so highly capital intensive. How many of the new service jobs will be well paid will depend on the pace of economic growth. Where growth is rapid, the quality of jobs is likely to improve. Where it is slow, many of the service-sector jobs will pay badly and will be a substitute for unemployment. Undoubtedly, many people will continue to work in unsatisfactory service jobs for a long time to come.

FURTHER READINGS

Augustin, B. and Kohout, M., "The Border Industries Program: A Geography of Maquiladoras." In *A Geographic Glimpse of Central Texas and the Borderlands: Images and Encounters*, edited by J. F. Peterson and J. A. Tuason. Indiana, PA: National Council of Geographic Education, 1995, pp. 75–84.

Dunkerley, J. "Political Transition and Economic Stabilization: Bolivia, 1982–1989." London: University of London, Institute of Latin American Studies Research Paper No. 22, 1990.

Gwynne, R. N. *Industrialization and Urbanization in Latin America*. Beckenham, England: Croom Hill, 1985.

Rowland, A., and P. Gordon. "Mexico City: No Longer a Leviathan?" In *Megacities in Latin America*, edited by A. G. Gilbert. Tokyo, Japan: United Nations University Press, 1996.

Shaiken, H. "Advanced Manufacturing and Mexico: A New International Division of Labor?" *Latin American Research Review* 29 (1994): 39–71.

Sklair, L. "The Maquilas in Mexico: A Global Perspective." *Bulletin of Latin American Research* 11 (1992): 91–108.

CHAPTER 8

Mexico

PETER REES

INTRODUCTION

In *The Latin-American Mind* (1963), the Mexican philosopher Leopoldo Zea wrote:

> *our past still has not become a real past; it is still a present which does not choose to become history.*

An enduring theme in understanding Mexico has been the significance Mexicans attach to their past. Although *indio* (Indian) today is often a term of abuse, nostalgia for the pre-Hispanic remains. Moctezuma and Cuauhtémoc are popular given names and, while Mexico is overwhelmingly Spanish-speaking, Catholic, and *mestizo* (a mixture of Spaniard and Indian), it still struggles to digest the original Spanish conquest.

Since independence from Spain (1810–1824), Mexican national identity has been shaped by resentment of foreign intrusion. France occupied Mexico (1862–1867). The United States, to whom it forfeited half its territory, invaded three times (1848, 1914, and 1916). After the Mexican Revolution (1910–1917), its oil industry was dominated by American and British interests until nationalization in 1938.

From 1929, Mexico has been governed by a ruling oligarchy exercising power through the PRI (*Partido Revolucionario Institucional*). For most of its 65 years in power, PRI economic policy was statist (largely controlled by state-owned corporations) and protectionist. Mexican self-definition since the Revolution has been driven by resentment of the "colossus" to the north while at the same time Mexico has become economically and culturally ever more closely bound with the United States. Today, around 14 million people of Mexican descent, equivalent to 15 percent of Mexico's 95 million population, are legally resident in the United States. The United States is Mexico's most important trading partner, absorbing more than 80 percent of its exports and providing 70 percent of its imports.

In the 1990s, attachments to the past have begun to weaken, at least in the visual landscape and in Mexico's institutions, if not in its cultural outlook. Economic nationalism has been tempered by a policy of free-market competition and foreign investment. The number of state-owned enterprises has been reduced from almost 1,200 in 1982 to 200 today. Mexico has entered an economic alliance with the United States and Canada, the North American Free Trade Agreement (NAFTA), signed in 1993. The political control of the PRI has been loosened as opposition parties gain ground and Mexico edges uncertainly toward a democratic system.

Yet, vestiges of the past remain. Living in extreme poverty are 7 million Indians. Their struggle for self-government most recently found expression in the uprising in Chiapas (1994–1995), which helped undermine foreign investor confidence in Mexico's economy, causing the value of the Mexican peso to fall drastically against other currencies. In late 1994, Mexico nearly defaulted on its foreign debts. This crisis was only the most recent manifestation of Mexico's economic instability. In 1982, the collapse of oil prices provoked a similar crisis.

Mexico's long history of concentrated social power, running uninterrupted from precolonial days, still perpetuates an extreme imbalance in the distribution of wealth. In 1992, the richest 10 percent of the population received 38 percent of the country's total income, while the poorest half earned only 18 percent. Although a middle class has emerged in Mexico, it constitutes only about a quarter to a third of the population, compared with two thirds in the United States (Casteneda, 1995). The poor still abandon the countryside for the major cities and contribute to swollen populations that cause extreme environmental stress, as in the capital of Mexico City. Finally, Mexico suffers from a long tradition of regional contrasts. Mexico's correct name, the United Mexican States (*Estados Unidos Mexicanos*) is a reminder of the country's regional diversity. The term *patria chica* (little country) still has relevance in parts of the country distant from Mexico City, where allegiance to a region's cultural traditions can transcend attachment to national identity. Consequently, as in many countries, Mexico exhibits a core–periphery tension in which the center, dominated by Mexico City, seeks to impose its economic, political, and cultural will on distinctly different and often independent regions whose origins are traceable to the pre-Hispanic and colonial past.

This chapter reviews the forces shaping Mexico's physical landscapes. It then charts the layers of human occupance—pre-Hispanic, colonial, and neocolonial— that diffused spatially to varying extents across the country. Contemporary sketches of Mexico's distinctive regions follow. Finally, an assessment is made of national trends and issues that may alter Mexico's future geography. Throughout, two questions should be kept in mind: Is Mexico's past becoming history? and Can the regional contrasts, especially in the periphery, be woven with the core to achieve a coherent and progressive nation-state?

PHYSICAL ENVIRONMENT

► Landforms

Mexico's position occupying a part of three crustal plates helps explain its landform diversity. Most of the country comprises a southern extension of the North American plate, ending at the Isthmus of Tehuántepec (Figure 2.16). Extending south of the U.S. border, a block of the earth's crust thrusts upward to form the great Mexican plateau, an elevated platform of land comprising the Mesa del Norte and Mesa Central. Ranging from under 4,000 feet (1,212 m) on the border at El Paso to 7,347 feet (2,226 m) at its southernmost point in Mexico City, the plateau is a series of flat-floored basins lined with low hills, often of volcanic origin. Some basins are tapped by rivers flowing to the ocean, such as the Lerma River that crosses much of the Mesa Central to Lake Chapala. In turn, the lake is drained by the Rio Grande de Santiago, which has cut a mountain gorge to reach the Pacific north of Tepic. Many other basins drain internally with no outlet to the sea and contain shallow lakes or dry salt pans.

There are many active volcanoes in Mexico. When El Paricutín erupted in the 1940s, it buried surrounding fields and villages.

Two mountain chains border the plateau: the Sierra Madre Occidental, which drops precipitously to a narrow plain along the Pacific, and the Sierra Madre Oriental, which grades to a wider Gulf coast plain (Figure 8.1). An east–west ridge of active and inactive volcanoes, the Transverse Volcanic Axis, defines the southern

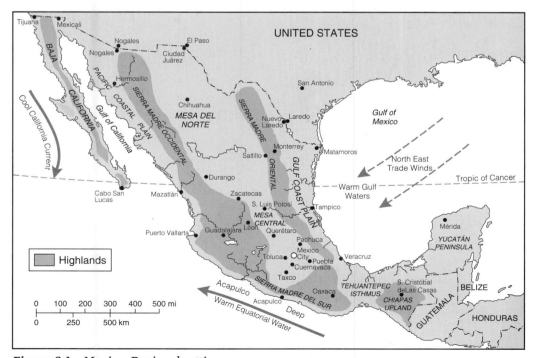

Figure 8.1   Mexico: Regional setting.

border of the plateau, and includes Mt. Orizaba (18,965 feet; 5,747 m), Mexico's highest peak. South of this axis, a structural depression is drained by the Balsas River, while south and east of the plateau lie the stream-dissected stumps of resistant crystalline outcrops that form the southern highlands of Oaxaca and Guerrero.

The Tehuántepec isthmus is the boundary between the North American and Caribbean crustal plates (Figure 8.2). Although they tend to slide past each other, nearby sites can experience some earthquake and volcanic activity. In 1982, the top of long-dormant El Chichón blew off in spectacular fashion, sending enormous dust clouds into the upper atmosphere, which contributed to a worldwide temporary cooling of 0.5 to 1°F. The Chiapas highlands form the western end of a spine of mountains that extend across the Caribbean plate to comprise many islands of the Greater Antilles. The Yucatán Peninsula is a limestone platform, thinly covered with soils and few surface streams. Recently emerged from the ocean, it is part of the undisturbed strata fringing the northern edge of the Caribbean plate and reappears to the east as the Bahamian islands.

Both the North American and Caribbean plates converge more violently with the Cocos plate underlying the Pacific, creating deep oceanic trenches offshore and fault zones that spawn severe earthquakes. Few towns in the Pacific half of Mexico have avoided earthquake damage. A 7.8 Richter-scale quake in a fault west of Acapulco generated crustal waves that devastated older sections of Mexico City in 1985, killed more than 5,000, destroyed buildings, and left 150,000 homeless.

The Baja California peninsula, a thin mountain spine fringed with dry, sandy shores is the only part of Mexico within the Pacific plate. At its junction with the North American plate, a set of double parallel faults, or a rift valley, has flooded to form the Gulf of California. North across the border, it becomes California's Imperial Valley.

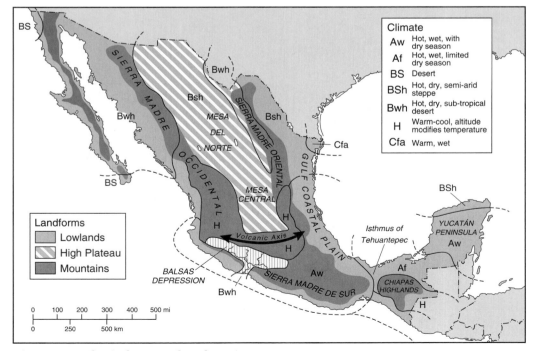

Figure 8.2   Physical geography of Mexico.

Colonial-style churches dominate the human landscape of Puebla, Mexico. The volcano Popocatepetl stands above all.

► *Climate and Vegetation*

Mexico's climatic variations are derived from the interplay of three influences. First is the Pacific, low-latitude, subtropical high pressure area that is strongest in the northwest of Mexico. It brings dry desert conditions to Baja California and the coastal lowlands surrounding the Gulf of California, but its influence affects two thirds of Mexico's land area. Second is the northeast tradewinds that blow off the Gulf and bring most of the country's rainfall. Finally is altitude, which modifies the otherwise warm temperatures that prevail throughout Mexico, produces rain-shadow effects that reduce rainfall on the Mexican plateau and Pacific-facing slopes, and creates stratified ecologic zones up the sides of steep mountains.

Unlike countries in temperate latitudes, the seasons in Mexico are defined by rainfall rather than temperature, since temperatures experience little seasonal variation. In summer, the subtropical high weakens on its southern side, the Intertropical Convergence Zone (ITCZ) moves farther north, and the northeast trade winds intensify to bring the rainy season. In winter and spring, the trades weaken as the Atlantic subtropical high pressure cell builds greater stability southward and the dry season prevails (Figure 2.9).

As the trades sweep across the warm waters of the Gulf of Mexico and the Caribbean, they absorb moisture. The low-lying Gulf coast and the flat Yucatán Peninsula have insufficient relief to produce orographic uplift and much precipitation. The coastal lowlands thus receive modest rainfall and even support semiarid vegetation along their coastal margins (Figure 8.2). Inland and farther south, away from the strong influence of the subtropical high, the lowlands of southern Veracruz, Tabasco, and southern Campeche receive rainfall in excess of 100 inches, and support tropical rain-forest vegetation.

The heaviest amounts (sometimes exceeding 200 inches per annum) drench the mountains of the Sierra Madre Oriental and Chiapas highlands, as the trades rise up

Much small-scale agriculture in Mexico takes place on steep hillsides. Here the farmer has mounded the soil across the slope to reduce erosion.

steep escarpments that reach over 8,000 feet. Dense stands of rain forest at lower altitudes give way to conifers as elevation cools temperatures. Once across these mountain barriers, the now drier air crosses the Mexican plateau where rainfall totals drop dramatically, ranging from 41 inches per annum in Mexico City to 17 inches per annum in Chihuahua to the north. Only dry, xerophytic vegetation can be supported by these semiarid conditions. Crossing the windward slopes of the Sierra Madre Occidental and Sierra Madre del Sur, the trade winds give up additional moisture, and the higher ground has a covering of conifers. But as the air descends to the Pacific, it releases little moisture. The west sides of mountains are therefore drier and obtain their rainfall from Pacific storms, whose frequency increases southward.

Overlaying these broad climatic patterns are day-to-day weather conditions that result from the interplay of tropical and temperate global air masses. No part of Mexico is spared the atmospheric violence that sometimes results. In winter, cold polar air surges into tropical latitudes, producing fierce storms (*nortes*), which may damage vessels at sea and prevent tropical plants such as bananas from bearing fruit. Temperatures can drop 15 to 20 degrees Fahrenheit in a few hours, and frosts may occur along the coasts of northern Mexico. In summer and fall, Atlantic hurricanes often hit the Yucatán Peninsula and Gulf coast lowlands, producing severe flooding and property destruction. In 1988, Hurricane Gilbert with 175 m.p.h. winds left a path of destruction across the Yucatán, curtailing the area's tourist industry for months.

Less destructive but still powerful Pacific storms batter the region's west coast and sometimes bring fierce summer downpours to the otherwise arid lowlands of Sonora and Sinaloa. Prolonged, severe droughts can occur, especially in the arid and semiarid higher interior of the northern Mexican plateau and the Baja California and Sonoran deserts, which collectively make up two thirds of Mexico's land area.

Since 1993, an extended drought in the north has cut the supply of water in dams in Chihuahnua and Coahuila to between 14 percent and 26 percent of capacity.

ORIGINS OF THE CONTEMPORARY LANDSCAPE

► *The Pre-Hispanic Legacy*

The present geography of states, cities, and people in Mexico evolved from earlier patterns of occupation.[1] Before 5000 B.C., a common language reflecting a relatively uniform culture stretched from northern Mexico to Colombia. Subsequent waves of newcomers disrupted this linguistic harmony, producing a mosaic of over 250 distinct languages. Many of these, such as Mayan, Zapotec, and Nahuatl, the language of the Aztecs, can still be heard today in the isolated valleys and highlands of the region.

A highly productive agriculture was developed, founded on combinations of maize, beans, and chilies or squash. This plant combination represented both a botanical and a dietary symbiosis. Corn provided a climbing pole for the bean plant; the bean drew nitrogen from the air through its leaves and deposited it in the soil, thereby contributing fertility; and chilies or squashes, as a low ground cover, prevented soil erosion. Impressively, this prehistoric plant assemblage offered a balanced human diet: corn provides carbohydrates, beans contribute protein, peppers or squash offer critical vitamins and trace minerals. Still a traditional diet for many rural folk today, this plant combination supported dense indigenous populations, particularly in the drier highland basins where elaborate systems of irrigation and terracing were created. Elsewhere in tropical coastal lowlands, population density was lower and swidden (shifting agriculture) prevailed (Chapter 4).

The appearance of political states imposed a degree of cultural uniformity on central and southern Mexico, but the geographic extent of states varied with the time and place of cultural flowering. The first were the Olmec peoples (900 B.C.) whose sites are now being revealed in the forests of lowland Veracruz and Tabasco. They left little evidence besides enormous heads carved from basaltic rock. Later, Totonac and Huastec culture diffused through lowland Veracruz at sites such as Tajín between A.D. 400 and 1200; Zapotecs spread their influence through Oaxaca, Guerrero, and southern Puebla from Monte Albán from 500 B.C. to A.D. 1000; Teotihuacán dominated highland central Mexico from A.D. 300 to 900; Mayan culture developed first among tribal groups in highland Guatemala, achieved its greatest advances in lowland Guatemala from sites such as Tikal (A.D. 250–600), and later diffused into subsidiary locations throughout the Yucatán; and Toltecs consolidated their power at Tula (A.D. 950–1250), from which their influence extended as far as northern Yucatán where they built Chichén Itzá as a Toltec colony. When Europeans arrived, a flourishing Tarascan state was centered on Tzintzuntzan beside Lake Pátzcuaro, and the Aztecs were rapidly consolidating power in Central Mexico and the Gulf lowlands from their capital in the Valley of Mexico.

A line running roughly across the Mexican plateau from the Panuco River to modern-day Zacatecas marked the northern limit of Meso-American high culture. North of the line was what the Spaniards called the "Chichimec frontier," a region inhabited by seminomadic peoples whose culture represented an extension of the American plains Indians (Figure 3.1). Lacking a complex social organization and permanent settlement, the people of the lightly populated Chichimec frontier, although initially resistant to the Spanish advance, ultimately vanished. Only remnants of simple slash and burn farmers in the remote northern Sierra Madre Occidental remain today.

---

[1]See Chapter 3 for more details.

By contrast, throughout central and southern Mexico numerous cultural differences among rural folk have pre-European origins. The people of Veracruz (called *jarochos*) owe much of their distinctive dress, diet, and tradition to Totonac forebears. Metal workers among the market craftsmen in Morelia trace their specialty to Tarascan ancestors. And Yucatecans still speak Mayan and practice community life much as did their ancestors.

The most visible expressions of the pre-European past are the ceremonial centers of earlier cultures, which were foci of religion and political power and repositories of art, architecture, and science on a scale that frequently surpassed anything in Europe at the time. Teotihuacán (100 B.C.–A.D. 900), for instance, extended over 8 square miles (20 square kilometers) and contained more than 4,000 residences and as many as 100,000 people. The center was occupied by two huge temple pyramids, the largest of which had a base greater than that of any Egyptian pyramid. Now unoccupied, these ceremonial urban centers attract a lucrative tourist industry. But their presence is testimony to the complexity of agriculture, trade, and political power inherent in once vibrant native cultures that European conquest suppressed, pushed to the remote recesses of the region, but never fully obliterated.

► The Colonial Legacy
Settlement Pattern

If today's rural ways and attitudes to life in Mexico reflect the pre-Hispanic past, the pattern of cities, towns, highways, and production owes more to European colonial influence. The Spaniards founded Mexico City, the capital of New Spain, on the rubble of Tenochtitlán, the conquered Aztec capital. The pattern of settlement in Mexico that followed was dictated by the location of silver, gold, and Indians, as well as the Spanish desire to centralize administrative power. The distribution of towns and connecting highways (*caminos reales*) reflected the geography of colonial impact.

In the first phase of settlement, Spaniards spread across the Mesa Central of Mexico, which contained the heaviest concentration of Indians. The colonists' route from Mexico City to Spain ran eastward to the harbor at Veracruz, a poorly protected roadstead on the leeward side of the island of San Juan de Ulúa where vessels of the Spanish fleet (*flota*) were often subject to *nortes* and tropical diseases might kill half a ship's crew. Carried by lighters to the mouth of the Antigua River, the goods brought by the *flota* were transferred to mule trains (*requas*). These wound up the steep inclines of the Sierra Madre Oriental to Mexico City on the plateau, where a trade fair (*feria*) took place. Prices at the fair were often set in secret by representatives of the *consulados* of Mexico City and Seville, thereby restricting the participation of smaller merchant houses located elsewhere in New Spain.

The fair focused economic activity on the capital, already the political center of New Spain and site of the colony's only mint. Under the Spaniards, Mexico City maintained the dominant role in the urban hierarchy, as had Tenochtitlán under the Aztecs, and remained the focal point of the route network. Goods brought from Spain and silver from the mines to pay for those goods were all assembled in Mexico City instead of following direct routes to the coast.

The earliest highway connecting Veracruz with Mexico City at first followed no specific path once the traffic had negotiated the Sierra Madre Oriental through the Jalapa Pass. Instead, travelers picked their own way across the flat-floored basins of the Mexican plateau to the capital. But, as colonial settlement was established and the route acquired a degree of permanence, *ventas* (inns) and rudimentary bridges over the more difficult *barrancas* fixed certain points of the highway. By 1530, documentary evidence indicates that stretches of the road were being constructed to hasten the movement of wagons across the plateau. As it evolved,

Delivering jars of water or pulque (an alcoholic beverage), Mexico, c. 1890.
Notice the two-wheel cart of the type used in the colonial era.

the highway proved typical of subsequent colonial transportation routes with an emphasis on the long-distance connection of terminal points over the shortest practical path, rather than adding deviations to pick up local traffic. Throughout the sixteenth century, the government refused to authorize the diversion of the official Mexico City–Veracruz highway to pass through the new Spanish city of Puebla, located only a short distance south of the *camino real*, in spite of pressure from citizens. By the time the government finally relented and the Mexico City–Puebla segment became recognized, the original alignment north of the city through the plains of Apan was so established that it continued to be used as an alternate highway by direct cart traffic throughout the colonial period (Rees, 1976).

Other highways were single arteries of commerce, but all fanned out from the capital and reflected the dominance of Mexico City. Across the southern border of the Valley of Mexico and down the escarpment of the east–west volcanic axis, an early route ran to Cuernavaca to serve the sugar *haciendas* surrounding the town, including the extensive estate of Hernán Cortés. From Cuernavaca, the highway continued south to Acapulco, with a branch leading to the rich silver mines of Taxco. To the west, another sixteenth-century road linked the agricultural basins of Toluca and Valladolid (Morelia) to Guadalajara, the colony's second city.

Guadalajara (founded in 1548) had developed as a supply point for the northern mining camps following the discovery of silver in Zacatecas in 1546. This second major phase of colonial settlement into the arid north beyond Querétaro depended on pushing back the frontier between Spanish settlement and the Chichimec Indian tribes inhabiting the plateau. By 1551, the main Zacatecas–Mexico City highway, or *Camino Real de la Tierra Adentro*, was a well-traveled route passing through a series of interconnected, flat-floored basins that allowed the use of two-wheeled carts (*carretas*). However, the Zacatecan mines were still beyond the settlement frontier at this time. The growth of the mining economy required a protected means of communication through hostile territory to carry silver ore to Mexico City and move agricultural supplies north to the mining camps. Consequently, in the latter half of the sixteenth century the highway was lined with military *presidios* at San Miguel, San Felipe, Ojuelos, and Cienega Grande. The discovery of silver in Guanajuato added to the importance of this highway, and Celaya was similarly founded as a *presidio* town, protecting shipments from the new mines as well as the Basque ranchers who populated the surrounding fertile lands of the upper Río Lerma.

Later in the seventeenth century, the highway was pushed north along the western edge of the plateau as silver was discovered near Sombrerete, Durango, Parral, and Chihuahua, while branches to Saltillo from Zacatecas and San Luis Potosí served mines on the eastern side. By the eighteenth century, mining gave way to religion as the major force extending the frontier, and the two northern highways on either side of the Mesa del Norte now reached more than 900 miles to the missions of Santa Fé and Taos (New Mexico) and San Antonio (Texas).

Missionaries pioneered a third route northwest along the Pacific coast into California. Behind priests came prospectors seeking minerals, and grain farmers and stockmen to supply the mines. A highway was opened along the coastal base of the Sierra Madre Occidental, and by the middle of the eighteenth century, this road connected a chain of frontier forts, missions, and silver-mining communities that lead to Baja and Alta California.

The three routes still define the major axes of northern Mexico's settlement and economy. The Indian population was sparse, and the region has a Hispanic character. The opposite is found in Chiapas and the Yucatán. Here, fierce aboriginal resistance to the early Spaniards, together with a lack of commercial resources, left these areas remote from Spanish influence except for the indefatigable priests. Today these distant regions remain strong centers of Indian heritage.

The Cultural Contribution

European colonialism was not a patina laid upon an undisturbed indigenous culture that would reemerge as foreign influence was peeled away after Independence. Rather, the colonial period created a new, culturally diverse mixture of people and places. When the Spaniards arrived, an aboriginal population of perhaps 17.2 million was present in Mexico (Denevan, 1992). Within a century, disease and hardship had reduced the Indian population by nearly 90 percent. Replacement by Europeans was gradual, and Africans, who were brought to work in tropical fields and mines, never reached the numbers seen in Brazil. When population growth began again in the eighteenth century, *mestizos*, a mixing of mostly Hispanic and Indian genes, became the ethnic majority in the region. Today, "mestizo" has a cultural connotation, referring to those who have adopted European ways. The European perception of social status associated with lightness of skin still prevails, despite a tendency to glorify the Indian past while denigrating the Indian present.

Hispanic culture also introduced Catholicism, but again, a distinctive modification of religious practice evolved. Especially in rural churches, it is not uncommon today to encounter supplicants appealing to a pantheon of gods and spirits in addition to those worshiped by Christians. In the nineteenth century, Mennonite colonies settled in northern Mexico and Belize, and recent inroads by fundamentalist Protestant churches have further enriched the variety of religious practices in the region.

Among the most lasting colonial influences have been those of landholding and settlement form. The indigenous system of collective land use organized around dispersed settlements was supplanted by the *hacienda,* a great estate owned by the colonial elite and worked with native labor. The Indian population was moved to concentrated agro-villages under a system called *congregación.* Land wealth still imparts social status and has led to a great imbalance in property ownership in all

Middle American nations. Inadequate supplies of peasant farmland continue to undermine the political foundation of Mexico, as the recent Chiapas uprising demonstrates.

Despite huge landholdings, the colonists preferred urban living. A majority of cities and towns in the region date from the sixteenth century. Many reflect common design principles dictated by the crown: a grid street plan organized around a central plaza and dominated on one side by the church. Plazas served as both markets and places for socializing. Today plazas remain spaces of central importance in numerous towns and small communities, often graced with shade trees and a central bandstand.

Finally, the Spaniards left an administrative legacy that helps to define the modern political geography of the region (Figures 8.3 and 8.4). From the early years of European settlement, the Viceroyalty of New Spain was administered from Mexico City. Administrative subregions evolved, reflecting dispersed pockets of Spanish settlement. In 1786, the crown restructured the colonial administration. New Spain was divided into twelve *intendencies*, each under the control of an intendent whose authority was strengthened by reporting directly to Spain rather than through the viceroy. When the independence movement swept Mexico from 1810 to 1821, the intendencies of New Spain remained centers of regional power and culture, although agreeing to unite as the United States of Mexico (*Estados Unidos Mexicanos*).

In Central America, a similar union of intendencies formed the United Provinces of Central America and at first joined independent Mexico in 1822. A year later, with the exception of the former Intendency of Chiapas, which elected to remain part of the Mexican union, the United Provinces separated from Mexico.

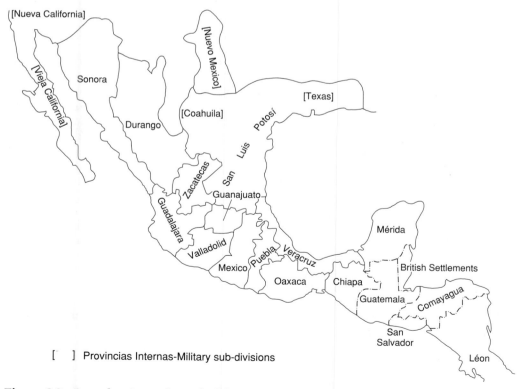

[ ] Provincias Internas-Military sub-divisions

Figure 8.3   Intendencies at the end of the colonial period.

Figure 8.4 Present day boundaries of Mexican states and Central American countries.

► *The Neocolonial Legacy*

The patterns of settlement established in the colonial period were intensified in the nineteenth century by neocolonial economic interests from the United States and northern Europe. New sources of mineral wealth were discovered. U.S. capital developed the rich Mexican copper ores found in Cananea in northern Sonora and deposits of iron ore near Durango. In the early twentieth century, British and U.S. interests opened the first oil fields on the Gulf coast near Tampico and Veracruz, and U.S. influence laid the groundwork for Mexico's early iron and steel manufacturing in Monterrey, based on local iron ore desposits and coal from nearby Saltillo. Germans developed the Mexican brewing industry with plants in Mexico City, Orizaba, and Guadalajara, while in retailing, French capital started a network of department stores in principal Mexican cities.

The major changes in agriculture occurred in the tropical lowlands. A network of light railroads fanned out from ports such as Progreso, helping to transform the interior of northern Yucatán into henequin plantations, which produced sisal rope and twine to tie the bales of U.S. prairie reapers.

Roads deteriorated throughout the first half of the nineteenth century. The English company that tried to run the Real del Monte silver mine at Pachuca between 1824 and 1849 needed 6 to 8 months to get supplies from England. Furthermore, one sixth of the delivered price of its chief import, mercury, was the cost of transport from Veracruz.

► *Railroads*

After 1873, the transport system improved through the development of railroads. Like much of nineteenth-century economic development, railroads were built by

British and U.S. companies. Mostly, railroads handled long-distance commerce, reflecting the desire of Mexico's elite to use raw material exports to pay for imported manufactured goods. The railroads generally followed colonial route alignments and thus reinforced the centrality of Mexico City (Figure 8.5). British-built lines imitated the earlier duplication of colonial highways between Mexico City and Veracruz, one line passing through Orizaba and the other through Jalapa. To the north, U.S. companies built three railroad lines: (1) along the northwest coast to Nogales, (2) through Chihuahua to Ciudad Juárez, and (3) San Luis Potosí and Monterrey to Nuevo Laredo. Each approximated the routes of the three principal northern colonial highways and tied Mexico to the expanding U.S. railroad network.

The socioeconomic consequences of the neocolonial period resulted in an intensification of colonial patterns. At the end of the nineteenth century, wealth and landownership were more concentrated than they had been in 1810. In the state of Mexico, one hacienda covered more than 500 square miles. Three hacienda owners in Colima claimed one third of the land in the state. Four haciendas in Michoacán owned one fourth of all the property. Meanwhile, in some areas nearly 90 percent of the peasantry did not own land, and the 1910 census reported that half the population lived in *chozos*, shacks virtually unfit for human habitation (Cumberland, 1968). The Mexican Revolution (1910–1917) was the result of an uprising among the peasantry in the remote periphery of the southern Sierra Madre del Sur under the charismatic leadership of Emiliano Zapata.

► The Modern Era
Land Tenure

A nationalistic phase of Mexican economic development followed the Revolution. Economic and political power remained concentrated among the elite, but land reform under Article 27 of the Mexican Constitution transferred land to the peasantry under the communal *ejido* system. Although slow to occur, land reform resulted in redistribution of about half of Mexico's arable land under the *ejido*, in which property ownership remained with the state but peasants, acting as a cooperative, acquired a protected right to its use. This form of communal property reflected pre-Hispanic systems of landholding where access to land was dictated by use rather than legal title.

By 1930, *ejidos* occupied only 13 percent of cultivated land. However, under the administration of Lázaro Cárdenas (1934–1940) some 17 million hectares of land were expropriated from former hacienda owners and redistributed to more than 8,000 new villages occupied by some 2 million peasants. By 1940, nearly 15,000 villages containing a quarter of Mexico's population had access to slightly less than half the cropland. In 1988, 70 percent of the country's farmers were working communal land where none existed at the time of the Revolution (Cumberland, 1968; De Janvry, 1995). A profound redistribution of property had occurred, and the *ejido* system had become enshrined as an almost inviolate principle of Mexican society. Nevertheless, much of this land was marginal. Only 16 percent is irrigated today, and 64 percent of *ejidatarios* farm less than 5 hectares, generally considered insufficient to maintain a family. Much of the best bottom valley lands remained in commercial agriculture; *ejido* lands were illegally bought by larger landowners; and today, many *ejidatarios* tend dispersed plots that cling precariously to steep, wooded mountain slopes.

In 1993, Article 27 was amended. Land redistribution, previously enshrined in the Constitution, was abandoned. *Ejidatarios* were free to sell their land to commercial farmers and land developers, retaining the proceeds. A previously illegal practice was now legal. One result has been to speed the expansion of the urban

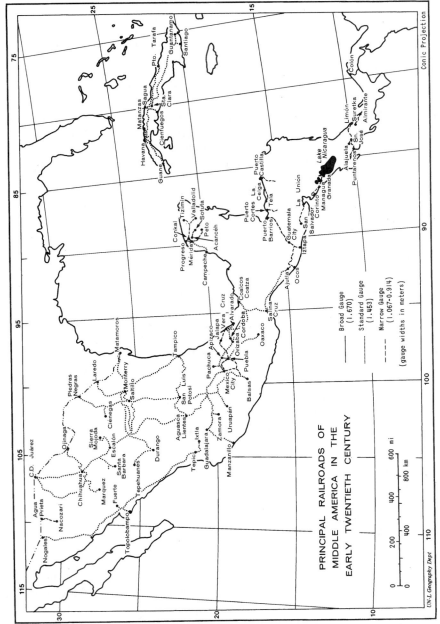

Figure 8.5 Principal railroads of Middle America in the early twentieth century.

PRINCIPAL RAILROADS OF
MIDDLE AMERICA IN THE
EARLY TWENTIETH CENTURY

Broad Gauge
(1.670)

Standard Gauge
(1.453)

Narrow Gauge
(1.067-0.914)

(gauge widths in meters)

Conic Projection

UN-L Geography Dept

Mounted guard at Amparo mine, c. 1890, in front of habitations occupied by miners. The arrogance of foreign mining companies was a cause of the unrest leading to the Mexican revolution.

fringe onto former *ejido* properties in major metropolitan centers such as Mexico City and Guadalajara. In remote rural areas such as Chiapas, where land reform had hardly begun, the change in Article 27 was greeted with dismay and helped fuel the recent uprising against central government authority.

Industrial Development

The Mexican Revolution unleashed a powerful economic nationalism, at first directed at oil companies. Extensive oil deposits along the northeast Gulf shore around Tampico had been developed by U.S. and British companies. In 1921, Mexico produced a fourth of the world's oil. The oil companies' laissez-faire defense of their right to pump unlimited quantities clashed with the Mexican government's desire to limit exploitation of what it considered its national wealth. After 15 years of legal conflict, the Cárdenas government in 1938 nationalized the oil industry. The new government oil company, PEMEX (*Petróleos Mexicanos*) became, like the *ejido*, an icon of Mexican identity. Today, Mexico remains resistant to foreign exploitation of its oil reserves.

The government's economic nationalism later extended to much of the industrial and manufacturing sector. After World War II, Mexico, like other Latin American nations, adopted a policy of import substitution (Chapter 7), protecting domestic industries with tariffs against imported foreign goods. The government was heavily involved in economic policy. Cement, steel, and chemicals were dominated by state-owned enterprises. Many industries located close to the government decision-making bureaucracy, and as a consequence, by 1960 Mexico City had a fifth of the population but nearly 50 percent of the manufacturing activity. With growing employment opportunities, the capital attracted young, ambitious migrants from rural areas.

Since 1990, Mexico has reversed its policy of economic nationalism in favor of private sector growth. Foreign nationals can now exercise majority ownership of Mexican businesses. A flood of foreign capital investment resulted which overwhelmed the economy and led to the recent peso crisis. However, long-term growth can be anticipated in *maquila* industries along the 2000-mile border with the United States, agribusiness, mining industries, particularly silver, bismuth, and celestite in which Mexico is the world's leading producer, and numerous manufacturing products from autos and parts to televisions.

Rural-to-Urban Migration

As the dominant center of power, as well as of news and information, Mexico City was portrayed in the media as a place of material well-being and prosperity. The federal government's tendency to invest in the capital's schools and health care, combined with the economies of scale that Mexico City's large population offered manufacturers, produced opportunities and better conditions for hopefuls from the countryside. Beginning in the 1950s, highways were paved, and a cheap, subsidized long-distance bus system provided the means to reach Mexico City from all parts of the country within 48 hours.

Rural-to-urban migration turned Mexico City into Latin America's biggest city. In 1900, Mexico City's half million people represented 2.5 percent of the country's population. By 1990, the capital's urbanized area, swollen by natural increase and in-migration, contained over 20 million, nearly a quarter of Mexico's total of 95 million people. Other cities, such as Guadalajara and Monterrey, have also become migration targets as population shifts from the countryside. Today, Mexico is 71 percent urban (versus 75 percent in the United States). With 36 percent of the population under 15 years of age, numbers will double in about 40 years.

Since the 1950s, the government (an essentially one-party system ruled by the *Partido Revolucionario Institucional* or PRI) has sought to divert economic growth away from Mexico City to other parts of the country. This policy of decentralization has met with little success. The federal government refuses to yield political power to state and especially local governments in outlying regions, which often view Mexico City with skepticism and animosity. Although the center may never be so dominant as to blur the distinctive regions that make up the country, a tension between core and periphery remains an important feature of Mexico's changing political geography.

REGIONS OF MEXICO

At its origin as an independent country in 1821, Mexico was a collection of individual intendencies. For many years, each of these former colonial administrative units acted like an individual country, or *patria chica* (little country). Despite the dominance of Mexico City and the power of the federal government, Mexico is still in the process of forging a national identity. In peripheral parts of the country distant from Mexico City, where allegiance to a region's cultural traditions can transcend nationalism, the term patria chica still has relevance. An appreciation of the distinctive regions of Mexico (Figure 8.6) remains significant.

► *The Independent North*

San Luis Potosí (founded in 1592) is Mexico's gateway to the north. The wide streets are clogged with long-distance trucks whose smoke and dust cast a grey plume across the townscape. For most of its history, San Luis has been a primary transport node. Highways intersect here forming a giant crossroad at the country's center, linking Guadalajara and Mexico City, the nation's two largest cities, with Ciudad Juárez and Nuevo Laredo on the U.S.–Mexico border. North of San Luis, the great Mexican

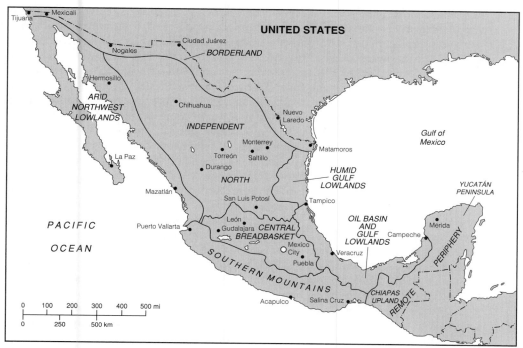

Figure 8.6 Regions of Mexico.

plateau expands in scale and becomes increasingly arid. In many ways it has the frontier feel of the American West, a land of wide vistas and dispersed settlement clusters, resource-rich and given over to mining, cattle ranching, and grain farming.

Regionally proud, the people of northern Mexico have a reputation for political independence and conservatism. In 1989, when the business-oriented opposition party PAN (*Partido Acción Nacional*) first captured a governorship from the dominant PRI, it was in Baja California. Chihuahua, also in the North, was the second of the four states that PAN presently controls (the others being Jalisco and Aguascalientes). The source of this independent character is found in the region's distinctive historical geography. The North was only lightly populated with nomadic Indians at the time of European contact, and cultural influences are more strongly Hispanic than the mestizo-Indian mix of central and southern Mexico. Wheat and barley, rather than maize, are the predominant commercial grains, and it was from here that Spanish ranchers spread the horse, lariat, rodeo, roundup, and other elements of the cowboy culture into the prairies and plains of what are now the United States and Canada.

Today, agricultural settlement avoids the semiarid basins in favor of the plateau edges. Large cattle ranches occupy grassy slopes at the base of the mountains. Where rivers and streams flow out onto the plateau, irrigated oases support grain, alfalfa, and vegetable farms. Among the earliest and, for modern Mexican history, the most important of these irrigation districts was the Laguna area surrounding Torreón. Here, Cárdenas expropriated over 600,000 acres held in large estates and made cultivation rights available to 30,000 peasant families under the *ejido* system of land tenure. The Laguna project turned a dry, flat basin into an irrigated oasis. Using both pumped groundwater and diverted flow from the Nazas River, the area became Mexico's largest producer of cotton. More recently, productivity has been limited by soil salinization, a problem that eventually afflicts most of Mexico's dry-land irrigation projects when downward flow of irrigation water in the soil is not sufficient to offset upward flow through evapotranspiration.

Mineral deposits and mining were a major stimulus to urban growth and industrialization in the north. Mexico's first modern integrated iron and steel plant opened in Monterrey in 1903, and the city has continued to play an important role in manufacturing. Sited in the foothills of the Sierra Madre Oriental, Monterrey benefited from Latin America's best coking coal in the nearby Sabinas basin, iron ore deposits in the Hercules–La Perla region, and rail access to both U.S. and central Mexican markets. The mining of copper deposits stimulated the rise of Chihuahua; lead and zinc found in association with silver ores revived the fortunes of old silver towns such as Aguascalientes and Parral; and Durango benefited from iron ore deposits in nearby Cerro el Mercado.

Most recently, two waves of manufacturing have dramatically altered the cities of the North. The first, encouraged in part by the Mexican government's turn toward a deregulated market, involves multinational firms putting down deeper roots in major northern cities and developing complete manufacturing processes. Auto manufacturers, in particular, have made significant investments. Ford opened an engine plant in Chihuahua in 1983. Nissan produces autos in Aguascalientes, and General Motors has engine plants in Monterrey. Other multinationals like IBM, Whirlpool, Kodak, and Caterpillar are investing in up-to-date plants, laying the foundation for a high-tech industrial zone in northern Mexico, which within NAFTA could link up with the southern California and Texas manufacturing zones to become one of the world's leading industrial regions. Already, Monterrey is tilting north. Mexico's third largest city, it is the center of the country's wealthiest state, Nuevo León. Only 150 miles from the U.S. border (but 600 from Mexico City) Monterrey has many of the surface features of a Texas city: broad, well-traveled boulevards, modern regional shopping centers, and elite suburban neighborhoods. Increasingly, its Mexican companies in cement, metals, and agribusiness are becoming international, regarding Dallas and San Antonio with the same attention as traditional markets in Guadalajara and Mexico City.

A second wave of manufacturing in the border towns from Tijuana to Matamoros, collectively now Mexico's fastest-growing cities, has transformed these communities into a distinctive regional landscape.

The Borderland Subregion

Significantly, Mexicans call this area *la frontera* (the frontier), implying a place of opportunity and progress. This view contrasts sharply with the traditional American perception of dusty border towns with lurid, honky-tonk tourist sections that attract quick visits by military personnel and students seeking a lower legal drinking age. These starkly different perspectives find expression in the townscapes of the border towns (Arreola and Curtis, 1993). Americans encounter the tourist district close to the border crossing, populated with curio shops and roving street merchants offering goods that appeal to an image of Mexico that owes much to the serape, the burro, and the broad brimmed mariachi hat. Beyond the few blocks of the tourist section, and rarely penetrated by visitors, is the border town the residents know: a vibrant commercial center, outlying retail strips with an increasingly American visual appearance, small elite neighborhoods and much larger sprawling residential areas from middle-class *colonias* (developments) to squatter settlements on nearby hillslopes (Arreola and Curtis, 1993, pp. 218–219). The latter result from the buildup of population associated with industrial growth.

After World War II, the border towns became attractive as conduits of legal, if temporary, immigration under the 1942 U.S. *bracero* program. The plan involved encouraging temporary agricultural labor from Mexico to work in the fields of

California, Arizona, and Texas, but also resulted in much illegal immigration. When the bracero program ended, the Mexican government in 1965 began the Border Industrialization Program, designed to siphon the streams of migrants heading north into employment in Mexico. The result is the *maquiladora* factories on the Mexican side of the border that import raw materials and components duty free, utilize low-cost labor for assembly, and export finished products to the United States. Duty is paid only on the value added by manufacture.

The program appeals primarily to U.S. manufacturers in labor-intensive industries such as clothing and textiles, electrical and electronic goods, and furniture and wood products, all of which take advantage of Mexican labor rates, which are presently four to five times lower than those in the United States. The number of *maquiladora* plants grew rapidly in the late 1980s. There are now more than 2,000 of them, employing nearly half a million workers (Figure 7.4). Only the oil industry produces more foreign-exchange revenue for Mexico.

European and Asian companies, especially Japanese manufacturers, have joined U.S. companies in opening *maquiladoras*. Once confined to the border towns, the program is spreading to other parts of the North, such as Ensenada in Baja California, Hermosillo in Sonora, Monterrey, Chihuahua, Saltillo, and even to Mexico City and Mérida in the Yucatán.

Maquiladoras have had a controversial effect on the border communities. Plants have stimulated construction and created factory jobs, a majority held by young women. Cross-border economies are increasingly integrated through retailing patterns, since Mexican workers spend 40–60 percent of their wages in U.S. retail outlets (Augustin and Kohout, 1995). As the focus of growth, the border cities benefit from infrastructure investments in roads, utilities, and modern industrial parks. A current proposal is to create a new $2 billion international airport astride the border, serving Tijuana and the San Diego metropolitan area, binding ever more closely the economies of these two large urban areas into a single market of 3 million people, which some now call "San Tijuana."

However, *maquiladoras* pollute air and water supplies. Unsafe working conditions exist from the excessive use of chemical solvents. Child labor is common. Towns with new factories have attracted migrants, resulting in overcrowding and a housing crisis. The masses of crudely built shacks, often lacking electricity and piped water, that sprawl around border towns demonstrate the dangers of rapid, unplanned growth. And the physical containment of illegal immigration has resulted in construction by the United States of fences between El Paso and Ciudad Juárez, and a ten-foot high steel wall along the edge of Tijuana.

Despite these real problems, the Borderland is a vibrant region. Economically, it barely seems to have noticed the 1994 drop in the value of the peso and subsequent economic stagnation throughout much of the rest of the country. Investment continues, and a new $600 million Samsung television assembly plant in Tijuana will soon employ another 5,000 workers (*Economist*, October 28, 1995).

The Arid Northwest Lowlands

If the plateau is the heart of the Independent North, the surrounding coastal plains share its personality and trends, yet reflect distinctive landscapes of past history and recent government action (West, 1993). In the arid northwest, early settlement focused in the subhumid mountain valleys where Jesuit missionaries were attracted to the dispersed communities of native farmers. Spanish ranchers followed, and a mining economy evolved based on silver ore in towns such as Alamos. Meanwhile, the desert lowlands remained a cultural backwater, barely inhabited except by small native groups such as the Yaqui in coastal deltas of rivers that issued from the Sierra Madre Occidental.

Market scene, Guanajuato, Mexico, c. 1890. Traditional markets have changed little in the last century.

The roles of these two areas reversed in the present century. Early attempts to settle the coastal lowlands based on irrigation in the lower Yaqui, Mayo, and Fuerte rivers were by Americans and Europeans. Following the Revolution and its attendant concern for land reform, control of agricultural development shifted from the private to the public sector. In the late 1930s, the government began a number of large-scale irrigation projects. The Fuerte, Mayo, and Yaqui were dammed as they emerged from the Sierra Madre Occidental, and the flat delta lands bordering the Gulf of California were irrigated. To the north, in the Sonora and Concepción–Magdalena rivers, where water naturally rarely reaches the Gulf, irrigation systems based on deep well pumping were employed.

Although much of this improved acreage was assigned to ejidos, large foreign and domestic commercial farm operations with modern farming techniques and equipment have often coopted production as well as marketing. In Sonora, winter wheat occupies 30 to 50 percent of the land, and cotton has been the traditional summer crop. More recently, oilseed has replaced cotton while in the deltas using well water, overpumping and decreased flow have encouraged farmers to shift into perennial fruit trees and alfalfa. In Sinaloa, where warmer temperatures reduce the danger of frost and the border is still close enough, the coastal oases have become major suppliers of winter vegetables, especially tomatoes, to the U.S. market. Farther south, sugarcane and rice are common crops. Today, the northwest coast has supplanted the Bajío, near Guanajuato, as the principal center of modern agriculture in Mexico.

A spectacular growth of population in the principal cities has followed, particularly in Hermosillo (now approaching 0.5 million), Ciudad Obregón, and Navajoa. Initially agricultural service centers, these cities are now attracting a manufacturing base. In 1985, Ford opened an auto plant in Hermosillo.

A further stimulus to the northwest and Baja California is development of the fishing industry. Shrimp from the Gulf of California, and tuna and sardines from the Baja California coast are sent from fishing ports such as Matzatlán, Guaymas,

Ensenada, and La Paz to markets in North America. Mexico has Latin America's third largest fishing industry, following Chile and Peru.

Humid Gulf Lowlands

Irrigation projects similar to those in the Arid Northwest are found along the northeast Gulf coast, mainly in the delta of the Río Bravo del Norte (named the Rio Grande in the United States) between Reynosa and Matamoros. Here in a more humid region, irrigation involves flood control as well as water distribution. Sorghum for livestock feed and cotton are important crops. Fishing is well developed, focused on the main port of Tampico whose fleets bring red snapper, sea bass, groupers, and lobsters from the Gulf of Mexico.

► *The Central Breadbasket*

Extending south from San Luis Potosí to the Cuernavaca Valley and west from Guadalajara to the Veracruz lowlands, the Mesa Central of Mexico has been the traditional center of Indian population and the main region of food production. Today, some 60 percent of Mexico's population resides on the central plateau. Mestizo culture dominates the region. Maize, beans, and chilies form the basis of an ancient diet. Before the Mexican Revolution, land within the flat-floored basins was mostly held in haciendas. These large estates favored wheat cultivation, except in the Cuernavaca and Cuautla valleys, where sugarcane at lower, warmer, altitudes prevailed. Maize cultivation was found on the poorer soils of surrounding higher ground, where peasants cultivated small plots on a subsistence basis.

After the Revolution, marginal upland farming remained the only support of many of Mexico's desperately rural poor. The better bottom lands were partially redistributed and irrigated, creating a mixture of ejidos and mid-sized private farms,

Flower market in Mexico City, c. 1890. In the previous picture women are selling vegetables, just as they do today in Latin American markets. The male flower sellers may indicate that the produce is being transported a considerable distance to market.

which became the basis for modern agricultural development. Among the most productive of these areas is the Bajío of Guanajuato, a lowland region partly watered by the Lerma River. The once-prevalent wheat fields are now giving way to fresh vegetables. Broccoli, cauliflower, and snow peas are sold to packing plants owned by Bird's Eye, Green Giant, and Del Monte in the agricultural town of Irapuato and dispatched to the U.S. market. Tomatoes, garlic, chilies, and beans move to the Mexico City market. In the basins east of Puebla, maize predominates, grown in fields surrounded by agave cactus, which in colonial times preserved Indian fields from wandering cattle.

Manufacturing Towns

Set within this agricultural heartland are two distinct types of commercial communities: modern manufacturing towns and craft centers. León is the center of Mexico's tanning industry and is the country's leading shoe producer; Guadalajara, the country's second largest city, distributes food and manufactures textiles; Querétaro has become the center of Mexico's electronics industry; Toluca, Cuernavaca, and Puebla support automobile plants.

At the heart of the region is Mexico City with over a third of all manufacturing employment in Mexico. Although the government has had limited success in dispersing economic growth to other cities, the basis exists for an emerging metropolitan axis extending from Guadalajara and its Pacific port of Manzanillo, through the Bajío towns of León and Querétaro, to Mexico City and its satellites, Toluca, Pachuca, and Cuernavaca, and on to Puebla and the Gulf port of Veracruz. This axis is presently connected by Mexico's densest, but aging, rail network and an upgraded highway system. Parts of this region of Mexico may eventually fuse, creating a megalopolis similar to the area between San Diego and San Francisco, and between Boston and Washington, D.C.

Craft Centers

A second distinctive commercial community is found in remote valleys, or sometimes the suburbs of major cities, where traditional craft centers have developed a thriving trade for the art and tourist markets. Communities such as Tlaqueplaque, outside Guadalajara, and Tonalá, are known for pottery, Olinalá and Uraupan for lacquered boxes, Santa Clara del Cobre near Lake Pátzcuaro for copper ware, and Tlaxcala for woolens. Peasant crafts people, drawing on ancient traditions and using modern techniques, are supported by government programs and have enhanced the value of local economies.

▶ *The Oil Basin of the Gulf Lowlands*

From Tampico to Campeche, an arc of forested and lightly settled lowland has become the source of Mexico's oil wealth. Offshore in the Bay of Campeche, and increasingly inland in the vicinity of Villahermosa, pools of oil and natural gas have been discovered in deep sedimentary deposits. The earliest wells were brought in by foreign capital, beginning with British and American development off the coast of Tampico and Poza Rica in the first decade of this century. Estimated reserves grew dramatically in the 1970s with the opening of the Akal, Chac, and Ixtoc wells east of Campeche, as well as the rich Reforma field in central Tabasco and northern Chiapas states. With these finds, oil and gas went from being 2 percent of the country's export earnings to 80 percent at the height of the world oil price rise around 1980. Mexico borrowed heavily against the value of future oil earnings, leading to a debt crisis when fuel conservation and recession in the developed world caused a collapse of prices. Today oil accounts for only 25 to 30 percent of export value.

Before the 1970s, stock raising on inland savannas and sugarcane in the coastal lowlands were the prevailing activities of the sparsely populated Gulf lowlands. The government attempted to settle farmers from the overcrowded highlands in the tropical lowlands on irrigation projects, such as the system of dams and canals built on the lower Papaloapan River, beginning in 1947. Tropical, lateritic soils and the lack of lowland farming experience by pioneers from the uplands limited productivity, but such projects led to substantial deforestation and settlement. Oil has since transformed the economic landscape of the tropical lowlands. Pipelines connect both the Campeche and Reforma fields to Coatzacoalcos, and thence northward along the Gulf lowlands to the United States as well as to Mexico City on the plateau. Other lines cross the Isthmus of Tehuántepec to Salina Cruz.

Given the sacred place of oil in Mexican nationalistic thought, government policy has been to process as much of the resource in Mexico as possible. At present, about 40 percent of crude oil production is refined within Mexico. Refineries are found in Reynosa, Tampico, Tuxpan, and Minatitlán along the coast, as well as at Salamanca in the Bajío. A major refinery existed at Mexico City until it was closed for gross air pollution violations. New refinery capacity has been constructed in the 1980s at Salina Cruz, Tula, and Cadereyta near Monterrey. Two giant petrochemical complexes to produce fertilizers and fibers are coming on stream in Veracruz state at Parajitos and La Cangrejera.

► *The Southern Mountains*

Stretching from Puerto Vallarta to Salina Cruz, a band of rugged mountains separates much of the southern edge of the great plateau from the Pacific. The remote, southern mountains have been a refuge for outlaws and insurgents, as well as home to isolated, self-sufficient communities. Most of the people are subsistence farmers of Indian origin, especially Mixtecs and Zapotecs, who cultivate maize fields (*milpas*) on the mountain slopes (Chapter 3). Wheat is sometimes raised as a cash crop, along with honey from beekeeping and wool from sheep rearing. Such products find their way to local periodic markets in nearby towns. Overgrazed and eroded slopes are common, especially in Oaxaca, where deep gullies scar the red earth.

Economic growth in the region is hindered by a lack of transportation. Few roads cross the rough topography, and only one rail spur connects the coast with the national rail system. The line, completed in 1979, terminates at Lázarao Cárdenas, site of an ambitious plan to turn a sleepy port community into Latin America's largest steel complex, based on local iron ore deposits and imported coal. The government's intention was to create a new industrial growth pole, dispersing investment from Mexico City, and creating a community of 850,000 by 1990. Supported by oil revenues in the late 1970s and early 1980s, the project briefly blossomed before revenues and the rate of investment declined. Today, the Sicartsa mill, although producing some steel, has yet to turn this coastal area into a major industrial complex.

Along the Pacific, a new coastal highway links a number of government-planned resort centers. Beginning with the old colonial port of Acapulco, whose bay is now polluted with sewage from beach front hotels and shantytowns on the hills beyond, tourist centers have been developed in Puerto Vallarta, Manzanillo, Zihuatanejo, Ixtapa, Puerto Escondido, and Huatulco. The key to development of these and other Pacific coast sites is proximity to international airports, which provide connections to the North American market. The southern highlands remain a land of starkly contrasting worlds, one populated by peasant farmers, often little changed from pre-European times, and the other by visiting North Americans. The two cultures rarely interact.

► *Chiapas and the Yucatán: The Remote Periphery*

Mexicans who have felt most psychologically distant from the federal capital are those in Chiapas and the Yucatán Peninsula, areas of traditional Mayan culture. Together with the Southern Highlands, this region constitutes Mexico's poverty belt, a stark contrast with the growing wealth of the Independent North. Recently, parts of urban and coastal Yucatán have begun to modernize from directed central government investment. Chiapas turned to armed uprising in the 1990s. The isolated region had protected its identity for decades, but as the forces of modernization intruded into the area, violence resulted.

Chiapas

Southern Chiapas attracted some Spanish settlement and has a mestizo character. The narrow Pacific slopes of the Sierra Madre de Chiapas support a band of coffee, cacao, and cotton production that extends into western Guatemala. Pastures of the Río Chiapa valley, once given to stock raising, are now used for intensive irrigated agriculture as flood control systems, built in the 1980s, have filled the valley with a 60-mile long lake behind the Angostura Dam.

In the hinterland of the colonial capital and market center of San Cristóbal de las Casas is an area of forested mountains as remote as any in the country. Little altered from its pre-Hispanic form, the area is scattered with Indian villages and populated with many of the 1 million Mexicans who speak no Spanish. Four-hour bus rides to reach the regional market in San Cristóbal are not unusual for these peasant farmers, who have resisted modernization through inaccessibility. The heart of this remote area is the Lacondón tropical forest in northern Chiapas, and the refuge of the Chiapas guerrillas.

Migrating to this area in the 1980s have been thousands of Central American refugees fleeing El Salvador's civil war and attacks by the Guatemalan army against its Indian population. Mexico now has its own severe illegal immigration problem, with many refugees living in impoverished camps beside the border with Guatemala. Under U.N. observation and with U.S. aid, these Central Americans remain isolated and restricted to work projects such as excavating ancient Mayan ruins.

The outcome of the Chiapas revolt remains ambivalent and unresolved. Although the flashpoint was a stolen election for governor of the state, the underlying cause was the poverty of native farmers whose precarious lands on steep slopes remain too small to support their families and whose communities are in conflict with cattle ranchers and logging companies. The central government has long ignored the welfare of Indian communities and failed to protect them when outsiders arrived to exploit resources formerly the preserve of traditional villagers.

Yucatán

The Yucatán Peninsula may be considered as two distinctive areas. Southern Yucatán remains as isolated as northern Chiapas. Lightly populated with subsistence farmers, the area's forests have recently begun to reveal new Mayan sites. In turn, this has attracted environmentally sensitive tourists, who with local and foreign archaeologists seek to persuade the Mexican and Guatemalan governments to establish national parks to preserve the region's heritage. Opposition to the idea from farmers in Quintana Roo resulted in the burning of forests in the region to prevent restrictions on traditional practices.

While Yucatecans retain a strong sense of independence from the central government, northern Yucatán has lost much of its isolation in recent years. Government development of a coastal tourist strip from Isla Mujeres to Isla Cozumel, particularly after the construction of an international airport at Cancún, has brought international investment and tourists from North America and Europe. In turn, remittances from the large pool of workers employed in tourism is a major economic support of the dense network of villages in the immediate area. Rapid growth has occurred around Mérida, now exceeding 700,000 people, fueled by the government's decision in 1989 to expand the *maquiladora* program to areas beyond the northern border cities. Some 50 *maquiladoras* in the Mérida suburbs now seek to shift their early focus on textiles and dental-care products to higher-value sectors such as jewelry and electronics. Other industries process natural fibers from local sisal (henequin) and convert soybeans imported through the port of Progreso to oil for the domestic market.

► *Mexico City: The Dominant Metropolis*

Mexico City contains a fifth of the country's population. With a population of 20 million, it is the world's largest city, but it is not a world city in terms of international finance. Mexico City has grown as a result of national centrality. Since the Aztecs, political administration has been concentrated in Mexico City. Similarly, it has been the focus of the country's road and rail network. In 1960, it manufactured over 40 percent of Mexico's industrial products. Since then, there has been a relative decline in manufacturing, but Mexico City now produces half of all of Mexico's services (Ward, 1990). In addition, the country's media are concentrated in Mexico City and disseminate an image of urban opportunity and affluence to the rest of the country.

Since World War II, there has been a massive migration from the impoverished countryside to the capital city. The early migrants came from communities within 75 miles. By 1989, the migration field extended an average of 880 miles and involved some 700,000 people a year. Population densities in the city are higher than Tokyo, twice those of metropolitan New York, three times those of Paris, and four times those of London. In the past two decades, urban expansion has spilled beyond the Valley of Mexico into surrounding parts of the State of Mexico. Once migration accounted for 60 percent of Mexico City's metropolitan growth; now this figure has dropped to 38 percent, as natural increase fuels most growth and the pool of potential migrants in rural areas is reduced with Mexico's growing urban majority.

The consequences of population growth have been shattering. The demand for employment has far outstripped available jobs, forcing reliance on the informal economy. While fluctuating with the country's economic fortunes, the informal sector of shoeshine entrepreneurs, Chiclet sellers, hawkers of newspapers, washers of car windows, street artists, and balloon sellers ranges between 34 percent in 1981 to 40 percent of all employment in 1987. Many underemployed are among the 5 million Indians who live in the city.

Government and private housing markets fail to provide for up to a third of the city's people, and new immigrants must fend for themselves. The result has been the growth of huge squatter settlements. For example, over 1 million people live in Nezahualcóyotl on the dry lake bed east of downtown. In the rainy season, the soil of this huge squatter settlement turns into a swamp, making movement difficult over the 750 miles of streets, only half of which are paved. The early squatters built their own shacks on land with dubious titles, tapped illegally into nearby electric distribution lines, created rudimentary sewage disposal systems that often failed, and purchased water from roving tank trucks. Services have slowly been provided by the government as the population grew to a density of over 100,000 per square mile. Meanwhile, the outward spiral of new squatting settlements with no services was repeated in more distant areas such as the Chalco valley in the southeast, where population has grown from 90,000 to nearly 750,000 in recent years.

Traffic congestion, Mexico City.

Not all of Mexico City is poor. Elite residential districts extend from the city center along axes to the west (Lomas de Chapúltepec and Polanco districts) and southwest (San Angel, Coyoacán, and Pedregal districts), following the model of the Latin American city described in Chapter 5. Large middle class developments extend to the northwest (Ciudad Satellite). In such districts smart shopping malls and multiservice hypermarkets dot the landscape. In the urban center, the Zona Rosa beside the broad Parisian-like Paseo de la Reforma boasts stores and restaurants as exclusive as any encountered on Park Avenue or the Champs Elysées.

Mexico City's hyper-urbanization has resulted in severe environmental degradation and acrid air pollution that respects no income limits. The city is 1.5 miles above sea level in a steep-sided basin, which provides excellent conditions for temperature inversions that trap pollutants. At this altitude there is a 23 percent reduction in oxygen content, which increases ozone-producing photochemical reactions. When the exhaust from 10,000 factories and 2.8 million vehicles is added, the pollutant levels of ozone, nitrogen dioxide, and toxins in the air in winter exceed by four times the tolerable limits set by the World Health Organization. Because 30 percent of the city's residents lack any sewage service, the atmosphere is further contaminated by fecal dust that contains infectious microorganisms such as salmonella, streptococcus, staphylococcus, shigella, and amoeba.

Although only 34 percent of urban journeys are by bus and car, 85 percent of pollution is thought to come from vehicular traffic. Mexico City authorities expressed an early construction bias toward building expressways, and a metro (subway) was not begun until the 1960s. The original track has been extended, but it is still overcrowded and insufficient. Attempts have been made recently to promote the wider use of unleaded gasoline and control auto emissions by forbidding 20 percent of cars to use the roads each day. But many wealthier Mexicans have bought additional old cars to circumvent the ban. Plans call to extend the subway, replace 3,500 smoke-belching buses, and retire taxis built before 1985, all in an effort to reduce air pollution. Factories, such as the oil refinery at Atzcapotzalco, a notorious polluter, have been closed and others fined for excess emissions.

Environmental controls may stabilize the situation, but the city will remain hostage to region-wide temperature inversions. Moreover, a recent reputable study suggests that the wrong causes may have been targeted. Rather than auto and factory emissions, the major component of Mexico City's smog may be leaking liquid propane from faulty gas tanks used almost universally within the city for cooking and heating.

The solution to the high rates of in-migration and natural increase is deflection of growth away from Mexico City to surrounding urban centers. Such programs have not been successful to date, involving major infrastructural development such as ring highways and railroads around the Mexico City basin to connect the capital to a wider polycentric metropolitan region including Cuernavaca, Toluca, Pachuca, Tlaxcala, and Puebla. Beside cost, such a program would involve cooperation among the five state governments that adjoin the Federal District.

SUMMARY

Since oil prices plunged and inflation surged in the mid-1980s and Mexico labored under a crushing foreign debt, the government has abandoned numerous nationalistic policies that had made it one of the world's most protected economies. Between 1988 and 1994 under the Salinas government, opportunities for foreign ownership of Mexican companies, reduction in tariff barriers, and privatization of government-owned businesses have helped slow the country's inflation, reduced unemployment rates, and improved the climate for foreign investment. These changes laid the foundation for the adoption of NAFTA. Although the peso crisis of 1994 revealed remaining weaknesses in Mexican economic policy, recovery has been far more rapid than after 1982. Exports grew rapidly in 1995 as the peso declined in value and Mexican exports became cheaper in export markets. Mexico had a large positive trade balance in 1995.

Mexico now needs significant domestic economic expansion. Fertility rates have declined in the past 20 years to an average of three children per family, but the workforce is still growing by 900,000 per year. An economic growth rate of 5 percent will be needed to absorb such numbers. To accomplish this objective, Mexico will need to address the following potentially destabilizing issues:

1. *Income distribution.* Mexico remains socially and geographically unequal. A great gulf in income and opportunity remains between the vibrant North and the poverty of the South. The Chiapas rebellion was a reminder of this geographic divide. A similar disproportionate income distribution, a feature of Mexico since colonial times, remains stubbornly in place. While the richest 10 percent of Mexican families earn 40 percent of the country's household income, the poorest 20 percent gain a mere 3 percent (*Statistical Abstract of Latin America*, 1993, p. 441).

2. *Drugs.* Mexico is rapidly acquiring drug cartels to rival those of Colombia in scale and income, in the process corrupting the political process with bribery and violence. Although not major producers, Mexicans are acting as middlemen for producers in several parts of the world, absorbing the risks involved in delivery to the U.S. market (see Figure 8.7). Recent reports suggest that Colombians are paying Mexican traffickers in drugs rather than cash. One ton of Colombian cocaine delivered to the Bronx in the past might have fetched a $2 million transport fee. A half share of the load yields $10 million in the New York market (*New York Times*, July 30, 1995).

3. *Migration.* Emigration to the United States, both legal and illegal, remains an important safety valve for Mexico as long as the population and economic growth rates remain imbalanced. Immigrants have always come from the poorer parts of Mexico, but in California's fields today Mixtec Indians from Oaxaca, one

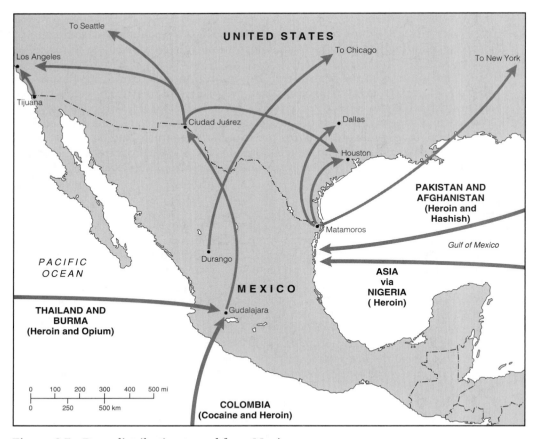

Figure 8.7 Drug distribution to and from Mexico.

third of whom are thought to be illegal, are rapidly replacing earlier migrants, who
originated from the country's north and center. Over 200 Oaxaca villages are now
represented on farms from San Diego to Salinas. Despite low wages, these new
arrivals send dollars back to their home communities. Consequently, a more restrictive U.S. policy toward immigration, especially from Mexico, could have political
and economic consequences in the impoverished southern highlands and remote
periphery.

4. *NAFTA.* The free trade agreement, implemented on January 1, 1994, presents contrasting opportunities and dangers. Mexico offers a young and fast-growing
population, a substantial potential market of new consumers. It stands to attract
foreign capital and businesses seeking lower labor costs and (although Mexican
officials deny it) looser environmental regulations. To the extent that new investments can be diverted to secondary urban areas, which seems likely, more balanced
economic growth may result.

On the other hand, reducing most tariffs across its North American border
immediately and phasing all others out over a 15-year period expose small Mexican
companies that previously depended on protected markets to competition from
multinational firms (Grayson, 1995). Most controversial is NAFTA's expected
impact on peasant agriculture and the maize crop. In 1992, 50 percent of arable
land was in maize and 78 percent of Mexico's farmers were maize producers.

However, of the 2.4 million producers, 2.2 million had farms that averaged only 2.3 ha in size (De Janvry, 1995). Most maize farmers use primitive technology, achieve low yields, and have properties insufficient to support a family. As a consequence, the Mexican government has long subsidized corn producers and the cost of maize flour for consumers. In 1991, Mexican yellow maize was supported at the equivalent of $71 per ton, compared to a U.S. price of $28 per ton and $21 per ton in Canada.

In 2009 tariffs and subsidies on corn will be eliminated, and Mexican peasant farmers will face competition from larger and more efficient producers in Iowa and Manitoba. The Mexican government's modification of Article 27 of the Constitution to permit the gradual privatization of ejido land was a response to the need to modernize agriculture. But many feel that the combined effect of cheaper maize imports and more flexible landownership will result in agricultural consolidation and the eviction of a large number of peasants, who will swell the numbers moving to the cities.

A contrary view is implied by the case of La Moderna, Mexico's largest tobacco company, which is working with 7,000 *ejidatarios* around Tepic (*Wall Street Journal*, July 26, 1995). The company has persuaded the farmers to pool their land into large farms while retaining individual land rights, and to specialize in the tasks of producing tobacco. The company provides advice, plants, seeds, and machinery and splits profits evenly with the farmers. Such a system preserves traditional relationships within the land and village communities while permitting farmers to produce a diversity of crops on larger farms with more productive technology.

5. *Transport infrastructure.* A factor limiting national economic integration is Mexico's aging transport infrastructure. Three quarters of the country's 16,000 miles of railroad predate 1910. More than 20 percent of locomotives are out of service at any one time. Track is so poorly maintained that speeds above 20 m.p.h. are rarely achieved. The only new recent additions to the system are: (1) the link between Chihuahua and the Pacific port of Topolobampo, which carries more tourists through the famed Copper Canyon than products of the northern hinterland destined for export, the original purpose of the line; (2) the 100-mile extension from Apatzingán to the Lázaro Cárdenas steel mill on the Pacific coast; and (3) electrification of the much-used line between Mexico City and Querétaro.

In the absence of a good rail system, most goods move by truck. Mexico lacks a network of interstate highways common in the United States, and public investment in superhighways as well as rural roads declined after the 1970s. But since 1989, Mexico has offered private companies the right to bid on construction of toll highways in exchange for revenues collected. By 1994, some 3,700 miles of new highways had been completed at a cost of $14 billion. Unfortunately, only 3 of 53 concessions are producing a profit. High tolls and low traffic volume have put many highway projects on the verge of bankruptcy. On the Cuernavaca–Acapulco highway, for example, tolls are only 13 percent of revenues needed to meet the construction debt.

Before NAFTA, all goods crossing the border overland (90 percent of total trade) had to transfer their cargoes and passengers to or from Mexican vehicles. NAFTA abolished transfer for rail and will do the same for buses in 1996 and trucks in 1999. However, modern intermodal forms of transport will be substantially impeded unless major infrastructural reforms in transportation can be accomplished. Bringing the transport network to twenty-first century standards remains the ultimate challenge if Mexico is to integrate its *patria chica* fully and finally loosen its ties to the past.

FURTHER READINGS

Arreola, D. D., and R. J. Curtis. *The Mexican Border Cities: Landscape Anatomy and Place Personality*. Tucson: University of Arizona Press, 1993.

Augustin, B., and M. Kohout. "The Border Industries Program: A Geography of *Maquiladoras*." In *A Geographic Glimpse of Central Texas and the Borderlands: Images and Encounters*, edited by J. F. Petersen and J. A. Tuason. Indiana, Pa: National Council of Geographic Education, 1995, pp. 75–84.

Doolittle, W. E. "Agricultural Expansion in a Marginal Area of Mexico." *Geographical Review* 73 (1983): 301–313.

Liverman, Diana M., "Drought Impacts in Mexico: Climate, Agriculture, Technology, and Land Tenure in Sonora and Puebla." Annals of the Association of American Geographers, 80 (1990): 49–72.

National Geographic. "Emerging Mexico: A Special Issue." Vol. 190, No. 3, August 1996.

West, R. C. *Sonora: Its Geographical Personality*. Austin: University of Texas Press, 1993.

———— and J. P. Augelli. *Middle America: Its Lands and Peoples*, 3d ed. Englewood Cliffs, N.J.: Prentice Hall, 1989.

CHAPTER 9

![decorative border pattern]

Central America

PETER REES

INTRODUCTION

Fragmented and marginalized, Central America is an area whose name mocks its struggle for unity. Since the short-lived United Provinces of Central America (1823–1839), following independence from Spain, 24 attempts have been made to achieve unity or economic integration (Melmed-Sanjak, 1993). All have failed.

The formation of the Central American Common Market (CACM) in 1960 raised hopes of regional integration with the creation of a central development bank, completion of the Central American highway from Guatemala to Panama, and attempts to allocate manufacturing among specific countries and reduce tariffs. Intraregional trade increased tenfold in the 1960s, and dependence on foreign trade lessened. However, Panama and Belize were never members. Furthermore, old national animosities undermined the advances, epitomized by a brief, bloody war between El Salvador and Honduras in 1969 that was sparked by a soccer match between the two nations, although the real cause was squatting by landless Salvadorians across the Honduras border.

For the seven republics of Guatemala (9.9 million), El Salvador (5.9 million), Honduras (5.6 million), Nicaragua (4.6 million), Costa Rica (3.6 million), Panama (2.7 million), and Belize (200,000) (Table 9.1), not only has regional unity remained elusive but internal political and cultural differences have often hampered achievement of national identity. An ongoing 30-year period of violence between peasants and the military elite in Guatemala and fierce civil wars in Nicaragua (1979–1990) and El Salvador (1980–1991) with spillover effects into Honduras, led observers to describe the 1980s as Central America's "lost decade" (Vilas, 1995). This domestic unrest and conflict within countries further undermined the CACM.

Regional disunity allowed the destiny of Central American nations to be shaped by external forces. To the outside world, Central America became noteworthy as the barrier to easy movement between the Atlantic and Pacific oceans. Its strategic importance tempted foreign nations, particularly the United States, to interfere in the area's political development. The country of Panama and the canal were creations of U.S. foreign policy in the beginning of this century. Later, U.S. troops occupied Nicaragua (1927) and Panama (1991), and the U.S. Central Intelligence Agency (C.I.A.) initiated the coup that overthrew the elected government of Guatemala in 1954. The U.S. government sided with Contra rebels against

TABLE 9.1 Central American Nations: Demographic Profiles

Country	Total Population (1000s)	Birth Rate (per 1,000)	Death Rate (per 1,000)	Natural Increase (%)	Doubling Time (years)	Infant Mortality (per 1,000 live births)	% Urban	GNP per Capita ($ U.S.)
Belize	.2	38.0	5.0	3.3	21	34.0	48	2,550
Costa Rica	3.6	26.0	4.0	2.2	31	13.0	44	2,380
El Salvador	5.9	32.0	6.0	2.6	27	41.0	45	1,480
Guatemala	9.9	36.0	7.0	2.9	24	50.0	47	580
Honduras	5.6	34.0	6.0	2.8	25	51.0	39	1,190
Nicaragua	4.6	33.0	6.0	2.7	26	49.0	63	330
Panama	2.7	22.0	4.0	1.8	39	18.0	55	2,670

Source: Population Reference Bureau *1996 World Population Data Sheet* (Washington, D.C.: Author, 1996).

Nicaragua's Sandinista government during the 1980s. Moreover, foreign enterprises often dictated Central American economies. U.S. banana companies "owned" the Honduran government in the 1930s and 1940s.

As a result, each Central American nation remains a classic dependent economy. Reliant on one or two export products and at the mercy of the vagaries of world markets, each must earn hard currency to pay for imported goods. Imports are necessary because a minuscule domestic market cannot attract capital or support competitive local industries capable of making products as cheaply as those imported from developed nations. Alternatively, protecting local manufacturers with a high tariff wall only drives up prices of finished goods, encourages smuggling, and restricts improvement of living standards. Economic diversification, together with a common solution to problems of land hunger among peasants, seems only achievable with a coordinated Central American approach. Today, the need for economic and political cooperation has never been more necessary as nations form regional groupings in response to the altered conditions of a new world economic order. NAFTA, by favoring Mexico, may disadvantage Central American nations in the North American market. Elsewhere, Argentina, Brazil, Paraguay, and Uruguay have formed Mercosur, while Andean nations seek to revive the Andean Pact. If Central America, together with the Caribbean island nations, is not to become the Balkans of the region, remnants in a shatter zone blown about before the winds of external political and economic forces, some form of regional unity must be achieved. The theme of this chapter explores whether a geographical coherence exists to support that unity.

PHYSICAL ENVIRONMENT

Put aside the region's frequent human hardship and conflict and Central America appears as a land of often startling beauty and tranquil, near-perfect climate. Majestic peaks and deep valleys characterize upland Guatemala. Richly diverse rain forests make Costa Rica a world destination of eco-tourism. The tropical waters on the Caribbean coasts of Honduras and Nicaragua hint at a largely isolated paradise. But even the physical environment is subject to turmoil from environmental hazards. Earthquakes and volcanic eruptions frequently bring violence to this geologically young region (Figures 9.1 and 9.2). Fierce winter storms from the disruption of tropical air masses by polar outbreaks can damage crops and disrupt shipping

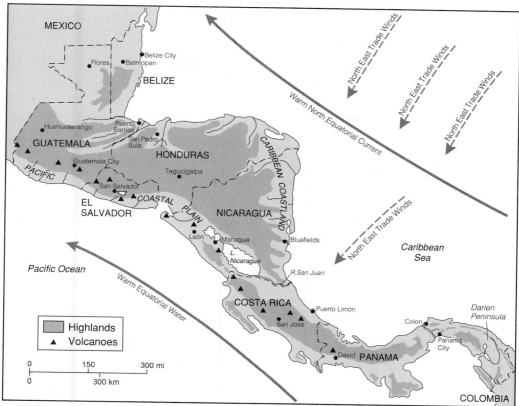

Figure 9.1 Central America: Regional setting.

along Caribbean shores. Atlantic hurricanes and Pacific typhoons have pursued destructive paths into isthmian coasts and their mountain borders. Consequently, Central America is a land of both fertile promise and considerable risk.

► Landforms

Geologically, Central America begins at the Isthmus of Tehuantepec in Mexico (Figure 2.16). From the isthmus to Panama, the Pacific coastal area is dominated by a spine of steep volcanic mountains and narrow coastal lowlands where the western edge of the Caribbean crustal plate collides with the Cocos plate underlying part of the eastern Pacific ocean. The result is a series of deep ocean trenches and fault zones off the Central American coast that spawn earthquakes as the plates lurch past each other. In 1972 an earthquake leveled Managua, Nicaragua's capital.

As the denser Cocos plate is forced down into the earth's mantle, its edges liquify from the pressure and fuel the upward movement of magma through crustal cracks to form the chains of volcanic mountains lining the Pacific. These volcanoes intermittently throw lava, ash, and dust across their slopes. Farmers are attracted to the fertile, well-drained volcanic soils that result but risk calamity. Villages and crops can be buried by lava or mud flows, roofs collapse from the weight of volcanic ash, and hot toxic clouds of gas can rush downslope with explosive force. More than 25 active volcanoes exist in the zone from Guatemala to western Panama, some rising to 14,000 feet. The region is one of the world's most susceptible to volcanic hazards.

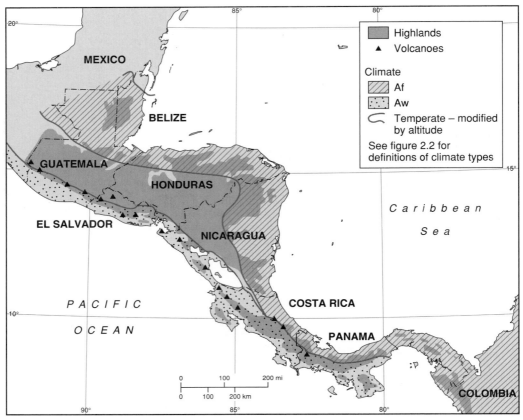

Figure 9.2 Regions of Central America.

The rest of Central America is a remnant of "Old Antillia," a geologic region formed when the submerged crystalline rocks of the Caribbean Plate were overlain with deposits of limestone. Subsequently, Old Antillia was buckled into a series of parallel ridges and valleys, before most of it was drowned once again by the Caribbean Sea. Erosion removed some of the limestone caps from ridges still above sea level, revealing the crystalline core beneath. These crystalline rocks provided the source of gold found by the Spaniards in northern Honduras. The porous limestone supports a semiarid vegetation cover or rich farmland where the limestone has washed down to fill valleys and depressions. But in Central America (particularly the southern highlands of Honduras and northern Nicaragua), the limestone areas have often been covered with thick deposits of volcanic lava and ash from the Pacific volcanic axis.

Elsewhere in Central America, the western, higher portion of Old Antillia is manifest in a series of mountain ridges extending generally eastward (for example, the highlands of central Guatemala, northern Honduras, and eastern Nicaragua), separated by intervening valleys (for example, the Motagua Valley in Guatemala). These ridges and valleys are traceable across the Caribbean as emergent islands or deep oceanic troughs, all the way to the arc of the Lesser Antilles that marks the eastern side of the Caribbean plate. The northern edge of Old Antillia escaped compression and folding, resulting in flat platforms of limestone strata that form northern, lowland Guatemala and the Yucatán Peninsula, as well as the many cays (islands) that line the Belize coast.

Extending from the northwest to the southeast through Nicaragua and Costa Rica is a great crustal crack, or rift valley. Huge deposits of volcanic ash and material have filled much of this fault-formed depression, disrupting drainage and empounding lakes Managua and Nicaragua, the largest bodies of fresh water in the region. A shallow river, which can cease to flow in the dry season, drains surplus water from Lake Managua into Lake Nicaragua. The San Juan River drains water from Lake Nicaragua to the Caribbean Sea. The land around the lakes has rich volcanic soils and is well populated. In the northwest, the rift valley has been drowned by the Pacific, producing the Gulf of Fonseca. Not only has this great valley, connecting both oceans, supported considerable settlement on its fertile and well-watered soils, but it also attracted attention after the California Gold Rush (1849) as a transportation artery across the isthmus and was a rival with Panama for the canal route at the end of the nineteenth century.

► Climate

Central America lies completely within the tropics, and annual temperatures vary less than 10° Fahrenheit. Temperatures that average 80–85° Fahrenheit are significantly modified by altitude, creating the ecological zones of *tierra caliente, tierra templada, tierra fría,* and *tierra helada* common to much of tropical Latin America (Figure 2.8). This modification of temperature with elevation is best seen along the Pacific slopes of the volcanic axis and is closely associated with the changing agricultural regimes from sugarcane, cotton, and bananas near the coast through tropical fruits such as oranges, and coffee, to wheat, corn, and pasture grasses in the highlands.

Seasonality in Central America is expressed by rainfall, with a dry season occurring in the early months of the year. The dry season is more evident on the Pacific side of the region than it is on the Caribbean coast. The dry season is shorter in Panama than it is farther north, as in Costa Rica. These contrasts can be seen by comparing rainfall at Gatun in Panama (Figure 9.3) with precipitation at San José

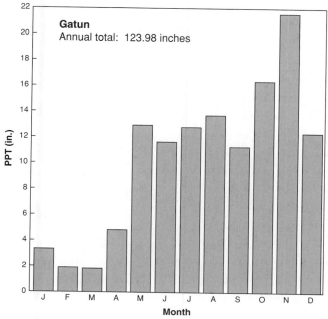

Figure 9.3 Rainfall of Gatún.

in Costa Rica (Figure 9.4). During the dry season, the northeast trade winds are more stable as the subtropical high pressure cell in the Atlantic shifts southwest toward the tropics. In summer, the subtropical high—the Bermuda high—moves north away from the tropics. The region then is increasingly influenced by the Intertropical Convergence Zone (ITCZ), a narrow zone of instability into which the northeast and southeast trades converge. In July, the ITCZ lies across Panama and Costa Rica (Figure 2.9).

As the trade winds sweep across the warm, enclosed waters of the Gulf of Mexico and the Caribbean, they absorb huge quantities of moisture. The low-lying Yucatán Peninsula has insufficient relief to produce orographic precipitation. Consequently, this flat plateau receives 40–80 inches of rainfall per annum and, combined with its dry limestone surface, supports only semiarid vegetation. Farther south and closer to the equator, on the Caribbean coasts of Nicaragua, Costa Rica, and Panama, rainfall totals are higher, and the dry season, in the early months of the year, is short. Annual rainfall totals along this coastline are often in excess of 100 inches. The heaviest amounts (sometimes exceeding 200 inches per annum) drench inland, eastward-facing mountain slopes, as the trades rise up steep escarpments over 8,000 feet. Dense stands of rain forest at lower altitudes give way to conifers as elevation cools temperatures. Descending the Pacific slopes, the warming trade winds release little moisture. Consequently, the west sides of mountains are less wet (under 80 inches per annum) and obtain most of their rainfall from Pacific storms, whose frequency increases southward to Panama.

Overlaying these broad climatic patterns are numerous microclimatic niches formed from a combination of elevation and orientation to prevailing winds, as well as day-to-day weather conditions. The latter may include hurricanes that attack the Caribbean coastline north of Panama from July to October and Pacific coast typhoons in September and October. In winter, blasts of Arctic air from mid-latitude

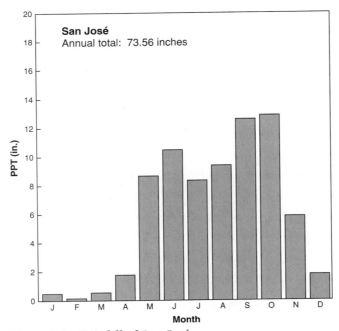

Figure 9.4   Rainfall of San José.

cyclones bring stormy conditions that can damage fragile banana plantsas far south as eastern Nicaragua and produce overcast skies and a fine, cold drizzle in the more populated centers of the interior highlands.

► *Soils and Vegetation*

Soils and vegetation derive from the interplay of climate and surface landforms. Soils in Central America vary greatly in their capacity to support human endeavor. Newly created volcanic material breaks down rapidly. Volcanic soils are rich in mineral nutrients and are well drained. Within three years, coffee plants can thrive in areas devastated by volcanic action. Crystalline rock breaks down only slowly and yields thin soils while limestone contains valuable calcium minerals for plants, however the rock's porosity and lack of surface water can limit its value for farmers. Where eroded limestone collects in depressions and stream valleys, the result is a mineral-rich, fertile, sticky clay such as that found in the Guatemala Petén, where, perhaps not coincidentally, the Mayan sites of the Classic period are found.

Elsewhere, red, leached, lateritic soils lie beneath rain forests, although much forest has been cleared, exposing the infertile soil to colonization by grasses. The region's soils have suffered degradation from centuries of woodcutting and over-grazing. Studies indicate that the erosive effect of running water increases exponentially: If stream flow, unimpeded by forest cover, doubles, scouring capacity quadruples, carrying capacity increases 32 times, and the size of particles able to be moved increases 64 times. Thus, in a forested area with an average annual rainfall of 85 inches per year, approximately one half ton of soil per acre would be removed from a 12–15 percent slope. However, on an unforested, cleared slope only half as steep, the soil eroded would be 15 tons per acre (Meggers, 1971). Soil loss, perhaps more than the widely publicized disappearance of tropical forests, is Central America's most pressing environmental problem.

Much of Central America was once covered with dense tropical forests, particularly on the eastward-facing slopes. But as with soils, the natural forest cover has been altered by human action. Much of the tropical dyewoods that lined the coastal lowlands from Belize to southern Nicaragua and attracted British incursions to the colonial Spanish domain have been culled. In eastern Nicaragua and northeastern Honduras, extensive areas of savanna grassland punctuated with North American pines prevail, despite rainfall of 80–100 inches per year, and may be the result of either prehistoric burning or the inability of the soils to retain moisture. Elsewhere, woodlands, particularly deciduous oaks and pines in the *tierra templada*, have given way to pastures and agricultural settlement, and natural plant cover is increasingly restricted to the more isolated, higher elevations where cloud forests prevail. On the drier Pacific slopes and in isolated micro-rainshadows, where moisture levels decline, semiarid, xerophytic plants and even cactuses are not uncommon. Beside both the Pacific and Caribbean shores, mangrove swamps of salt-tolerant plants present a tangled and forbidding facade but contain within them enormous varieties of animal and fish life. Mangroves are the wetlands of the tropics, and their destruction is a net environmental loss.

ORIGINS OF THE CONTEMPORARY LANDSCAPE

Most of Central America is a culturally *peripheral* area, a fractured buffer zone on the land bridge connecting North and South America. As such, it straddles many divides. Some date from the prehistoric, such as the boundary between the seed-based agriculturalists of highland Mexico and South American root-based cultures. Others have colonial origins, such as the cultural and social divide that separates

Hispanic, mestizo populations in the densely settled highlands, from isolated Indian and Caribbean black coastal communities, whose contemporary descendants still speak an indigenous language or an English patois and feel a distinct lack of affinity for their Spanish-speaking countrymen to the west. Still other divisions have neo-colonial roots, such as the uneasy relationships that developed between foreign entrepreneurs, particularly on commercial plantations on either coast, and the traditional landowning classes whose cattle ranches spread across highland pastures. These divisions contribute to contemporary Central America, which continues to reflect its past.

▶ The Pre-Hispanic Legacy

The Mayan civilization, the major pre-European cultural florescence, had already declined by the time the Spaniards arrived. Evolving in highland Guatemala, the Mayan culture area spread downslope to the forested Petén lowlands, where from A.D. 300–900 city-states of the Classic Mayan period flourished at sites such as Tikal. The Mayans relied on seed-based agriculture, including intensive forms of irrigation. They constructed elaborate temple complexes in the center of major urbanized zones. At Tikal, archaeologists measure an urbanized area with a diameter of 30 miles. The Mayans were traders by land and water, establishing their cultural influence at sites on the Belize coast and in northern Honduras at Copán, and spreading farming techniques as far south as Lake Nicaragua. The collapse of the network of Mayan states within a century is incompletely understood but appears to have an ecologic cause, in which farming techniques were insufficient to support an increasingly large nonagricultural population. Modern scientific examination of this question is important because of the claim that indigenous, rather than modern, agricultural methods offer the most productive ways to support today's dense populations.

Seed-based cultures in pre-Hispanic times spread southward from Guatemala along the Pacific coast to the Nicoya Peninsula. Nahuatl (Aztec) speakers established trading outposts as far south as the San Juan River (Newson, 1987). In the central highlands of Honduras and Nicaragua, South American root-cultivators based on shifting cultivation predominated, spreading into the eastern savannas toward the Costa de Mosquitos. The people spoke Sumu, Matagalpa, and Rama languages and lived in self-contained villages. More advanced chiefdoms of Huerta culture occupied the Costa Rica highlands, developing skills in gold metallurgy. Most of these indigenous groups vanished with the onslaught of European disease and slavery, but along the isolated Caribbean coast of Nicaragua, indigenous communities survived. In Panama's western periphery, around Cerro Colorado and down toward the coast of Bocas del Toro, Guaymí forest farmers in villages of extended families live today much as they have since before the Spaniards. To the east, Cuna speakers survive among the San Blas Islands, producing colorful *molas* for the tourist trade, while canoe-bound Chocó occupy riverine environments in the remote Darién Peninsula. However, the most populated areas of contemporary indigenous settlement are in Guatemala, especially in the remote northwest Cuchumatán highlands, west of the Chixoy River. There, Mayan languages and corn-centered culture that resisted colonial Spanish acculturation still shape community life and subsistence practice (Lovell, 1992).

▶ The Colonial Legacy

At the time of Spanish conquest, Central America's indigenous population may have stood at 5.650 million (Denevan, 1992, p. 291), although some estimates reach as high as 13.5 million (Dobyns, 1966). Whatever the total, the native population declined very rapidly. In the first half of the sixteenth century, over half a million Indians were exported as slaves from Nicaragua and Honduras, bound for

Panama and the Peruvian mines, and slaving was an early economic activity of the Spaniards until the practice exhausted the native populations. What slavery failed to do, disease accomplished, especially in the tropical lowlands where depopulation rates may have reached 25 to 1 by the end of the sixteenth century.

Indian population density and the location of mineral deposits determined the early direction of Spanish settlement. The Pacific coast was the principal point of entry, and the central highlands drew European settlement inland. Gold mining developed around Nueva Segovia and was administered from León, and western Honduras experienced an additional boom in the early 1800s with the discovery of rich silver veins. But mining was never on the scale of Mexico and Peru, and soon Spanish interest turned to land.

Large grants of land in savannas formerly farmed by the now-decimated native population attracted continued Spanish settlement. Haciendas rejected cacao, a common crop in pre-Hispanic times along the Pacific coast, because of the need for large amounts of now-scarce labor, and instead turned to indigo (especially in El Salvador in the eighteenth century) and later tobacco. But neither was as successful as stock raising for local meat consumption and more importantly for hides. In Costa Rica and Panama, raising mules for the trans-isthmian trade further attracted Spaniards to ranching. Administrative settlements such as Antigua Guatemala, San Salvador, León, and San José developed by the end of the sixteenth century, while ports such as Realejo and Nicoya supported shipbuilding. In the highlands, and especially in areas of dense native populations, the Spanish practice of *congregación*, assembling dispersed Indian populations into larger, centralized agricultural communities, accounts for much of the network of rural towns found today.

Overall, Spanish Central America was oriented toward the Pacific and toward the south. Few Spaniards penetrated the river valleys to the Caribbean coast, considered to be the region's "back door." As a poorly defended fringe of empire, it was vulnerable to intrusion by rival European powers. British and Dutch privateers established beachheads at communities such as Bluefields, and semipermanent English settlement along Nicaragua's Costa de Mosquitos was underway by 1620. In 1641, a ship carrying a large number of black slaves was shipwrecked. Other blacks arrived along the coast as imported labor from Jamaica to work on plantations established by the English or as runaways escaping from the mines of Tegucigalpa and Comayagua. Blacks settled, intermarried with Indians, and traded with the English in dyewoods and mahogany. By the end of the eighteenth century, these Zambo-Miskito people dominated eastern Nicaragua and parts of eastern Honduras (Newson, 1987). Their influence had spread northward to Belize, where the English also established settlements, supplied from Jamaica. The Spaniards never successfully gained control over this eastern region. Although nominally part of the Spanish-speaking realm, the coastal communities continued to speak a blend of English-native patois and to reject cultural imports from the highlands to the west. Such was the gulf between the two populations that, following independence from Spain, the Miskito Kingdom was recognized by Britain as distinct from the Central American nations (Naylor, 1989).

Although administered from Guatemala, Central America in the colonial era was economically oriented southward to Panama. The land bridge between the two oceans was made more significant because of the crown's insistence on trying to control trade and make it flow along a limited number of routes. The network of commercial links between Spain and its American colonies, the *Carrera de Indias*, consisted of two annual protected convoys (*flotas*) to and from Spain, one destined for New Spain and the other to Panama, although Lima was its real terminus. The South America *flota* departed Spain in August and reached Nombre de Dios in late

September. After 1597, the Caribbean terminus was moved to Puerto Bello, near present-day Colón, a better-protected port. There, the largest fair in the Spanish colonies developed, primarily because of the restriction the crown imposed that merchants from Spain could only trade on their own account as far as Panama while Peruvian merchants were limited in the opposite direction. The fair at Puerto Bello thrived for 14 to 40 days, after which the *flota* sailed for the fortified harbor of Cartagena for supplies and overhaul. In January, the fleet sailed for Havana where it linked up with the convoys from Veracruz for the return voyage to Spain. The thousands of mules used annually to carry goods across Panama, as well as provisions for the *flota*, spread ripples of demand across all of Central America. Consequently, when the mid-seventeenth century brought a decline in the *flota* system, Central America retreated to a backwater status, limiting the growth of Spanish settlement.

► *The Neocolonial Legacy*

The wave of independence movements that swept Spanish America in the early nineteenth century separated Central America from Spain in 1822. In 1786, Spain had restructured its political administration, dividing Central America into five intendencies, reflecting the major pockets of Spanish settlement in the region. At independence, the former intendencies joined to become the United Provinces of Central America and were associated with Mexico. In 1823, with the exception of the former intendency of Chiapas, which elected to remain part of the Mexican union, the United Provinces separated from Mexico. Central American unity was short-lived, however, and by 1838, the former intendencies (see Figure 8.3) had become the independent nations of Guatemala (including Belize), El Salvador, Nicaragua, Honduras, and Costa Rica (Panama became independent of Colombia in 1903).

Independence brought few changes to the social organization of these nations. Although the link with Spain was severed, people of European origin continued to control the economies and hold power over those with native roots. The vacuum left by Spain was filled by economic interests from the United States and northern Europe, particularly Britain. Their activities modified the landscape patterns established in the colonial period, but their motivations were similar, hence the use of the term *neocolonial*.

The initial interest in Central America in the nineteenth century was prompted by the continuing need to find a cheaper crossing between the Atlantic and Pacific that would avoid the long journey around Tierra del Fuego. The pressure for a shorter, trans-isthmian route was intensified after the discovery of gold in California in 1848. The earliest attempt to connect the two Central American coasts followed the San Juan River to Lake Nicaragua where a U.S.–owned steamship line ferried prospectors across the lake to a trail that ran down to the Pacific coast. Attempts to supplant this route with an all-rail link through Honduras were begun in 1853, but efforts lasted only until the first U.S. transcontinental railroad was completed in 1869. At this point, the project was abandoned, leaving 57 miles of track running inland to a point a little beyond San Pedro Sula (Clayton, 1987).

In 1879, a railroad was built in sections from the Caribbean port of Limón toward San José in Costa Rica. The project, initiated by Minor Keith, encountered delays because of a lack of capital. Keith, recognizing the need for freight to support income on the unfinished line, conceived the idea of raising bananas beside the track for export to the growing industrial populations of U.S. east coast port cities. The Costa Rican railroad was completed through San José to Alajuela in 1890 and extended to Puntarenas in 1910 to complete the transoceanic link. By that time, construction of the Panama Canal was underway, and the Costa Rica project's original purpose was eclipsed. But the association of railroad construction and banana cultivation using a plantation system was established. Keith's operation had become the United Fruit Company, and bananas had spread to the Caribbean coasts of other Central American nations. In Guatemala, in 1884 a railroad was started from Escuintla on the Pacific to the Caribbean along the Motagua Valley, which it reached in 1908. The United Fruit Company planted over 1 million banana trees beside its tracks, resulting in immense plantations along the Motagua and the company's ownership of the docks at Guatemala's port of Puerto Barrios.

When, in the present century, sigatoka and leafspot disease afflicted banana production in the Caribbean valleys of Central America, the only solution was to abandon the infected soils, and banana production spread to the Pacific coast, which had been less intensively planted in the nineteenth century. Previously existing railroad spurs, originally part of earlier trans-isthmian projects, were adapted to facilitate local raw material exports.

The greatest of all the projects to connect the two coasts was the Panama Canal. In 1855, a railroad between Colón on the Atlantic and Panamá City on the Pacific coast was completed, but its advantages were diminished by customs duties levied by Colombia and high trans-shipment costs at the two terminals. In 1882, a French company began construction of a canal. The Panamanian isthmus rises to 275 feet at this point, is covered with tropical forest, and is subject to heavy rainfall. Making the 190-foot-deep Gaillard Cut through the higher ground presented enormous construction problems, and the French company did not have sufficient capital to solve them. Meanwhile, American investors were exploring what was perceived to be an easier path through Nicaragua, using the old gold rush route: up the San Juan River, ferries across Lakes Nicaragua and Managua, and overland down to the Pacific shore. The Nicaragua versus Panama route question became mired in

Panama Canal; the Galliard Cut. In the picture is a vessel, 85 feet above sea level, moving between the Pacific and Atlantic Oceans. Locks raise vessels up and down to the level of the Gaillard Cut and Lake Gatún for the crossing of the continental divide.

U.S. congressional politics, stimulated by rival factions seeking U.S. government support. Ultimately, the Nicaragua faction lost through a scare resulting from the eruption of Mt. Momotombo, north of Lake Managua, that occurred only a few days after the devastating loss of 30,000 lives from the explosion of Mt. Peleé in Martinique. Proponents of the Panama route managed to arrange publicity of Nicaragua's volcanic violence on a U.S. postage stamp!

In 1905, the United States took over the failed French company and completed the Panama Canal, which opened for traffic in 1914 (McCullough, 1977). Thousands of contract laborers were brought in to dig the canal. Jamaica rejected efforts to recruit its people because so many had perished working for the French. Instead, the canal company established a recruitment office in Barbados, and the island sent nearly half of the 45,000 contract workers the canal used. It is estimated that another 25,000 Barbadians arrived in Panama informally during construction (Richardson, 1985). Their impact was substantial, not only for the money they sent back to their island home, but for the many who remained in Panama, significantly transforming the ethnic composition of the population, especially in Colón.

Completion of the canal in 1914 brought major benefits to Pacific countries exporting to U.S. and European markets. Growth occurred in the production of sugar in Peru, ores in Chile and Bolivia, coffee in Colombia, and later, in the 1950s, bananas in Ecuador. The United States regarded the canal as a security issue. It encouraged Panama to become independent from Colombia in 1903, and the new country granted the United States sovereignty over a zone paralleling the canal. Panama was transformed by the canal, which attracted an international mixture of immigrants and substantial income from the transit operations. But elsewhere in the country, traditional patterns of colonial landownership overlain by nineteenth century banana plantations and mining operations prevailed, with pockets of traditional native populations in the periphery. The canal reinforced and broadened a dual society and economy in Panama.

In addition to grand schemes to connect the two coasts, international investors in neocolonial Central America were motivated by export opportunities in agriculture and mining. Banana plantations spread from Guatemala and Costa Rica to Nicaragua and especially to Honduras, where, at the turn of this century, the government sought increased income through offering generous concessions to banana companies. Beginning in the 1890s, vast estates served by feeder railroads were developed on the northern coast and around the eastern port of La Ceiba, later to become the property of Standard Fruit Company. Rival United Fruit (known locally as *el pulpo*—the octopus) acquired huge estates along the Cuyamel River near Tela and by 1914 controlled nearly 1 million acres of Honduran land. By that time, bananas accounted for about two thirds of the country's total exports, and the banana companies in the 1930s ultimately came to control the Honduras government. Local labor in the low-density tropical valleys of the Caribbean coast was scarce, and plantations relied mostly on imported Afro-Caribbean workers from English-speaking islands—a practice repeated the length of Caribbean Central America. This Afro-Caribbean population further reinforced ethnic differences between the eastern lowlands and Spanish-speaking highlands.

The other major nineteenth century export crop that transformed Central America was coffee, which began to be grown in the *meseta* of Costa Rica in the 1830s. From there it spread to the forests around San Salvador in the 1840s and along the Pacific slope of Guatemala in the 1850s and 1860s. In Nicaragua, coffee was being grown in the hill areas around Matagalpa by the 1850s, but in Honduras, the rugged topography inhibited transportation of the product to export markets, and coffee growing expanded only in the 1940s (Weaver, 1994). Elsewhere, railroad construction aided the export of the crop to Europe and the

United States, while well-drained, productive, volcanic soils and ideal climatic conditions between 500 and 1,700 meters supported its growth.

Ultimately, coffee had significant social and ecological consequences (Faber, 1993). Its ideal altitude coincided with rich oak-pine forests, which were rapidly replaced by both large and medium-sized coffee estates (*fincas*). The cleared land was planted in rows of coffee bushes, which take three to five years to mature, and shaded with a canopy of cacao trees. Coffee harvesting requires a substantial supply of labor to pick the berries and process the coffee beans. The plant is also a heavy depleter of soil productivity (witness the hollow frontier in Brazil; see Chapter 12). These two factors, intensified by growing foreign demand, caused expansion of lands under coffee cultivation. In Costa Rica, relatively large areas of unpopulated virgin forest were cleared for estates. But in El Salvador, Guatemala, and Nicaragua, governments conspired to take Indian communal lands, forcing indigenous people to higher elevations in the central and western mountain slopes where soils were thinner and less productive, and self-sufficiency more difficult to achieve. In El Salvador, the northern mountains became denuded of woods and soils and now represent the country's poor, back country. Displaced Indians, in turn, became the source of seasonal labor required on the coffee plantations. In Guatemala and Nicaragua, governments passed vagrancy laws and forced labor drafts to assure that indigenous communities yielded sufficient workers at harvest time for estates that were increasingly owned, not only by wealthy criollos and mestizos, but by European immigrants, especially Germans. The potential social conflict resulting from geographically interwoven patterns of wealthy coffee growers and impoverished Indian communities is perpetuated to the present and underlay much of the civil unrest during the 1980s.

In the early twentieth century, Central American countries were concentrating on coffee and banana exports and reflected classic dependency economies. By 1913, coffee represented 85 percent of Guatemala's total exports by value and 80 percent of El Salvador's exports. Bananas accounted for half the exports of Costa Rica and Honduras (Weaver, 1994). Domestic manufacturing lacked the demand that could produce economies of scale that might compete with foreign imports. Consequently, agricultural exports paid for imported finished goods for a small, wealthy elite.

In the worldwide depression in the 1930s, Central America experienced declines in production of the major export staples, although the effect was cushioned for a while in some countries such as Honduras where mining of precious metals sustained the economy as foreign countries abandoned the gold standard and the price of gold rose. Disease in banana plantations intensified and extended the region's economic decline, while competition from inferior but far larger suppliers of coffee, such as Brazil, lessened Central America's market share. Over time, some smaller plantations were consolidated into larger estates. Yet, in the 1960s in the five Central American countries (excluding Panama and Belize), only 26 percent of agricultural properties were considered large enough to generate income above self-sufficiency (Vilas, 1995).

THE MODERN ERA

The legacy of the past continues to shape the Central American present. The neocolonial period introduced new export staples but reinforced and intensified colonial social patterns. Thus, income distribution became more extreme. By the 1970s, the top 20 percent of families earned 51 percent of income in Costa Rica, 64 percent in El Salvador, 29 percent in Guatemala, 52 percent in Honduras, and 60 percent in Nicaragua.

► *Diversification of Exports*

Although agricultural exports have been diversified, coffee remains the leading export by value in El Salvador (72 percent of total exports), Nicaragua (50 percent), Guatemala (49 percent), Honduras (39 percent), and Costa Rica (36 percent). However, after the post–World War II boom, coffee prices fell in the late 1950s, and banana production suffered significant competition from newly established plantations in Ecuador. Central America's response was to diversify the agricultural sector and produce more cotton, cattle, sugarcane, tobacco, and irrigated rice. In Costa Rica, land under sugarcane doubled between 1950 and 1973, and production tripled as Cuba's sugar quota was curtailed by the United States following the Cuban Revolution. Beef exports from Costa Rica increased rapidly, and acreage in pasture doubled as demand from fast-food chains stimulated cattle production. In Guatemala, cotton production grew tenfold between 1950 and 1963 in response to a rapid growth in world prices, while the area under sugarcane expanded by a factor of 12 and beef exports grew 8 times. In Nicaragua, cotton acreage expanded tenfold between 1950 and 1963 while land under cattle ranching doubled. In El Salvador, cotton production expanded fourfold. Finally, irrigated rice in Panama grew to become the country's largest agricultural product (Vilas, 1995).

In 1960, the initiation of the Central American Common Market (CACM) promised the beginnings of a region-wide rise in industrial capacity. Although focused on processing of food, drink, and tobacco, and development of a textile industry (ironically based on imported synthetics), industry thrived during the CACM. So-called "finishing-touch" industries—making shoes, pharmaceuticals, cosmetics—also expanded (Weaver, 1994). Industrial capacity was most strongly developed in Guatemala and El Salvador, especially in the respective capital cities of Guatemala City and San Salvador. However, despite these initial advances, together with those noted earlier in infrastructure and capital organization, Central America

Pesticides are used heavily on export crops in Central America, as on this cotton field. Small farmers try to cultivate the steeper slopes running to the valley floor.

failed to solve its twin historic problems: deep divisions between the oligarchical leadership of each country and its people, and a failure to enact land reform, which might make more land available to peasants. The result was the turmoil of civil war that engulfed the region in the 1980s.

The beginning of the 1990s has seen a reduction in the internal violence of the previous decade. Renewed cooperation among governments offers a glimmer of hope that the promising start to economic development achieved by the CACM might be revived. However, as the following sketches of each country indicate, fundamental problems of social inequality and environmental deterioration restrain optimism for Central America's future.

► Guatemala

Guatemala is a country of sharp human and environmental contrasts. Physiographically, it can be divided into three regions: (1) the Central Highlands; (2) the Pacific Slopes and the Pacific Coastal Plain; and (3) the Petén (Figure 9.5).

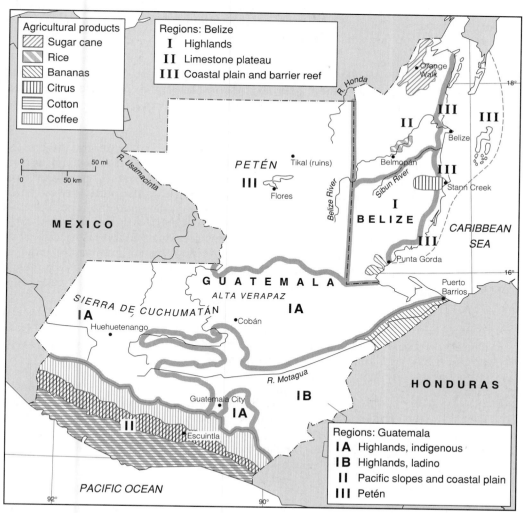

Figure 9.5 Map of Belize and Guatemala.

1. The Central Highlands dominate the country and consist of two mountain chains separated by the valley of the Motagua River. To the north of the river, the Cuchumatánes mountains form a series of limestone ridges and basins between 9,000 and 11,000 feet, which can be traced across the Caribbean. Similarly, the Motagua depression reappears in the Caribbean in the Cul-de-Sac of lowland Haiti. Running northwest to southeast, on the Pacific margin of the highland, is the volcanic ridge where active and inactive volcanic peaks rise to over 12,000 feet. The highest peak is Volcán Tajumulco (13,816 feet). Densely settled high basins and plateaus are dispersed throughout this region.

2. The Pacific Slopes and Pacific Coastal Plain represent the second physical region. The higher slopes are covered with a layer of ash and lava into which rivers flowing to the Pacific have cut steep valleys. Altitude, modifying temperature, produces distinct environmental bands from highland to the coastal lowland. In the cool high mountain basins from 5,000 to 8,000 feet, traditional agriculture based on corn, beans, and squash or chilies prevails. Downslope in the Boca Costa, elevations between 1,500 and 5,000 feet (the *tierra templada*) are well suited to coffee cultivation. Below 1,500 feet, in the *tierra caliente*, cotton and bananas predominate. The narrow Pacific Coastal Plain is edged with mangrove swamps and lagoons in which shrimp farms have been established.

3. A third region, about half of the country, lies north of the central highlands, and in Guatemala is called *El Petén*. Tropical forest covers much of this lightly settled frontier region. Outliers of the highlands poke through as low foothills, but the Petén is mostly a flat lowland, a southern extension of Mexico's Yucatán Peninsula.

Guatemala's human landscape is full of contrasts. It has Central America's largest population (9.9 million) with a doubling rate of 24 years, second fastest after Belize. Infant mortality, a measure of overall health care, is 51 per 1,000 live births, worst in the region. Life expectancy, at 65 years, is jointly lowest with Nicaragua. The overall figures mask differences within the population, both demographically and geographically. Guatemala contains Central America's greatest concentration of Indians, mostly of Mayan descent, speaking Mam, Quiché, Ixil, and Chuj. Among Indians, life expectancy is only 44 years. Small numbers of Indians live throughout the Petén, but the great majority, nearly half the country's population, live west and north of Guatemala City in highland country rightly called the "Alps of Central America." The area's physical beauty masks misery. Indians have long cultivated fertile valleys and steep, terraced hillslopes, raising maize on a subsistence basis and funneling surpluses into a self-contained system of local markets. But 80 percent of Guatemala's land is in the hands of 2 percent of the population and over half a million farmers own no land.

Poverty and land hunger sparked resistance among the Indian population and since 1961, a civil war has raged intermittently across the country. Military crackdowns on the Indian population have been terroristic. During a flareup in 1980–1981, thousands of Indian civilians were killed in a scorched-earth policy that destroyed 440 villages. More than 150,000 Indians fled to Mexico and helped cause problems in Chiapas. Some 60,000 survivors were forced off their lands to live in 70 newly constructed model villages: dingy cinderblock houses lining muddy streets in which the population is under military oversight and forced to join local militia patrols (Faber, 1993).

In many ways, this recent policy of population concentration mirrors similar efforts by the Spanish. In the early colonial period, priests sought to convert natives

Lake Atitlan in western Guatemala is
surrounded by volcanic peaks.

by congregating them in agricultural villages. But the Indians have an intense relationship with their traditional lands, and, in the eighteenth century, their settlement
pattern began to drift back into a more dispersed, precolonial form reflected in local
administrative units known as *municipios* (Lovell, 1992). Similar resistance has met
the present government policy of model villages. Furthermore, as travel has become
more dangerous, traditional market systems among towns have been disrupted. The
Indian area still reflects conflict and destitution, although a relative political calm
has recently been imposed.

Comparative peace prevails in the highlands east of Guatemala City, through
the Motagua Valley, and along the slopes of the Boca Costa and Pacific Coast.
These areas are occupied by *ladinos*, the term used in Guatemala, El Salvador, and
Honduras for peoples of mixed Indian and Spanish ancestry who adopted Hispanic
ways. Commercial agriculture dominates areas of ladino settlement. Along the
Motagua Valley leading to the Caribbean, traditional banana plantations are now
giving way to sugarcane cultivation. But the country's most important agriculture is
found in parallel bands rising up the fertile volcanic slopes that face the Pacific. At
the highest elevations, spilling over into the highlands, are the coffee fincas, which
produce half of Guatemala's export earnings. Coffee spread across this region in the
1870s, promoted especially by immigrants from Germany as well as local
Guatemalan families. Estates range in size today from small units of no more than
25 acres to some 1,500 large plantations of up to 4,500 acres in size, which produce
80 percent of the crop (West and Augelli, 1989). Downslope is sugarcane, a crop
raised since colonial times in the area. The coastal plain is occupied by cotton fields
interspersed with banana plantations. Cotton production, introduced in the 1950s,
has declined recently. The heat and humidity encourage pests, and with no winter
frost to keep insect populations in check, yields decline dramatically after four or
five years. Heavy use of pesticides in response have produced high levels of DDT in
the body fat of local Guatemalan peasants (Weinberg, 1991). In mangrove swamps
along the coast investment in shrimp mariculture is growing, despite the loss of
mangroves and their essential role in tropical wetland habitat.

The Indian-ladino geographical divide has generally been complete. Indians and ladinos kept to their respective areas, and interpenetration was limited even under the Spanish. With the arrival of coffee, and more recently sugarcane and cotton, a seasonal labor force demand at harvest time has resulted in highland Indians migrating temporarily to the ladino agricultural areas. Sometimes voluntarily, at other times coerced by law or by impoverishment, Indian laborers have worked for low wages in circumstances that have exposed them to environmentally dangerous conditions. Instructions on imported U.S. pesticides are often in English. Most laborers speak native languages and cannot understand Spanish, much less English. Many bathe in streams or irrigation canals contaminated with agro-chemicals or breathe dust from aerial spraying (Weinberg, 1991).

In the late 1950s and early 1960s, the growth of cattle ranching stimulated the expansion of ladino settlement into highland Indian lands across the traditional cultural divide. Around towns such as Huehuetenango, previously untitled land was conveyed to private owners, the forest was cleared, African grasses replaced the native varieties, and Santa Gertrudis, Brahman, Zebu, and other breeds of beefier, faster-maturing animals were introduced. A similar process spread across the Pacific Coast lowlands around Escuintla, resulting in the clearing of the forest. By 1960, frozen beef exports to U.S. fast-food and pet food manufacturers left Guatemala at the rate of one ship every five days (Faber, 1993).

Other more recent trends are slowly changing the character of the indigenous highlands from a subsistence to an agro-export economy. Thirty years ago, a German immigrant brought cardamom seeds from India, where they were used as an ingredient in curries. In the department of Alta Verapaz, around Cobán, conditions are ideal for cardamom growth. Today, some 250,000 Keq'chi Maya Indians produce over two thirds of the world's cardamom, a plant that grows nearly wild, needing little attention and fertilizer until harvesting. Today, 85 percent of the cardamom is exported to the Middle East, where in Arab countries, principally Saudi Arabia, the spice is mixed with coffee to produce a pungent, aromatic drink called *gahwa*. In areas within reach of Guatemala City, the cultivation of vegetables is beginning to transform the traditional maize culture, as roads to the highlands are opened and air freight costs from Guatemala City to U.S. markets decline. Broccoli, mange-tout (a type of sweet pea), french beans, and radiccio, all labor intensive crops that can be grown on small plots of land, allow highland Indians to be competitive in U.S. and European markets.

Land hunger still plagues impoverished highland communities. Traditional indigenous people relied on a collective occupancy of land, foreign to Western views of property ownership. Many who could not show title became landless, which has encouraged out-migration in two directions: the city and the frontier.

While 60 percent of Guatemala's population is still rural, Guatemala City (1.5 million) has grown rapidly. Employment is focused on service and food-processing industries, as well as agro-chemical processing, reflecting the dominant activities in the countryside. More recently, as in other Central American countries, the large floating employment pool has attracted importation of garment parts for final assembly. Woven apparel has become the country's leading export by value to the United States. However, domestic strife hampers the growth of this sector. One importer takes extraordinary measures, moving products by truck convoys guarded by a private security force to prevent theft and hijacking. Such procedures, together with frequent port strikes, slow the flow of materials and hamper the just-in-time structure of modern manufacturing. Furthermore, *maquila* industries in the city are feeling the competitive effects of Mexican industries operating under NAFTA's more favorable terms. Thus, although it is the country's principal manufacturing center, Guatemala City does not attract sufficient employment to accommodate the needs of domestic rural-to-urban migration. Up to half its workforce is unemployed, and huge numbers occupy squatter settlements that lack sewage and electricity, repeating a social condition common to Latin America's primate cities and other large urban centers.

Guatemala's frontier region, the Petén, offers an alternative opportunity to the city. This lightly populated and heavily forested lowland hides numerous Mayan archaeological sites, shelters guerrillas, and supports isolated pockets of pioneer farmers who moved down from the highlands, as well as *chicleros* who collect the sap of chicle trees from which chewing gum is made. But the Petén remains undeveloped, and its future symbolizes the dilemma facing other Latin American frontier areas. Below its surface lies an extension of Mexico's rich La Reforma oil field with reserves that may place Guatemala among the world's top 20 oil nations. Oil could underwrite the costs of economic development. Yet to date, government desire for resource control and guerrilla harassment have restrained oil exploration and production in the Petén. Furthermore, oil drilling and the associated support infrastructure of roads conflict with strong international pressure to preserve the region's tropical forest.

Over the past 30 years, 40 percent of the forest has been logged for mahogany and cedar and then burned. Equally disruptive of forest cover is the Usumacinta River dam, a joint project of Mexico and Guatemala, which will flood 500 square miles of rain forest and drown two classic Mayan sites. Much of the forest cover has already been cleared on the Mexican side of the border. The dam project now threatens incursions into Guatemala. Environmentalists have fought back, prevailing upon Guatemala to designate a third of the Petén as a national park and forest reserve, and advocating eco-tourism to Mayan sites as the base for the region's economic development. But policy made in Guatemala City is rarely enforced in the lowlands, and loggers and cattle ranchers will probably set the terms of the Petén's future landscape.

► *El Salvador*

El Salvador (5.9 million population), a Massachusetts-sized country and the smallest in Central America, lies on the Pacific volcanic axis, which runs west to east across the country (Figure 9.6) An older volcanic range forms the northern back country, while a geologically recent range, dotted with active cones, occupies the southern slopes leading down to the Pacific. The Lempe River flows from the northern highlands to the Pacific, cutting the country into two unequal parts. Between the two volcanic ranges is a series of intermontane basins, filled with ash, punctuated with cinder cones, and frequently occupied by internally drained lakes. In these basins are found the densest pockets of population in what is already the most densely populated nation in Latin America outside a few of the Caribbean islands. Demographic indices like infant mortality and life expectancy place El Salvador near the median for Central American countries. But a high doubling rate of 27 years puts severe pressure on already heavily settled land and makes the country an extreme example of the region's ills.

Economically, El Salvador is dependent on a single product, coffee, which brings 72 percent of export earnings. Coffee occupies the intermontane basins and upper volcanic slopes above a coastal strip of cotton production, just as in Guatemala. But a history of colonial stock raising, which is still prevalent in the interior highlands, together with intense demand for land has left the country with only 3 percent of its forest cover. Erosion is so severe that 77 percent of the soil has been damaged or rendered useless, and a local joke calls soil flowing to the sea the

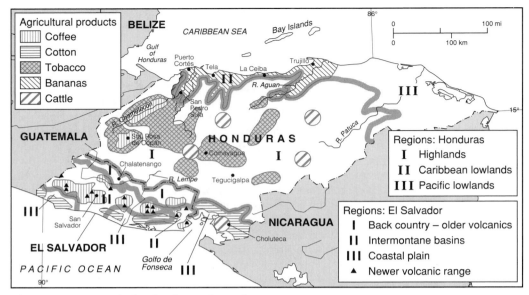

Figure 9.6 Map of El Salvador and Honduras..

country's chief export. Applications of pesticides, including DDT, are uncontrolled, killing off plant and bird species. One San Salvador hospital alone experiences the death of 50 children a year from pesticide poisoning.

El Salvador's ills have been accentuated by an extreme imbalance in landownership. Today, more than 70 percent of the land is owned by 1 percent of the population, who rule the country as a powerful oligarchy. Traditional peasant views of land occupancy emphasize use rather than legal title, and encroachment on lightly worked estate lands has grown, producing conflict with the landowning oligarchy (Browning, 1971). Land shortage prompted Salvadorean peasants to cross the northern border into Honduras and led to the "soccer war" between the two countries in 1969. Part of El Salvador's motivation in joining the Central American Common Market was to promote the free movement of labor within the region to ease the pressure on land resources. The soccer war ended that hope.

In 1980, responding to peasant unrest in the eastern part of El Salvador, a reformist government undertook an ambitious program of land redistribution, expropriating farms larger than 1,235 acres (22 percent of the agricultural land) and redistributing the land to peasant cooperatives. Perhaps a quarter of the rural poor received title to land for the first time. Yet much of the distributed land was marginal, and by locking peasants into specific plots, the process restricted the ability of subsistence farmers to apply shifting agricultural techniques to disperse the impact of farming degraded land. However, when land reform efforts attempted to dismantle the larger coffee estates to gain peasant access to the better soils in the intermontane basins, the event touched off a legal challenge from landowners and quickly degenerated into a violent civil war between leftist peasants and the right-wing oligarchy, supported by the military. El Salvador was ravaged. More than 100 towns were destroyed. Agricultural land was abandoned as tens of thousands fled the country or moved to the capital city of San Salvador.

The capital had succeeded, during the years when the Common Market thrived, in developing one of Central America's strongest industrial manufacturing complexes, especially in cotton textiles. This industrial base was undercut by a disrupted domestic and regional market. Unemployment grew as a result of industrial decline and the influx of migrants from the countryside. The housing situation became desperate, and squatter settlements grew within the city's garbage dumps. Demand for water caused the water table to drop by 3 feet per year through the 1980s, resulting in the need to import water from rivers as far as 80 miles away. Inadequate sanitation and polluted drinking water today make San Salvador a prime target for a cholera epidemic.

In the 1980s, agriculture suffered as coffee production dropped by 17 percent and cotton by 85 percent. The end of the civil war in 1992 was the only positive note in a country whose environmental and social fabric had been wrecked by the civil war and economic collapse of the previous decade.

Rescuing El Salvador is an extreme challenge. The country's rapid rate of population growth, extreme social inequality (which the civil war did little to dent), and lack of environmental sensitivity all place huge burdens in the path of reversing the country's downward spiral. Some glimpses of prosperity suggest a way back from the social and environmental abyss that El Salvador now faces. Ironically, these hints of progress are consequences of the civil war. Chalantenango Province, in the northern range, was one of the strongest areas of guerrilla support in the war and successfully resisted the worst of its effects. Traditions of community farming remain strong in this area, and agricultural cooperatives can be organized if land-ownership issues can be resolved. Additionally, because there is a small niche for organic coffee in world markets (especially in California and Germany), small peasant producers north of San Salvador are taking advantage of abandoned lands that lay fallow and were leached of pesticides and herbicides during 12 years of warfare. These lands are now ideal for organic farming, capable of producing chemical-free coffee as well as organic cashews, soybeans, vegetables, and even bananas. In El Salvador, organic products are as yet a minuscule part of total agricultural production, but given the heavy labor inputs necessary for organic farming and a premium of up to 30 percent on the world market for its products, this farming niche could

Volcán San Vicente, El Salvador. The volcanic soils make fertile fields for small farmers.

well provide a modern basis for peasant farming in the country. Another consequence of the war was the significant number of El Salvadorians who escaped the country, especially to North America. Although some returned, many were granted asylum and have remained. The repatriated funds these emigrants send back to relatives helps undergird the El Salvador economy.

► *Honduras*

Most of Honduras is part of the Old Antillian mountain chain of core crystalline rocks overlain with limestone and sandstone sediments (Figure 9.6). Only the southwest is close enough to the volcanic axis to receive coverings of ash and lava that bring enhanced soil fertility. For most of the country, thin acidic soils of low productivity on steep, rocky slopes (75 percent of the land surface has slopes exceeding 25 percent) are easily subject to erosion when the native oak and pine forests are cleared. The remarkably fractured terrain has long isolated Honduran settlement in fragmented pockets. Only since the 1960s has a network of paved roads linked the country's dispersed parts and provided an infrastructure for national development. Along the coast of the Gulf of Honduras and the Costa de Mosquitos that Honduras shares with Nicaragua, the mountains drop down into hot, moist lowlands of tropical rain forest. This Caribbean lowland fringe is bisected by a series of narrow river valleys from the Chamelecón to the Coco that drain north and east from the highlands.

Honduras remained outside the orbit of most Central American cultural influences. Mesoamerican high culture touched only the western fringe around Copán, and the dispersed pre-Hispanic settlement in the mountains was quickly erased under the Spanish by disease and slave expeditions. Pockets of gold and silver production along the Guayape River, around Tegucigalpa, and in western Honduras dictated patterns of Spanish settlement, which was generally light elsewhere. Land granted to ex-conquistadors formed the basis for cattle estates, but by the end of the colonial period, much original natural vegetation cover still stood untouched.

The Spaniards generally kept to the highlands and never established themselves on the tropical Caribbean shore. Infested with black sandflies, mosquitos, and other insects, these humid lowlands were highly susceptible to yellow fever and malaria. This area was originally sparsely settled by Sumu-speaking Indian tribes of hunters and gatherers who had migrated here originally from the coast of Colombia. The Miskitos, one Sumu group occupying the isolated coastal area from the mouth of the Cocos River south, gave their name to the coastal lowlands. From the 1630s, British privateers made contact along the coast, introducing a few semi-permanent settlements populated with Afro-Caribbean peoples whose subsequent mixing with local Indians led to the Zambo-Miskito population.

By the 1730s, English dyewood cutters were scattered around the Gulf of Honduras, on both the Belize and north Honduran coasts as well as along the Costa de Mosquitos lowlands, and were actively trading with Jamaica. After 1786, the English formally withdrew except from the Belize coast but continued to have political contact with Zambo-Miskito leaders. In the 1840s, Britain recognized a Miskito Indian kingdom on the shores of Honduras and Nicaragua. Significant English-speaking populations existed all the way south to Bluefields and the San Juan River. By then, the attraction was extraction of mahogany. Only United States pressure claiming infringement of the Monroe Doctrine (1823) led to British withdrawal from the area and acknowledgment of Honduran claims to its north shore region in 1859 (Naylor, 1989). Today, a significant ethnic and socioeconomic divide still exists between the Caribbean coastal communities, with their historic cultural attachments to the Caribbean islands, and the Hispanic highlands where the country's power center resides.

After British withdrawal, U.S. banana companies established plantations in the lush river valleys running to the Caribbean coast. Puerto Cortés, Trujillo, and La Ceiba were developed as company-owned port facilities, local labor from Honduras was supplemented with contract workers from Caribbean islands, and by the 1930s, bananas dominated the economy. The companies, primarily the United Fruit Company, controlled the government. But banana production was an enclave economy, restricted to the Caribbean lowlands. The companies contributed little to the national economy, and in the highlands in the early twentieth century, cattle ranching, mining, and subsistence peasant farming prevailed.

Banana production peaked before the Depression of the 1930s and thereafter succumbed to disease. Today, the industry is changed. Disease-resistant varieties are produced by smaller domestic growers, with the large foreign companies limited to handling exports. Moreover, bananas now account for only a third of export earnings, being superseded by coffee, which entered Honduras with Salvadoran immigrants in the 1960s and now makes up 39 percent of export earnings. Stock raising is still dominant. Pastures often occupy the fertile valley bottoms while subsistence farmers are driven to the poorer upper slopes. Tobacco cultivation north and east of Santa Rosa de Copán dates from the eighteenth century. More recently, land on the Pacific slopes south of Tegucigalpa has been opened to cotton. Another recent industry, accounting for 6 percent of exports, is shrimp farming developed in areas once occupied by mangrove swamps around the Gulf of Fonseca on the Pacific coast. Stimulated by a growth of demand for seafood in North America, investors have been expanding production in the region at the expense of mangrove forest, which performs an important ecological role of nurturing fish populations.

Honduras escaped the civil unrest that touched all its neighbors in the 1980s, but not its consequences. The best land remains concentrated in a few hands, and while Honduras has a law permitting peasants to claim uncultivated tracts, procedures for land titling take eight or nine years. Although Honduras is still lightly populated (5.6 million in a country five times the size of El Salvador), only about a quarter of its land is agriculturally productive, and the country has suffered considerable illegal immigration from peasants fleeing violence in surrounding countries. Consequently, land hunger is a growing issue, and Honduras remains a poor nation, with a per capita G.N.P. half that of Guatemala and exceeding only that of Nicaragua in Central America. Two thirds of the peasant families cannot meet their own subsistence needs, and the poor constitute 70 percent of a population that will double in 25 years, given present growth rates. Sorghum, which can grow on thin, degraded soils, now constitutes nearly one half of the country's food production, supplanting corn as the dominant crop of peasant farmers despite the low status that sorghum has as a food source for local people (Faber, 1993).

Honduras contains considerable natural resources, especially its inland forests. The original cover of mahogany and cedar has been logged, but an extensive second growth of pine forests has spread across the interior. Unfortunately, clearing land for cattle ranches, cutting by commercial sawmills, expansion of shifting agriculture, and demands for charcoal fuel have reduced the forest cover by 26 percent in the past two decades and threaten what could have become, with careful management, an important feature of a more diversified Honduran economy. Moreover, with the land stripped of vegetative cover, soil loss and river siltation are growing. For example, in the El Cajón Dam and Reservoir, 100 miles northwest of Tegucigalpa, which supplies 70 percent of the country's electrical power, over half the reservoir has become a giant mud flat, leading to diminished capacity to generate power. The frequent blackouts in the nation's capital interfere with the manufacturing and service sectors, driving some desperate producers to spend foreign exchange on imported generators and requiring Honduras to import more oil.

▶ *Nicaragua*

In the 1970s, Nicaragua (4.6 million population) had a relatively balanced economy with both agriculture and industry contributing equally. But the country had the same inequality of wealth and poverty that characterizes much of Central America. Severe social conditions prompted the Marxist-oriented Sandinista revolution in 1979, followed by a debilitating civil war waged by ousted right-wing insurgents supported by the United States. Together with a trade embargo against Nicaragua by most Western nations, the economy was devastated, and comparative peace that came in 1990 has left the country deeply in debt and starved of investment. Today, it has the lowest per capita G.N.P. of any Central American nation (a tenth that of Mexico) and faces perhaps the longest journey of recovery.

Nicaragua, which borders the Atlantic and Pacific oceans, can be divided into three physical regions (Figure 9.7).

1. A majority of the population lives in the western structural trench that contains Lakes Nicaragua and Managua, as well as a line of active volcanoes along the Pacific coast. Parts of the coast have been drowned to form inlets, lagoons, and bays, such as the Gulf of Fonseca shared with Honduras. This western region represented the southernmost extension of the Mesoamerican high cultures and also attracted Spanish settlement. Fertile volcanic soils and well-watered valleys supported indigo and cattle ranching, with exports through the old port of Realejo.

2. To the east lie the Central Highlands, not unlike the highlands of Honduras, particularly in the north, where pine forests prevail. Elsewhere, the highlands were once covered with rain forest on east-facing slopes and drier scrub on the rain-shadow sides. Much forest was cleared as colonial settlement moved inland from the west, invading lightly populated areas of indigenous subsistence farmers. As in Honduras, small pockets of mineral resources have been found among the crystalline rock core of this region. Gold and more recently copper have been uncovered around Bonanza at the headwaters of the Wawa River and nearby Siuna and Rosita.

3. A third region contains the Costa de Mosquitos lowlands, again cut by eastward-flowing rivers from the highlands. Culturally and physically, this lowland region is one with northeast Honduras. It shares its history from early hunting and gathering through eighteenth- and nineteenth-century English influences, the introduction of Afro-Caribbean peoples, and the early twentieth-century banana economy. Banana plantations were established inland up the Matagalpa and Escondido rivers, but most have succumbed to disease, and today the fruit contributes less to export earnings than sugarcane. The Sandinista government, wishing to end the historic isolation of the Mosquitos region, sought to reorganize the area along revolutionary principles. That resulted in the migration of 40,000 Miskito Indians across the border to the relative safety of Honduras. A further 5,000 fled to Costa Rica (Neitschmann, 1989).

The neocolonial and modern periods brought the now-familiar commercial crop combinations to overlay those of the earlier colonial period. Coffee reached Nicaragua in the 1860s, spread by German and British immigrants in the Diramba highlands south of Managua where large estates were laid out. In the twentieth century, a more varied mixture of local middle-class ladinos and European immigrants extended coffee production to the moist, lightly settled central highlands east of Jinotega and Matagalpa. By the early twentieth century, coffee accounted for two thirds of Nicaragua's export earnings and underpinned the wealth of the ruling elite.

In the 1950s, cotton spread along the Pacific lowlands west of Managua to the Gulf of Fonseca and especially around the old colonial town of León and today is

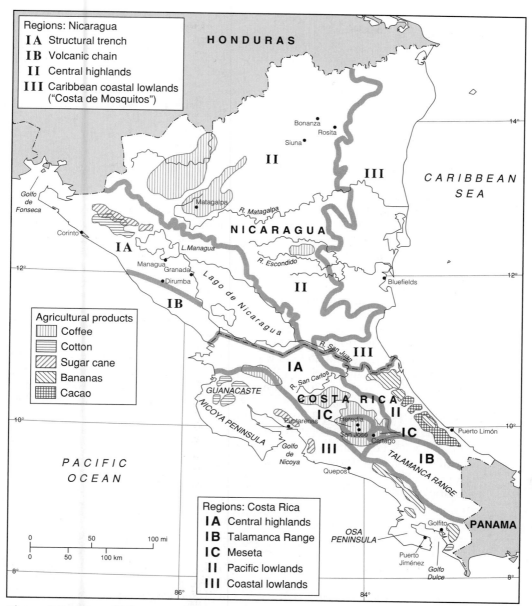

Figure 9.7 Map of Nicaragua and Costa Rica.

Nicaragua's second most important export crop. But the pesticide problems found in Guatemala were equaled in Nicaragua. In the 1960s and 1970s, Nicaragua (together with Honduras) was the world leader in per capita illness and death from pesticide poisonings. Breast milk among the women of León contained DDT 45 times higher than the safe level set by the World Health Organization (Weinberg, 1991). Northwest of Lake Nicaragua, cattle herds roamed the savannas around the town of Granada, as well as the middle and southern drier valleys of the central highlands. Trucked to Managua, they contributed to a chilled beef market developed in response to the demands of the U.S. fast-food industry.

The population (63 percent urban) remains concentrated in the western structural depression and the Pacific coast, areas subject to earthquakes and volcanic eruptions. The capital has twice been leveled in this century, and natural hazards pose obstacles to the development of an industrial base. Manufacturing in Managua focuses on consumer goods such as chemicals, plastics, electrical products, and food processing. The revolution starting in 1979 destroyed more than a third of Nicaragua's industrial capacity. The Sandinista government hoped to reconstruct industry based on processing the country's forest and mineral products, as well as to disperse production geographically to regions outside the western highlands. The expansion of manufacturing activity has not happened.

The 1980s did bring some changes. The Sandinista government achieved a large-scale reallocation of land to more than 100,000 poor peasants, although it bungled attempts to incorporate the Miskito Indian communities along the Caribbean coast with the rest of the country. In 1984, the government launched a project to rebuild the railroad from the Pacific port of Corinto to Managua, and then extend it to Bluefields on the Caribbean, but construction is unfinished for lack of external funding. While coffee remained the mainstay of the economy, earning almost half the country's export income, cattle ranching suffered a severe blow. As in Honduras, it traditionally had been a favored activity of the elite. With the revolution, wealthy owners slaughtered up to a quarter of the herds, attempting to convert cattle to exportable cash, and thus crippled a growing business in the shipment of chilled beef to North American markets.

Although the Contra war ended in 1990, the new Chamorro government has failed to resolve the conflicts between left and right that provoked hostilities in the 1980s. Land redistribution schemes under the Sandinistas put many land titles in question. Foreign investment and aid were slow to return. Nicaragua remained devastated. The government inherited a $10.8 billion debt; per capita output and living standards dropped in the 1980s and have not recovered in the 1990s. Hurricane Joan devastated 22 percent of Nicaragua's rain forest in 1988, destroying 29,000 homes, killing 114 people, and injuring 178.

Nicaragua's recovery has thus been slow and replete with unusual schemes to generate income. For instance, the government is soliciting proposals from foreign companies to build a 400-km high-speed rail line to link container ports on the Atlantic and Pacific coasts. However, the traffic on the proposed railroad would be insufficient to pay interest on the construction debt. Such schemes appear to be more fantasy than reality, despite their attraction as a way to achieve the long-sought integration of the Caribbean lowlands with the rest of the country.

► Costa Rica

Costa Rica (3.6 million population), whose Pacific and Caribbean coasts are separated by only 130 miles, has had rail connections between Puerto Limón and Puntarenas since the nineteenth century, although service eastward has now ceased in favor of highways. Nevertheless, most of the population remains in the west, centered in the central plateau basin of San José.

Costa Rica's physical environment can be divided into three parts (Figure 9.7)

1. Central highlands flanked by Pacific and Caribbean lowlands. The central highlands are further subdivided into a northern range of still active volcanoes, whose peaks and high valleys are covered by a tropical rain forest that conservation efforts seek to preserve through national park designation. The rich diversity of species, especially bird and mammal populations, make this area one of Costa Rica's major centers of eco-tourism. To the south, a huge, batholithic, basaltic intrusion forms the Talamanca Range. Highly resistant to erosion, the area has thin soils and isolated settlements. Between these two mountainous areas within the central highlands lies the Meseta Central, composed of two basins centered on Cartago and San José. Filled with volcanic ash and lava that produce highly fertile soils, and lying at an elevation of 3,000–5,000 feet within the *tierra templada* in close to ideal climatic conditions for humans, this area has attracted dense populations.

2. The Pacific lowlands are formed by the extension of the rift valley that runs northwest to southeast through Nicaragua into Costa Rica. Much of the rift has been drowned to form bays such as the Golfo de Nicoya and the Golfo Dulce. Rivers from the highlands deposit alluvium in the Pacific coastal areas, especially in the Guanacaste region. Beyond the depression, the Nicoya and Osa peninsulas form mountainous outposts beside the Pacific. The Pacific lowlands were once covered in a tropical deciduous forest; however, clearing for banana plantations in the 1930s and 1940s, together with burning to facilitate settlement, especially cattle ranching, resulted in the transformation of much of this region to tropical grasslands and savanna.

3. The Caribbean lowlands are an extension of similar physical environments to the north. The conditions of the Costa de Mosquitos extend along the eastern margins of Costa Rica and into the Boca del Toro area of Panama. As in Honduras and Nicaragua, rivers bisect this region as they run from the central highlands to the Caribbean. In Costa Rica, the San Juan River creates a wide alluvial swath, providing access to the west and an extensive area for banana plantations. The Matina Valley with its port of Limón likewise became an area for banana production in the 1880s in association with Minor Keith's construction of the trans-isthmian railroad. Rainfall levels increase southward down the coast, and the Caribbean coastal region supports a rich tropical forest.

Costa Rica was lightly settled by aboriginal peoples, and their numbers were rapidly depleted by disease in the highlands. Small numbers of Miskito Indians remained in the Caribbean lowland forests, surviving and adapting to European contacts in a process already described for Honduras and Nicaragua. Despite Costa Rica's Spanish name ("Rich Coast"), the early Spaniards found no gold and few clusters of native population, and the country's distance from major markets in Mexico and Peru limited its function as a supply base. Unlike most of Latin America, the usual economic and religious motives that shaped colonization were absent, and institutions such as *encomienda* and *repartimiento* never gained a foothold. The few Spaniards who remained after discovery settled around the colonial capital of Cartago, as well as in nearby San José and Heredia in the Meseta Central. Today, four fifths of Costa Ricans claim their descent from these early colonists, rather than the traditional mestizo and ladino cultures of other Central American countries. Moreover, alone in the region, Costa Rica appears as a stable democracy with a more broadly developed middle class, higher rates of literacy, and lower levels of infant mortality. These conditions have led to a view of Costa Rica's different colonial experience that has acquired mythical dimensions. It is argued that prior to the introduction of a coffee culture, colonial settlement was characterized by a pattern of rural, dispersed, subsistent peasant households. This homogeneous community is regarded as the source of Costa Rica's relative egalitarian social order and democratic tradition. Recent analysis of pre-coffee Costa Rica indicates that this interpretation of the colonial past is a myth. City populations in San José, Cartago, and Heredia dominated the countryside through an internal market of artisan and agricultural exchanges that preempted any sort of egalitarian homogeneity of the region's population.

Depending on the author, this inequality was either weakened or reaffirmed by the introduction of coffee in the 1830s and 1840s, in response to growing European demand. For some, coffee brought rapid population growth, spread

Coffee Finca, Costa Rica.

settlement into more distant parts of the highlands, and reinforced inequality between city and countryside. For others, coffee's commercialization of mostly small and medium-scale producers (over half of today's fincas are less than 25 acres) occurred because it was a crop that did not require a large plantation system to generate adequate income, and insufficient native labor pools discouraged large farms. Consequently, the benefits of coffee culture have been spread more broadly through the population, mirroring similar transitions in other coffee-culture economies in Colombia, Venezuela, and Puerto Rico (Gudmundson, 1986). Today, coffee accounts for more than a third of the value of Costa Rica's exports.

Unlike coffee, Costa Rica's other primary agricultural export—bananas—evolved under a commercial plantation system controlled by foreign rather than domestic producers, who initially sought freight for newly constructed railroads. Once exclusively grown along the Caribbean coast, production in the 1940s moved to the Pacific coast to escape sigatoka disease. Many Jamaican workers who had previously labored on the Caribbean plantations were induced to migrate to the Pacific. After construction of a network of estates and feeder railroads focused on the new Pacific ports of Quepos and Golfito that duplicated the infrastructure previously established around Puerto Limón, the Pacific estates also began to succumb to disease. On the Caribbean side, old banana land has been converted to the once historically important cultivation of cacao, while Pacific lands are being turned into stands of African palm oil. These crops now coexist with a reduced production of disease-resistant strains of banana, raised mostly by small farmers. Bananas still rank second after coffee, earning 22.5 percent of exports.

Over the past 125 years, expansion of settlement has spread population out from the central plateau into the highland tropical forests and San Juan valley to the north, east toward the Caribbean around Puerto Limón, and along both sides of the mountains bordering the Pacific. Little vacant land now remains except for a few sections of prime tropical rain forest preserved from squatting, not always successfully, by designation as national parkland. As in many other pockets of Central America, with proximity to European and North American markets provided by a network of locally paved roads and international air freight capacity, farmers are increasingly moving toward nontraditional crops, such as tropical fruits, ornamental plants, fresh-cut flowers, and macadamia nuts.

Political stability has stimulated manufacturing in San José. Multinational businesses operate in seven free-trade zones, attracted by a skilled workforce and duty-free exports to the United States, under the Caribbean Basin Initiative. Textiles and pharmaceuticals are exported, while cement, tire, and auto assembly plants serve the domestic market. Costa Rica possesses attractive beaches, especially around the Nicoya Peninsula, that complement the remarkable tropical forests. The potential for tourism is considerable. There is a significant population of retirees from North America who find the climate and amiable life-style attractive.

Costa Rica's fortunes have slipped recently. G.N.P. per capita is now behind Panama and Belize, and government efforts to nationalize banking are scaring away foreign investment. The country's famed environments are under threat from an explosion of cattle ranching, illegal loggers, landless peasants, and mining interests. Between 1961 and 1987, cattle pastures in Costa Rica, planted in aggressive African grasses, grew from 19 percent to 44 percent of the land area, with growth directed especially toward the San Carlos Plain. Meanwhile, despite a national park system that covers 22 percent of the country, Costa Rica has lost forest cover at a higher *rate* (6.6 percent per year) than any Latin American country, including Brazil. Between 1961 and 1991, deforestation reduced forest cover from 55 percent to 22 percent of the land area (Faber, 1993). Symptomatic of the invasion of national parks is the Corcovado rain forest which covers the Osa Peninsula. Since 1968, a boom town has developed at Puerto Jiménez on the inland coast, serving a flood of *oreros* (prospectors) seeking fortunes from newly discovered gold deposits. By the 1980s, 1,500 *oreros* were working illegally in the park while 3,500 mineral concessions, some held by large companies, surrounded the park's borders. In 1989, roads were pushed through to the Osa Peninsula from the main

Banana plantation, Costa Rica.

Pan-American Highway, linking the previously isolated peninsula with the heavily populated highlands. Further encroachment on the national park has followed.

► *Panama*

Panama (2.7 million population) differs from its Central American neighbors. Before 1903, Panama was a province of Colombia, and its culture retains a South American flavor. For most of its history, Panama's significance resulted from the narrow isthmus that separates two of the world's major oceans. The United States created Panama as a client state and financed the canal. Until the treaty of 1977, when the United States agreed to return the canal to full domestic Panamanian authority in 1999, Panama was a virtual colony of the United States. Americans ran the canal. Built at great human cost by mostly Jamaican and Barbadian labor, who contributed a strong Afro-Caribbean element to Panama's population, the canal also drew Chinese, Italians, Greeks, Arabs, as well as those of mixed Spanish and Indian blood to Panama's major cities. The U.S. dollar is the currency of Panama, and the country is a major center of banking and financial services for Latin America. The contribution of financial services to the gross domestic product is more than twice that of agriculture. A recent contributor to the economy is shrimp harvesting in the Golfo de Panamá, which has had extraordinary success and has now made shrimp the leading export, providing 21 percent of Panama's foreign earnings.

Despite the economic setbacks engendered by the U.S. invasion in 1989, Panama's economy remains tied to its international role. Colón (established in 1948) is one of the world's largest free-trade zones, performing warehouse and distribution functions. Hundreds of companies now import electronics, pharmaceuticals, and consumer goods, reshipping to the rest of Central and South America without payment of any customs duties to Panama. Over 90 percent of all consumer goods from the Far East destined for Latin America pass through Colón. Today, the duty-free zone is a walled maze of streets and wholesale shops owned principally by Lebanese, Chinese, East Indian, and recently arrived Arab businessmen. Busier than any place outside Hong Kong, Colón is doing $11 billion of business a year; yet it is also an island of wealth in a sea of poverty where crack cocaine consumption is rampant. The capital Panamá City, at the Pacific end of the canal, is equally cosmopolitan in character and houses international banks and trading companies, reflecting its role as a center for both legal and illegal capital flows into and out of Latin America.

International trade, however, is threatened by an aging canal. Its 27-foot draft cannot accommodate the world's largest merchant ships. At present, the largest volume of goods passing through the canal is grain (24 percent), followed by petroleum (16 percent) and containers (14 percent). Traffic is growing at a sluggish 2 percent per year, producing income from tolls of $605 million. These conditions have prompted suggestions for a new sea-level canal west of the existing route. However, such proposals have lost favor in the face of enormous construction costs and the ecological uncertainty of connecting two dissimilar oceanic environments. Some 10,000 species of marine organisms populate the oceans around Panama, although opinions over what proportions inhabit both oceans vary from 10 percent to 50 percent. At the very least, Pacific predators such as the crown-of-thorns starfish and the venomous yellow-bellied snake have not yet found their way to the Atlantic, mainly because freshwater Lake Gatún is a barrier in the center of the present canal. However, the Pacific mean sea level is one foot higher than that of the Atlantic, while tides off the Pacific coast vary by 21 feet compared with two feet on the Atlantic side, almost assuring species transfer in a sea-level waterway. Alternatives involve increasing lock width and widening the Gaillard Cut, now underway, reducing a major bottleneck that has prevented two-way passage. Enlarging the locks, however, would increase the demand for water to assure efficient operation, thereby increasing the saltwater pumped into Lake Gatún, the artificial "holding tank" at the highest point in the crossing (85 feet). In turn, dilution of the lake's freshwater content would decrease its effect as an ecological barrier between the oceans.

Rural Panama contrasts greatly with its urban counterpart, taking on a more typically Central American appearance. To the west, mountain ranges extend to the Costa Rica border, dividing the country into a series of lowland-fringed slopes facing the Atlantic and Pacific (Figure 9.8). Banana plantations are found in western Chiriquí province beside both coasts, interspersed with rice and palm oil cultivation. Coffee is grown on the higher slopes of Volcán Barú. Chiriquí, with its provincial capital at David, displays a streak of independence within Panama. The new oil pipeline from Puerto Armuelles on the Pacific to Chiriquí Grande on the Atlantic coast allows tankers carrying Alaska crude to avoid the canal and adds to the western region's sense of economic self-sufficiency.

The western region contains significant indigenous populations. In particular, the 40,000 Guaymí are growing in numbers and are the largest native group in Panama. Traditionally shifting cultivators in the tropical forests around the slopes and base of Cerro Colorado, the Guaymí live in dispersed villages composed of a network of extended families. They grow a wide variety of subsistence and commercial products, from rice and corn to coffee and bananas. Population densities are increasing and fallows are shortening in the shifting cultivation system. Pressure on land resources is growing. The land shortage cannot be relieved by extending the area occupied because commercial farmers cultivate on the periphery of traditional Guaymí lands. One consequence is the growth of Guaymí seasonal labor migrations to coffee and vegetable harvests in western Chiriquí, or cutting sugarcane in Veraguas Province and working on banana plantations in the Bocas del Toro region. Such cases represent the traditional association of indigenous contract wage labor and commercial plantation agriculture common to many other parts of Central America.

Another potential basis for conflict with the Guaymí is the introduction of extractive mining industries between 1970 and 1981. Exploration by foreign

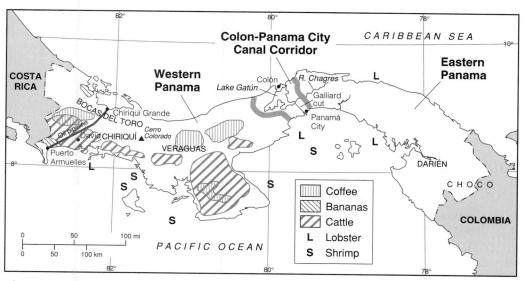

Figure 9.8   Map of Panama.

companies beneath the surface of Cerro Colorado uncovered a large deposit of low-grade copper, together with commercial quantities of molybdenum, gold, and silver. A proposal was made to create a huge 4-km by 5-km open-pit copper mine; a concentrator at the mine site; and a smelter at a new, deep-water harbor beside the Pacific, which would be connected to the mine by a 65-km slurry pipeline (8 km equals 5 miles). Some 7,000 construction workers and up to 3,500 permanent workers would have been needed, together with a city of 30,000 to support the operation. The plan would have involved a radical rearrangement of Guaymí living space and the invasion of thousands of foreigners. Ultimately, the transnational mining interests pulled out of the project, considering the price of copper in the world market to be insufficient to justify development costs. Pressure exerted on the Panamanian government by the Guaymí, linked to a worldwide environmental coalition, also played a role (Gjording, 1991) and marked an early example of indigenous activism that was later duplicated among Mayan peoples suffering from Guatemalan military oppression and Chiapas Indians seeking greater equity from the Mexican government.

East of the Canal Zone lies the third and most remote of Panama's three regions, including the San Blas coast and the poorly-drained Darién Peninsula. Traditionally, this region has been composed of Indian settlements, based on subsistence agriculture and fishing. Since the 1970s, however, as a highway has been pushed into the Darién, cattle ranchers have moved in behind the construction engineers, clearing the forests and threatening one of the last remaining stands of primary rain forest in Central America. Placing cultural parks around the highway route is being tried as a way of preserving the forest and the cultural life of the Indian villages it supports (Herlihy, 1989).

Panama's close political relationship with the United States is expected to end in 1999 when full ownership and operation of the canal passes to the Panamanian government. Already, many vestiges of an "American culture strip," separated from Panama by tall chainlink fences, have passed. Evidence on the ground of the Zone is fast disappearing. The United States has begun to move troops and dependents back to home bases. This early withdrawal has prompted second thoughts from Panamanians, given the 16,000 jobs and $330 million (8 percent of Panama's budget) that the military presence brings. Discreetly, Panama has begun to negotiate for a continued smaller U.S. presence into the next century. Such a reversal of nationalistic passions by Panamanians represents a way to cushion the economic loss and provide the stability required by international investors in a region of fragile peace.

► *Belize*

Belize (200,000 population) stands apart from the rest of Central America, a distinct enclave of English-speaking, Afro-Caribbean culture. Its origins as part of the English incursion on the Spanish domain, beginning with piracy and dyewood logging and shifting in the nineteenth century to mahogany extraction, have already been detailed. At one time in the late eighteenth century, Britain formally acknowledged Spain's claim to most of the Caribbean coastline of Central America, but never to the community around Stann Creek. Guatemala based its claim to Belize on the *uti possidetis* principle of prior Spanish authority over Central America (see Chapter 14). In 1993, Guatemala renounced its claim to Belize, which had achieved independence from Britain in 1981. The small British garrison, whose presence had deterred any Guatemalan threat, was sent home to mixed feelings by Belizeans, since the military contributed 10 percent of the economy.

Belize is Central America's wealthiest nation after Panama, but it has the most rapid population doubling rate (21 years) of any country in the region. About half the population is urbanized, many living in the vicinity of the Stann Creek valley and

Belize City (45,600), as well as smaller numbers along the New River between Orange Walk (12,100) and Corozal (7,100) to the north, Dangriga (6,400) and Punta Gorda (3,900) on the south coast, and San Ignacio (10,600) and Benque Viejo (4,100) in the interior (Camille, 1994). Belmopán (3,800), 30 miles inland from the coast in the center of the country, is the new forward capital, begun in 1970 after Hurricane Hattie (1961) destroyed most of Belize City. The vulnerability of coastal locations to hurricanes, which often track close to the Belize shore, together with an emerging nationalism prompted the decision to establish the new capital. To date, the city has grown slowly, and coastal communities continue to dominate the population distribution.

Belize divides into three regions (Figure 9.5).

1. South of the Sibun River, the country is an extension of the Guatemalan uplands that extend across the southern Petén. Stann Creek, the focus of early British settlement from the 1700s, cuts a steep trench through the edge of the Maya Mountains.

2. North of the Sibun, the land is rolling to flat, part of the great limestone plateau that forms the Yucatán Peninsula. Rivers, particularly the Belize and the New, bring watered strips of fertile alluvium as they drain north and east of the Maya Mountains, and with Stann Creek, have been the main centers of agricultural development.

3. The third region is the narrow lowland plain bordering the Caribbean, accessible mostly by coastal ferries. Since 1966, an inland highway has stimulated some settlement southward from Dangriga to Punta Gorda. Paralleling the coast from 10 to 20 miles offshore is a dense strip of mostly tiny islands (*cays*), part of the longest barrier reef in the Western Hemisphere. Together with the clear water and coral formations that surround them, the cays offer idyllic remoteness and some of the world's finest scuba diving. In the offshore waters, a growing commercial fishery is emerging, while along the coast for the past 15 years, more than 600 hectares of shrimp farms have been developed.

At one time, tropical forests covered more than 70 percent of the land surface. Today, that figure has dropped to 42 percent as loggers and cattle ranchers have made significant inroads. But for most of its history, forests have been the principal attraction of settlement and human enterprise to Belize. In the colonial period, logwood found mostly in the swampy lowlands of northern Belize provided vivid dark blue, black, and deep red dyes for European textile producers. Logwood exports peaked in the mid-eighteenth century, after which chemical aniline dyes began to replace natural products. The focus on lumber then shifted to hardwoods, especially mahogany. Woodcutters soon recognized that the wood found in northern Belize is of superior quality to that in the south. Lower rainfall in the north produces a closer grain, although the trees are not as large and grow more slowly than in the moister forests of southern Belize (Camille, 1994). By 1940, overcutting in the northern forests, mostly accessible by rivers, was exhausting supplies. However, a system of logging roads and the introduction of large logging trucks invigorated the industry, which made inroads into the previously inaccessible southern mahogany forests. Mahogany extraction lasted well into the 1950s, despite fluctuations in price and redirection of the market from Europe to the United States and, more recently, Japan.

Since 1960, except for a brief period (1949–1959) when pine destined for the West Indies achieved over 20 percent of Belizean exports, lumber products have been supplanted by sugar, citrus, bananas, and coconuts. Sugar was a modest export in the late nineteenth century, but the industry floundered through much of the present century from inefficient production and lack of available labor. Guaranteed markets under the Commonwealth Sugar Agreement of 1951 became the modern stimulus to sugarcane cultivation, mainly on the alluvial lands along the New River in the north. Bananas were attempted in the southern valleys, particularly Stann Creek, but the lack of flat land and the frequency of wind damage, together with the usual problems of sigatoka and Panama disease, limited production. Much more significant in recent years has been production of citrus crops, which are becoming a mainstay of the Belize economy. Grapefruit and oranges grow well on the black alluvial soils of the upper reaches of many rivers, particularly in the Stann Creek area (Camille, 1994).

Belize's population density (24 per square mile), by far the lowest in Central America, has attracted both the landless and the oppressed from outside the country. Black Caribs, or *Garifuna*, were brought from Dominica in the late eighteenth century to coastal communities in southern Belize; from there they have spread into Honduras and Nicaragua. A mixture of African and Carib, the Garifuna speak a form of Arawak language and practice subsistence agriculture based on South American manioc. In 1958, Mennonites from Canada and Mexico established communities in the Orange Walk area that now produce rice and beans for the domestic market. More recently, refugees from Guatemala and El Salvador brought a renewed presence of Maya language to modern Belize. Maya speakers were once common in Belize, and archaeological sites have been uncovered along the north shore, but at the end of the seventeenth century, the Spanish forcibly removed the dispersed indigenous communities to Alta Verapaz and the Petén in Guatemala. Into this vacuum came the predominantly English-speaking Afro-Caribbean and Anglo populations.

Today, Belize stands on the brink of substantial growth. The cays on the Caribbean coast are attracting the interest of the tourist industry, in which the use of English is an advantage. Citrus cultivation is poised for major expansion, given proximity to the U.S. market and fine growing conditions including the absence of the frosts that damage Florida and Texas groves. Democratic traditions and a lack of serious land tenure issues give a political stability uncommon in Central America. Yet, Belize is a small country vulnerable to changing external circumstances. Economic prosperity owes much to links with the British Caribbean, especially Jamaica, and markets in Britain through the Commonwealth affiliation and access to the U.S. market via the Caribbean Basin Initiative (CBI). Recently, Caribbean links have weakened, and Belize turns more toward Central American neighbors, epitomized by its forward capital of Belmopán, with its Mayan building motifs that echo a pre-British past (Camille, 1994). As Britain becomes more closely tied to the European Union, preferential trading terms for Commonwealth countries are disappearing, and NAFTA raises questions about the prospect of continued favorable treatment in the U.S. market when the CBI ends in 1997.

Consequently, Belize has begun to seek alternatives and turn itself into an offshore financial center. In 1989, it established a flag of convenience for registration of world shipping. More recently, it has authorized the establishment of international business companies (IBCs): anonymous, tax-exempt entities that can be formed for $100. In 1995, an offshore banking law was passed, allowing Belize to compete with the Bahamas, Cayman Islands, and Panama as centers for legal—and often illegal—international business.

Belizean attempts to develop alternative enterprises are also prompted by a desire to limit environmental damage from a raw material and agricultural export economy. The government remains sympathetic to a strong local environmental movement and has designated 28 percent of the total land area as park and forest reserves. Aerial spraying of the herbicide paraquat was discontinued in 1983 due to concern that honeybees were being killed. In 1985, a huge, undeveloped interior tropical forest was purchased by Coca-Cola for citrus cultivation to supply the North American orange juice market. However, the environmental movement inside and outside the country forced the company to curtail its plans. In 1988, the company began to break up its holdings, passing some to small farmers and others to environmental organizations such as the Belize Audubon Society, but retaining substantial lands for future development. The growing political power of environmental movements cannot be discounted as a factor shaping Central America's landscapes.

SUMMARY

Central America is a peripheral region of fragmented political units with little power to resist outside economic and political forces. It is suggested that Central America might prosper if it manages to achieve some degree of unity. The sketches of each country describe common problems afflicting the region: excessive reliance on a few raw material exports; lack of land for the large subsistent population, which is growing in numbers; a major imbalance in the distribution of wealth and power; severe environmental degradation; and a fundamental cultural divide between what Augelli has called the Rimland (the Caribbean culture region) and the Mainland (the Spanish-speaking mestizo and highland Indian region). At the same time, Central America also possesses attributes that include tropical climates, fertile soils—albeit in proximity to locations of high risk from volcanoes and earthquakes—and the potential for a wide diversity of sustainable products from cut flowers to macadamia nuts and pesticide-free coffee. Moreover, the proximity of Central America to North America, together with the decline in ground and air freight rates, puts much of the region far closer to its market than South American competitors. What is needed is the will to put aside national conflicts, such as those caused by migrants from El Salvador fleeing to Honduras or Nicaraguans settling in Costa Rica, and instead begin to appreciate the cultural affinities of language, history, and environment that once, briefly, bound Central America into a single entity.

FURTHER READINGS

Augelli, J. P. "Costa Rica's Frontier Legacy." *Geographical Review* 77 (1987): 1–15.

Browning, D. O. "Agrarian Reform in El Salvador." *Journal of Latin American Studies* 15 (1983): 399–426.

McCullough, D. *The Path Between the Seas: The Creation of the Panama Canal, 1870–1914*. New York: Simon and Schuster, 1977.

Wall, D. L. "Spatial Inequalities in Sandinista Nicaragua," *Geographical Review* 83 (1993): 1–13.

CHAPTER 10

The West Indies

OLWYN M. BLOUET

INTRODUCTION

The West Indian islands have been more important in world affairs than their size suggests. Europeans brought the islands into the trans-Atlantic economy by developing sugar plantations, based on transported, enslaved African labor. During the seventeenth and eighteenth centuries, European colonial rivalry involved intermittent warfare in the Caribbean. The islands were considered so important, because of sugar production, that at the close of the French and Indian Wars the British pondered whether to retain Canada or Guadeloupe (Peace of Paris, 1763). They settled for Canada. Part of the reason why Napoleon decided to sell the Louisiana Territory to the United States for $15 million in 1803 was that, with Haiti lost to the French empire, in a slave revolution, the significance of Louisiana as a supply base was diminished.

During the nineteenth century, the islands, especially Cuba, became important to the United States. As Europeans gained new empires in Africa and Asia, the United States began to regard the Caribbean as its sphere of influence. In the twentieth century, the United States has developed political, economic, and military dominance in the region, despite lingering European connections. American culture is pervasive.

Why are the West Indies important to the United States? First, their strategic location is crucial. In enemy hands, they pose a threat to national security, hence the quick action in Grenada in 1983. The islands can control trade routes to the Americas, involving critical supplies such as petroleum from Venezuela or the Middle East.

Second, although now widely grown, tropical agricultural items such as sugar, coffee, and bananas, and mineral resources such as bauxite and nickel come from the Caribbean. Third, the West Indies produce migrants. Currently, about 5 million people of Caribbean extraction live in the United States. Conditions in the islands affect rates of migration, as the recent exodus from Haiti and Cuba demonstrates. People with Caribbean links, ranging from Marcus Garvey (Jamaica), to Malcolm X (mother from Grenada), to Louis Farakhan (mother from the West Indies), to Colin Powell (parents from Jamaica) have an impact on culture and events in the United States.

Finally, the islands attract U.S. attention because of illegal activities, such as drug running and money laundering. The United States has an excuse and an opportunity for surveillance and intervention in the region.

THE WEST INDIAN REGION

The West Indies[1] include the islands between the Florida Peninsula and the northern shore of South America—from the Bahamas to Trinidad. The land area is approximately 91,000 square miles, roughly the same as the United Kingdom and a little larger than Minnesota. Population totals about 36 million. The sizes of islands and their population numbers vary greatly. Cuba has almost half the land area of the Antilles with over 44,000 square miles, and almost a third of the population with 11.2 million. At the other end of the scale, Anguilla has only 35 square miles and about 9,000 people.

The fifty or so islands, scattered over more than 2,000 miles of sea, can be divided into four groups.

1. *The Bahamas* consists of more than 700 islands. Most are small, low-lying, limestone, dry and uninhabited.

2. *The Greater Antilles* include the islands of Cuba, Hispaniola, Puerto Rico, and Jamaica. They are part of a submerged continental tract with high mountain peaks. The four islands comprise over 80 percent of the land area of the West Indies (Figure 10.1).

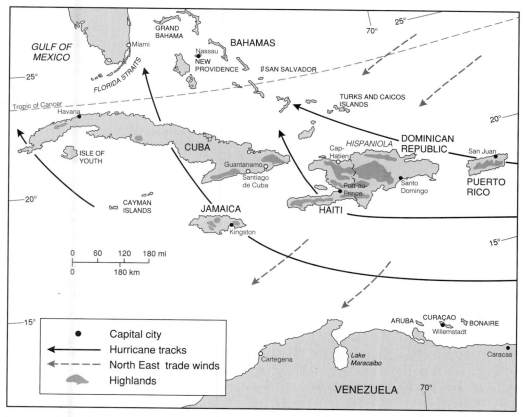

Figure 10.1 The Greater Antilles.

---

[1]The terms West Indies, Antilles, and Caribbean are used interchangeably in this chapter although there is no general agreement on definitions of these terms.

3. *The Lesser Antilles* are composed of two parallel chains. Islands in the inner arc have formed around volcanic peaks. Five of the islands—Guadeloupe, Martinique, St. Kitts, Montserrat, and St. Vincent—have active or potentially active volcanic cones. Coral limestone islands, such as Antigua and Barbados, form the outer arc of the Lesser Antilles (Figure 10.2).

4. *South American offshore islands* include Aruba, Bonaire, Curaçao, Trinidad, and Tobago. Venezuelan offshore islands, like Margarita, are not included in this discussion of the West Indies.

The West Indies is a diverse region based on history, culture, and physical characteristics. People of African, European, and Asian extraction mingle in a Caribbean setting, producing one of the most ethnically and culturally diverse areas of the world. A tourist traveling to the Dominican Republic is in an Afro-Hispanic-Caribbean realm, where Spanish is spoken and Roman Catholicism practiced. The same tourist who crosses the boundary into Haiti discovers an Afro-Franco-Caribbean world where the local creole language predominates, but French is spoken by an elite. Roman Catholicism and voodoo exist side by side. The tourist visiting Barbados is in an Afro-Anglo-Caribbean area where English is spoken and Anglican churches dot the landscape. Bridgetown, the Barbadian capital, has the flavor of an English county town with its government buildings, cathedral, and Kensington Oval cricket ground. Trinidad, a former Spanish and then British colony, presents a different face displaying African, Spanish, French, English, Chinese, and East Indian cultural enclaves. In Trinidad, the mosque has a place beside the Roman Catholic church, the Protestant chapel, and the Hindu temple. Afro-Caribbean influences, such as calypso, steel bands, and reggae are ubiquitous.

French language and culture predominate in Guadeloupe and Martinique and are significant in Haiti, St. Lucia, Dominica, and St. Vincent. The Netherlands Antilles have a distinct cultural flavor, displayed by the Dutch language and architecture. As tourism expands in the Caribbean, developers attempt to project island identities and frequently emphasize the colonial heritage. In Aruba, for example, new shopping malls are constructed in Dutch architectural styles, and a windmill has been erected to underline the Dutch connection.

Islanders have a sharp sense of identity and a strong sense of place. Even within the English-speaking Caribbean there is a marked difference from island to island. A Jamaican, for instance, knows a Barbadian from a Trinidadian, and vice versa. A great deal of island particularism is resistant to change. People born in Tobago, for example, guard their identity against the central Trinidadian government. Similarly, inhabitants of tiny Barbuda resent control from Antigua. Democratic, representative governments exist in the former French, British, and Dutch colonies. Cuba is under Fidel Castro's control, and Haiti struggles to understand democracy.

Everywhere in the Caribbean, creole cultures and local patois coexist within a Euro-American-African milieu. Everywhere Third World images face First World scenes. A tethered cow grazes beside a brand new McDonald's restaurant. Small, wooden shacks are around the corner from high-rise Hiltons. Peasant farmers work small plots, global companies run plantations to produce sugar and fruit for world markets, and modern container ports function alongside women carrying produce on their heads. The Americanization of the region has recently proceeded at a fast pace.

Given the diversity and particularism, how can the West Indies be discussed as a region? John Augelli proposed a Mainland–Rimland division of Middle America that sought to differentiate culture realms (West and Augelli, 1989, pp. 12–16). His Mainland consisted of Mexico and most of the Central American Republics, and

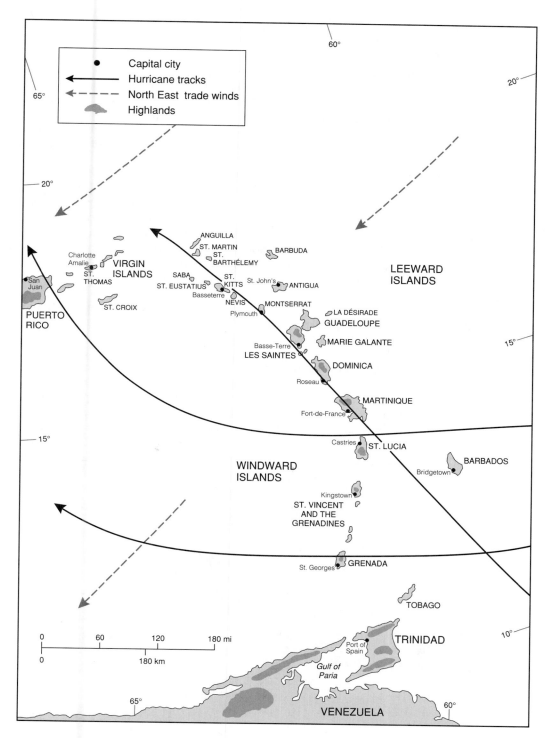

Figure 10.2 The Lesser Antilles.

the Rimland included all Caribbean islands and the Atlantic margin of Central America, which had been penetrated by Caribbean influences. Augelli noted significant differences between the Mainland and Rimland. While the Mainland displayed strong indigenous and Hispanic influences, the Rimland was influenced by several European groups and had few indigenous elements. Conversely, the Rimland showed a very strong African presence but the Mainland less so. Most Rimland inhabitants live on tropical coastal plains, susceptible to outside influences, but Mainland populations generally live in interior uplands. The Mainland and Rimland differ in land use (*hacienda* versus plantation), types of crops, and material culture.

Augelli subdivided the Rimland into a Central American (Coastland) and a West Indian (Island) sector, differentiated by a number of factors. First, the plantation has a long slave-based tradition in the West Indies, rather than a postcolonial development as on the Caribbean coastlands of Central America. Second, North Europeans, rather than North Americans, initially developed plantation agriculture in the West Indies. Third, sugar has traditionally been the prime commercial crop in the West Indies and bananas in Central America. Finally, the West Indies lacks Amerindian and mestizo labor pools found on the Central American Rimland (Table 10.1).

Therefore, three distinct realms can be identified: Mainland, Coastland, and Island (Figure 10.3). Let us discuss in more detail the common geographical and historical characteristics of the West Indian Island realm.

► *Geographical Characteristics*

The *insular nature* of the West Indies has been a significant factor in historical development. Living on islands with limited refuge areas, indigenous populations were vulnerable to demographic destruction by Europeans through warfare, enslavement, labor practices, deportation, and diseases. The Indians were virtually

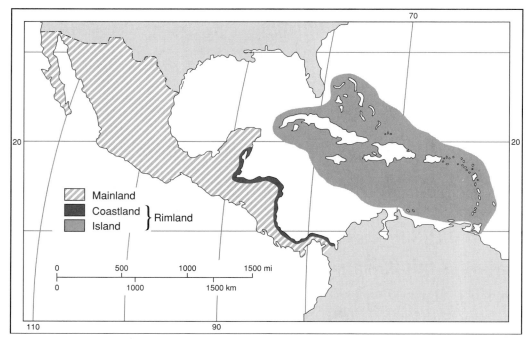

Figure 10.3   Middle America: Mainland, coastland, island.

TABLE 10.1 Middle America

Mainland	Coastland	Island
Mexico and Most of Central America	Caribbean Coastlands of Central America	Islands of the Greater and Lesser Antilles
High, relatively remote environment	Tropical coastal plains, susceptible to interaction	Tropical, strategic islands, susceptible to interaction
Strong indigenous influences	Some indigenous influences	Weak indigenous influences
Weak African influences	Some African influences	Strong African influences
Strong Hispanic influences	Hispanic and other European influences	Hispanic and strong northern European influences
Traditionally hacienda	Post-colonial plantation development by North Americans	Long colonial plantation tradition, based on slavery
		Developed by Europeans and then North Americans
Mixed farming	Bananas	Sugar

eradicated from the islands; in contrast, most Latin American Indian populations were depleted but not eliminated (see Chapter 3).

The large number of islands permitted different European countries, such as Spain, France, England, and Holland, to colonize the area and mold distinctive cultural patterns in unique West Indian forms. More contact and interaction took place with metropolitan powers in Europe than with Caribbean neighbors. The island nature of the Antilles militates against control by one power. Insularity intensifies fragmentation, although the Caribbean Sea has acted as a transportation and communication conduit.

The second common geographical characteristic of the West Indies is the *small size* of most islands, especially when their large populations are taken into account. The Caribbean is the most densely populated region of the Americas. Small states face common economic problems, such as land shortage and limited environmental variability. Most islands have small internal markets and rely heavily on imports. Transportation costs are high; economies of scale are weak (Clarke and Payne, 1987).

The numerous small states in the Caribbean find it expensive to operate their governments, educational systems, and health and welfare services. Political independence is costly, and the smaller the state, the higher the per capita cost of many services. For example, each West Indian country, regardless of size, employs an extensive diplomatic corps on the world scene, and costs are high.

Third, the West Indian islands lie in the *maritime tropical* air mass, although the Bahamas and northern areas of the Greater Antilles can experience incursions of cold air from the North American landmass. Average annual temperatures are about 80°F. with little seasonal temperature variation. The northeast trade winds modify the tropical heat year round and produce most precipitation. Rainfall is highest in the summer months, with a drier season, when sugarcane is usually harvested, between December and April.

Precipitation varies markedly from island to island, with many arid or semiarid environments. Generally speaking, low-lying islands receive little rainfall because the moisture-laden air does not rise high enough to cool and produce precipitation. For example, the flat Netherlands Antilles of Aruba, Bonaire, and Curaçao have a semiarid climate with about 20 inches of rain per year. The vegetation is composed of cacti, scrub, and aloe. Aruba has a large desalinization plant to produce its water supply. Conversely, islands such as Dominica, with high volcanic peaks, receive over 100 inches of rain and support tropical rain forest.

Precipitation varies markedly even within islands, with heavier rain generally on the northeast side of islands (windward slopes) and lighter rain on the southwest side (leeward slopes). For example, Montego Bay north of the Blue Mountains in Jamaica, gets on average 51 inches of rain per annum, but Kingston, on the southern coast, receives only 31 inches per year (Figure 10.4). Irrigation is necessary for agriculture in southern Jamaica, which lies in a rain shadow. Similar rainfall variability is common in many islands.

The Caribbean is regularly visited by *hurricanes*, many of which proceed on to threaten the United States. Sir William Reid, a nineteenth-century pioneer meteorologist, began research on hurricanes after assisting with the cleanup of "The Great Hurricane" in Barbados (1831), which killed over 1,500 slaves. Reid wrote *The Law of Storms* (London, 1838) in which he attempted to explain how hurricanes develop. Hurricanes form over the Atlantic off West Africa and sweep across the Caribbean between June and November, damaging crops, livestock, people, and property. Recent severe hurricanes have included David (1979), Gilbert (1988), Hugo (1989), and Andrew (1992). The 1995 hurricane season was particularly active, producing numerous tropical storms and hurricanes. Among the most destructive hurricanes

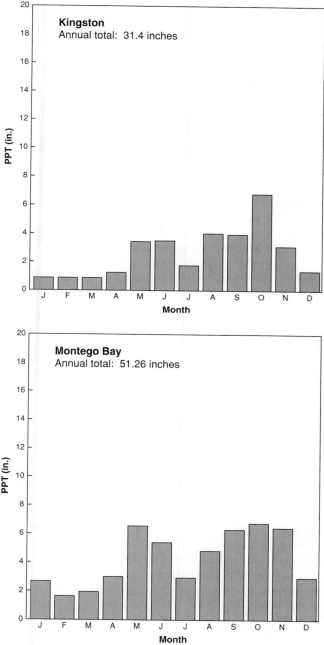

Figure 10.4   Annual rainfall: Montego Bay and Kingston.

were Felix, Iris, Luis, and Marilyn, producing considerable damage in Antigua, Barbuda, Montserrat, the Virgin Islands, Puerto Rico, and St. Martin.

In addition, several islands face the possibility of earthquakes (Port Royal, Jamaica, 1692) and volcanic eruptions (Mt. Pelée, Martinique, 1902). Mt. Soufrière on St. Vincent erupted in 1979. In 1995 inhabitants of Montserrat were evacuated because of volcanic threat.

The West Indies share a legacy of *environmental degradation* that gathered momentum after the introduction of sugarcane plantation agriculture in the seventeenth century. The production of tropical crops transformed the Caribbean landscape. Native flora and fauna were replaced by introduced species, as forests were cleared for agriculture. Just as forests are being cut down in Amazonia today, so mahogany and other woods were largely eradicated from the Caribbean landscape. The pattern was set in Barbados where, between 1645 and 1665, nearly all the native woodland, with small exceptions like Turners Hall Wood, was cleared to make way for sugar plantations. Deforestation increased the risk of erosion and drought, and changed the nature of the land resources (Watts, 1987). Once vegetation is cleared, excessive heat destroys humus, rainfall leaches out mineral nutrients, and fertility declines. Impoverished peasantries, most notably in Haiti, exacerbate environmental destruction. Forest is cut to provide firewood, and cropping on steep slopes exposes the land to erosion, especially after heavy rainfall.

The development of tourism, which is important from an economic standpoint, threatens the environment in the long term by putting pressure on resources. Tourist hotels contribute to pollution and waste disposal problems, and can increase beach erosion.

Yet another common feature of the West Indies is their *strategic location*, as American offshore islands, midway between North and South America. Caribbean waterways give access to the Gulf of Mexico and the Panama Canal, link Europe with Latin America, the east coast of the United States with the west coast and Asia, and South America with North America.

The strategic significance of the islands was obvious to the Spaniards, especially after the discovery of large silver deposits at Zacatecas and Potosí on the mainland. Havana became the main port of the *flota* on the homeward voyage, as the city lay on the Gulf Stream route, through the Florida Strait to the North Atlantic and Europe. Havana was heavily fortified by the Spaniards. Other colonial powers later established naval bases in the West Indies, such as Nelson's dockyard at English Harbour, Antigua, which today is a tourist attraction. The French had a naval base on Martinique from the late eighteenth century.

Since the late nineteenth century, the United States has maintained a dominant presence in the region. After the Spanish-American War (1898–1902), the United States annexed Puerto Rico and secured a naval base at Guantánamo Bay, Cuba, in 1902. The completion of the Panama Canal in 1914 increased the strategic significance of the Caribbean. In World War I (1914–1918) the United States, fearing the establishment of German submarine bases, occupied Haiti and the Dominican Republic, and purchased the U.S. Virgin Islands from Denmark (1917). During World War II (1939–1945) the United States gained several bases in the Caribbean from Britain on 99-year leases.

The strategic importance of the region was highlighted in 1962 with the Cuban missile crisis, when the United States was prepared to go to war rather than allow Soviet missiles on Cuba. In addition, in 1965 and 1983 the United States intervened, respectively, in the Dominican Republic and Grenada because of strategic interests. Currently, the United States has a strong military presence in Puerto Rico, bases in the Lesser Antilles, the naval base at Guantánamo, missile tracking stations in the Bahamas, and drug surveillance radar in the western Caribbean. U.S. military presence has grown because the Caribbean is one routeway for illegal drugs entering the United States.

► *Historical Characteristics*

The West Indies share a colonial history of slavery and sugar plantation agriculture. Although some territories, such as the Bahamas and Aruba, did not have sugar plantations, slavery was common to all islands. Historical characteristics form the foundation of a sense of regional identity.

The Caribbean islands were shaped as European colonial possessions (Table 10.2). Arawak and Carib cultures did not withstand European contact, despite large pre-Columbian populations (see Chapter 3). All domestic animals and many plants, with the important exceptions of maize, manioc (cassava), sweet potato, peanuts, beans, tobacco, avocado, and pineapple, are Old World imports. The Spaniards introduced sugarcane, citrus, bananas, and rice.

Spanish settlers spread out from Hispaniola to Jamaica (1509), Cuba (1511), and Puerto Rico (1512) in search of precious metals. Expeditions raided other Caribbean islands for slaves, leaving pigs and cattle to breed in the wild. In the 1520s and 1530s gold production in Hispaniola and Cuba declined, and Indian labor was depleted. Enslaved Africans were introduced as a labor force.

After Cortés invaded Mexico in 1519, many settlers moved to the mainland. The Greater Antilles remained Spanish colonies and were important from several perspectives. Large ranches acted as provision grounds for mainland Spanish America, supplying meat, hides, and tallow. Some sugar and tobacco were exported to Europe. Fortified ports like Santo Domingo, Havana, and San Juan guarded the trade routes between the New World and Spain.

In the buccaneering seventeenth century the Dutch, French, and British entered the colonial race. The Dutch concentrated on acquiring trading bases, and in the 1630s took the islands of Curaçao, Aruba, Bonaire, Saba, St. Martin, and St. Eustatius. Curaçao, with valuable salt deposits, was, and continues to be, the major Dutch possession. Today Willemstad is a free port with oil refining facilities and serves as an offshore financial center.

The French and English settled for the Lesser Antilles, where Spain was weak and the Caribs lingered. In the 1620s and 1630s the English colonized St. Kitts, Barbados, Nevis, Antigua, and Montserrat. The French settled on part of St. Kitts (1624), Martinique, and Guadeloupe (1635). Initial settlement in the English and French colonies was by white freeholders and indentured servants who hoped for land after serving their contracts. Venture capital was supplied by merchants. Early export crops included tobacco, indigo, ginger, and cotton.

In the 1630s, the islands in the Lesser Antilles were fledgling colonies, economically vulnerable and not vital interests of European governments. The development of plantation sugar production after mid-century changed the colonial relationship, making the small islands profitable overseas possessions of England and France. The islands were linked into the expanding Atlantic economy that involved Europe, Africa, and the Americas in a trade network that helped to fuel the development of capitalism and the Industrial Revolution.

In the second half of the seventeenth century, Spain lost Caribbean territory. The English captured Jamaica between 1655 and 1660, and Spain recognized French control of western Hispaniola at the Treaty of Rywick (1697). During the eighteenth century, England and France vied for dominance in the Caribbean. The islands became chips in the international war game, several changing hands numerous times. At the close of the Napoleonic Wars (1815), the British had a prime position in the Caribbean, and the Royal Navy was supreme. Trinidad (originally Spanish) and many formerly French islands were in British hands. The French were left with Martinique and Guadeloupe, having lost western Hispaniola (Haiti) to slave rebellion and independence (1804).

During the nineteenth century, as Spain and Portugal lost control of the mainland, British, Dutch, Danish, French, and Spanish colonial control continued in the island Caribbean. Only Haiti and the Dominican Republic (intermittently) had independent status, but neither flourished economically. The British and French

TABLE 10.2   The West Indies: Colonial Zones

U.S. Zone	Hispanic Zone	British Zone	French Zone	Netherlands Zone
U.S. Virgin Islands (St. Croix, St. Thomas, St. John)	Cuba (1902)*	Jamaica (1962)	Martinique	Aruba (due 1996)
	Dominican Republic (1844)	Trinidad & Tobago (1962)	Guadeloupe	Bonaire
	Puerto Rico**	Barbados (1966)	Haiti (1804)	Curaçao
		Bahamas (1973)	St. Martin (Northern)	St. Martin (Southern)
		Antigua–Barbuda (1981)		St. Eustatius
		St. Kitts–Nevis (1983)		Saba
		Montserrat		
		Anguilla		
		British Virgin Islands (Tortola, Virgin Gorda, Annegada)		
		Caymans		
		Turks & Caicos		
		Grenada (1974)		
		St. Vincent and the Grenadines (1979)		
		St. Lucia (1979)		
		Dominica (1978)		

* Dates indicate independence.

** Politically, Puerto Rico is in the U.S. zone, but culturally and historically, it is part of the Hispanic zone.

colonies became less important in the imperial scheme as the islands adjusted to the abolition of slavery, the end of protected European markets, and competition from other areas of production and from sugar beets. Conversely, the Spanish islands, especially Cuba, became more economically important as extensive areas were opened up for sugar production, using mechanized techniques that were difficult to introduce profitably into the smaller islands.

Toward the end of the nineteenth century, Cuba and Puerto Rico sought independence from Spain. During the Spanish-American War (1898–1902) Spain was removed as a Caribbean power. In the twentieth century, the British colonies became independent, and the United States assumed a dominant role in the Caribbean. The United States controls Puerto Rico and part of the Virgin Islands, and exerts influence in the region by economic penetration, political pressure, and direct intervention. Economically and geopolitically, the United States controls the region.

What has been the impact of the United States and Europe in the area? Politically, major decisions were made outside the region, with imperial considerations taking precedence over West Indian concerns. Economically, the islands have looked to outside powers for governance, markets, imports, and investments. Agricultural economies were geared to export markets. Socially, a white elite controlled political and economic power, a brown group struggled for recognition, and a large, former slave, black population was at the bottom of the social hierarchy. Culturally, European and increasingly American influences are strong.

Sugar, slavery, and the plantation are the historical keys to understanding the region. Without sugar production on a large, capitalist scale, the Antilles would have looked different. The sugar industry was a reason for and an instrument of colonial expansion. Plantation labor was performed by enslaved Africans, but the wealth accrued to a white elite of merchants, financiers, and planters—many of them absentees who lived in Europe or the United States.

Sugarcane originated in New Guinea, spread westward across Asia, and became an important crop on irrigated land in the Mediterranean area. In the fifteenth century, sugar cultivation expanded to the Atlantic islands of Madeira, the Canaries, and São Tomé, and traveled west to Hispaniola with Columbus on his second voyage in 1493. The crop diffused into Puerto Rico, Cuba, Jamaica, and Mexico. Some sugar was produced for export in the sixteenth century, but it was one crop among many and not a dominant monoculture in the Spanish islands until the nineteenth and twentieth centuries.

The Portuguese developed the sugar industry in Brazil, where they established a thriving export trade to Europe (see Chapter 12). Shipping sugar across the Atlantic was costly, but Brazilian sugar could be grown without irrigation, was favored by abundant timber for fuel, and had a long growing season. By 1600, commercial sugarcane cultivation was no longer economically viable in the Mediterranean because of New World competition (Galloway, 1989).

Brazil became the major New World sugar producer and, from Brazil, the Dutch[2] diffused the industry into Barbados, providing the technology, equipment, capital, and credit. Between 1640 and 1680, landholdings grew larger in Barbados, woodland was cleared, and enslaved Africans provided labor on sugar plantations (Dunn, 1972). Many white small holders left Barbados, unable to compete with large landowners and slave labor. Whites migrated to the mainland North American colonies, especially South Carolina. In the late seventeenth century, Barbados was the wealthiest, most populous English colony in the Americas, surpassing both

---

[2]In the early seventeenth century, the Dutch controlled part of northern Brazil until the Portuguese regained it in 1654.

Sugar cane was propagated by planting cuttings in holes. The arduous work of digging the holes was done by slaves.

*   *   *

Massachusetts and Virginia. Barbados was a center of diffusion for innovations such as cane-holing (digging holes rather than trenches for young cane plants, to minimize erosion), windmills (to press the cane), and the burning of bagasse (cane trash) for fuel (Galloway, 1989).

Other British and French territories in the Lesser Antilles converted to sugar production for metropolitan markets. In the eighteenth century, Jamaica succeeded Barbados as the richest British sugar island, and French Saint Domingue became the wealthiest sugar area in the whole Caribbean.

Sugar production was an agricultural and industrial operation combining field and factory. Sugar is produced in four stages:

1. Cultivation and harvesting of cane
2. Extraction of cane juice by pressing or milling

Cane cutting had to be done rapidly when the crop was ready.

Cut cane had to be brought promptly to a crushing plant where the sugar juice was extracted. The crushing mill in the picture is wind powered. Other forms of power included mules, oxen, and water. Later, steam and diesel engines were used.

*   *   *

3. Boiling the juice to crystallize sugar
4. Refining

The first three operations occurred at the plantation site. Minimal refining was sometimes performed at source to produce clayed or white sugar. But refining was usually done, as it is today, at the point of consumption because sugar can become a solid block without complete control of humidity. Amsterdam and London were, and continue to be, important refining centers.

The sugar industry required large labor inputs, especially at harvest time, when ripe cane had to be cut promptly, carted to the mill, and crushed to extract

In the boiling house on the plantation the cane juice was concentrated.

Molasses, a syrup separated from raw sugar in sugar manufacturing, and rum were shipped in barrels to North America and Europe.

the juice. Once cut, sugar rapidly ferments and will not crystalize. Because of the critical demand for labor, a demographic revolution ensued. It is estimated that nearly 10 million enslaved Africans were imported into the Americas between 1451 and 1870 (Table 10.3). Brazil took approximately 39 percent of the slaves; nearly 50 percent went to the Spanish, British, French, and Dutch Caribbean; almost 8 percent went to the Spanish mainland; and less than 5 percent of slaves were imported into British North America. The labor demands of the sugar industry resulted in large and continuous slave imports into the Caribbean, since, unlike in the United States South, there was no natural increase in total slave numbers. Researchers suggest several reasons for the lack of natural increase in West Indian slave populations, notably the severe labor regime, food deficits, unbalanced sex ratios, climate and disease, lactation practices, and the use of abortifacients.

TABLE 10.3   Estimated Numbers of Slaves Imported into the Americas, 1451–1870

Country	Number	%
Spanish Caribbean	809,100	c 8.60
Spanish Mainland	743,000	c. 7.90
Brazil	3,646,800	c. 38.80
British Caribbean	1,655,000	c. 17.80
French Caribbean	1,600,200	c. 17.00
Dutch Caribbean	500,000	c. 5.30
Danish Caribbean	28,000	c. 0.30
British North America	399,000	c. 4.25
Total	9,391,100	99.95

A more recent estimate of 9,778,500 slaves imported into the Americas is very close to Curtin's original figure (Lovejoy 1982).
Source: Curtin, 1969, p. 268.

Africans brought a rich culture complex to the West Indies, including agricultural and handicraft skill, folk tales, music, dance, adornment, medicine, and belief systems. The role of African women in rural life, agriculture, and marketing continued in modified form. The Caribbean region shares a deep African heritage, strongly displayed in musical rhythms and carnival traditions.

Sugar and slavery led to a revolution in land use. The plantation, geared to external markets and profits, developed in the West Indies, whereas the *hacienda*, designed for self-sufficiency, developed on the mainland. Early settlement in the British and French colonies consisted of dispersed farms and small holdings, but when sugar monoculture developed, landholdings were consolidated, and the estate complex of plantation house, provision grounds, sugar mill, boiling house, and slave village emerged (Figure 10.5).

Plantations and population were mostly located on coastal plains and lowland areas, monopolizing the best land. Plantation land use was originally of four main types:

1. Sugarcane fields
2. Forest for timber and fuel
3. Provision grounds for slave food
4. Pasture for livestock

Later, land use was modified. Forest was cleared and bagasse used as fuel. Pasture was reduced as wind power replaced animal power in the mills. In some islands sugar expanded onto provision grounds, and food was imported. Reliance on food imports continues to the present.

The colonial plantation, frequently 200 acres with 200 slaves producing 200 tons of sugar, was at its height in the British and French islands during the eighteenth century. Then slavery came under pressure, first in Haiti with the successful slave revolt that began in the 1790s. Next, Denmark outlawed the slave trade in 1802, followed in 1807 by the British, who attempted to enforce the ban internationally. Metropolitan governments later abolished the institution of slavery. Slaves were emancipated in the British islands in the 1830s. Slavery was abolished in the French islands in 1848 and in the Dutch areas in 1863. Slavery ended in Puerto Rico in 1876 and in Cuba in 1886 (Table 10.4).

The end of slavery in the British and French islands had an impact on population and settlement. Out of slavery, black population numbers increased. Where land was available, as in Jamaica, ex-slaves moved from plantations and settled on dispersed plots, often in hilly, interior areas. Sharecropping, small holdings, and squatting developed on bankrupt or subdivided plantations. Peasant villages appeared, the number of plantations declined, and the value of plantation land dropped. Many ex-slaves became part-time peasants and part-time wage laborers.

Not all islands with interior uncultivated land followed the Jamaican pattern. In the French islands of Martinique and Guadeloupe, planters were more successful in retaining labor by using sharecropping (*métayage*) and by selling peripheral plantation land to former slaves.

Where all land was in cultivation, as in Barbados, Antigua, and St. Kitts, ex-slaves continued to work plantations or migrated to other islands. After emancipation, sugar production in Barbados remained stable, since ex-slaves had little choice but to continue to labor on sugar estates. Areas facing a labor shortage in the sugar industry recruited indentured workers, mainly from the Indian subcontinent and China, who added to the cultural diversity of the region. The majority of indentured laborers went to new plantation areas, such as Trinidad and British Guiana, where

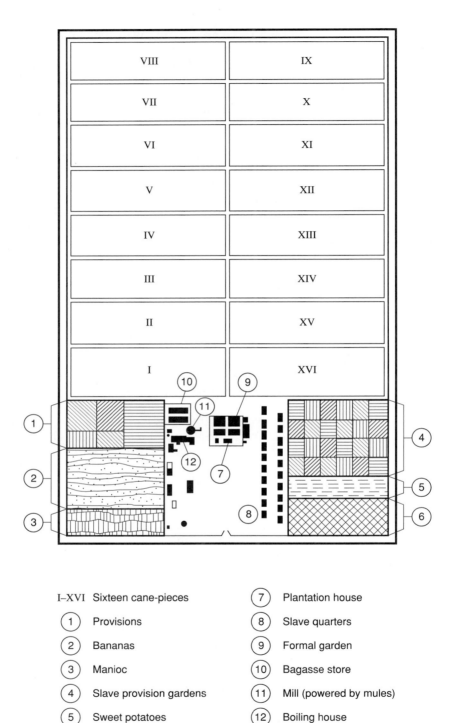

Figure 10.5 Layout of sugar plantation (after Avalle).

I–XVI Sixteen cane-pieces

(1) Provisions

(2) Bananas

(3) Manioc

(4) Slave provision gardens

(5) Sweet potatoes

(6) Guinea grass

(7) Plantation house

(8) Slave quarters

(9) Formal garden

(10) Bagasse store

(11) Mill (powered by mules)

(12) Boiling house

TABLE 10.4   End of the Atlantic Slave Trade and Slavery

Metropolitan Country	Year: Legislation Passed to Bar Nationals from the Slave Trade	Year: Emancipation of Slaves
Denmark	1803	1848
Britain	1807	1833*
U.S.A.	1807	1865
France	1831	Haiti 1794
Spain	1845	Guadeloupe & Martinique 1848**
		Puerto Rico 1873–1876
		Cuba 1880–1886
Portugal	1850	Brazil 1888
Netherlands	1814	1863

*The Emancipation Act was passed by the British parliament in 1833, to come into effect in 1834. Compensation was paid to planters, and ex-slaves were to serve an apprenticeship period that was curtailed in 1838 with "full freedom."

**Slavery was finally abolished in Guadeloupe and Martinique in 1848, although it had been temporarily suspended in 1794 by the French Revolutionary government.

Source: Conniff and Davis, 1994.

they worked for 5 to 10 years before hoping to acquire crown land. Between 1838 and the close of the indentured labor scheme in 1917, over 400,000 people traveled from Madras and Calcutta to the British Caribbean (Richardson, 1992).

In contrast to struggling economies in the British and French islands, Cuba and Puerto Rico increased sugar production in the nineteenth century. Developments in Cuba mirrored earlier events in the British and French islands:

1. Deforestation occurred.
2. Sugar acreage increased rapidly.
3. Large landholdings dominated.
4. Slave labor increased.
5. Black population numbers increased.
6. Population density increased.
7. Dealer dictated to producer.
8. Foreign capital replaced local capital.
9. Absentee control replaced local ownership. (Parry and Sherlock)

Cuba became the world's largest sugar producer in the second half of the nineteenth century. This development was linked to an abundance of fertile land, slave and indentured labor, proximity to the U.S. market, U.S. investment and tariff concessions, adoption of new technologies (steam mills), and the movement to free trade that opened up some European markets. Rail construction helped the geographical expansion of sugar. It is no coincidence that the first Latin American railroad was built in Cuba in 1837.

In the second half of the nineteenth century, sugar producers in the Caribbean employed new technologies, including steam power, horizontal rollers, the vacuum pan, and centrifugal machinery. These innovations increased efficiency and allowed cane to compete with beet sugar by World War I. The new, expensive technology favored the development of larger factories (*centrales*) and bigger corporate land-holdings. North American companies increasingly owned estates and *centrales*, supplied machinery, and controlled the marketing of sugar on islands such as Cuba and the Dominican Republic.

The twentieth century has seen the displacement of the family-run plantation by the corporate operation, such as Tate and Lyle in Jamaica, or Gulf and Western in the Dominican Republic. Recently, transnational corporations have tended to relinquish ownership of plantation land to island governments. Corporations retain control of manufacturing and marketing. In Cuba and St. Kitts state-run industries have developed. Sugar, slavery, and the plantation continue to leave an imprint on the Caribbean region.

THE CONTEMPORARY CARIBBEAN

▶ *Population*

Between 1960 and 1990, despite significant emigration, the population of the West Indies doubled from approximately 17 million to 34 million. In 1995, the population was 36 million. At the present rate of increase (1.5 percent per year), the population will double again in 46 years (Table 10.5). The West Indian islands face severe population pressure in relation to resources. There are too many people for too few jobs, resulting in high unemployment and underemployment. Population density to cropland is the highest in the Americas.

Population growth rates vary from island to island. Grenada has the highest rate of increase at 2.4 percent per year. The annual increase is 2.1 percent in the

TABLE 10.5 Caribbean Population Growth Characteristics

Country	Population Estimate Mid-1995 (millions)	Birth Rate per 1,000	Death Rate per 1,000	Natural Increase (Annual %)	Population Doubling Time in Years at Current Rate
Total Caribbean	36.0	23	8	1.5	46
Antigua and Barbuda	0.1	18	6	1.2	58
Bahamas	0.3	20	5	1.5	47
Barbados	0.3	16	9	0.7	98
Cuba	11.2	14	7	0.7	102
Dominica	0.1	20	7	1.3	55
Dominican Republic	7.8	27	6	2.1	32
Grenada	0.1	29	6	2.4	29
Guadeloupe	0.4	18	6	1.2	56
Haiti	7.2	35	12	2.3	30
Jamaica	2.4	25	6	2.0	35
Martinique	0.4	17	6	1.1	62
Netherlands Antilles	0.2	19	6	1.3	55
Puerto Rico	3.7	18	8	1.0	67
St. Kitts–Nevis	0.04	23	9	1.4	50
St. Lucia	0.1	27	6	2.0	34
St. Vincent & The Grenadines	0.1	25	7	1.8	38
Trinidad & Tobago	1.3	17	7	1.1	64

Source: Population Reference Bureau, *1995 World Population Data Sheet.*

Dominican Republic and 2.3 percent in Haiti. At present rates, the 7.8 million people of the Dominican Republic will double in 32 years, and 7.2 million Haitians will double in 30 years. Overall, the rate of natural increase is decreasing in the Caribbean, and at 1.5 percent is less than South America (1.8) and Central America (2.3). Barbados, the island with the highest population density (1,548 per square mile), has the lowest rate of increase at 0.7 percent, the same as Cuba, the United States, and Canada. In common with other regions of rapid population growth, the West Indies has a high percentage of young people (Table 10.6). For example, in Grenada 43 percent of the population is under age 15, and in Haiti 40 percent is under 15 years of age. The young are an expensive, nonproducing group who place a heavy burden on the rest of society in terms of health, welfare, and education costs. The young do, however, produce children. Recent evidence suggests that anticipated offspring loss to migration may sustain high fertility rates in some populations, especially the smaller islands of the eastern Caribbean.

With declining infant mortality and better health care, life expectancy at birth has risen to over 70 years in many West Indian countries, and it reaches over 75 in Barbados, Cuba, Dominica, Guadeloupe, Martinique, and the Netherlands Antilles. But Haiti has a Third World life expectancy of 57 and a high infant mortality rate of 74 per 1,000 live births.

The Antilles, with the exception of a couple of thousand Caribs on Dominica and an Arawak presence on Aruba, have no indigenous people, which provides a significant contrast to most of Latin America. Due to sugar and slavery, a large

TABLE 10.6   Caribbean Population Measures of Well Being

Country	Infant Mortality Rate*	Total Fertility Rate**	Percent Population Under Age 15/65+	Life Expectancy at Birth
Total Caribbean	39.0	2.9	31/7	70
Antigua & Barbuda	18.0	1.7	25/6	73
Bahamas	23.8	2.0	29/5	73
Barbados	9.1	1.8	24/12	76
Cuba	9.4	1.8	22/9	75
Dominica	18.4	2.5	29/8	77
Dominican Republic	42.0	3.3	35/4	70
Grenada	12.0	3.8	43/5	71
Guadeloupe	10.3	2.0	26/8	75
Haiti	74.0	4.8	40/4	57
Jamaica	13.2	2.4	33/8	74
Martinique	8.0	2.0	23/10	76
Netherlands Antilles	6.3	2.0	26/7	76
Puerto Rico	12.7	2.2	27/10	74
St. Kitts–Nevis	19.0	2.5	32/9	69
St. Lucia	18.5	3.1	37/7	72
St. Vincent & The Grenadines	16.0	3.1	37/6	73
Trinidad & Tobago	10.5	2.7	31/6	71

*Infant deaths per 1,000 live births.

**Average number of children born to a woman in her lifetime.

Source: Population Reference Bureau, *1995 World Population Data Sheet*.

Afro-Caribbean population is common to almost all islands.[3] In general, the Hispanic islands have a larger proportion of Euro-Caribbeans than the areas with French, British, and Dutch connections. Approximately four of five Puerto Ricans and two of three Cubans are classified as white. Whites comprise between 5 and 10 percent of the total population in most French- (including Haiti) and British-connected islands.

Miscegenation (racial mixing) has occurred for centuries; consequently, mulatto populations (black and white) are present in the Caribbean. Asian minorities exist in several islands, for example Jamaica, and are significant in Trinidad and Guyana. Racial distinctions continue to be important in the Caribbean, although race and class can no longer always be correlated.

► *Migration*

Migration is an important element in population change. There are four main types of international migration: in-migration, out-migration, return migration, and repeat migration. All four modes have been important in the Caribbean context, where migration and circulation are clearly perceived as adaptive strategies in the face of economic, political, and/or ecological pressures. Some commentators view Caribbean labor as a reserve pool for developing the global economy, dominated by U.S. and European capital.

Until 1900 migration into the Caribbean (in-migration) predominated. The Spanish islands developed a strong base of Hispanic settlers before African numbers grew in the nineteenth century, due to the development of large-scale sugar production. Overall, a relatively small number of Europeans (including indentured laborers and convicts) was overshadowed by 4 to 5 million enslaved Africans moving to the region. Significant numbers of East Indian (500,000) and Chinese (150,000) indentured laborers migrated in the nineteenth and early twentieth centuries to fill the labor shortage created by emancipation. They went mainly to Trinidad and the Guianas, plus a significant number of Chinese went to Cuba. Europeans, mostly Spaniards, moved to Cuba early in the twentieth century to work in the expanding sugar industry.

During the twentieth century, migration from the West Indies (out-migration) has been the major trend, reflecting population pressure and perceived opportunities elsewhere. Large numbers of West Indians have moved in response to internal and external economic and political forces. Destinations and rates of emigration have been affected by job possibilities and government regulations in receiving countries (Table 10.7). The migrants have had three main destinations:

1. Circum-Caribbean (including Central America)
2. Europe
3. North America

Circum-Caribbean (including Central America)

After emancipation, ex-slaves migrated from British islands in the Caribbean to areas with expanding sugar frontiers, principally Trinidad and British Guiana. Although some migrants later returned home, a regional migration tradition evolved (such as from Barbados to Guiana) that continues to the present. There is a great deal of interisland circulation on a daily, seasonal, and permanent basis.

---

[3]Whites predominate in the small French islands of Les Saintes and St. Barthélemy (St. Barts), and the small Dutch island named Saba. Les Saintes is off the coast of Guadeloupe, and St. Barts and Saba are off the coast of St. Martin.

TABLE 10.7   Major Migration Flows from and to the Caribbean

Periods (approximate date)	Migration from	Migration to
I. Before 1492 Preencounter	South America (Arawaks, Caribs)	Caribbean Islands
II. 1492–1640 Early plantation economy	a. Forced movement of Caribbean Indians from islands	Inter-island and mainland
	b. Europe: Spain, Netherlands, England, France (many indentured laborers)	Caribbean Islands
	c. Africa	Hispanic Caribbean
III. 1640–1830 Development of full plantation economy	a. Europe (e.g., continuing small number of indentured servants and circulation of military administration, and planters)	Caribbean Islands
	b. Africa (large number of forced migrants)	Caribbean (first to Lesser Antilles and Jamaica in quantity before 1807; then in late eighteenth and early nineteenth centuries to Cuba and Puerto Rico)
	c. British Caribbean (small number of whites due to economic problems)	North American mainland (e.g., South Carolina)
IV. 1834–1880 Post-Emancipation in British and French islands	a. Small Caribbean islands such as Barbados, St. Vincent, Grenada, St. Kitts, Nevis	Trinidad and British Guiana
	b. Indentured laborers from Indian sub-continent, China, and Java	Trinidad, British Guiana, Dutch Guiana (continued to 1917)
	c. China	Cuba, Dominican Republic, Puerto Rico
	d. Europe (mainly Spain)	Cuba and Puerto Rico
	e. Portugal, Madiera, Syria	British West Indies
	f. Jamaica	Panama Railroad
V. 1880–1924 Age of Imperialism (1924 U.S. restrictive Immigration Act)	a. Small Caribbean islands such as Barbados, Grenada, St. Vincent, St. Lucia	Trinidad, British Guiana and Dominican Republic
	b. Jamaica	Central America (1870s) Costa Rica railroad for banana plantations; 1880s Panama Canal (French effort)
	c. Jamaica and Barbados	U.S. Panama Canal Zone (1904–1914)
	d. Jamaica, Barbados, and Trinidad (relatively affluent people)	U.S. (mainly New York and Boston)
	e. British West Indies (especially Jamaica) and Haiti (contract sugar labor) (1895 U.S. gave preference to sugar from Cuba and Dominican Republic)	Cuba and Dominican Republic

TABLE 10.7   *Continued*

Periods (approximate date)	Migration from	Migration to
V. 1880–1924 (cont.)	f. Spain	Cuba
	g. Cuba	U.S. (New York, Florida)
	h. Puerto Rico	Hawaii, U.S. South, New York
	i. British West Indies and U.S. Virgin Islands	U.S. (e.g., Bahamas to Miami; Jamaica to New York; Barbados to Boston)
VI. 1924–1940 Depression and Interwar Years	a. British West Indies	Venezuela, Trinidad, Aruba, Curaçao (oil industry)
	b. Puerto Rico	U.S. Virgin Islands and New York
	c. Cuba	Miami
VII. 1940–1965 WWII and Post-war Expansion (1952 U.S. Immigration and Naturalization Act limited British West Indies to U.S.) (1962 Canadian Act opens up immigration to educational and occupational groups) (1962 U.K. Commonwealth Immigration Act closed door to Great Britain)	a. Small British West Indies, e.g., Barbados, Grenada, St. Vincent	Trinidad and Curaçao
	b. Barbados and Jamaica (short-term workers and emigrants)	U.S. (Bracero Program)
	c. Puerto Rico (especially after 1946)	U.S. (mostly to New York)
	d. Cuba (1952 Batista coup; 1959 Castro coup)	U.S. (mostly Miami)
	e. Trinidad, Jamaica, Barbados, British Guiana, and Haiti	Canada
	f. British, French, and Dutch West Indies	Metropoles
VIII. 1965–1994 (1965 U.S. Immigration Act opens up Caribbean migration to U.S.) (1980 U.S. Refugee Act restricted Cuban entry)	a. Whole Caribbean (e.g., especially Dominican Republic, Haiti, Jamaica, Cuba, and CARICOM countries)	U.S. (mainly New York and Miami)
	b. British West Indies, Haiti	Canada
	c. Suriname (Javanese, 1975–1976)	Netherlands
	d. Caribbean (inter-regional movement)	Caribbean, e.g., Haiti to Dominican Republic and Bahamas; Guyana to Suriname
	e. U.K. (return migrants)	Commonwealth Caribbean

Note: Many migrants move on a seasonal and/or fluctuating basis.
Source: Adapted from Conway, 1992, and Richardson, 1989.

The sugar industry has affected interisland migration patterns, frequently offering opportunities for contract labor. For instance, Jamaicans, Haitians, and people from the Lesser Antilles migrated to Cuba and the Dominican Republic during the early years of this century, after the United States gave preference to sugar imports from the Hispanic Caribbean. In 1919 and 1920 nearly 50,000 Jamaicans went to Cuba to cut cane (Richardson, 1985, p. 139). Thousands of Haitians have traditionally crossed into the Dominican Republic for the sugar harvest.

The Isthmus of Panama has been a major destination for Caribbean migrants. Jamaicans labored on the Panama Railway, which opened in 1855. Caribbean labor built railroads associated with banana plantations in Costa Rica. Jamaicans and other islanders worked on the French canal project in the 1880s, which failed and took a heavy toll due to accidents, malaria, and yellow fever. When the United States revived the Panama Canal scheme early this century, the Jamaican government forbade labor contractors to recruit on the island. Barbados became the headquarters for labor recruitment, and thousands of West Indians (approximately 150,000), especially Barbadians and Jamaicans, migrated to the Canal Zone between 1904 and 1914. Some later moved to the coastal banana plantations of Central America, where blacks now comprise a significant percentage of the population.

As the oil industry has developed in Trinidad and Venezuela (production and refining) and Curaçao and Aruba (refining only), inhabitants of the Antilles have moved south to work in the oil fields and refineries. During World War II, U.S. bases in St. Thomas, Antigua, Trinidad, and British Guiana attracted thousands of West Indians to work in construction and services. There is a reservoir of West Indians ready to migrate for economic reasons. Some governments within the region, such as Jamaica, Trinidad, and Barbados, where economies are growing or stable, currently restrict immigration to skilled and professional groups.

Europe

After World War II, West Indians migrated to Europe to fill a labor vacuum. Thousands of West Indians from British islands went to the United Kingdom in the 1950s[4], where they settled in industrial working-class areas of London and the Midlands, particularly engaged in transport and health sectors of the economy. Fearing economic and race problems, the British government passed the Commonwealth Immigration Act of 1962 that reduced the flow of migrants (especially from the West Indies, India, and Pakistan) to Britain. Many Caribbean migrants switched destination to the United States, where immigration rules were relaxed in 1965. Today approximately 650,000 people of West Indian extraction are U.K. citizens. Currently, there is significant return migration to the Caribbean, as 1950s migrants reach retirement age. Since World War II, French West Indians have migrated to France and Dutch West Indians to the Netherlands. Many people from Suriname, with an Indian or Java heritage, have fled to Holland for political reasons.

North America

According to Richardson, there may be more than 5 million people of Caribbean origin in the United States (1985, p. 145). Approximately 2 million are Puerto Ricans, and as U.S. citizens, should not be considered foreign migrants. Puerto Ricans migrated to the United States mainland during the twentieth century, and in large numbers after World War II. Their destination was overwhelmingly New York City.

Cubans have migrated to the United States in waves over a longer period. For example, Cuban exiles left the island as a result of the Ten Years War between 1868 and 1878, and movement continued until the restrictive National Origins Law (Johnson-Reed Act) of 1924 put a brake on general Caribbean migration to the United States. After the dictator Juan Batista took power in 1952, Cubans left for Miami. About 1 million Cubans have migrated to the United States since Castro's 1959 Revolution, ostensibly for political reasons. A large Cuban group (115,000–125,000), including some criminals, arrived in Miami in 1980, when Castro allowed them to leave, from the port of Mariel. Most Cubans live in the Miami district called Little Havana. Recently, people of Cuban ancestry were allowed to visit Cuba from the United States, taking consumer items and dollars with them.

---

[4]Estimates range from 230,000 to 280,000 (Peach, 1968).

An estimated 800,000 people from the Dominican Republic (many via Puerto Rico) and a similar number from Haiti (many via the Bahamas) live in the United States, mostly in the New York City area. They are joined by about 600,000 British West Indians (many originally from Jamaica) who have consistently moved to the United States when restrictive legislation has not been enforced. Migrants from the Anglo-Caribbean went mainly to cities in the Northeast.

West Indians, mostly from British- or French-connected islands, have migrated to Canada since the 1960s, some directly and some from the metropoles. There are approximately 350,000 people of Caribbean extraction in Canada.

In 1991–1992 thousands of Haitian boat people attempted to sail to the United States in the wake of a military coup. It was difficult to determine to what extent they were political or economic refugees, and the majority were returned to Haiti by the U.S. Coast Guard. Cuban refugees faced no such problems, until the numbers of would-be immigrants (many in hastily made rafts) mushroomed in 1994. Despite pressure from Cuban-American groups, the United States decided to restrict Cuban migration, and an agreement (allowing 20,000 migrants per year) was reached with Cuba. For a time in 1994–1995, Cuban and Haitian boat people were both taken to Guantánamo Bay for processing before many were returned home.

What has been the impact of out-migration on the West Indies? Population pressure, unemployment, and poverty have been relieved. It is estimated that emigrants living outside the West Indies (including children born in Europe and North America) equal almost one quarter of the resident West Indian population (West and Augelli, 1989, p. 111). Emigrants send money and goods (remittances) to relatives in the islands, and many migrants return home with wealth acquired abroad. Currently, return migration is a feature of many West Indian islands, although more research needs to be done on this topic.

Nevertheless, in total, a skill, youth, and ambition drain has resulted. Frequently, the old and very young remain. Males tend to be more migratory than females, especially in seasonal migration chains. Since the 1940s, for example, thousands of West Indian males, mainly Jamaicans, have seasonally cut cane in Florida. Haitian men cross the border into the Dominican Republic to cut sugar cane, and travel to the United States to pick apples. Women are left as heads of household, although skilled females, such as nurses, find jobs elsewhere. Migration, legal and otherwise, will continue to be a Caribbean strategy. The United States appears to offer material and skill opportunities, but the islands kindle culture and identity.

► *Urbanization*

Increasing urbanization is a common feature of the Caribbean and is linked to the processes of economic change and globalization (Table 10.8). In 1995, 62.8 percent of the total Caribbean population lived in urban areas (Potter, 1995), up from approximately one third in 1960. People move to the city, usually the primate city, for a variety of reasons. Urban areas are perceived to offer better educational and social services, more job opportunities and higher incomes, better amenities and life chances. Young people leave the countryside because of poor job prospects, especially where mechanization has decreased agricultural labor needs. Rural population growth rates have declined significantly in recent years.

Urbanization continues to increase despite high unemployment and abysmal housing conditions in cities. Many cities simply cannot cope with the in-migration and high fertility rates of migrants, who tend to be in their prime child-bearing years. Urban population growth rates in the Caribbean averaged 2.4 percent between 1990 and 1995. This is significantly higher than in North American cities. The increasing urban population results in congestion, shantytowns, pollution, and

TABLE 10.8 Caribbean Population—Density and Urbanization

Country	Total Area (sq. miles)	Capital City	Urban Population (%)	Per Capita GNP, 1993 (U.S. $)
Antigua & Barbuda	170	St. John's	31	6,390
Bahamas	5,380	Nassau	84	11,500
Barbados	170	Bridgetown	38	6,240
Cuba	44,220	Havana	74	2,644*
Dominica	290	Roseau	—	2,680
Dominican Republic	18,810	Santo Domingo	61	1,080
Grenada	130	St. George's	—	2,410
Guadeloupe	690	Basse-Terre	48	—
Haiti	10,710	Port-au-Prince	31	—
Jamaica	4,240	Kingston	53	1,390
Martinique	420	Fort-de-France	81	—
Netherlands Antilles	300	Willemstad	92	—
Puerto Rico	3,400	San Juan	73	7,020
St. Kitts–Nevis	140	Basseterre	42	4,470
St. Lucia	240	Castries	48	3,040
St. Vincent & The Grenadines	150	Kingston	25	2,130
Trinidad & Tobago	1,980	Port of Spain	65	3,730

*World Almanac, 1992, figure for 1990.

Source: Population Reference Bureau, 1995 *World Population Data Sheet*.

crime. As urban services fail to meet the demands of uncontrolled urban growth, the bright city lights evaporate into cardboard squalor.

Within the region, levels of urban primacy are generally high, higher than in North America. In most islands the primate city—San Juan (1 million), Santo Domingo (2 million), and Kingston (750,000)—has the fastest population growth. Even in Cuba, where Castro has favored rural development, Havana (2.1 million) remains dominant. Haiti is the least urbanized of the Greater Antilles, even though the capital, Port-au-Prince, has more than 1 million people. Primate cities would be larger if not for international migration, which usually starts from the capital.

Capitals of small island states in the eastern Caribbean, such as Port of Spain, Bridgetown, Castries, and Fort-de-France (Table 10.8) are growing rapidly, due especially to the development of tourism. They not only act as political and economic centers for each island but also link the islands into the global network.

It is expected that in countries with high levels of urbanization (over 50 percent), there will be a reduction in growth rates for three reasons (Boswell, 1992). Fewer people will move to the cities as rural population increase declines. General fertility rates appear to be decreasing in the Caribbean. Furthermore, people will move to the suburbs, as they have increasingly done in North America.

▶ Settlement Pattern

A three-stage settlement pattern—the Plantopolis model—has been identified for some former British Caribbean islands (Potter, 1995). This is a variant of the Mercantile model referred to in Chapter 6. Stage 1 of the Plantopolis model represents the period before Emancipation (1830s) when one major town and a number of relatively independent plantations existed. Stage 2 occurred from Emancipation until the 1950s. In this phase, small farming communities, practicing subsistence agriculture and supplying occasional plantation labor, were added to the previous settlement pattern.

Stage 3, the modern era since the 1950s, has involved the expansion of the major urban zone, with suburban and tourist development, some enclave industrialization, and retail development. Flows between town and country are pervasive. The modern era has seen an extension of the polarized pattern of development that has focused on the existing urban cores.

▶ Economy

West Indian economies are small, undiversified, and dependent. Typically, a limited range of primary products is exported, and manufactured goods are imported. In addition, the West Indies imports most of its food and energy supplies. Puerto Rico, Barbados, and Trinidad, for example, import over three quarters of their food requirements. Industrialization and tourism are seen as the road to development.

International prices and terms of trade are determined outside the region. Transnational corporations (for example Gulf and Western and Tate and Lyle in sugar; Geest in bananas; Amoco in oil; Reynolds in bauxite; and many others) provide the marketing and refining of raw materials. Commodity prices rise and fall, producing boom and bust conditions. At present, the world market is glutted with sugar and bananas. Prices for bauxite are low. Conversely, manufactured goods are relatively expensive. The structure of trade results in deficits, indebtedness, and shortage of hard currency that hinders economic growth. A large proportion of export earnings services debt, and capital is drained from economic expansion.

The West Indies (periphery) is dependent on the United States and Europe (core) for markets, technology, investment, credit, and aid. Because the cycle of dependency is difficult to overcome, West Indian governments have experimented with two strategies to counteract dependency and expand the manufacturing sector. A policy of *import substitution* was designed to promote manufacturing for domestic consumption by protecting nascent industries with tariff walls (see Chapter 7). But competition was reduced, prices increased, and domestic profits were disappointing. Another strategy of *industrialization by invitation* or export-oriented industrialization (supported by Sir Arthur Lewis, the West Indian economist) encouraged manufacturing for export. Governments frequently offered incentives, such as tax breaks and duty-free zones, to attract foreign manufacturing and investment. Taking advantage of low labor costs, some U.S. assembly plants (principally garments and sporting goods) located in the West Indies. The finished products are reexported to North American markets. Although providing some jobs, these initiatives have not changed the structure of dependency.

▶ Agriculture

Although declining, agriculture remains a significant activity in the Caribbean. A large percentage of the labor force is involved in agriculture in, for example, Jamaica, the Dominican Republic, and Haiti (Figure 10.6). Agriculture is polarized. On one hand, large plantations, usually on the best land, produce tropical cash crops for an international market. On the other hand, small peasant holdings (frequently less than 5 acres) produce for local markets and subsistence. Peasants supplement income with plantation work. Many Caribbean countries have preferential trade agreements, including price supports, with European Union members. Without guaranteed markets, many West Indian producers would find it difficult to compete internationally.

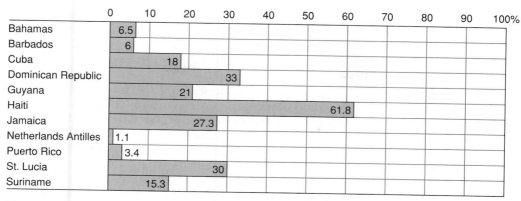

Figure 10.6   Percentage of labor force in agriculture. Data from *Europa World Yearbook,* 1995

Sugar continues to be the main cash crop. It is the leading export in the Dominican Republic, Guadeloupe, St. Kitts, and Cuba and is important in terms of land and labor in Jamaica, Trinidad, and Barbados.

Sugar acreage, yields, and production have declined as a result of high production costs involving increased fertilizer inputs on exhausted soils, higher labor costs, and low world prices. Sugar from Brazil, Hawaii, and the Philippines has increased world supply. Beet sugar and sugar substitutes have meant further pressure. United States and European Community quotas were cut in the 1980s. The prospects for sugar are not good unless alternative uses, such as the production of ethanol, are developed.

After sugar, *bananas* are the leading agricultural export. Jamaica, the Windwards, Guadeloupe, and Martinique export bananas to the European Community, where the crop is given preference over cheaper Central American bananas by former colonial powers. In France, bananas most likely come from Martinique or Guadeloupe. Bananas from the British Windwards and Jamaica dominate the British market. If protection ends, Caribbean bananas and fruit will face competition from Central America, where fewer hurricanes, rich soils, and modern production techniques provide advantages. Bananas (both sweet and plantains) are grown for local consumption in the West Indies.

Marijuana, or *ganja*, as it is sometimes called in the Caribbean, was introduced by Hindu East Indians in the nineteenth century and used in a religious context. It has attracted attention as an internationally traded illegal substance. Most ganja is grown by small farmers in Jamaica, especially in the central Cockpit region, the traditional home of the maroons (runaway slaves). Jamaica is a significant supplier of marijuana to the United States. Marijuana generates foreign currency, but it provides an additional reason for U.S. intervention.

The West Indies acts as a conduit through which South American drugs (mainly cocaine) pass. The Bahamas is a distribution center for drugs bound for U.S. markets. Haiti and the Dominican Republic are important as drug transshipment areas.

The decline of *tobacco* is widespread. It is hard on the environment and contributes to soil exhaustion. Drying and curing require high energy inputs, which are expensive. In addition, U.S. competition is stiff on the world market. Today tobacco is a minor crop in Puerto Rico, Jamaica, and the Dominican Republic but more significant in Cuba. Hand-rolled Havana cigars remain world famous.

Coffee, grown in the Greater Antilles, faces competition from larger and more fertile growing areas in Brazil and Colombia. An important cash crop for peasants, coffee is exported from Haiti and the Dominican Republic. It is grown in Cuba and Puerto Rico for domestic consumption. A small amount of good-quality Jamaican Blue Mountain coffee commands high prices in Europe and Japan. Other cash crops exported from the region include cacao, citrus, spices, pineapples, coconuts, rice, and flowers.

The prospects for agriculture are bleak, for most tropical soils are exhausted and eroded, fertilizers and pesticides are expensive, and world markets are extremely competitive. Peasant farms are too small for commercial agriculture and are usually located on marginal land. The decline in sugar production has not resulted in increased subsistence agriculture. The Caribbean is dependent on food imports, and agricultural productivity is declining on many islands.

On some islands with disproportionate male migration, such as Montserrat, Grenada, and Nevis, agriculture is feminized. Women heads of household, unable to get credit, have reverted to subsistence farming to feed families. Land has gone out of cultivation, resulting in declining production.

► *Minerals*

The Greater Antilles are richer in minerals than the Lesser Antilles. Cuba has the most diversified mineral resources including iron, chrome, cobalt, copper, oil, manganese, marble, and nickel. Used in the production of military hardware, nickel is Cuba's most important mineral. It is also mined in the Dominican Republic, where it is an important export.

Bauxite, the ore used for aluminum production, was until recently widely mined in the Caribbean—in Jamaica, Haiti, and the Dominican Republic. In 1990 only Jamaican mines operated. Despite a recent decline in production, Jamaica continues to export bauxite, most of which goes to the United States and Canada. Because of high energy requirements, much refining occurs in North America.

Trinidadian petroleum is the exception to a general lack of domestic energy in the Caribbean. Trinidad's petroleum and natural gas reserves brought prosperity in the 1970s. Since oil prices dipped in the 1980s, the pace of development has slowed, and GNP per capita has dropped abruptly. Barbados and Cuba produce small amounts of petroleum.

► *Industry and Manufacturing*

Many islands have small processing plants—breweries, cement works, food processing—to serve local needs. Large plants are associated with refining of bauxite to alumina (Jamaica) and sugarcane (several countries including Puerto Rico, the Dominican Republic, Cuba, Jamaica, and Barbados). Much sugar refining takes place outside the region.

Molasses and rum are locally produced from sugarcane. Molasses is a thick syrup that remains after sugarcane juice has been crystallized by boiling. It is used as table syrup, cattle feed, and flavoring agent, and can be distilled to make rum (although juice and other residues may be used). Some molasses is shipped to New England for rum production, continuing a trade pattern of the colonial era. Caribbean rums vary in taste and color. Darker, heavier rums are produced in Jamaica and Barbados. Lighter, drier rums, such as Bacardi, come from Puerto Rico and the United States Virgin Islands. Large quantities of rum are shipped to North America and Britain. Another important by-product of sugarcane is bagasse (the fiber left after squeezing cane), which is used to fuel boilers or make paper and wallboard. About one ton of wet bagasse is equivalent to a barrel of oil.

The Caribbean region is a big refiner and shipper of world oil for the United States market. Location close to North America and the Panama Canal, deep harbors, access to Venezuelan oil, and weak environmental rules have contributed to this development. Oil is refined in Puerto Rico, Trinidad, Curaçao and St. Croix.

Transshipment terminals, where Middle Eastern crude oil is unloaded from supertankers and transferred to smaller vessels, are found in several islands including St. Lucia, St. Eustatius, and Bonaire. Environmental pollution is a growing threat.

Puerto Rico, because of its commonwealth relationship with the United States, has the most industrialized economy in the Caribbean. Tax laws have encouraged U.S. industry to locate plants in Puerto Rico, resulting in a wide range of manufacturers including heavy industry, electronics, and, most recently, pharmaceuticals (Johnson and Johnson, Bristol-Myers). There is a very large container port in San Juan. Assembly operations, particularly involving garments, which predominantly employ females, are important in Puerto Rico, the Dominican Republic, Barbados, and elsewhere.

► Trade and Transportation

The United States is the West Indies' most significant trade partner. For example, 64 percent of the Dominican Republic's exports go to the United States, and 35 percent of imports are from there. Nearly 50 percent of Barbadian exports and imports involve the United States. Normally, the United States provides two thirds of Haitian imports and takes 85 percent of exports. Other trade partners include Canada, the European Union (EU), and Latin American neighbors. Guadeloupe and Martinique are part of the European Union and the direction of their trade is with France. Until 1990, Cuba's trade was almost exclusively with the former Soviet Union and Eastern Europe. In the 1990s, Cuba has looked for new trade and investment partners such as Canada, France, Spain, China, Mexico, and other Latin American countries.

Transportation patterns are externally oriented rather than regional. Nevertheless, in recent years, an interisland air transportation system has developed, which has attracted both freight and passenger traffic. Most islands have a good, low-cost bus service, with numerous routes radiating from the capital into the countryside.

► Tourism

Travel and tourism is the biggest industry in the world (*Economist*, March 23, 1991, p. 3). More than 50 percent of all international travel and tourist expenditure is generated by the nationals of five countries: Germany, United States, Britain, Japan, and France. The United States provides two thirds of all tourists to the West Indies, and Canadians and Europeans comprise the rest. Just as southern Europe is a resort area for northern Europe, so the Caribbean is a convenient destination for North Americans.

Before World War II, U.S. tourists traveled to Cuba and the Bahamas, but mass tourism did not take off until the 1960s. Increasingly, sunshine is used to attract visitors rather than produce food. Caribbean economies rely more and more on tourism which is subject to external economic forces. Puerto Rico and the Bahamas dominate in number of arrivals. Tourism is increasing to the British and United States Virgin Islands, Antigua, and Martinique. Cruises are important, although they inject less money into local economies than hotel tourism and can overwhelm port areas when five or six cruise ships call in one day. Up-market, purpose-built resorts, away from local settlements, are fashionable.

Tour operators, capitalizing on Caribbean sun, surf, sail, sex, scenery, and proximity to the United States, have rushed to develop parts of the West Indies. Some island governments, such as those of the Bahamas, Barbados, and Antigua, have promoted tourism to provide jobs in hotels, restaurants, retailing, and trans-

Luxury resort, St. Barts.

portation. Tourism is viewed as a foreign exchange source and as a way to stimulate economic growth.

The tourist industry, however, can present problems:

1. It relies on the economic situation in the source areas. Recession in the United States, for example, has a big impact on tourism to the West Indies.

2. It is sensitive to local factors such as adverse weather and energy shortages.

3. Employment in tourist industries is seasonal, with most visitors arriving in the Northern Hemisphere winter months, which coincides with cane cutting.

4. Wages are generally low for locals.

5. Tourists can kill tourism by polluting sea, beaches, and countryside. Small islands especially experience difficulty with water demands and garbage problems. The environmental impact of tourism needs to be carefully assessed.

6. Tourism is capital intensive (usually foreign capital), involving complex investments and infrastructure.

7. Foreign exchange frequently "leaks." It goes to buy imported merchandise, including food served in tourist hotels, furniture, and even gifts. Earnings are frequently repatriated by foreign corporations, many of which own hotels.

8. Tourism is subject to changes in fashion. A prime destination can quickly lose its appeal.

The question of image is important, since the islands have similar attractions. Some, such as the Bahamas opt for mass tourism, offering package tours, complete with casinos. Others provide luxury vacations. Still others seek to develop forms of eco-tourism or nature tourism. Most offer duty-free shopping and promote water activities such as snorkeling and windsurfing. Different approaches to tourism can, for example, be found on the island of St. Martin, which is half French and half Dutch. Tourism on the French side is regulated and caters to high-income travelers. The Dutch part, the "wild west" of the Caribbean, has unplanned development that threatens the environment with overcrowding and high rates of pollution.

Islands that have successfully encouraged tourism have experienced economic growth, together with environmental and social costs. Economic development and environmental protection are not easy to balance. For the large islands, like Puerto Rico and Jamaica, tourism helps to diversify the economy. For small islands in the Leewards and Windwards, tourism dominates the economy. The challenge is to create linkages, such as food production, between tourism and the local economy. There is evidence that this can occur with maturity in the industry (Momsen, 1986a).

In addition to tourism, another service industry that has expanded since the 1960s is government employment. In most islands a high percentage of the workforce is in government, as police officers, teachers, bureaucrats, and inspectors. In general, governments, desperate to reduce unemployment, have not resisted the trend.

► *Economic Integration*

Given the small size of island economies and markets, regional economic cooperation appears attractive for the West Indies. But, economic integration is difficult because:

1. Islands have competitive rather than complementary economies.
2. Levels of economic development differ between islands.
3. Island particularism makes cooperation difficult.

An attempt at political federation of the British West Indies (1958–1962) failed, although it should be noted that at independence, most British West Indian colonies opted to remain in the Commonwealth (a global association of former British colonial territories). Economic integration of the former British territories has been relatively successful. Beginning in 1965 with the formation of CARIFTA (Caribbean Free Trade Association), tariff barriers between member countries were eliminated and efforts made to develop a common external tariff. In 1973 CARICOM (Caribbean Community and Common Market) superseded CARIFTA and tried to create a common market, hoping to exert more bargaining power in world markets. Thirteen former British colonies with a market of approximately 5.5 million people are in CARICOM. The members are Antigua-Barbuda, the Bahamas, Barbados, Belize, Dominica, Grenada, Guyana, Jamaica, Montserrat, St. Kitts–Nevis, St. Lucia, St. Vincent and the Grenadines, and Trinidad and Tobago. In 1995 Suriname became the first non-British connected territory to join the group. Haiti and the Dominican Republic have observer status. CARICOM has two subregional groups—the Less Developed Countries (LDCs) and the more developed countries (MDCs) of Jamaica, Trinidad, Barbados, the Bahamas, and Guyana.

CARICOM has produced some economic integration, such as the creation of the Caribbean Food Corporation to improve marketing of local food and the West Indies Shipping Corporation to coordinate transportation of West Indian goods. But CARICOM has failed to narrow the economic gap between the MDCs and the LDCs. Jamaica and Trinidad, CARICOM's bigger members, gain most from the organization. Economic integration appears to be more successful among countries that are closer together than those that are far apart. Insularism continues to impede integration, and the small size of the total market remains a disadvantage.

The *Caribbean Basin Initiative* (CBI) went into effect on January 1, 1984, and was designed to promote economic growth in the Caribbean and Central America, while safeguarding the United States' economic and political interests in the region. Cuba, Grenada, and Nicaragua were not included in CBI for political reasons. Under CBI, $125 million was loaned to Caribbean territories, specified products can enter the United States duty-free (80 percent of exports to the United States were already duty-free), tax exemptions are allowed for business conventions held in the West Indies, and investment and tax treaties are negotiated bilaterally rather than multilaterally. Excluded from duty-free status are leather goods, traditional agricultural commodities, and petroleum products. Garments assembled in the region from fabric produced and cut in the United States, pay only a value-added tax upon reentering the United States. They are subject to quotas. Participants in CBI were required to conclude tax and information exchange agreements with the United States; these agreements are designed to control money laundering and tax avoidance. In the Caribbean region, the Dominican Republic and Jamaica have been leading beneficiaries of CBI, with most investment going to garment assembly plants. Total investment, however, has been disappointing.

CBI has a number of disadvantages. For example, CBI militates against economic cooperation between West Indian neighbors and does nothing to promote regional cooperation. Inexpensive United States imports hurt local industries. In addition, CBI promotes dependence on United States markets, capital, and investments.

In 1994 the Association of Caribbean States (ACS) was established to promote economic integration and cooperation in an extended Caribbean region. The ACS includes members of CARICOM plus Mexico and Central America, the Dominican Republic, Haiti and even Cuba. Puerto Rico is not a full member.

THE GREATER ANTILLES

▶ Cuba

Cuba, the most extensive island in the Antilles, is 90 miles south of Florida. The first windsurfer made it from Cuba to Florida in 1994. The population is predominantly Hispanic and Spanish speaking, with a significant Afro-Cuban population. Since 1959, a communist government has been in power.

The plantation sugar economy developed strongly in Cuba in the late eighteenth and early nineteenth centuries, aided by steam and rail technology, abundant fertile land, and the African slave trade. Between 1780 and 1840, Cuba was transformed from a colony of predominantly white, free settlers to one with a large, enslaved black population (Knight, 1970). During the third quarter of the nineteenth century more than 100,000 Chinese laborers migrated to Cuba. Euro-Cubans maintained their strong population numbers by immigration from Spain and a relatively high birth rate.

Toward the end of the nineteenth century, a group of creole landowners pushed for independence from Spain. A bitter Ten Years War (1868–1878) ensued that gave Cuba some revolutionary heroes, such as José Martí, but was not successful. The war led to the demise of many Cuban sugar planters. This aided U.S. interests in acquiring Cuban sugar plantations and factories.

Cuba also faced severe readjustments as a result of slave emancipation (1880–1886) and a decline in sugar prices. Subsequently, the Cuban economy came to rely increasingly on the United States for markets and capital. After the Spanish-American War of 1898, the United States occupied Cuba until 1902. The Platt Amendment, incorporated in the Cuban-American Treaty of 1903, gave the United States the Guantánamo naval base and a right to intervene in Cuba (the right was relinquished in 1934 as a result of Franklin Roosevelt's "Good Neighbor Policy").

Hurricane Hugo hits the West Indies, 1989.

The United States leases Guantánamo for just over U.S. $4,000 per annum, but the Cuban government, wishing to end the agreement, refuses to cash the checks.

In the first half of the twentieth century, American involvement in Cuba increased with U.S. interests (Hershey's, for example), buying up sugar estates and mills. U.S. companies also built highways and electricity-generating plants. The tobacco manufacturing industry declined, as the United States protected its own manufacturers with tariffs on imported cigars and cigarettes. Cuba, though politically independent, was economically dependent on the United States.

In 1959, when Castro took power, ties with Western capitalists were severed and banks and industrial concerns nationalized. Large foreign-owned estates were confiscated, many without compensation. Much of the land was consolidated into state farms. After the Bay of Pigs (1961) and the Cuban missile crisis (1962), the United States placed an economic embargo on Cuba and prohibited U.S. citizens from traveling to the island. Castro responded by concluding trade agreements with the Soviet Union, under which Cuba supplied sugar, tobacco, and minerals to the Soviet Union and received petroleum, wheat, fertilizer, and mechanized equipment in return. Cuba became economically dependent on the Soviet bloc.

Cuba achieved some success in diversifying its economy, although sugar continues to be the economic mainstay. The production of food crops, such as rice and beans, for domestic consumption increased. Citrus fruit was exported to Eastern Europe. Cattle and dairy herds improved, and the fishing industry expanded. Mining of nickel, cobalt, iron ore, chrome, and copper was significant.

Castro's socialist policies increased health, housing, and education services provided by the state. Rural living standards went up. Cuba boasts 40,000 doctors and 35 universities. Literacy levels are high (96 percent), the death rate is low (7 per 1,000), and life expectancy good (75 years). But there is little freedom of thought and action, and living conditions are basic. Many essentials, such as shoes, clothing, and toiletries, are expensive.

The collapse of the Soviet bloc has had a devastating effect on the Cuban economy, depriving it of cheap oil, markets, machinery, imports, and aid. Production of commodities such as sugar, tobacco, and nickel has dropped dramatically since 1989. For example, the sugar harvest fell from 7.5 million metric tons in 1989 to 3.3 million metric tons in 1994 (*Washington Post*, November 17, 1995).

The Cuban government has responded by courting foreign investment (from such countries as Spain, France, and Canada), authorizing limited self employment, allowing farmers to sell surplus produce on the open market, and by paying workers, over their work quotas, with now-legal dollar incentives that can be spent in special stores. In an attempt to revitalize the sugar industry, the government has borrowed heavily to provide fertilizers and spare parts, because increased sugar production is crucial to economic recovery. Tobacco and nickel production (with the aid of foreign capital) appears to be rising. Tourism, higher education for foreign students, and biotechnology are viewed as important growth sectors of the economy, and there have been efforts to expand these activities.

In 1996, after Cuban jets shot down two planes flown by a Miami-based Cuban exile group, President Clinton signed the Helms-Burton Act. The Act was designed to strengthen the embargo against Cuba by penalizing foreign companies that do business with Cuba.

► *Puerto Rico*

Puerto Rico, the smallest of the Greater Antilles, is dominated by a central mountain region that rises to over 4,000 feet. Three quarters of the island is mountainous, with fertile coastal plains. The north-facing slopes receive plentiful rainfall, while the south and west are in a rain shadow and dry. The population of 3.7 million is predominantly Hispanic, Spanish-speaking, and Roman Catholic. During the Spanish colonial period, sugar plantations developed modestly. Agriculture was characterized by a few large landowners and numerous tenant farmers on small holdings. The plantation, sugar and slavery developed between 1815 and 1850. Slaves and forced laborers (*agregados*) worked side by side on plantations.

At the close of the Spanish-American War, Puerto Rico was ceded by Spain to the United States. Today, Puerto Ricans are United States citizens (since 1917) with internal self-government. The United States is responsible for Puerto Rico's foreign affairs, defense, customs, and mail. Puerto Ricans are eligible for federal aid and the military draft, do not pay federal income taxes, and are not directly represented in Congress, although they do lobby (and so do U.S. companies with operations on the island). Goods produced in Puerto Rico are not subject to tariffs on entering the United States (section 936 of the federal tax code), and Puerto Ricans can settle in the United States if they wish.

In the early twentieth century coffee was Puerto Rico's most important commercial crop, followed by sugar and tobacco. But sugar quickly became the major crop, grown on large, U.S.-owned corporate plantations, despite efforts in 1917 to limit size to 500 acres. Dry land in the south was irrigated for sugar. At the same time, coffee growing declined.

In 1940, when Muñoz Marín's Popular Democratic party came to power, land reform was begun. Plantations in excess of 500 acres were purchased by the government and leased to small cane farmers as proportional profit farms, but large corporations retained control of the *centrales* (sugar mills).

Since World War II, the Puerto Rican economy has been transformed. Farming has declined in importance, and today less than 5 percent of the labor force is in agriculture. Dairy cattle and livestock are important, however. Although Puerto Rico is the world's largest producer of rum, the island is a net importer of sugar, tobacco, and coffee—items that used to be exported. More land is again in coffee than is in sugar. Puerto Rico imports much of its food from the United States.

In the late 1940s the U.S. and Puerto Rican governments inaugurated "Operation Bootstrap," an industrialization program built on tax exemptions, low interest loans, and cheap labor. Manufacturing is currently the leading sector of the economy and focuses on heavy industries, such as machinery and metal products, chemicals, pharmaceuticals, petroleum refining, rubber, and plastics. Assembly operations and the garment industry are important, although light manufacturing is moving to cheaper labor areas, such as the Dominican Republic, where wages do not conform to United States minimum wage laws, as they now do in Puerto Rico.

Tourism is a leading economic sector in Puerto Rico, with more than three quarters of all tourists coming from the United States. Historical attractions include old San Juan and the Ponce Museum of Art.

Migration to the United States has acted as a safety valve for population pressure, and more Puerto Ricans now live in New York than in San Juan. Although the island has benefited from economic integration with the United States, U.S. corporations control a large proportion of its manufacturing, banking, construction, and communications industries. Many Puerto Ricans receive federal aid, including food stamps.

Whether Puerto Rico will become the fifty-first state is a charged political issue in an island with a Latin American identity. Only about 20 percent of the population speaks English with ease, and nearly 60 percent do not speak English at all. In 1991 Spanish became the official language, reversing the 1902 law by which Spanish and English were dual languages. The two main Puerto Rican political parties are split on the issue of statehood or continuing commonwealth status. A third, independence party gets less than 10 percent of the vote. Statehood would make Puerto Ricans eligible for federal taxes and more federal funds, but it could hurt manufacturing jobs and business, because tax advantages could end. Tax exemptions for U.S. companies' operations in Puerto Rico are under review in Washington. In 1993, a referendum was held in Puerto Rico regarding political status. It produced a slight edge for the status quo over statehood (48.6 percent against 46.3 percent). Support for statehood has risen significantly in recent years.

After several decades of apparent advantages (such as duty-free access to U.S. markets), Puerto Rico has lower living standards and higher unemployment and crime rates than other islands, such as Barbados. This should give pause to those who think that the creation of *maquiladoras* in Mexican border towns will solve economic problems in Mexico and curtail Mexican immigration to the United States.

► *Haiti*

The Republic of Haiti occupies the western third of Hispaniola. It consists of two mountainous peninsulas that enclose the Gulf of Gonaïve, as well as the two inhabited islands of Gonâve in the gulf and Tortue (Tortuga) off the north coast. About two thirds of Haiti is mountainous.

Early Spanish settlement of Hispaniola concentrated on the eastern part of the island; later (1697) the French gained the western section from Spain, naming it Saint Domingue. Sugar plantations were developed first on the north coast lowlands around Cap Haitien and then in the south, where irrigation was needed. In the 1740s and 1750s about 100,000 acres of sugar land was irrigated (Watts, 1987). Other commercial crops, such as cotton, indigo, cacao, and especially coffee were grown. The total value of agricultural exports in the French Revolutionary year of 1789 superseded their value in the mid-1980s, without taking account of inflation (Boswell in West and Augelli, 1989, p. 166).

In 1804, after a successful slave insurrection, the newly named Haiti (an Indian name meaning "mountainous") became the second independent country in the Americas. Instability and economic stagnation followed, as sugar and coffee plantations deteriorated and collapsed. Irrigation works and dams were neglected. Haiti faced an international economic embargo. Haiti has never recovered from this economic catastrophe.

In 1915 the United States occupied Haiti to quell a revolt and safeguard the approaches to the Panama Canal. U.S. rule lasted until 1934. The dictatorial Duvalier family (first Papa Doc and then Baby Doc) ruled Haiti from 1957 to 1986, when Baby Doc was forced into exile. The regime was corrupt and repressive. In 1991 elected President Jean-Bertrand Aristide was forced out of office and out of Haiti by a military coup. At this point, Haitian boat people attempted to flee to the United States, but most were turned back by U.S. authorities. In 1994, with the assistance of U.S. Marines and a U.N. force, Aristide was returned to power in a bloodless action. In 1996, presidential power passed from Aristide to René Preval after a peaceful election.

Haiti is the poorest country in the Western Hemisphere, and it has the highest infant mortality rate and the lowest life expectancy in the Caribbean, as well as an illiteracy rate of over 60 percent. The population growth rate is high, and there are too many people (7.2 million) for the weak resource base. Approximately 75 percent of the population is rural (the highest in the Caribbean) and involved in peasant agriculture. Coffee is the most important cash crop. Title to land is frequently ambiguous and difficult to obtain.

Haitian agriculture is not able to feed the population of the country, for land is impoverished by overcropping and burning. Forest covers less than 2 percent of the land. Moreover, since trees are cut down for fuel, massive erosion has ensued. Haiti's agricultural yields are therefore low, and food imports are expensive. Haiti is a hungry land. Women play an important role in agriculture. Whereas men do the heavy work of clearing land, women, as is common in parts of Africa, tend the cultivation plots and sell produce, frequently walking miles to small town markets that dot the countryside. Coffee has traditionally been the leading cash crop for peasants.

Land hunger has encouraged Haitians to go east into the Dominican Republic. Haitian men seasonally cross the border to cut cane. Poor economic circumstances and political instability have fueled emigration overseas. Many Haitians have moved to other islands (such as Cuba and the Bahamas), to France, Canada, and the United States.

In the 1970s tax incentives led foreign companies to build factories in Haiti. The mostly U.S. plants employed cheap female labor. The country's manufactured exports included clothes, electronics, and sports equipment. Factories are predominantly located in and around the primate city of Port-au-Prince, but in 1995 only a fraction of the light-manufacturing plants were operating. Tourism in Haiti is handicapped by poverty, disturbing slum conditions, political instability, and the high incidence of AIDS. Haiti's mineral resources, including bauxite, copper, gold, iron, silver, sulphur, and tin are underdeveloped. In fact, the bauxite mine closed in the 1980s.

Haitian society has a marked color and cultural divide. The population is predominantly black, rural, illiterate, and Creole-speaking. African influences are strong. The elite is light-skinned (descendants of French fathers and African mothers), urban, educated, and bilingual (French and Creole). The new constitution of 1986–1987 for the first time granted official status to Creole, placing it on a par with French, the language of roughly 7 percent of the population.

In a country where the majority struggle to survive and where unemployment is well above 50 percent, development prospects are bleak. When Aristide was returned to power, the United States and allied nations promised substantial financial aid, much of it to come from international lending agencies. But promised privatization programs inside Haiti are slow, some aid has been withheld, and foreign investment is weak. Attaining political equilibrium and international confidence are imperative.

► *Dominican Republic*

The Dominican Republic, formerly Santo Domingo, occupies the eastern two thirds of Hispaniola. With a population of 7.8 million, largely Hispanic and mulatto, pressure on the land is not as intense as it is in Haiti. The frontier with Haiti is officially closed. The Cordillera Central runs northwest to southeast, with Pico Duarte rising to 10,417 feet, the highest point in the Caribbean. To the north and east lie the rich Cibao lowlands, especially fertile in eastern Vega Real. Population density is high in this region. The other major area of agricultural production is the southeastern Caribbean coastal plain, where sugar plantations cluster (Chardon, 1984). The capital, Santo Domingo, the oldest European city in the Americas, is located on the western edge of this coastal region.

Initially, the Spaniards mined gold but soon shifted attention to raising sugar, tobacco, cacao, and cattle. During the eighteenth century, in contrast to intensive sugar production in Saint Domingue, agriculture in the Spanish colony was extensive and diversified. Spanish Hispaniola complemented the sugar industry of Saint Domingue by producing cattle and beef. Population numbers, especially of African slaves, were relatively low.

Santo Domingo became independent in 1821 and was promptly occupied by Haitians until 1844. Economic stagnation followed, and many affluent Caucasian families left. In 1844 the Spanish-speaking easterners again proclaimed independence as the Dominican Republic. The Haitians were expelled, leaving a legacy of distrust. In the latter part of the century, the United States got financially involved, and Cuban planters, fleeing the Ten Years War (1868–1878), helped to develop the sugar industry.

Twice in the twentieth century the United States has sent troops to the Dominican Republic. The first instance occurred between 1916 and 1924 when U.S. Marines occupied the country. During this period foreign debt was reduced, the infrastructure improved, and U.S. interests invested in the sugar industry. By 1925 the U.S.-owned West Indies Sugar Company controlled 40 percent of Dominican production (Galloway, 1989, p. 173). The second intervention occurred in 1965, during a period of civil unrest amidst fear of a communist takeover. Once again the United States sent the Marines, this time for only a two-month occupation.

Unlike Haiti, the Dominican Republic has seen periods of economic development, albeit at the cost of human rights. Under the rule of General Rafael Trujillo (1930–1961), new land was brought under cultivation, irrigation projects started, roads built, and new industries formed. Foreign investment, much of it North American, was encouraged. As a result of Trujillo's brutal border policy, thousands of Haitian squatters were rounded up and massacred in 1937, adding to the tension between the two neighbors.

During the 1970s another growth phase occurred, based on rising sugar prices and U.S. loans. Commentators began to talk of a "Dominican Miracle" as the gross national product rose by 10 percent per annum. By the end of the decade, however, economic conditions had deteriorated. The 1980s saw declining sugar prices, a drop in U.S. sugar quotas, and mounting foreign debt. The economy declined, and though more prosperous than Haiti, the Dominican Republic remains a poor country.

Despite manufacturing and tourism, agriculture is the Dominican Republic's major activity and sugar the main cash crop. Mechanized, irrigated sugar plantations are located on the southeast coastal plain. The Standard Fruit Company operates high-yield banana plantations in the western Cibao which are irrigated and sprayed for disease. In addition, Dole has a pineapple operation in the Dominican Republic. In contrast, small peasant holdings in the Vega Real concentrate on cacao for cash and produce a variety of foods. Numerous small holders grow coffee on lower mountain slopes (as a cash crop). Other crops include tobacco, rice, and peanuts. As one of the few Caribbean countries with a successful cattle raising industry, the Dominican Republic is an exporter of meat.

Mining figures in the Dominican economy. Ferro-nickel and gold are mined. Dore (an alloy of silver and gold) is produced and is a significant export item, as is amber. Oil in the southwest awaits exploitation.

Manufacturing, involving the assembly of imported components, has created jobs in the clothing, electrical, toy, and leather goods areas. Several tax-free industrial zones have been created. As is common in the Antilles, tourism is the growth industry, providing many service-sector jobs.

► *Jamaica*

Jamaica has approximately 2.4 million people, and four fifths of the island's area is mountainous and only sparsely settled. The Blue Mountains in the east are composed of igneous and metamorphic rocks and rise to over 7,000 feet. Elsewhere, in the interior, limestone plateaus range from 1,000 to 3,500 feet.

Jamaica has had a number of capitals. In 1509, the Spanish originally settled on the north coast around Seville, but in 1534, Villa de la Vega (today Spanish Town) was established. The site was chosen because it was less exposed, located in the most fertile agricultural region of the St. Catherine and Liguanea plains, and close to the fine Kingston harbor. When the English took Jamaica in 1655, during the rule of Oliver Cromwell, unplanned Port Royal grew rapidly at the end of the spit protecting the entrance to Kingston harbor. Port Royal was near sea level, and in the 1692 earthquake, most of it sank into the sea. The present capital of Kingston, on the landward side of the harbor, is a primate city, being ten times larger than any other place in Jamaica. The city sprawls along the coast and inland to the foothills. The shantytowns of West Kingston are among the worst in Latin America and have little hope of full integration into the urban system.

Under the Spaniards, Jamaica was a minor colony whose main economic activity was cattle raising. After Britain took the island, it became an international pirate base. Plantation sugar production developed in the 1670s along coastal areas, typically using enslaved African labor. By the 1720s, Jamaica had outstripped Barbados as the leading British sugar producer.

Resistance to slavery, notably in the 1831 Jamaican Insurrection, contributed to the parliamentary act abolishing slavery in the British islands (1833), following which the Jamaican sugar industry went into sharp decline. The early nineteenth century levels of sugar production were not reached again until the 1930s. Many ex-slaves moved off plantations, seeking land and establishing villages in the interior. A lasting impact on land use and landownership resulted. Today, large estates monopolize the fertile and well-watered areas and concentrate on export agriculture. Peasant farmers operate on small plots (frequently less than 2 acres) located on interior slopes. Coffee, citrus, cacao, ganja, and allspice are grown by peasants as cash crops. Food crops, such as yams, are widely grown for subsistence and sale in local markets, representing an important, though not easily documented, contribution to the economy. The acreage in sugar is declining, although it is still Jamaica's major agricultural export.

Since the late nineteenth century, bananas have been an important crop in Jamaica. (Shipments to the United States began in 1870.) Today, banana growing is mainly in the hands of small-scale, independent farmers, and the Banana Producers' Association organizes transportation and sale to Fyffes, a U.K. company. Exports go to the protected market of the United Kingdom. Though declining, agriculture remains an important economic activity, providing more employment than mining. Agriculture contributes less than 10 percent of gross domestic product but employs one third of the labor force.

Since World War II, in common with many Caribbean islands, Jamaica has attempted to diversify its economy. The country has one valuable mineral—bauxite, the raw material for aluminum production. During the 1960s Jamaica was the world's largest supplier of bauxite, with foreign companies such as Reynolds operating in the island. Jamaica has to compete with high-grade ores from Australia, Brazil, and Guinea, but bauxite and alumina remain its leading exports by value. The production and price of bauxite affects the Jamaican economy. Also mined are gypsum, marble, sand, gravel, and industrial lime. Of the former British islands, Jamaica has been the most successful in attracting light manufacturing plants.

Tourism provides most new jobs for Jamaicans, with resorts prominent on the cooler north coast from Negril to Montego Bay to Ocho Rios and Port Antonio. Cruise ship arrivals are increasing. Tourists encounter a variety of cultural experiences, including Jamaica Talk (the local dialect), Rastafarians, and reggae music. Approximately three quarters of all tourists are from the United States. The United States is also Jamaica's biggest trade partner, followed by the United Kingdom. Significant Jamaican communities live in the United States, Canada, and Britain.

THE LESSER ANTILLES

► *Trinidad and Tobago*

Trinidad is like no other Caribbean island. First, geologically, it resembles the Venezuelan mainland across the Gulf of Paria. Second, Trinidad has an oil economy. Commercial oil production began in 1909. Third, the racial and cultural composition of the population, with roughly equal proportions of blacks and East Indians (both about 40 percent), has no parallel in the islands but approximates the situation in mainland Guyana.

Three low mountain ranges cross Trinidad from east to west and account for the island's name. The Northern Range at just over 3,000 feet is the highest, and a well-watered, fertile plain intersects the North and Central ranges. Some coastal areas are swampy. Being farther south and closer to the Orinoco Basin heat island, Trinidad experiences higher average temperatures and more rainfall than other West Indian islands. It lies outside the hurricane track.

Initially part of the Spanish empire, Trinidad was an economic backwater. Fewer than 1,500 inhabitants lived on the island in 1777 (Wood, 1968, p. 32). Trinidad became a refuge for French and Spanish migrants fleeing war and the slave revolt in Hispaniola. By 1803 the population totaled about 28,000, of whom over 20,000 were slaves.

In 1797 the British captured the island from Spain, and soon after, plantation agriculture and sugar production began to develop. English settlers from other islands came in search of fertile land. But at the time of emancipation in 1834, only 209,000 acres had been appropriated, from a total of 1.4 million acres, and less than 44,000 acres were being cultivated (Johnson, 1972, p. 38). At this point Trinidad was ready for expansion of the sugar economy. The labor shortage was met by indentured workers, mostly from the Indian subcontinent. Sugar thrived on large estates and exports rose. Technological innovations and *centrales* were introduced relatively early in Trinidad.

After emancipation, many former slaves acquired land, established villages, and only worked on sugar estates if they desired. When the indentures of East Indians and Chinese expired, they often purchased land and operated small farms. Since the last quarter of the nineteenth century, small cane farmers have grown substantial amounts of sugar, which is processed by large *centrales*. In the early nineteenth century cacao was grown by small holders of Spanish descent. Then, in the 1870s, French creole planters, using indentured labor, developed cacao estates, especially in the Montserrat district. By 1905 more than three times the amount of land was under cacao than was in sugarcane (*Statesman's Year Book*, 1905), and cacao was the principal agricultural export. Today, a relatively small amount of cacao is grown, due partly to competition from African producers.

In recent years total agricultural production has declined sharply. Around 10 percent of the workforce is in agriculture, but agriculture contributes less than 5 percent of gross domestic product (GDP). Attempts have been made to expand the cultivation of rice (mostly by farmers of East Indian extraction), fruit, and food crops. Coconuts are grown in coastal areas.

Trinidad's economy runs on oil and fluctuates according to oil prices. Petroleum production centers in the southeastern area (around Pitch Lake) and offshore. A large amount of oil from the Middle East (formerly Venezuela was a major source) is refined in Trinidad and reexported. Asphalt production, from the world's largest natural source at La Brea Pitch Lake, is declining as international demand slows. Large natural gas reserves are assuming importance. Gas powers domestic industry and oil refining, and is exported in liquified form.

During the 1970s the Trinidadian government embarked on a policy of industrialization by invitation, concentrating on heavy industry. An industrial complex was built at Point Lisas in the south. Success has come in petrochemicals, ammonia, and fertilizers, but the Iron and Steel Company of Trinidad and Tobago has consistently lost money since opening in 1981. Manufacturing employs about 15 percent of the workforce. The service sector, consisting mostly of government jobs, employs almost half the workforce. Tourism, though not well developed, is growing. Despite economic fluctuations, Trinidad in the 1990s is prosperous by West Indian standards.

Cultural, religious, and racial diversity complicate politics. Geographically, blacks are concentrated in urban areas and petroleum districts, while people of East Indian descent predominate in rural areas, especially in the sugar regions. Tension between blacks and East Indians is sometimes intense, as the 1990 unsuccessful coup by Muslim extremists indicates.

Port of Spain, with geometric streets and spacious squares, resembles a typical colonial capital laid out by Europeans. The city focuses on Woodforde Square, around which are situated government offices. To the north, surrounded by late-nineteenth century buildings, is the Queen's Park Savannah with a racecourse. To the west of the Savannah is the cricket ground, and to the east, the botanical garden. The Anglican and Roman Catholic cathedrals are prominent in the urban core.

The urban area sprawls out along the Eastern Corridor—a nearly unbroken string of settlements including St. Joseph, Tunapuna, Arouca, and Arima—all linked by frequent, crowded buses. The Eastern Corridor settlements, in which ethnic groups cluster, are thriving, as migrants move in from rural areas. Meanwhile, the colonial core of Port of Spain is in decline.

Tobago, twenty miles northeast of Trinidad, has been politically united with its neighbor since 1888, but retains a strong sense of particularism. In the seventeenth and eighteenth centuries, after the Caribs were overcome, cotton and sugar were grown and exported under various European regimes. Britain finally won control of the island during the Napoleonic Wars, and sugar production remained buoyant until the end of slavery, after which economic decline set in. Today, Tobago, with its fine beaches and relaxed pace, relies on tourism. It is a favorite vacation spot for Trinidadians. Population movement from Tobago to Trinidad reflects job opportunities and the relative prosperity of the larger island. The main Tobagan town of Scarborough has a new deep-water harbor that opened in 1991.

► *Barbados*

Barbados lies 100 miles east of the arc of the Lesser Antilles. Relative proximity to Europe gave the island an early transportation advantage. Bridgetown was frequently the first port of call from Europe, and in the seventeenth century developed as a mercantile city. Barbados is a coral limestone island with the central uplands rising to about 1,100 feet (Mt. Hillaby). Most of the land has fertile red and black soils, with the exception of the clays of the less productive Scotland district in the northeast, which faces severe land erosion problems.

Lacking an indigenous population at encounter, Barbados was colonized relatively easily by the English in the 1620s. Between about 1640 and 1680 the so-called Sugar Revolution involved a rapid shift from white indentured labor to slave labor, from tobacco and cotton to sugar, and from small-scale farming to plantations. After the abolition of slavery, sugar cultivation remained the major economic activity. Most of the island was in plantations, and ex-slaves had little opportunity to acquire land. Labor was plentiful, and migrants left the island to find work, mainly in British Guiana and Trinidad.

During the last 30 years, Barbados sugar production and exports have declined. Rum continues to be produced. Sugar cultivation is concentrated in the estates of the central region, on land best suited to the crop. Agriculture as a whole employs about 5 percent of the workforce. Agriculture has been diversified, more vegetables are grown, and broiler chickens are produced, especially in marginal areas, such as St. Lucy parish in the north. Barbados continues to import much of its food.

More than 80 percent of the workforce is employed in the service industry, with tourism as the major sector. Most tourists come from England, not the United States. Economic development is concentrated in the southwest near Bridgetown where light manufacturing, involving textiles, electronics, precision instruments, and sporting goods, has been successful. Data processing firms from the United States operate on the island, taking advantage of the educated labor force (99 percent literacy rate). A small oil and gas field provides some local energy.

The Barbadian population is relatively homogeneous, about 80 percent black, less than 5 percent white (some poor whites, "Red Legs," descended from seventeenth-century indentured laborers), and the remainder mixed. Barbadians enjoy political stability, a sense of identity, and modest prosperity. Barbados has long had one of the highest population densities in the world (1,548 per square mile), and realizing the imbalance between population and resources, the Barbadian government was the second in the world (after India) to establish a National Family Planning Program in 1955. The result is one of the lowest birth rates in the hemisphere. At the current rate it will take a century to double the population. The average age of the population is increasing as a result of a low birth rate and return migrants from the U.K. Many West Indians who migrated to Britain in the 1950s and 1960s are now reaching retirement age, and a significant proportion are choosing to retire to the islands from which they came, often building retirement homes in the process. The suburbs of Bridgetown are growing, partially as a result of return migration.

► *Anglo–Leeward Islands*[5]

The islands in the AngloLeeward group are Antigua–Barbuda, St. Kitts–Nevis, Montserrat, and Anguilla. The Leewards are small, resource poor, and economically vulnerable. St. Kitts–Nevis and Montserrat have high central volcanic cores. Antigua, Barbuda, and Anguilla are predominantly flat, semiarid, limestone islands. Antigua, which has volcanic remnants in the southwest, frequently experiences drought conditions, and water is sometimes shipped in from St. Kitts. The group has little sense of identity. Barbudians, for example, are dissatisfied with government from Antigua. Anguilla retained its dependent status with Britain rather than be associated with St. Kitts–Nevis. Montserrat, with a legacy of early Irish Catholic settlement, continues as a British territory.

The islands share several characteristics. First, with the exceptions of Barbuda and Anguilla, they were agricultural colonies, reaching their commercial height during the eighteenth century. Sugarcane predominated in St. Kitts–Nevis, Antigua, and Montserrat. In Antigua more than 200 windmill ruins dot the landscape. Nevis, the birthplace of Alexander Hamilton, withdrew from the sugar industry early in the twentieth century. Sea-island cotton continues to be cultivated in Montserrat and Antigua.

Second, commercial agriculture has been declining for decades, although sugar, cotton, and coconuts continue to be commercially grown. Small farmers contribute to local food supplies with vegetables, corn, and fruits. In Antigua, corn acreage is increasing in the eastern part of the island. The St. Kitts sugar industry, nationalized in 1975, remains significant, with most of the coastal region in cane. St. Kitts has the only remaining export sugar industry in the Leewards, which has been harmed by shrinking U.S. sugar quotas.

The Leewards are attempting to promote tourism, which is most developed in Antigua, where about one quarter of the labor force is involved in the tourist industry. Light manufacturing is also encouraged as part of diversification. The United States has a naval station on Antigua. In relation to the resource base, the islands are overpopulated, and emigration continues to be a response. Remittances assist the remaining population.

► *Anglo-Windward Islands*

The Windward Islands of Dominica, St. Lucia, St. Vincent, and Grenada[6] lie south of the Leeward group. Caribs retarded European penetration. The last few surviving Caribs live on a reservation in the northeast of Dominica. The French and British traded overlordship in the eighteenth century until the Windwards became British colonies. The islands still retain a French flavor, illustrated by French patois and culture, although English is the official language. The population is black, young, and increasing. The Windwards are spectacularly mountainous, with steep volcanic slopes and numerous streams. The rugged terrain limits agriculture, and the islands remain wooded. Mt. Soufrière on St. Vincent is an active volcano that last erupted in 1995.

With the exception of Dominica, which has little level land, the islands were sugar producers. Today the chief cash crop of the group is bananas, mostly shipped

---

[5]The terms *Leewards* and *Windwards* are administrative terms used by the British. Confusingly, the *Leewards* are not to the leeward (west) of the Windwards but lie to the north. The Spanish originally called all the Lesser Antilles the Windward Islands (points from where the wind comes) and used the term *Leeward* for the islands off the South American coast. The Dutch use the term *Leeward* for the ABC islands and *Windward* for their other territories.

[6]Several small islands, including the Grenadines, Carriacou, and Petit Martinique, are in the Windward group.

to the protected U.K. market, although some diversification to dairy farming and fishing is taking place. Some of the islands grow a specialty crop. For instance, St. Vincent produces most of the world's arrowroot. Traditionally grown for its starch content and important in baby and invalid foods, arrowroot is now used to make computer paper. Dominica specializes in limes and has important timber reserves. Grenada is noted for exporting nutmeg, mace, cinnamon, and cloves, and has long been known as the "Spice Island." Grenada is second only to Indonesia in the production of nutmegs. Recently, the value of nutmeg exports from Grenada has declined, because Indonesia has flooded the world market. Spice trees flourish at relatively high altitudes and on volcanic soils in the Windwards. Other commercial crops include cacao, coffee, flowers, and coconuts. Food for local consumption is cultivated on peasant plots.

Although the economies are still heavily agricultural (particularly in Dominica), tourism and manufacturing have been added. Tourism is most developed in St. Lucia where the scenery is outstanding. Castries, the capital of St. Lucia, developed as an important coal bunker station on the Panama Canal route. Now the island is a large oil storage and transshipment point and leases military bases to the United States.

In 1979 Maurice Bishop, a Marxist, assumed leadership of Grenada unconstitutionally, and politically moved closer to Cuba. In 1983, after Bishop was overthrown and killed by a more radical Marxist group, the Organization of Eastern Caribbean States (with Dominica's prime minister, Eugenia Charles, in a prominent role) asked for U.S. military assistance, and received it. The island was taken in three days. Shortly afterwards, U.S. troops left, and an election was held in 1984. U.S. intervention was partially motivated by fears that Cuba was building an international airport on Grenada at Point Salines which might be used for military purposes. The incident illustrates the importance of every West Indian island to the strategic policy of the United States.

► *French Antilles*

Martinique and Guadeloupe[7] comprise the French Antilles. They are overseas departments of France (*départments d'outre mer*) and belong to the European Union. Residents are, therefore, French citizens. Martinique is volcanic and the largest of the Windward Islands. Its highest peak, Mt. Pelée, erupted in 1902, killing about 30,000 people and destroying the town of Saint-Pierre. Guadeloupe is divided into two—Basse-Terre and Grande Terre—by a swampy channel. The highest volcanic peak in the Lesser Antilles, Mt. Soufrière, is found on Basse-Terre (Low Ground).

Martinique and Guadeloupe were settled by the French in the 1630s. Plantation agriculture developed slowly, but in the eighteenth century the volcanic soils made them two of the richest sugar islands in the Caribbean. Slavery was abolished in the French islands in 1848, but sugar production continued. Sugar acreage has decreased in recent years, although sugar and rum continue to be exported. Because of French protection of domestic beet producers and European Union policies, a shift to fruits, especially bananas, melons, and pineapples, has occurred.

The largest employer is the French government, with a bureaucratic structure overseeing health, education, and social services. Tourism is increasing, especially to Martinique, although it has been hurt by pro-independence violence in Guadeloupe. Impressive scenery, French character, cuisine, wines, and perfumes are considerable tourist attractions.

---

[7]Guadeloupe has six small dependencies: La Désirade, Marie Galante, Petit-Terre, Iles des Saintes, St. Barthélemy, and the northern part of St. Martin (the rest of St. Martin is Dutch). Isles des Saintes and St. Barts have relatively large white populations, descendants of Breton and Norman fisherfolk.

Two Pitons, St. Lucia.

The French Antilles have little contact with other islands. Martinique and Guadeloupe are distinctively French in culture, language, and tradition, and use the French franc as their currency. They trade with France, importing far more than they export. Because of integration with France, inhabitants have one of the highest living standards in the developing world. As French citizens, they receive social security, health, family, and unemployment benefits. Emigration from Martinique and Guadeloupe to France has been high, easing population pressure on the islands. In return, some white French workers have moved to the islands.

► *Netherlands Antilles*

The ABC islands—Aruba, Bonaire and Curaçao—are located off the Venezuelan coast, outside the hurricane belt. They are low, barren, and arid with less than 25 inches of rain per annum. Low rainfall and high sunshine hours allow a large production of salt by evaporating seawater. But water supply for industry and tourism is short, requiring desalination plants to augment the water supply. Cacti and aloe vera thrive on the ABC islands. Agriculture is limited by lack of water, but small herds of goats are common and feed on the cacti and scrub vegetation. About 500 miles to the northeast (near Puerto Rico) lie St. Eustatius, Saba, and St. Martin,[8] the other Dutch territories. They have higher elevations and about twice as much rainfall as the southern group. With the exception of Aruba, the islands are administratively linked together as a union within the Netherlands. In 1986 Aruba gained separate status as an autonomous part of the Netherlands, because it resented the dominance of Curaçao. Aruba is scheduled to become independent in 1996. Willemstad, on Curaçao, is the major urban center and port, with a fine, deep harbor.

The inhabitants of the ABC islands are descendants of Amerindians, Africans, Dutch, Portuguese, Danish, and Jewish settlers. An old trading language

---

[8]The southern part of St. Martin is Dutch.

known as Papiamento is spoken, although Dutch and English are also used throughout the Netherlands Antilles. Aruba's population has a significant Amerindian intermixture.

Curaçao and Aruba have over 90 percent of the population of the Netherlands Antilles and are economically the most viable, owing to oil refining. The Dutch developed the Maracaibo oil fields in Venezuela, and during World War I Royal Dutch Shell built a refinery on politically stable Curaçao, using Venezuelan oil. Standard Oil of New Jersey later opened a refinery on Aruba.

The decade of the 1980s was not kind to the oil industry in Aruba and Curaçao. Prices declined and profits dropped. Venezuela refined more of its own oil, and in 1985 the Aruba refinery closed. In order to prevent a similar occurrence on Curaçao, the Dutch government bought the Shell refinery and leased it to the Venezuelan State Oil Company. A Texas company then reopened the Aruba plant.

Oil refining on Aruba and Curaçao has had a demographic impact, attracting migrants from other islands in search of employment. Bonaire, although larger than Aruba, has significantly fewer people and more males than females, owing to emigration patterns.

Tourism and manufacturing provide alternatives to oil refining. Resort hotels have been built on Aruba, Curaçao, and St. Martin, and attract tourists mainly from the United States and Venezuela. Duty-free shopping plazas, displaying merchandise from all over the world, crowd the downtowns, such as Willemstad. Tax incentives and duty-free zones encourage manufacturing. Offshore banking developed in the Netherlands Antilles during World War II when major Dutch companies moved assets to protect them after the German invasion of the Netherlands in 1940. Today offshore banking is a significant economic activity.

► The Bahamas

The Commonwealth of the Bahamas is a group of about 700 islands off the Florida coast. Most of the islands are uninhabited, while two—New Providence and Grand Bahama—have approximately four fifths of the population. British settlement began in the seventeenth century, and Africans were imported. The population is approximately 85 percent black. The Bahamas gained independence from Britain in 1973.

The islands are low (highest elevation about 200 feet), with poor limestone soils surrounded by shallow sea. In contrast to most Caribbean islands, plantation agriculture never was an important activity.

Tourism is the backbone of the Bahamian economy. Proximity to east coast urban areas of the United States is an advantage. Gambling grew rapidly as a tourist attraction when Cuban casinos were closed by the revolution. Oil refining and oil transshipment provide jobs and revenue.

International banking and investment are substantial service industries in the Bahamas. Over 150 major banks (and trust departments) operate in this area. Laws relating to taxation and trusts have been arranged to allow the development of offshore banking services, and international companies often find it advantageous to deposit funds offshore, where profits can be tax sheltered until repatriated. Most banks offering offshore services are major North American or European institutions, bound by banking laws in their own countries in relation to the laundering of drug money. Abuses exist, but the main concern is to attract legitimate funds. The advantage to the host country is the creation of jobs in banking, finance, and law. Another service offered in the Bahamas is international ship registration.

The Bahamas is a center for the smuggling of marijuana and cocaine into the United States (just as it was for alcohol during the Prohibition years of 1920 to 1933). The United States Drug Enforcement Agency assists the Bahamas authorities in trying to curb illegal activities.

▶ *The Virgin Islands*

The Virgin Islands are located east of Puerto Rico. St. Thomas, St. Croix, and St. John are the major American islands with a population of over 100,000. People born on these islands are U.S. citizens. The United States purchased the islands from Denmark in 1917 for strategic reasons. Anegada, Virgin Gorda, and Tortola are the major British islands with about 10,000 inhabitants. Most Virgin Islanders are descendants of Africans.

The islands have poor agricultural land and insufficient water supply. There are large concrete rainwater collection areas on the hillsides. Some water is shipped into the American Virgin Islands from Puerto Rico, and seawater distilleries have been built. Export agriculture has given way to food production for local markets, especially for the tourist industry.

Economic connections are close within the entire group, and the U.S. dollar is the common currency. Charlotte Amalie, on St. Thomas, is the major urban and commercial center in the islands. Industrial development focuses on St. Croix where oil and bauxite are processed, rum is distilled from imported sugar, and textiles and pharmaceuticals manufactured.

The Virgin Islands National Park, composed of land given by the Rockefeller family, covers two thirds of St. John and is a tourist destination. Tourism, however, is most developed on St. Thomas. Migrants from other Caribbean islands, the United States, and Britain have moved to the Virgin Islands in search of jobs and a relatively high standard of living.

▶ *The Cayman, Turks, and Caicos Islands*

Like the British Virgin Islands, the Caymans (Grand Cayman, Little Cayman, and Cayman Brac) and the Turks and Caicos Islands (eight inhabited islands) are British crown colonies. They are similar to the Bahamas in three respects. First, they are low-lying limestone islands, with inadequate rainfall and little agriculture. Second, tourism provides most jobs and income. Third, offshore banking has developed, and money laundering causes concern. The Cayman Islands is one of the largest offshore financial centers in the world (with many corporate offices), and financial services employ about 10 percent of the labor force.

GUYANA, SURINAME, AND FRENCH GUIANA (GUYANE)

For historical, economic, and cultural reasons, it is appropriate to include the Guianas in this chapter concerning the West Indies. Historically, the Guianas were first colonized by North Europeans—the Dutch, British, and French. Sugar plantations were developed in the Guianas. Culturally, large numbers of people of African and Asian ethnic heritage were imported to satisfy the labor demands of sugar plantation agriculture. In addition, Guyana has been a member of CARICOM since its inception, and Suriname has recently joined the organization.

The Guianas lie on the Atlantic flank of the Guiana Highlands of Venezuela and Brazil. Along the Atlantic coast is a low-lying, ill-drained, coastal plain (Figure 10.7) consisting of alluvium deposited by rivers flowing out of the highlands and material moved along the shore by the North Equatorial current from the mouth of the Amazon. The major rivers, such as the Essequibo, are navigable inland, but as their valleys penetrate into the highlands, there are cataracts, and eventually precipitous waterfalls that prevent continuous navigation.

The coastal region is characterized by mangrove swamps, tidal marshland vegetation, and some savanna grassland. In the coastal region, temperatures vary

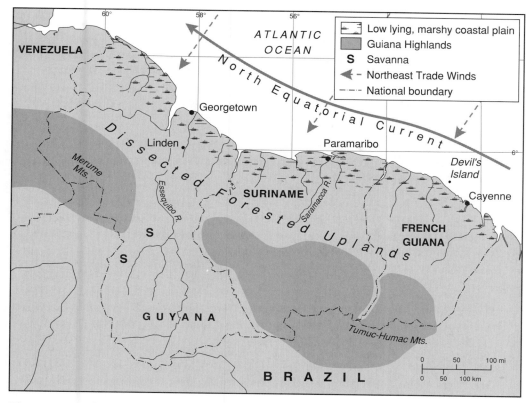

Figure 10.7 The Guianas.

little from month to month. Daytime highs are in the high 80s; mean monthly temperatures are around 80°F. There is little cooling in the hours of darkness. Inland, tropical rain forest predominates, but in Guyana there are extensive tropical savanna grasslands where stands of trees are found only along the water courses. Even in Guyana, more than 80 percent of the land is forested.

The rainfall regime is more variable than the temperature regime. At Georgetown, for example, where the total annual rainfall is around 90 inches, some months are much wetter than others. There are heavy rains in May, June, and July when the ITCZ of low pressure, cloud, and precipitation lies over the area. As the ITCZ retreats south, the northeast trades reassert themselves. August, September, and October are less rainy, but November through January is another wetter period. Rainfall decreases in February to less than 5 inches before building up again as the ITCZ courses north once more. The rainfall distribution is similar in Cayenne and Paramaribo.

European settlement on the coast was begun by the French in 1604, a few years before France established settlements on the St. Lawrence starting in 1609. Subsequently, the Dutch and English established colonies. The coastal settlements changed hands several times. The Dutch acquired sole possession of Surinam (Surreyham) from the English in 1667 in exchange for Manhattan Island. The British took the territory that became British Guiana from the Dutch during the Napoleonic Wars.

The early settlers grew tobacco and cotton on the coastal plain and obtained dyes and wood from the local Indians in exchange for trade goods. Subsequently, sugar became the major crop as on the West Indian islands. Enslaved Africans were introduced as a labor force. When well drained (and the Dutch were expert at water management), the coastal plain was fertile, having deep soils, rich in organic matter. By the end of the nineteenth century, coastal plantations producing sugar, cotton, coffee, bananas, and citrus were common.

After the abolition of slavery, sugar remained the major crop. Indentured laborers were imported into British Guiana from the Indian subcontinent and, to a much lesser degree, from China. The Dutch imported laborers from India and from Java in the Dutch East Indies (Indonesia).

In the postemancipation period, former African slaves tended to move into towns, and indentured laborers, after completing their contracts, stayed in the countryside and grew rice on the well-watered coastal plain, in addition to providing labor on plantations. Rice remains a major crop in the Guianas, but sugar cultivation is much reduced everywhere, although still important in Guyana. In 1992, Guyana produced 257,000 metric tons of sugar, representing a decline since 1960 when 340,000 metric tons were produced. Other crops cultivated for sale and subsistence include bananas, cassava, corn, coffee, cacao, and citrus fruits.

Away from the coastal plain, the landscape rises inland, culminating at peaks in the Merume Mountains in the west and the Tumuc-Humac Mountains to the south. Away from the coastal plain, forest is the dominant vegetation type. Interspersed are shifting cultivation plots—the *abattis* system as it is called in French Guiana—and some livestock raising in zones of savanna grassland.

The forests are a major resource, and all the Guianas cut and export hardwoods and make wood products. The exploitation of the forests is limited to areas where passable roads link exploitable woods to navigable water. The estuaries of the rivers are full of shrimp, which support fishing fleets and an export industry.

► *Mineral Resources*

The Guiana Highlands, which consist largely of the eroded igneous stumps of once-impressive mountains, are rich in minerals. Gold, in placer deposits, is worked in Guyana and French Guiana. Bauxite, derived from the weathering of granite rocks in tropical conditions, is mined in Suriname to produce alumina and refined aluminum for export. Guyana has the greatest range of exploitable minerals, including gold, diamonds, bauxite, manganese, copper, and molybdenum.

Bauxite has been an important mineral in Guyana. The Demerara Bauxite Mining Company (DEMBA), a subsidiary of the Northern Aluminium Company of Canada (ALCAN), began operations in 1917. Reynolds Metal Company, another foreign-owned concern, was active from 1953. The center of bauxite mining developed around Linden, the only interior urban settlement of any size (about 40,000), situated about 100 kilometers from the capital of Georgetown.

In the 1950s and 1960s, bauxite mining was in high gear. After Independence in 1966, DEMBA was nationalized by the socialist government and, together with the Reynolds operation, run as a state-owned enterprise. In the 1970s and 1980s, due to a number of factors including high oil prices, inflation, and collapse of the aluminum market, production and efficiency declined. Alumina production stopped in 1983. In the 1990s, the bauxite industry in Guyana is in decline, facing stiff competition from more reliable and cheaper ores in China, Australia, and Brazil.

The Guianas, as a whole, contain extensive unexploited mineral deposits including iron ore, lead, silver, and platinum. Many of the resources lie well inland, and it is unlikely they can be exploited profitably because communications are bad, transport costs high, and distances long.

Many rivers in the Guiana Highlands offer potential for the generation of hydroelectric power as they descend from the upland by waterfalls and rapids. The best sites are far from centers of population, and the limited industrial activity in the coastal region does not provide a large enough market to justify investments in dams, generators, and power distribution lines.

► *Population and Settlement*

Historically, the population of the Guianas has lived predominantly on the coastal plain, particularly in towns built close to the estuaries of rivers that give some waterborne access to the interior. Despite many small rural places along the coastal plain, the coastal capital cities contain a high proportion of the total population. The population of Guyana is .8 million (Table 10.9), and approximately one in four Guyanese live in or around Georgetown. It is estimated that about 49 percent of the population is of East Indian heritage and 36 percent of African ancestry. There are .4 million inhabitants of Suriname and almost 50 percent live in or close to the capital of Paramaribo. The population of French Guiana was estimated at 132,500 in 1994. The capital, Cayenne, has many small town characteristics.

Suriname is home to several maroon (runaway slave) groups, such as the Saramaka, with villages along the Saramacca River deep in the rain forest. The maroon tribes total about 40,000 people (*Washington Post*, August 28, 1995). Their houses are similar to dwellings found in West African villages. Dress and dance follow West African styles, and the language incorporates many African words. Lately, tribal land has come under threat from Asian timber companies anxious to gain logging rights from the government of Suriname.

Population growth in the Guianas is around 2 percent per annum and would be greater if not for out-migration to France, the Netherlands, and in the case of Guyana, to Britain, the United States, and Canada. A high proportion of the Javanese, whose forebears came to Suriname as agricultural laborers in the second half of the nineteenth century, left Suriname for the Netherlands before and after independence in the 1970s, fearing that, as Muslims, they would be discriminated against by the Christians and Hindus.

SUMMARY

The Caribbean islands are characterized by diversity. The island arc stretches from the low coral islands of the Bahamas, to the highlands of the Greater Antilles, to the volcanic peaks of Martinique and St. Vincent, to Trinidad, which geologically is part of South America. Most islands share a history of colonialism, sugar plantation agriculture, and slavery. But because the Caribbean was a region of intense imperial rivalry, the islands display different European cultural traits—Hispanic, British, French, and Dutch—intermixed with a strong African component. Politically, too, diversity is apparent. Political systems range from stable independent democracies, such as in Barbados, to the communist dictatorship in Cuba, to departments of France (such as Guadeloupe), to associated U.S. states like Puerto Rico.

Yet, the countries of the region face common problems. Too many people hunt for too few jobs. Many people leave the islands in search of opportunity and work. In general, island economies are vulnerable, and a reliance on commodity exports, whether sugar in Cuba, alumina in Jamaica, or bananas in St. Lucia, persists. Most countries seek to attract tourists, encourage foreign investment, and diversify economies. To the tourist industry, the Caribbean is a commodity.

The Guianas, on the northeast coast of South America, share many characteristics of the West Indian islands. They were colonized by North Europeans. Sugar was a major cash crop. Enslaved African labor was introduced, followed by indentured workers from Asia, leading to multiethnic societies. The Guianas struggle to attain economic viability and political stability.

TABLE 10.9 The Guianas: Population, 1995

Country	Population	Birth Rate per 1,000	Death Rate per 1,000	Natural Increase (annual %)	Doubling Time (years)	Infant Mortality Rate	Life Expectancy at Birth	% Urban	Per Capita GNP, 1993 (U.S. $)
Guyana	0.8	25	7	1.8	39	48	65	33	350
Suriname	0.4	25	6	2.0	36	28	70	49	1,210
French Guiana (Guyane)	0.13*	—	—	—	—	—	—	—	—

* = *Statesman's Yearbook*, 1995–96, Estimate of Population for 1994.

— = no data.

Source: Population Reference Bureau, 1995 *World Population Data Sheet.*

322

FURTHER READINGS

Barker, D., and D. McGregor, eds. *Environment and Development in the Caribbean: Geographical Perspectives.* Kingston, Jamaica: University of the West Indies Press, 1995.

Barry, T., B. Wood, and D. Preusch. *The Other Side of Paradise: Foreign Control in the Caribbean.* New York: Grove Press, 1984.

Boswell, T. D. "Characteristics and Processes of Urbanization in the Caribbean." In Benchmark 1990: *Conference of Latin Americanist Geographers,* edited by T. L. Martinson, vol. 17/18, 1992, pp. 67–90.

Conway, D. "Migration in the Caribbean." In *Benchmark 1990: Conference of Latin Americanist Geographers,* vol. 17/18, edited by T. L. Martinson, 1992, pp. 91–98.

Galloway, J. H. *The Sugarcane Industry: An Historical Geography from its Origins to 1914.* Cambridge: Cambridge University Press, 1989.

Kimber, C., *Martinique Revisited: The Plant Geography of a West Indian Island.* College Station: Texas A&M University Press, 1988.

Knight, F. W. and C. Palmer, eds. *The Modern Caribbean.* Chapel Hill: University of North Carolina Press, 1989.

Meinig, D. W. *The Shaping of America: A Geographical Perspective on 500 Years of History,* Vol. 1, Atlantic America, 1492–1800. New Haven, Conn.: Yale University Press, 1986.

Mintz, S., and S. Price eds. *Caribbean Contours.* Baltimore, Md.: Johns Hopkins University Press, 1985.

Parry, J. H., and P. Sherlock. *A Short History of the West Indies,* 3d ed. New York: St. Martin's Press, 1971; and with A. P. Maingot, 4th ed., London: Macmillan Caribbean, 1987.

Potter, R. B. "Urbanization in the Caribbean and Trends of Global Convergence-Divergence." *Geographical Journal* 159 (1993): 1–21.

Richardson, B. C. *The Caribbean in the Wider World, 1492–1992.* Cambridge: Cambridge University Press, 1992.

Watts, D. *The West Indies: Patterns of Development, Culture and Environmental Change Since 1492.* Cambridge: Cambridge University Press, 1987.

CHAPTER 11

Andean America

BRIAN W. BLOUET

INTRODUCTION

The countries of Andean America discussed in this chapter are Colombia, Venezuela, Ecuador, Peru, and Bolivia. Although the Andes run through and separate Chile and Argentina, those countries are discussed in Chapter 13 on the Southern Cone.

By the early years of the sixteenth century, what is today the Caribbean coast of Venezuela, Colombia, and Panama was used by Spanish expeditions as a source of several commodities including slaves. In the east was the Pearl Coast, the area lying in and around the straits between the island of Margarita and the north coast of Venezuela. Much of the Caribbean coast of Tierra Firme (the Mainland) was a source of brazilwood, and the native inhabitants produced salt by evaporation, which they traded for inland products including objects made of gold. Spanish explorers traded and plundered on this coast for decades before permanent settlements were established, although Cartagena was named and used as a harbor by 1503 (Sauer, 1966, p. 170). The well-sheltered harbor was easy to sail into and out of and provided protection from hurricanes and the boisterous northeast trade winds. The northeast trades are powerful in the southwestern Caribbean because they attain maximum fetch blowing across a long, uninterrupted stretch of sea (Figure 11.1).

Early settlement on the Caribbean shores of present-day Colombia and Venezuela was developed from island bases in the Caribbean. Santa Marta, Colombia, the earliest Spanish town on the mainland of South America, was founded in 1525 at a well-protected harbor. The settlers raised livestock, acquired some gold, and traded with Indian communities. To the west in 1533 the town of Cartagena was established on a defensible, near-island site. Neither Santa Marta nor Cartagena was immediately prosperous, but in 1536 Jiménez de Quesada, starting from Santa Marta, went up the Magdalena Valley and into the Cordillera Oriental. In the upland basins, he encountered and subdued dense populations of Chibcha Indians. Santa Fe de Bogotá was established in 1538. The major foci for Spanish settlement were in the temperate upland basins of the Andes where populations of sedentary Indians lived. Despite problems of adjustment, sheep, cattle, and horses were introduced along with wheat and barley and other European crops. New World crops like corn and potatoes were included in Spanish agricultural

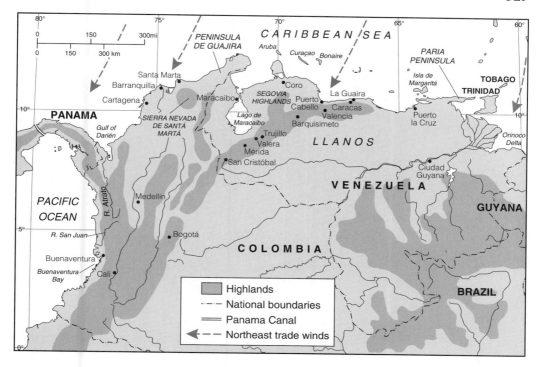

Figure 11.1 Settlement sites: Colombia, Venezuela.

activity. The original coastal settlements were fortified and retained as points of attachment to the Iberian Peninsula, but more colonists lived in the upland basins than in the tropical Caribbean coastal region.

The base for the settlement of Venezuela was the island of Curaçao, occupied around 1520. From there the town of Coro was established in 1527. Shortly afterward, the Welser family of Augsburg obtained a grant from the Spanish crown to develop Venezuela and in 1529 took over the municipality of Coro. Mérida was founded in 1558, San Cristóbal in 1561, and Caracas in 1567. All these towns were small settlements dependent upon unintensive farming and ranching. As in colonial Colombia, the major centers of settlement were in the upland basins. Coastal towns had relatively small populations but provided links with Spain.

The conquest of Ecuador, Peru, and Bolivia commenced in 1531 when the Francisco Pizarro expedition set out from Panama, journeyed south along the Pacific side of the Andes, ascended into the mountains and destroyed the Incan empire. Cuzco, the Inca capital, was captured in 1533. Quito was established in 1534, and Lima was founded in 1535. From Quito, settlers moved north to establish Popayán and Cali (1536) in the Cauca Valley of what is now Colombia.

THE ANDES

The Andean mountain chain runs like a spinal cord along the western margin of South America, from the Caribbean to Tierra del Fuego. The highest mountains in the Andes lie in Argentina (Aconcagua, 22,831 feet), Ecuador (Chimborazo, 20,577 feet), and Peru (Huascarán, 22,205 feet). The Andean chain is widest in Bolivia, where the Altiplano separates two distinct mountain ranges. Even in southern Chile peaks exceed 10,000 feet. A great part of the Andes lies in the tropics, but altitude

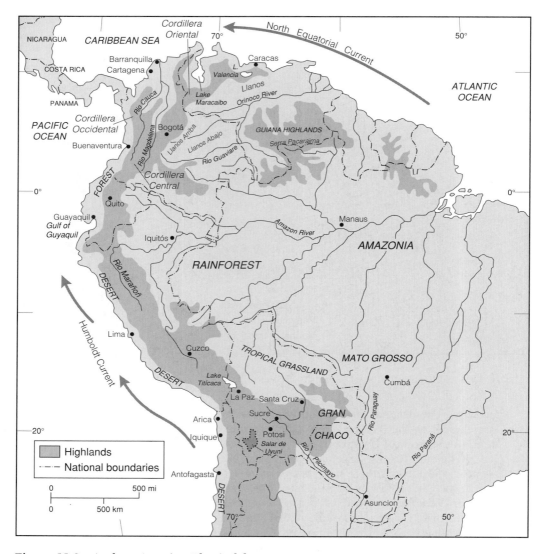

Figure 11.2 Andean America: Physical features.

modifies climate (Figures 2.8 and 11.2). At high levels are permanent snowfields and glaciers even at equatorial latitudes. High altitudes cause difficulties for humans, who may suffer from shortness of breath, nausea, and disorientation brought on by lack of oxygen.

Few highways link Andean America together in the way Alpine Europe is traversed by modern roads with reasonable grades. The Andes are more geologically active than the Alps (there are more earthquakes and landslides in Andean America), and slopes are less stable than in Alpine Europe.

COLOMBIA

In Colombia, the Andes divide into three ranges: the Cordilleras Occidental, Central, and Oriental. Between the western and central ranges flows the Cauca River. The Magdalena Valley, which the Cauca joins, separates the central and

Fortifications at the Port of Cartagena, Columbia.

eastern ranges and drains into the Caribbean. There are settlements in the valleys of the Cauca and Magdalena as well as in the basins of the cordilleras. East of the cordilleras is the *Oriente* region, part of which drains into the Orinoco. The more southerly streams enter the Amazon (Figure 11.2).

Colombia has Pacific and Caribbean shorelines. North from the border with Ecuador, the Pacific coastland is a low-lying, often marshy region. Around the latitude of the port of Buenaventura, the character of the Pacific coast changes and the Serranía de Baudó rises to over 5,000 feet.

The southern border of Colombia lies north of the equator, and the influence of the cold Peru (Humboldt) Current is not felt on the Colombian Pacific coast. Sea-surface temperatures are high, and the Pacific coast of Colombia receives warm, moist air from the ocean. Very high rainfall totals are recorded, particularly where the maritime tropical air is forced upward by the Serranía de Baudó. The port of Buenaventura near sea level receives 180 inches of rain well distributed throughout the year. Rainfall totals are greater at higher elevations in the Serranía de Baudó. The Serranía (mountain chain) rises north of Buenaventura Bay and runs in a roughly south-to-north direction, forming a coastal range in the Pacific lands of Colombia. The Serranía is separated from the Andes by the valleys of the Atrato and San Juan rivers. It is a land of perpetual rains, dense forests, few roads, isolation, and low population density. The lower valley of the Atrato is a swamp that impedes communication between Colombia and Panama. Before the Panama Canal was built, the Atrato Valley was considered as a route for a Caribbean–Pacific canal. The canal would have run from the Gulf of Darién, up the Atrato Valley, through the drainage divide between the Atrato and the San Juan rivers, and followed the San Juan Valley to the Pacific north of Buenaventura. Given the problems of digging the shorter Panama canal, it is unlikely that the Atrato–San Juan canal could have been completed. A major problem would have been controlling the river waters in the valleys through which the canal would be cut.

The Serranía does not prevent Pacific maritime tropical air from moving inland, and high rainfall totals are recorded in the Cordillera Occidental. During the summer rainy season, the valleys of the Cauca and Magdalena are also penetrated by rain-bearing winds from the Caribbean.

► The Cordilleras

In the cordilleras, temperatures are influenced by altitude. Tropical temperatures and crops extend up to about 3,000 feet (*tierra caliente*). In the next 3,000 feet (*tierra templada*) corn and coffee do well. From 6,000 to 10,000 feet (*tierra fría*) grains and tubers are the major crops. At higher altitudes, bleak pasture (*páramo*) predominates. The vegetation is grassland, and cold rain or drizzle is frequent. On the flanks of the volcano Cotopaxi, in Ecuador, on average 250 days a year receive rain (Parsons, 1982, p. 259). Snowfields are found above 14,000 feet (*tierra helada*). At Bogotá (8,355 feet), on the eastern edge of a large basin in the Cordillera Oriental, mean monthly temperatures are consistently in the high 50s. The city receives 40 inches of rain each year with no dry season. Rain or drizzle mark many days. Depending on taste, the climate of Bogotá can be described as perpetual spring—or damp and chilly.

► The Oriente

Both the *Oriente* and the Caribbean coastlands of Colombia have seasonal rainfall regimes. The regions lie in the tropical wet and dry climate zone. As the Intertropical Convergence Zone (ITCZ) migrates north in the Northern Hemisphere summer, rain falls across the Orinoco drainage basin of Venezuela and Colombia. When the rain belt retreats south, drier conditions prevail from November through the early months of the following year. In the southern part of the Colombian *Oriente*, which is in the rain belt for a greater part of the year, tropical forest (*selva*) is the dominant vegetation. Northward, the dry season becomes more pronounced, and rain forest gives way to grassland (*llanos*) where trees are confined to watercourses. In Colombia the boundary between the *llanos* and the *selvas* lies approximately along the line of the Río Guaviare, a tributary of the Orinoco.

In the north, the *llanos* of Colombia are drained by west-to-east flowing streams that empty into the Orinoco. The southern part of the *llanos* drains into the Amazon River system. From west to east, the Colombian *llanos* can be divided into two major regions—the high plains (*llanos arriba*) and the lower and drier *llanos abajo*. On the *llanos arriba*, close to the Cordillera Oriental, rainfall totals are higher and agriculture is practiced on the alluvial fans and valleys that issue from the mountains. The *llanos abajo* have less rainfall, and the region is used primarily for grazing in the north although there is enough precipitation to support forest in the south. In the northern areas with a pronounced dry season (December to March) rivers and lakes diminish, trees shed their leaves, and all vegetation is covered in dust. Grazing is scarce and animals are short of water. In the wet season (April through November) rivers overflow and inundate surrounding land, and cattle, confined to small islands above the flood, can then be short of fodder (Rausch, 1984, pp. 4–6). The economic value of the *Oriente* has been greatly enhanced by the discovery and exploitation of a number of oil fields.

► The Caribbean Coastlands

The Colombian Caribbean coastland is a predominantly low-lying, rolling landscape, composed of alluvium and the eroded remnants of drainage divides. To the east of the lowland, the granitic Sierra Nevada de Santa Marta rises to 18,947 feet and provides snow-covered peaks within view of the Caribbean Sea. The Caribbean shore of Colombia culminates in the east at the Guajira Peninsula, a semiarid region rich in coal, natural gas, and airstrips for the drug trade.

The Caribbean coastlands of Colombia receive most of their rainfall in the summer months. Precipitation totals decrease significantly from west to east. At the western end of the Caribbean coast of Colombia, around the Gulf of Darién, annual rainfall totals approach 80 inches per annum. Moisture is received from both Caribbean sources borne in by the northeast trades and from the Pacific Ocean. Rainfall is concentrated in the summer months when the intertropical front migrates north (Figure 2.9).

Moving northeast to Barranquilla, on the Colombian Caribbean coast, rainfall decreases to about half the totals experienced on the Gulf of Darién. Farther east and north, the Guajira Peninsula is semiarid. The vegetation reflects the change in rainfall regime, grading from *selva* on the Gulf of Darién to xerophytic scrub on the Guajira Peninsula. Little rain is received between late December and April. Rains commence in May, becoming heavier in June before trailing off toward the end of the year.

► *Colombian Agriculture*

Environmentally, Colombia is a varied country, a diversity that is reflected in the country's agricultural activity. The upland basins were early centers of Spanish settlement, and a number of crops imported from Europe could be grown in the cooler highland basin environments. Present-day agriculture in the Andean basins is a mixture of European and American crops: wheat, barley, corn, and potatoes. At higher elevations livestock grazing is the predominant activity. Livestock, notably cattle and sheep, are of European origin.

Cash crops, including cotton, marijuana, sugarcane, bananas, cacao, and rice, are grown on the Caribbean lowlands. Subsistence crops include manioc, corn, and vegetables. In the west, where rainfall is higher, crops grow without watering, but eastward, as precipitation totals decline, irrigation is required. A great part of the Caribbean lowland east of the Magdalena River is used for low-intensity livestock raising.

Colombia is the second biggest producer of coffee in the world after Brazil. Coffee is Colombia's largest legal agricultural export. The value of marijuana and cocaine, both of which are illegal, exceeds that of coffee, but illegal agricultural commodities do not appear in the official export figures.

On the flanks of the cordilleras, in the *tierras templadas* (3,000–6,000 feet), are found optimum temperature conditions for coffee. At these elevations rainfall is adequate, and the well-drained slopes are too steep for mechanized agriculture (Parsons, 1968, p. 142). Colombian coffee production involves large inputs of hand labor. The shrubs are well tended, and the berries carefully picked. The most productive Colombian coffee region is Antioquia, but the crop is widely grown.

In the cordilleras, the valleys of the Cauca and the Magdalena are steep-sided. The areas suitable for arable farming are limited but highly productive in such places as the vicinity of the river town of Girardot on the Magdalena, which ships coffee, hides, and cotton when the water level is high enough to permit navigation. On the Río Cauca, north of Cali, is the fertile Valle del Cauca. Here the Cauca Valley widens out in an old lake bed, and excellent crops of sugarcane, cotton, and tobacco are grown, although much land is used for cattle raising.

The *Oriente* covers more than 50 percent of the territory of Colombia, but little land away from the alluvial fans issuing from the mountains is cultivated. In the *llanos* there is low-intensity cattle raising. On the flanks of the Andes, where there is sufficient rainfall to support forest, coca is grown. Because the locally grown raw

TABLE 11.1   Andean Countries: Demographic Data

Country	Population 1995 (millions)	Birth Rate (per 1,000)	Death Rate (per 1,000)	Infant Mortality (per 1,000 live births)	% Urban	GNP per Capita (U.S. $)
Bolivia	7.4	36	10	71	58	770
Colombia	37.7	24	6	37	50	1,400
Ecuador	11.5	28	6	50	58	1,170
Peru	24.0	29	7	60	70	1,490
Venezuela	21.8	30	5	20	84	2,840

Source: Population Reference Bureau, *1995 World Population Data Sheet* (Washington, D.C.: Population Reference Bureau, 1995).

material is insufficient, coca paste is imported from Bolivia, Peru, and Brazil to make cocaine.

About a third of the Colombian labor force is in farming, and just half the population still lives in rural areas (Table 11.1). Over half the exports, by value, are agricultural products, excluding nonreported exports such as cocaine, marijuana, and poppy products.

▶ Urban Areas in Colombia

Colombia offers a contrast to most Latin American countries in possessing numerous large urban centers and no primate city (Figure 11.3). The Bogotá metropolitan area has 5.3 million inhabitants. Medellín, the capital of the department of Antioquia, has a population of 2.5 million, followed by Cali with 2.0 million and Barranquilla with 1.9 million. Both Cartagena and Bucaramanga have nearly .75 million inhabitants.

The cities contain numerous manufacturing industries, and Colombia has a more diverse economy than other Andean countries. Physical geography has played a part in this diversity. The cost of importing goods into Bogotá and Medellín has historically been high. Once industries were started in the cities, the high transport costs for imports gave some protection to local manufacturers. Similarly, it has been difficult for Colombian industries to serve national markets, and this has helped promote the growth of regional manufacturing centers.

Commentators attribute the relatively early development of manufacturing industry in Colombia to the nature of coffee farming in the country. From the mid-nineteenth century to the present, coffee growing has expanded with occasional pauses brought on, for example, by the depressions of the 1930s and 1980s. Unlike coffee production in Brazil, most Colombian coffee is produced on small family farms, particularly in Antioquia (Bergquist, 1978, pp. 259–260). By the end of the nineteenth century, Colombian foundries were producing simple agricultural equipment for which there was an expanding family farm market. The small farms produced and sold coffee, bringing cash into the countryside and helping create sufficient purchasing power to support the regional manufacturing of textiles and other items for domestic consumption. By the early years of the twentieth century, Medellín was a manufacturing center for textiles, clothing, shoes, soap, foodstuffs, glass, beer, and iron and steel products (Eder, 1913, p. 202). Bogotá had similar industries. Cali developed later, but today it too is a major manufacturing center.

Cali is linked by road and rail to the Pacific port of Buenaventura, through which most Colombian exports pass. The growth of Buenaventura accelerated after the opening of the Panama Canal (1914). Both Cali and Buenaventura are growing

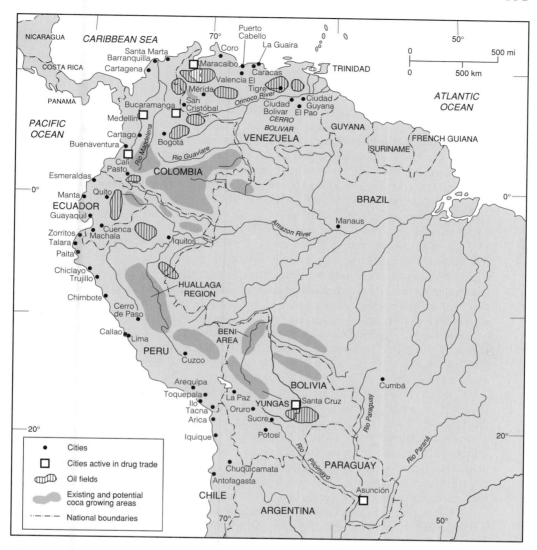

Figure 11.3 Andean America: Countries and cities.

rapidly and possess extensive shantytowns. Cali has grown to rival Medellín in the drug trade. The city is a center of "guerrilla and counterinsurgency activity and drug and anti-drug operations. It is best not to arrive after dark . . . and be prepared for frequent police checks" (*South American Handbook*, 1990, p. 697).

Colombia has a violent history. In 1840 there was a civil war. Toward the end of the century, there was civil unrest culminating in the War of a Thousand Days (1899–1903). After World War II the country was engulfed in *La Violenca* (1947–1958), resulting in numerous deaths. Today Colombians are battered by civil unrest, guerrilla warfare, and battles between the drug cartels.

Colombia exploits an array of mineral resources including coal, oil, gold, iron, emeralds, natural gas, nonferrous ores, and semiprecious stones. Agriculture is productive and is based on small units, which are adaptable. A lack of political stability prevents Colombia from enjoying those improvements in living standards and health care that would bring the country into the developed world.

VENEZUELA

Venezuela can be divided into four major physical regions: Lake Maracaibo and the lowlands along the Caribbean coast; the Andes; the Orinoco basin; and the Guiana Highlands.

► *Maracaibo Lowlands and the Caribbean Coast*

The Maracaibo lowlands occupy a depression framed to the west by the Cordillera Oriental and to the south and east by the Cordillera de Mérida. Forest covers the flanks of the mountains and much of the lowland surrounding the southern part of Lake Maracaibo. To the north and along the Caribbean coast, the climate becomes drier and a major climatic anomaly is experienced (Trewartha, 1966, pp. 59–64). In theory the Caribbean coasts of Venezuela and Colombia should be bathed in warm, moisture-laden air carried onshore from the Carribean Sea by the northeast trade winds. There should be substantial precipitation in summer with enough rainfall to support tropical forest, as is the case on most large Caribbean islands exposed to the northeast trades. In fact, along much of the Colombian and Venezuelan Caribbean coast, conditions are dry. Tropical grassland and thorny scrub are common vegetation. Factors contributing to the creation of the dry zone include subsidence of air (subsiding air warms, which decreases its relative humidity), and low sea-surface temperatures at the coast. The relatively cool sea surface cools the air blowing onshore. When the air arrives ashore, it is warmed by the land surface, and warming air absorbs moisture. Figure 2.12 displays the annual monthly temperature and rainfall statistics for the weather station at Maracaibo. On average, the city receives precipitation on only 65 days a year. The early months of the year are dry. In May, over 2 inches of rain are received, and a wetter season continues through summer and fall, with the highest monthly rainfall totals coming in October and November, before the dry season returns in December.

On the slopes of the Cordillera de Mérida overlooking the Maracaibo lowlands, coffee is grown. In the wetter portions of the lowland, toward the south, sugarcane and cacao are produced. In drier areas tropical grassland provides grazing for beef and dairy cattle. The scrublands support goats. Overall, the Maracaibo region is not a rich agricultural area.

The major economic activity in the Maracaibo basin is the oil industry. Oil has been extracted from the area since World War I, and the shallow lake has had many wells drilled into its bed. In spite of oil finds in the Orinoco basin, the Maracaibo fields remain the most productive in Venezuela. The first oil refineries were constructed on the Dutch islands of Aruba and Curaçao. The oil companies believed that Netherlands territory provided a secure base for investment and operations. Now there are many refineries in Venezuela, and the Curaçao refinery, formerly owned by Shell, is leased to the Venezuelan industry.

The oil extraction industry grew rapidly in the 1920s and 1930s. The industry was given impetus by the nationalization of Mexican oil in 1938, which had the effect of removing Mexican oil from international markets. World War II (1939–1945) boosted demand for oil, and Venezuela again benefited. In an effort to add value to the export of oil products, wartime legislation required that a percentage of oil be refined in Venezuela prior to export. The oil industry was nationalized in 1976 in an effort to keep more oil profits in the country. One result of nationalization was that Venezuela, rather than multinational oil companies, had to provide capital for development. Capital was raised by loans on international markets, which helped to create a Venezuelan foreign debt problem. The costs of nationalization were incurred in the late 1970s when oil prices were high. The loans came due in the 1980s when oil prices slumped.

Until the twentieth century, most Venezuelans regarded the Maracaibo region as hot, humid, and unhealthy. Much land was occupied by Indian groups who still live in the area, particularly in the swamplands west of Lake Maracaibo, and seek to preserve traditional culture. The rise of the oil industry resulted in the rapid growth of the city of Maracaibo and several other towns on the low-lying shores of the lake. In the 1950s a channel was dredged through the bar at the mouth of the lake, and vessels in excess of 20,000 tons can berth at Maracaibo. Much of the oil, however, is piped rather than shipped out of the lake region to refineries and terminals lying beside deep water on the east side of the Gulf of Venezuela.

The Maracaibo metropolitan area, with a population approaching 2 million inhabitants, is a manufacturing and service center. Employment in the tertiary sector has grown rapidly. Maracaibo is the Houston of Venezuela, although it is hotter and has no cool season. Maracaibo does not, however, serve as a center of the world oil industry the way Houston does. Traditionally, coffee grown in the Cordillera Oriental of Colombia was exported via Maracaibo; today cocaine follows the same route.

► Andean Venezuela

North of Bogotá, Colombia, the eastern cordillera of the Andes divides. A branch of the Andes runs northeastward and then eastward, across Venezuela to the Paria Peninsula, and out into the Caribbean Sea toward Trinidad. In the Venezuelan Andes are a series of upland basins, which attracted early Hispanic settlement. To the present, most major Venezuelan cities, with the exceptions of Maracaibo and Ciudad Guayana, lie in the Andean highlands, and most of the population lives in the cooler upland basins in the Andes, the floors of which are 2,000–4,000 feet above sea level.

The Venezuelan Andes are highest in the west in the Cordillera de Mérida, where slopes are steep and mountain peaks rugged. The upland decreases in altitude to the east. The Cordillera de Mérida rises to 16,411 feet, and a number of peaks are snow covered including the Sierra Nevada de Mérida, which forms a backdrop to the city of Mérida. Crops include coffee and corn, with wheat and barley at higher elevations. Potatoes can be grown up to around 10,000 feet. Above that altitude the slopes of the Cordillera de Mérida are used for pasture. Major towns in the cordillera include San Cristóbal (250,000), the university city of Mérida (170,000), Valera (120,000), and Trujillo. All are colonial foundations. The scenery surrounding the towns of the Cordillera de Mérida is spectacular; granitic peaks dominate steep, fault-formed slopes that rise up from the fertile upland basins. The region is healthy, rainfall is plentiful, and agriculture is productive, but many young people leave the area to find work in Maracaibo or Caracas.

The Andes are lower toward the city of Barquisimeto (population more than 600,000), which stands at an altitude of just over 1,500 feet. In this area the mountain chain swings due east to run parallel to the Caribbean shore. To the north of Barquisimeto are the Segovia Highlands, a heavily dissected upland. The Segovia area is close enough to the coast to be in the low-rainfall zone. Population is sparse, and agricultural activity tends to be confined to irrigated river valleys.

The central highlands, lying to the east of Barquisimeto, contain the major centers of population, economic activity, and political power. The extensive Valencia basin is an important agricultural region producing sugar, coffee, cotton, corn, rice, vegetables, and cattle. The region is linked to Caracas by road and rail. Puerto Cabello on the Caribbean coast 35 miles north of Valencia is the port for the Valencia basin. The city of Valencia (population 900,000) is a center of industry, commerce, and administration. Lake Valencia is polluted and shrunken as a result of deforestation and the extraction of irrigation water.

The Venezuelan capital, Caracas, fills a narrow, fault-formed, steep-sided basin to the east of Valencia. The city stands at around 3,000 feet, and temperatures are moderate. The annual rainfall is 32 inches, with May through November being the wettest months. The city was founded in 1567 and became the most important urban center in this part of Spanish America. Colonial Caracas had the advantage of occupying a temperate environment less than 10 miles from the Caribbean. It is separated from the sea by the Serranía de la Costa, but a pass gives access to the port of La Guaira, which has a hot, dry climate. At La Guaira, temperatures average over 80°F throughout the year, and the annual rainfall on the Caribbean coast is only 11 inches. The mountains come close to this part of the Caribbean coast. La Guaira is built on steep slopes, and breakwaters have had to be built to create a sheltered port.

In a few decades in the middle of the twentieth century Caracas was transformed from a colonial city into a massive metropolitan area complete with high-rise buildings, freeways, pollution, and as a result of rapid in-migration from rural areas, sprawling *ranchos* to shelter the urban poor. The metropolitan region with more than 4 million inhabitants is the major center of administration, industry, commerce, and finance in Venezuela. Caracas contains the large government bureaucracy. The government is a major employer, and most public money is dispersed through agencies in the capital. The rate of population growth in Caracas has slowed, but the capital remains the focus of political and economic power in Venezuela.

East of Caracas, the highlands are interrupted by the Caribbean Sea but reappear close to Puerto La Cruz, an oil shipment and refining center that serves El Tigre and other oil fields to the south in the Orinoco basin.

► *The Orinoco River and the Llanos*

Tributaries of the Orinoco originate in the Andes to the north and west and in the Guiana Highlands to the south. The Orinoco lowland is broken by outcrops of older, harder rocks from the Guiana Highlands, and there are mesas (flat-topped hills) of the *llanos*, formed of sedimentary rocks, in the northern part of the basin. Overall, the Orinoco landscape is rolling, with low interfluves separating alluvium-filled watercourses. The vegetation of the *llanos* is predominantly savanna grassland, with woodlands along the watercourses.

The Orinoco basin experiences a marked seasonal climatic rhythm associated with the annual movement of the Intertropical Convergence Zone (ITCZ). During the Northern Hemisphere summer months, the rain belt moves northward over the Orinoco basin. By April, rains start to spread from the south across the region, with the heaviest precipitation in June, July, and August. By late November the rainy season is over, and clear skies with dry conditions predominate until March. Vegetation responds to the seasons. In the dry season the grass dies off, and the landscape away from the watercourses becomes brown. In the wet season the rivers rise, flooding is commonplace, and grasses green up and grow rapidly.

Spanish settlers released cattle and horses into the tropical grassland environment of the *llanos*. The animals became the basis of feral herds that were culled periodically by *llaneros*. Cattle were driven from the *llanos* to the upland basins around Mérida, Valencia, and Caracas for fattening and slaughter. Agriculturally, the *llanos* has the disadvantage that the dry season curtails grazing, and in the wettest months much pasture is flooded and not available to livestock. The seasonal drought limits arable farming. Some flood-control and irrigation schemes have been built, for example, on the Río Guárico, but much more could be done, particularly because Venezuela is a large importer of agricultural produce. Good crops of rice and cotton are obtained where water resources are well managed.

The Orinoco River has disadvantages as a navigation route. The rainfall regime results in changes in river levels between the wet and dry seasons. The head of navigation on the Orinoco is at Puerto Ayacucho on the Colombian border, more than 700 miles from the Atlantic coast. There are no regular, direct riverboat services to Puerto Ayacucho from ports downstream like Ciudad Bolívar and Ciudad Guayana. In the Orinoco Delta, the river branches into many distributary channels. Dredging is necessary to allow passage of ore carriers that export iron excavated at Cerro Bolívar and El Pao.

Until World War II, the *llanos* was a pastoral land with low population densities and limited economic activity in the Orinoco river towns. In the second half of the twentieth century, the eastern part of the basin has undergone economic change. Around El Tigre oil and gas are extracted. To the south of the river in the Guiana Highlands, deposits of bauxite and high-grade iron ore are exploited. Some ores are exported, but at Ciudad Guayana iron and bauxite are smelted into steel and aluminum. Hardly more than a village at the end of World War II, Cuidad Guayana now has half a million inhabitants. The city complex lies on the south shore of the Orinoco where the Río Caroní joins the main stream. The Caroní is a fast-flowing river that falls, via rapids, off the Guiana Highlands. It rises well to the south in a region with a longer rainfall season. A series of hydroelectric plants on the lower course of the river produces power for Ciudad Guayana industries.

► Guiana Highlands

South of the *llanos* lie the sparsely inhabited Guiana Highlands that comprise some 50 percent of the territory of Venezuela. The highlands consist of an ancient block of rocks that are heavily mineralized as a result of igneous and metamorphic processes. The highlands increase in altitude toward the south and culminate in the Serra Pacaraima that forms the border between Venezuela and Brazil and divides the Orinoco and Amazon drainage systems.

The Guiana Highlands of Venezuela are drained by north-flowing tributaries of the Orinoco that have dissected the plateau, creating a series of erosion surfaces on which remnants stand out as peaks. Valleys are steep sided and difficult to cross. The beds of rivers are broken by rapids and waterfalls that prevent navigation. The region is covered with forest and savanna grassland. Soils are thin and infertile, and erode easily if vegetation is cleared.

Resources in the Guiana Highlands include bauxite, high-grade iron ore, manganese, and gold. Many of the rivers have the potential to generate electricity. A tiny proportion of the resources has been exploited on the northern margin of the region, close to the Orinoco. The greatest part of the highlands is isolated and lacks roads. The existence of a road from Ciudad Guayana to the Brazilian border makes it possible to drive to Manaus on the Amazon. At the eastern edge of the Guiana Highlands the Pacaraima Mountains fall steeply down to the Atlantic coastal plain. The plain is not part of the territory of Venezuela and contains Guyana, Suriname, and French Guiana (Guyane). Venezuela claims much of the territory of Guyana.

BOLIVIA, PERU, AND ECUADOR

When Bolivia, Peru, and Ecuador emerged as independent states in the early nineteenth century, each possessed three recognizable natural regions:

1. An eastern region (*oriente*) consisting of tropical lowlands and the Andean slope leading down to them

2. The Andes (*sierra*)

3. A coastal zone (*costa*)

Bolivia lost its coastal region and outlet to the sea to Chile in the Pacific War of 1879–1883. A large part of Ecuador's *oriente* was annexed by Peru in 1941. Peruvians refer to their eastern region as *montaña* or *selva*.

► *The Oriente*

The *oriente* regions of Bolivia, Peru, and Ecuador (all with extensive oil deposits) are predominantly tropical lands covered by forests and drained by tributaries of the Amazon River that rise in the Andes. If we were to journey westward from the Peruvian port of Iquitos on the Amazon and ascend into the Andes, we would pass through a variety of environments—tropical rain forest (*selva*), and then mountain forest (*montaña*), before moving into high grassland above the tree line. At Iquitos the climate is equatorial. Humidity is high. The total annual rainfall is around 100 inches, more than at the downstream Brazilian port of Manaus in the heart of Amazonia. Between December and March Iquitos receives over 10 inches of rain each month. From April to November, when the intertropical front has migrated to the north, monthly totals are significantly less, but there is still sufficient precipitation to support the growth of tropical forest.

Iquitos (population 375,000) is 328 feet above sea level and 2,000 miles from the Atlantic. Yet, the city is an Atlantic port. It is only partially linked to other parts of Peru by river boats and air travel. There are no road or rail connections with other regions of the country. Products of the *costa* and *sierra* cannot easily be exported to Europe or North America on a route that would run from Iquitos down the Amazon to the Atlantic and on to Western Europe or the eastern seaboard of the United States. Iquitos is a center of timber, chemical, and petroleum industries. There is little commercial agriculture in the region. Subsistence cultivators practice shifting agriculture in the surrounding forests, but much food for Iquitos arrives by boat or air.

Iquitos grew in the Amazon rubber boom of the late nineteenth century and became the upper Amazon center of the rubber trade. The industry developed on the basis of collecting latex from wild trees scattered in the rain forest. Upstream of Iquitos, river boats, launches, and canoes navigate the rivers of the tropical lowlands and penetrate the eastern valleys of the Andes.

On the flanks of the Andes, the maritime tropical air, which originates over the Atlantic and flows in the trade winds across Amazonia, is pushed upward. As the air rises, it is cooled, condensation takes place, cloud cover increases, and heavy precipitation results. The rain forest stretches up the front range of the Andes. The mix of species changes with altitude, but woodland extends up to 10,000 or 11,000 feet in many places. (See Figure 2.8 for a generalized representation.)

The microclimates of the front range are complex. Valleys lying to the west of prominent relief features can be in a rain shadow. Slopes with an aspect exposing them to long hours of sunlight have high rates of evaporation and support grassland rather than forest.

Fast-flowing tributaries of the Amazon, rising in the Andes, cut through the front range to reach the tropical lowlands. For example, the Río Urubamba rises south of Cuzco in the *sierra*, flows north in a deeply incised valley past Machu Picchu, swings east, and cuts through the front range to join the Río Ucayali, another stream with headwaters in the Andes. The Ucayali forms a confluence with the Huallaga–Marañón upstream of Iquitos. In spite of steep gradients in the front range, all these rivers and many others allow passage upstream for canoes and small craft. Between Iquitos and the Atlantic the Amazon falls 328 feet. Some of the Andean tributaries of the river drop more than 10,000 feet falling from the *sierra* to the *oriente*. As a result of these deep cuts into the *sierra* made by eastward-flowing streams, tropical environments extend into the mountains and exist in the valleys of incised streams. In the Andes it is often possible to follow an incised stream down thousands of feet and move from a temperate to a tropical environment in a few hours.

Cultivated crops of the *oriente* include manioc, sugarcane, sweet potatoes, bananas, mangoes, cacao, and coca. The American geographer Isaiah Bowman (1878–1950) observed that few commercial crops could stand the cost of transportation out of the *oriente*. There are areas of great fertility, but the region was, and is, isolated. The sugar crop could be hauled out of eastern Peru as rum. Cacao and dried coca leaves could also withstand the transportation costs (Bowman, 1916, p. 74). Inaccessibility, poor infrastructure, and high transportation costs constrain development of the *oriente* region. Projects like the Marginal Highway bring minor improvements in road transportation but do not form a farm-to-market road system. However, isolation has not stopped oil companies getting into the *oriente* to exploit petroleum resources in forests that are the traditional habitat of Indians who live by hunting and gathering (Kane, 1995).

Coca Cultivation

From pre-Columbian times, the leaves of the coca plant have been chewed as a stimulant. The coca shrub, *Erythroxylon coca*, will grow widely in the tropical lands of South America (Gade, 1975, pp. 174–180). The plant does well on the eastern flanks of the Andes, where temperatures are warm and rainfall is plentiful. The plant prefers shade, and the forests of the Andean eastern flank provide good conditions. Coca does not like temperatures below 40°F., which places an upper limit on cultivation on the Andean slopes. Not all environments on the eastern slopes are suitable. In the northern part of the Colombian *oriente* the dry season inhibits growth and leaf is of poor quality. In Peru there are valleys in rain shadow or hillsides heavily exposed to sunlight that lack sufficient moisture and shade. The southern limit of coca cultivation occurs in Bolivia where rain forest gives way to the deciduous scrub of the Gran Chaco. The range of the plant could be extended north and south by providing irrigation water and shade. The best quality chewing coca is grown between 3,000 and 4,000 feet. If coca is converted into cocaine, the quality of the leaf is less important.

The longstanding cultivation of coca is documented in the geographical literature (Weil, 1983). Clements Markham of the Royal Geographical Society described coca cultivation in Peru in the 1850s. In the valleys of eastern Peru Markham saw terraces of coca plants, interspersed with coffee, rising up hillsides (Markham, 1862, p. 229). Markham chewed coca and found it fragrant, soothing, and invigorating, noting that the leaf suppressed fatigue and hunger. Markham was less breathless when ascending steep slopes. He recognized that chewing coca leaf could be abused but thought it less injurious than most narcotics (Markham, 1862, 238–239).

Isaiah Bowman, reporting on the Yale Peruvian Expedition of 1911, described coca cultivation on the eastern slope of the Andes in southern Peru. Bowman observed coca seedlings being raised in nursery beds and then planted out in orchards that yielded, in the best conditions, 2 tons of leaves per acre per year (Bowman, 1916, pp. 74–77). The British geographer Alan Ogilvie (1927, pp. 169–171) provided an account of coca cultivation in the Yungas region of Bolivia, on the eastern slope of the Andes. Seeds for a new coca field were sown in November in beds covered with cut grass. As the seedlings emerged, they were protected with banana leaves. After a year the 12-inch high plants were transplanted into deep trenches. When mature, the shrub yielded three to four leaf pickings a year in the Yungas region. The life of a coca plantation (*cocal*) was 20 to 40 years, with the plants producing longer in higher, cooler environments. Ogilvie noted that the climatic requirements of coca—warmth and moisture with cloud cover to provide protection from the sun—were similar to those for coffee. Coca and coffee could be grown together. The harvests of coffee berries and coca leaves came at slightly different times of the year and were dried on the same paved areas. The coca and coffee were grown on both haciendas and the plots of free Indian cultivators. Wet weather inhibits drying and reduces the production of coca leaves (McGlade et al., 1995).

From Bowman's and Ogilvie's accounts, we know that coca cultivation was a part of traditional farming systems on the eastern slopes of the Andes in Peru and Bolivia. The dried leaves were transported by porters and llamas from the *oriente* to the *sierra*. A tiny proportion of the traditional crop was made into cocaine and exported. The cultivation of coca attracted international interest when demand for drugs resulted in more leaves being processed into cocaine for export in the 1970s.

Today most coca is cultivated on the eastern slopes of the Andes in Peru, Bolivia, Colombia, and Ecuador, in output order. But coca can be grown in Amazon rain forests. The British geographer Henry Bates, who lived on the Amazon from 1851 to 1859, noted plots of coca grown by women in the forest. The leaves were oven-dried and ground up with ashes, the potash helping to release the cocaine. Bates noted that many middle-aged women appeared to be addicted to *ypadú* as the coca powder was known (Bates, 1892, p. 283).

The point is that coca will grow widely in tropical rain forests, and curtailing cultivation in Peru and Bolivia may only encourage production elsewhere. Ray Henkel (1988, p. 54) reported that coca cultivation was increasing in Brazil. The governments of Peru and Bolivia, with U.S. financial assistance, encourage farmers to substitute other crops for coca. In Peru loans have been made available to finance crop changes, but the loans are difficult to obtain and costly (Morales, 1990, pp. 91–98).

In Bolivia coca growers have been given cash incentives to switch to alternative crops, and in 1990 over 15,000 acres were converted from coca to other crops (Robinson, 1990). No statistics are available on new coca acreage brought into cultivation, but coca production rose in Bolivia in 1993 and 1994 after slowing when the incentives were introduced (Sims, 1995). Little progress has been made in

Coca cultivation, Bolivia.

Harvesting Coca, Bolivia, 1950. The picture was
taken long before cocaine entered world markets on
a large scale. The leaves were picked, spread out to
dry, and then sold for chewing. The embanked cultivation plots and intensive labor indicate that coca
growing was profitable before the cocaine boom.

the Chapare region of Bolivia to provide the infrastructure of rural development
that will be needed to sustain the development of alternative crops (Escobar, 1995).

As a result of external demand for cocaine, the area under coca cultivation has
increased. The Huallaga Valley in Peru is the major coca-growing region of South
America. In Bolivia the Yungas and Chapare regions are the main growing areas.

Coca leaves for sale in an Andean market.

Cultivation of coca extends across the Bolivian border into Brazilian rain forests. In parts of Peru and Bolivia, coca cultivation is so intensive that blights pass easily from *cocal* to *cocal*.

When the demand for cocaine grew in the United States, much coca paste from Peru and Bolivia moved by networks of river routes, forest trails, and airstrips to remote areas in Colombia for refining. Cocaine was then smuggled from Colombia through the Caribbean to the United States. In recent years refining has become more important in the coca-growing areas of Bolivia and Peru, and the routes to the United States and Europe have become more diverse as Atlantic coast and Chilean ports of South America have entered the cocaine export trade, as has Mexico.

Attempts to eradicate coca cultivation are futile. The plant has been grown widely since pre-Columbian times. It is a folk plant chewed to suppress hunger, cold, and fatigue. It has other uses, including the making of a tea (*maté de coca*). Suppressing coca cultivation in one part of the region will drive up the price paid to growers elsewhere, in the same way that a drought hitting corn fields in Ohio helps corn farmers in Kansas get a better price.

To small-scale, largely subsistence farmers, the economics of coca growing are attractive. The dried leaves from one acre of coca plants are estimated to yield the grower $1,000 a year (Sims, 1995). Few other crops will stand the cost of transportation out of the *oriente*, and bananas, citrus, and pineapples would rot before they could be hauled to already glutted markets. Coca cultivation provides a cash income, often in U.S. dollars, for small farmers who previously had little to sell. It will be difficult to find cash crops to replace coca, for there are few facilities to process perishable crops and preserve them for marketing.

There are reports of extensive poppy growing in Colombia. (Poppies are the plant from which heroin is derived.) If heroin becomes the next fashionable drug, the demand for cocaine might drop.

Isolation of the Oriente

The remote nature of the *oriente* region makes it ideal country for both the drug trade and guerrilla activity. The legendary Che Guevara (1928–1967), who was part of Fidel Castro's revolution in Cuba, chose the Bolivian *oriente* to start his unsuccessful South American campaign in 1966. The isolated, forested, terrain of the *oriente* remains an environment in which rebel groups can operate with cover. The problem is exacerbated because parts of *oriente* regions often lie beyond effective central government administration. The future of the *oriente* regions is uncertain politically and economically. Increasingly, the regions will be involved in environmental issues as indigenous Indians and environmental groups attempt to preserve habitats and traditional rights.

A threat to political stability in the Peruvian *oriente* is the Shining Path (*Sendero Luminoso*), a Maoist guerrilla force that began operations in 1980. The leader has been captured, but the organization still exists in many parts of the country, including Lima. There are other guerrilla organizations operating in Peru, including Tupac Amaru (named after the eighteenth-century revolutionary) and armed groups supporting the coca-cocaine industry.

It would be wrong to give the impression that the *oriente* regions of Bolivia, Peru, and Ecuador are similar. The Ecuadorian *oriente* lies closer to the equator and receives more rainfall. The Bolivian *oriente* in the south merges into the arid Gran Chaco, which is covered with deciduous scrub and savanna. In addition, national governments have different attitudes toward development. The Peruvians have expanded their *oriente* at the expense of Ecuador. The building of the *Carretera Marginale* and roads from the *sierra* to link into it reflect a more aggressive attitude in Peru to open the east than is present in Ecuador.

Eastern Bolivia has better prospects than the *sierra* and *altiplano* regions of the country. The importance of tin mining in the highlands has declined as tin cans are replaced by plastic containers in the packaging of food. In the *oriente*, agricultural settlement, ranching, and coca growing are on the increase. There are several important oil fields with refineries and pipelines that export to surrounding countries. Santa Cruz (700,000) is the second largest Bolivian city (after La Paz, 1.2 million) with the boomtown problems of high living costs, poor housing, growing crime rates, and inadequate or corrupt law enforcement (*South American Handbook*, 1990, p. 253). Road and rail links in and out of the eastern region remain poor, but trade connections are maintained with Brazil, Paraguay, and Argentina.

▶ *Sierra*

The Andean *sierra* stands in sharp physical contrast to the surrounding lowlands. The environment changes rapidly as the mountains are ascended from the *oriente* or *costa*. The German geographer, Alexander Von Humboldt (1769–1859), described the altitudinal zonation in the Andes during his Latin American travels in 1799–1804. Humboldt observed the Ecuadorian volcanoes Cotopaxi (19,347 feet) and Chimborazo (20,561 feet) and recorded vegetational changes as he ascended. Stadel (1992) provides an analysis of the literature relating to altitudinal zonation in the tropical Andes.

Descending from the sierra to the tropical lowlands are four major crop zones, roughly corresponding to the *páramo, tierra fría, tierra templada,* and *tierra caliente*. In the high grassland (*páramo*) are livestock such as horses, cattle, and sheep, together with the indigenous species, llama and alpaca. Frosts in the *páramo* confine cultivation to a few favorable areas. Below the *páramo,* in the *tierra fría,* the dominant crops are tubers, including the potato. In the *tierra templada,* barley, wheat, and maize are produced. In the *tierra caliente,* tropical crops including citrus, bananas, manioc, sweet potatoes, peppers, coca, sugarcane, and more maize are grown. Irrigation may be needed in areas of rain shadow (Brush, 1977, pp. 9–10).

In the Andes of Southern Peru, crops can be grown up to approximately the following heights: citrus and bananas to 6,000 feet, sugarcane to 8,000 feet (although cane takes months longer to ripen than at 3,000 feet), the vine up to 10,000 feet, corn to 11,000 feet, wheat to 12,000 feet, and barley to 13,000 feet. In favorable circumstances, the potato will grow at near 14,000 feet, but above that height there is no cultivation (Bowman, 1916, p. 62).

It is misleading to think of Andean altitudinal zonation as producing discrete crop and land use zones. The zones are complementary. Commodities (such as coca, cacao, and rum) are transported from tropical regions to the highlands. Products from mountain environments (such as tubers, meat, wool, and grains) are traded down to lower altitudes. Potatoes are naturally freeze-dried into *chuñu,* and the freezing can only be done at higher altitudes where frost occurs regularly (Murra, 1989, p. 208). *Chuñu* is consumed at lower levels as a food. The tropical environments of the *oriente* regions and the upland environments of the *sierra* are interconnected by production and exchange systems that result in communities having access to a range of temperate and tropical products. Communication is facilitated by the river systems of the Ucayali, Huallaga, Marañón, and Napo, which flow from the mountains into the lowlands. So deeply incised are these rivers in the *sierra* that tropical environments exist in their valleys.

Although a general correspondence between the altitudinal environmental zones and crop zones exists, in practice crop zones are more numerous. In the Uchucmarca and Vilcanota valleys of Peru, Stephen Brush (1977, pp. 74–75) and Daniel Gade (1975, p. 104) noted at least six crop regions where cultivators have many choices to make. Decisions on which crops to plant are influenced by population density, labor supply, rainfall, the availability of irrigation water, vegetation, aspect, accessibility, and slope. Altitude is not the sole consideration.

Settlements in which farmers live are found on the boundaries between crop zones. In Brush's and Gade's Uchucmarca and Vilcanota studies, villages were located on the junction between tuber and cereal zones. Communities do not necessarily control land in all production zones. Periodic markets facilitate interchange. Gade recorded the following plant products for sale in the Sunday market at the town of Pisac on the Urubamba River: barley, maize, wheat, chuñu, new potatoes, peas, beans, oca, ulluca, coca leaves, lemons, bananas, and tomatoes. At the larger Quillabamba market farther downstream on the Río Urubamba, tropical crops (citrus, bananas, pineapple, avocados, manioc, and sweet potatoes) were for sale in addition to tubers and cereals from higher levels. Different ethnic groups are often identified with particular products (Gade, 1975, pp. 52–54). Gregory Knapp (1991) discussed the complex adaptive strategies of Andean Ecuadorian farmers.

As they pass through Ecuador, Peru, and Bolivia, the Andes display topographic and climatic contrasts. In Ecuador the structure of the *sierra* is relatively easy to comprehend. Between two parallel mountain ranges lie a series of upland basins with floors ranging in height from 7,000 feet to nearly 10,000 feet. They are the setting for the major concentrations of population (Basile, 1974, pp. 5–18). The mountains flanking the basins contain active, dormant, or extinct volcanoes. Soils contain volcanic material.

Major crops in the Ecuador basins include wheat, barley, oats, corn, quinoa, beans, and potatoes. Much land is used as pasture for horses and cattle. The limits of many crops are encountered on the steep slopes that rise from the basin floors. At Quito (9,242 feet), for example, crops are grown on the flatter ground around the city, but on the slopes above the capital, cultivation gives way to pasture.

The Iberian conquest of South America brought into contact two intricate agricultural systems—Andean and Mediterranean farming. Many crops, including

Picking coffee, Colombia.

Small-scale coffee production, Colombia.

tomatoes, potatoes, squashes, and species of beans were introduced into Mediterranean lands. Livestock and crops were moved from the Old World to the New. The following plants were introduced into Ecuador after the conquest: "grapes, olives, figs, apples, quince, peaches, oranges, limes, lemons, pomegranates, citron, sugar cane, bananas, plantains, melons, eggplants, garbanzo beans, cabbage, lettuce, radish, and mustard" (Knapp, 1991, p. 120). Many of the crops were grown on irrigated oases in highland Ecuador. Although irrigation systems existed in pre-Columbian Ecuador, the Spanish introduced technical innovations, drawing on their Roman and Arab irrigation heritage.

Andean agricultural systems have many different varieties of crops. For example, several hundred varieties of potato are grown, which possess varying characteristics and allow differing environmental niches to be farmed.

In Mediterranean and Andean lands, traditional farming is in decline. Small field agriculture producing many crops makes large demands on labor. Varieties are ceasing to be cultivated and are in danger of being lost to the agricultural genetic bank. Agricultural prices are low worldwide. Moreover, Bolivia, Peru, and Ecuador can import food cheaply. As a result, high-intensity, high-cost traditional farming systems will contract.

A sample study revealed that the cultivation of native maize is very labor-intensive in southern Peru. Each hectare of maize required 121 person days for the "plowing, raking, planting, first mounding, second mounding, weeding, harvest, sheathing, husking, drying, shelling, and storage" (Zimmerer, 1991, p. 423). Land is going out of cultivation in the *sierra* regions of Bolivia, Ecuador, and Peru. Less labor-intensive livestock rearing, which was always important in the farming systems introduced from Iberia, is becoming more important.

Travel between the upland basins of Ecuador is not difficult, with roads rising up into the *páramo* zone and then descending into the next basin. Several of the *sierra* towns are linked by rail. Quito, with more than 1 million inhabitants, retains in places the air of a Spanish colonial town. It has not grown as rapidly as some capital cities in Latin America, partly because modern economic activity is concentrated at Guayaquil in the *costa*. Many of the *sierra* towns (such as Ambato) display characteristics of market towns serving surrounding agricultural areas (Stadel and Moya, 1988).

In Peru the structure of the Andes becomes more complex, and travel in the *sierra* is difficult. Though upland basins have extensive flat, alluvial surfaces, they have been deeply cut by streams draining into the Amazon River system to the east. The outlet gorges from the intermountain basins mark the start of the steep descent to the eastern lowlands.

The Altiplano

In southern Peru, the Cordillera Occidental and the Cordillera Oriental are well defined and spread apart around an extensive high plain—the Altiplano—which extends into Bolivia. The Altiplano is a region of internal drainage with many streams falling into Lake Titicaca (12,506 feet). Only at the north end of the Altiplano do the streams join the drainage system of the Amazon. Much of the Altiplano is over 12,000 feet above sea level and close to the growing limits of many crops. Tuber and livestock raising are important in Altiplano agriculture. Around Lake Titicaca, temperatures are moderated by the water body. Wheat and corn can be grown near the lake in favorable locations. Population densities are higher around the lake. Most of the Altiplano is surrounded by mountains, and access to the region is via high-level passes. Lake Titicaca does facilitate waterborne transportation around the lake.

The Altiplano is a cold, windswept environment that is dry from April to October and wet from November to March. It is marginal in agricultural terms, with a large number of days with frost. Low rainfall inhibits plant growth. The most important crops are hardy tubers—potato, oca, ulluco, ysano—and the chernopods quinoa and cañihua (see Chapter 4).

Bolivia: The road from the Altiplano to the Yungas.

There are many fatalities on mountain
roads without guardrails.

To outsiders the Altiplano, which stretches from southern Peru into Bolivia,
appears bleak and poorly endowed for agriculture. Historically, the region has supported larger populations than the tropical environments of the *oriente*. The use
of a wide range of plants, particularly tubers, at high altitudes suggests a considerable degree of adaptation in environments close to the limits of many crops.
Archaeological evidence indicates that the Altiplano was effectively utilized in pre-
Columbian times. For example, the Tiwanaku culture (A.D. 400–1000) cultivated
lands to the southeast of Lake Titicaca, utilizing raised fields and irrigation water

The Andes in Bolivia, Tierra Helada.

from the lake to improve yields. The raised fields helped drain off cold air. When the Tiwanaku methods were reintroduced to present day farmers, yields of potatoes increased sevenfold (Robinson, 1990b)! The use of raised fields has been extensive throughout the Andean region (Knapp, 1991, p. 159).

In the Andes there are significant climatic differences from Ecuador to Bolivia. Quito (9,242 feet) lies close to the equator, and the length of day, from sunrise to sunset, varies little through the year. Altitude modifies temperature, but from month to month temperatures at Quito are constant as can be seen in Figure 2.10. Annual rainfall totals approximately 49 inches. The mountains partly shield the Quito basin from moisture-bearing Atlantic air. Usually a short dry season from June through August requires irrigation water to sustain crop yields. Moving southward in the Andes, temperatures display more seasonal variation than at the equator. Rainfall totals decrease, and a marked dry season is encountered. La Paz (11,916 feet) receives 23 inches per annum (totals are less on the Altiplano), and there is a pronounced dry season from May to July. Just as the *oriente* of Bolivia merges into the arid Gran Chaco, so aridity increases to the south in the *sierra*. In the Salar de Uyuni in southern Bolivia, swirls of dust and salt pollute the air in windy weather.

Mining

The *sierra* region has been a rich source of minerals, including lead, copper, silver, tin, zinc, and tungsten. The Bolivian town of Potosí was founded in 1545 at 12,000 feet to exploit a silver lode. In the nineteenth and twentieth centuries, tin was important, but production is now reduced. The core of Potosí—around the plaza—retains a colonial air. Oruro, south of La Paz, still produces silver, tin, and tungsten, but it is a mining town on hard times. At the Peruvian town of Cerro de Pasco, copper, zinc, lead, silver, and gold are mined, but the settlement is rundown and unemployment rates are high. As a result of environmental conditions and transport costs to market, Andean minerals tend to be high-priced. During the 1980s the prices of tin and silver collapsed. When prices drop, the high-cost producers are the first to reduce output.

In terms of economic development the *sierra* has a high-cost, high-risk environment that lacks an integrated transportation network. The engineering problems are not insuperable, but the costs of road building are high in relation to returns. Many of the communication and integration problems were illustrated in 1986 when a volcano erupted east of Quito. Earthquakes occurred at the same time. The disaster left 20,000 people homeless, and 30 miles of the trans-Andean pipeline, which transported oil from the *oriente* to the port of Esmeraldas on the Pacific coast, was wrecked. Oil production was disrupted, and exports were reduced for two years.

▶ The Costa

The coastal zone of Ecuador and Peru is a land of contrasts. In northern Ecuador the climate is equatorial, and the natural vegetation is a tropical forest that extends southward, in less luxuriant form, to the Gulf of Guayaquil. At the city of Guayaquil and in the surrounding area there is a dry season from June to November, and agriculture in the Guayas Valley employs irrigation in these months. South of the Gulf of Guayaquil, the major climatic influence at the coast is the cold, north-flowing Peru (Humboldt) Current. A desert stretches along the coast of South America from north of Valparaíso in Chile, along the coast of Peru, and through Ecuador to the Gulf of Guayaquil. The cold current lowers temperatures throughout the year along the Peruvian coast. From late April through October, mists (*garúa*) obscure the sun and lower temperatures further. Figure 2.11 displays the average temperature and precipitation figures for Lima. Lima is a few miles inland on the Rimac Valley, and it shares the climatic characteristics of much of the coastal desert. Rainfall is less than 2 inches per annum, but on average, 137 days a year record a trace of precipitation. Most of the precipitation is light drizzle associated with the *garúa*.

Agriculture

Land use in the *costa* regions of Ecuador and Peru is related to the availability of water. In humid, northern Ecuador, sugarcane, bananas, coffee, and cacao are grown in the coastal lowlands and on the flanks of the Andes. Rice does well in the low-lying Guayas Valley, where the river waters can be diverted into agricultural land. In Peru agriculture is prolific where irrigation water is available in the valleys of streams that rise in the Andes and water the coastal desert. In 1852–1853, Clements Markham traveled along part of the Peruvian coast lying southeast of Lima and described agriculture in the Cañeta, Pisco, Ica, and Nazca valleys (Blanchard, 1991). The crops recorded included dates, alfalfa, vines, sugarcane, cotton, and many fruits and vegetables. In addition, Markham observed wine making, the refining of sugar, and the distillation of spirits.

The Spanish settlers introduced Mediterranean crops and irrigation techniques, but previous inhabitants had irrigated crops in the valleys running from the Andes to the sea, across the coastal desert. At Nazca, where the whole valley was covered with productive *haciendas* in vines and cotton, Markham noted that there was little water in the stream. Irrigation water was brought from higher ground inland in trenches (*puquios*) to water fields downstream. The irrigation system had been built by "ancient inhabitants" who predated the Spanish colonists (Blanchard, 1991, p. 50). Since World War II, the Peruvian government has financed irrigation schemes, bringing water from the *sierra* to the *costa* to enlarge the area in commercial crops.

► *Mining*

Bolivia lost access to the sea in the Pacific War (1879–1883) in which Chile defeated both Bolivia and Peru. Peru surrendered the southern ports of Arica and Iquique. Bolivia lost the port of Antofagasta and all other territory that fronted the Pacific Ocean. The territorial dispute between Bolivia and Chile began over nitrate deposits that lay in Bolivian territory but were exploited by Chileans. When Bolivia attempted to exert control over the resource, war started. Bolivia lost the war, the Pacific shore, and the nitrate deposits. In 1992 Peru agreed to allow Bolivia to build a free-port zone at Ilo on the south coast of Peru. Currently, most Bolivian exports go through the Chilean ports of Arica and Antofagasta. For the Ilo link to be practical, an Ilo–La Paz highway will have to be built.

In the Pacific War, Peru lost territory that contained the Chuquicamata copper deposits northeast of Antofagasta. Peru retained deposits of copper at Toquepala on the Peruvian side of the new border. The Peruvian *costa* also contains deposits of coal, iron (there is a small, high-cost, electric-powered iron and steel plant at Chimbote), phosphates, and potash. The oil field at Zorritos in northern Peru is one of the earliest tapped in Latin America. Across the Gulf of Guayaquil from Zorritos, the Ecuadorians exploit an oil field.

Fishing

The seas off Ecuador, Peru, and Chile contain resources. For example, guano (the phosphate-rich droppings of seabirds) is exported as a fertilizer from islands where the birds nest. In the warmer Ecuadorian waters tuna, white fish, and shrimp are caught, and in the cold Peru Current, shoals of anchovies thrive. Fishing ports dot the desert coasts of Peru and Chile. Because anchovies are a low-value fish, the catch is dried and turned into fish meal or fertilizer. Every few years the cold, north-flowing Peru Current is displaced by warmer water (El Niño) flowing south from equatorial regions. As a result, the anchovy shoals and the fishing industry are disrupted. *El Niño* alters the marine environment and brings heavy rainfall on shore. Early in 1992, prolonged rainfall resulted in the washing out of rice fields and shrimp farms in northern Peru and widespread flooding in the Guayas Valley of Ecuador (*Latin American Weekly Report*, April 2, 1992).

In the estuary of the Guayas River, mariculture techniques are used to produce shrimp, which are harvested, frozen, and shipped to North America. Shrimp are raised on the Gulf coast of Texas by mariculture, but the year-round higher temperatures at the equator give the Guayas producers an advantage to offset the higher transportation costs they incur to reach the North American market. Shrimp is the second largest Ecuadorian export by value.

The Distribution of Population

There is rapid population growth in the *costa* region of Ecuador and Peru. The fast-growing Ecuadorian coastal towns include: Guayaquil (population approaching 2 million), where industry is still based on food, raw-material processing, and the manufacture of clothing; Esmeraldas (close to the Pacific end of the trans-Andean oil pipeline); Limones (saw milling); Manta (agricultural exports); and Machala (agriculture and mariculture). The agricultural hinterlands of the coastal towns are innovative and respond to market opportunities.

In the Peruvian *costa*, the Spanish founded towns in the valleys of the streams that flow across the desert from the flanks of the Andes to the Pacific Ocean. Today the fastest growing Peruvian towns and cities, including Lima, are in the *costa*. Many migrants have moved to the *costa* from the *sierra*. Major urban centers include Chiclayo (400,000), Trujillo (800,000), and Chimbote (200,000). Tacna (180,000), close to the Chilean border, is a prosperous town based on fishing and a Mediterranean style of irrigation agriculture.

The Lima–Callao conurbation sprawls down the Rimac Valley some 20 miles from capital to coast. The central area of Lima has lost the battle against overcrowding and poor administration. When Charles Darwin visited Lima in 1835, he commented that a heavy bank of clouds constantly hung over the land and almost every day of his visit he saw a drizzling mist. Lima was in a wretched state of decay, with unpaved streets and heaps of filth lying around, which buzzards picked over for carrion. However, Darwin noted, it had been a fine city in the colonial era (Darwin, 1890).

The former capital of the viceroyalty of Peru did recover in the late nineteenth century as government revenues grew with export booms in guano, sugar, cotton, copper, and silver. Between 1890 and the end of World War I, economic activity began to concentrate in Lima. Although Peru suffered economically in the 1930s, it benefited from the World War II–generated demand for minerals, foodstuffs, petroleum, and rubber. Economic growth was strong during and after World War II. Lima, in particular, grew in population and as the center of government.

In the 1980s most of Latin America went into prolonged economic stagnation as a result of foreign debt. Peru was no exception, and the effects of decline were noticeable in the central area of Lima. Crime rates increased. Shops closed, forced out of business by street vendors who sold imported goods that had avoided duty and did not pay sales, property, or income taxes. Lack of revenue undermined the ability of government to maintain Lima. With the containment of the Shining Path bombing campaign in the 1990s, conditions in the city improved.

At the beginning of the twentieth century, the most heavily populated regions of Peru were in the Andes, where people made a living raising llama, alpaca, grains, quinoa, and potatoes. Most west coast towns—situated on river valleys that flowed across the coastal desert—were small. Lima, the Spanish colonial capital, was the largest city and commercial center but contained only 130,000 inhabitants in 1900.

In 1900, birth rates and death rates were high, and population growth was slow. By 1940, death rates had dropped to 14 per thousand and birth rates (27 per thousand) began to rise. In the 1950s, the birth rate exceeded 30 per thousand and in 1965 reached 45 births for every 1,000 Peruvians. In the 1960s, death rates continued to fall in spite of infant mortality in excess of 100 per 1,000 live births. At over 3 percent per annum, population growth was fast (Table 11.2).

In the 1950s and the 1960s, people left rural areas and migrated to urban places. In 1950, the most populous departments were in the Andes. Migrants did not flock to towns in the mountains but came down to the cities of the coastal plain. The urban places on the west coast were being linked to the global economy, and job opportunities were growing in the region.

Between 1950 and 1980, Peruvian cities gained population at nearly 6 percent per annum, and migrants coming from rural areas accounted for half the urban population increase. Lima was growing more rapidly than the average for all cities. In the 1960s, the population of Lima doubled, indicating a growth rate of around 7 percent per annum. The city could not easily absorb newcomers. By the 1980s, between a quarter and a third of the population of Lima lived in squatter settlements without proper services. Most squatter settlements eventually acquired electricity and standpipes to provide drinking water, but the sewer system was incomplete.

It was no surprise when the disease of unsanitary urban growth—cholera—appeared in Lima in 1991, just as it had devastated European and American cities in the nineteenth century. The bacterium (*Vibro cholerae*) that produces cholera proliferates in water contaminated by sewage. The bacterium infects the small intestine, causing diarrhea and dehydration. In the nineteenth century, the disease commonly killed 40 percent of those infected. In Peru in the 1990s, tens of thousands of

TABLE 11.2 Population Estimates of Lima–Callao, Peru

	Peru Total Pop. (in millions)	Lima–Callao Estimated Pop. (in millions)	% of Total Pop. in Lima–Callao
1900	3.0	0.150* (estimate)	
1910	4.0		
1920	4.83	0.255	5.3
1930	5.65		
1940	6.68	0.597	7.4
1950	7.97	1.010	13.2
1960	10.02	1.700	17.0
1970	13.45	3.850	21.5
1980	17.30	4.410	25.5
1993	22.1	6.500	

*Keltie, S. (ed.) *Statesman's Year Book*. London: Macmillan, 1905, p. 995.
Source: J. W. Wilke, ed., *Statistical Abstract of Latin America* (Los Angeles: University of California at Los Angeles Press, 1995).

people were infected in the coastal cities, but the death rate was less than 1 percent! Rehydration techniques allowed most people contracting cholera to survive.

In the 1980s, as the population of Peru urbanized, the birth rate came down and population growth slowed from the 3.5 percent in the 1960s to around 2 percent per annum in the 1990s. Population growth is still rapid, but the pace of growth is slowing. With around 70 percent of the population in urban areas, the greatest part of the rural-to-urban migration has taken place, but there are still flows of people from the Andes to the coastal cities.

In the space of a few decades, the distribution of population in Peru has altered. In 1930, two thirds of the population lived in the Andes and approximately one Peruvian in 20 lived in the capital at Lima. Today, the majority of the population lives in the crowded west coast cities of Peru, and one in four Peruvians lives in the Lima–Callao metropolitan area. Many smaller Andean towns and villages have lost population or have aging populations. There is not sufficient labor to run traditional farming activities. Much land in the Peruvian Andes formerly used to grow grains and root crops is now pasture for animals.

▶ *Export-Led Underdevelopment*

Boliva, Peru, and Ecuador have Third World economies. They are dependent on the export of primary products (Peru—copper, silver, iron, oil, and fish meal; Bolivia—tin and zinc; Ecuador—oil, fish, shrimp, and bananas). Prices of primary products fluctuate, markets tend to be glutted from time to time, and commodities are usually exported with little processing or manufacturing that might add value.

Bolivia, Ecuador, and Peru do not have large manufacturing sectors. Bolivia is isolated, and the population lacks purchasing power. The market for manufactured goods in Ecuador is small. Peru has had economic booms based upon commodity exports, which have stimulated export-import trade in the Pacific ports. It might be thought that manufacturing would develop at the ports but Peru has experienced "export-led underdevelopment" (Thorp, 1991, p. 23). As a commodity export boom gets underway, available capital is invested in the successful activity. The exports earn hard currency that is used to pay for imports. Historically, the Peruvian government has used export taxes as a source of revenue rather than place tariffs on imports. The result has been that the Peruvian economy has not diversified into manufacturing. When import-substitution policies were introduced to encourage local manufacturing, illegal, duty-free imports undercut the Peruvian-produced, high-priced product. Coca dollars are now used to finance smuggling. Even where manufacturing has developed, products are basic: textiles, food processing, plastics, oil refining, brewing, cement making, and chemicals.

▶ *Ethnic Groups*

Ethnic and cultural differences within the populations of Peru, Bolivia, and Ecuador are large. In the coastal region of Peru, for example, there are people of Japanese, Chinese, African, and European origins, in addition to the majority, who are of Indian stock. The *sierra* is still dominated by Indian populations speaking Quechua or Aya. Many dialects of Indian languages exist, and groups of villages exhibit distinct speech patterns and ethnic traits. The Indian tribes of the *oriente* are separate from those of the *sierra*; language, house types, clothing, agriculture, and diet are distinct.

In the decades after World War II, as economic development quickened and migration from rural to urban areas accelerated, it was assumed that Indian culture groups would be absorbed into modernizing societies. Indians left traditional villages in the *sierra* and moved to rapidly growing towns in the *costa*, abandoning traditional dress and using Spanish more frequently. It is now clear that traditional Indian life-styles are resistant and persistent, just as they have been since the intrusion of the

The road from Bogotá to the Colombian Oriente.

conquistadores nearly five centuries ago. Many Indian ways have survived migration from traditional villages to expanding urban centers. As migration from rural to urban areas tends to take place in chains established between groups of villages and known urban neighborhoods, the transference of cultural traits is not surprising.

Ethnic, cultural, and racial distinctions make it difficult to develop cohesive national political units in Andean countries. The election of Alberto Fujimori, the son of Japanese migrants, as president of Peru in 1990 showed how an outsider was able to develop support in the provinces and among minorities who felt left out of a political system dominated by an Hispanic elite. As president, Fujimori dissolved Congress, and placed himself at the head of an emergency government to fight the Shining Path (*Sendero Luminoso*), and won reelection to a second term.

The states of Andean America reviewed in this chapter—Colombia, Venezuela, Ecuador, Peru, and Bolivia—are part of the Andean Group of countries. The Andean Group attempts to coordinate economic policy and has agreed to free trade between its members. Efforts to create a common external tariff have failed, and looming over the region are long-standing territorial disputes. Venezuela and Colombia do not agree on their respective rights in the Gulf of Venezuela (Anderson, 1989); Peru and Ecuador have not settled their disputes in Amazonia and skirmished on their boundary in 1994. Bolivia still wants to regain Antofagasta from Chile, and Venezuela claims part of Guyana.

SUMMARY

The prospects for the Andean countries reviewed in this chapter are mixed. Colombia and Venezuela are rich in oil and other resources but have failed to establish the political stability on which all prosperity is ultimately based. Ecuador has been relatively stable and has succeeded in achieving economic growth in recent years. In both Peru and Bolivia, formerly important sectors of the mining industry are in decline. Traditional agriculture in the *sierra* and the Altiplano is running down, and the undoubted resources of the *oriente* regions are not being developed for commercial farming other than the production of coca leaves.

FURTHER READINGS

Bowman, I. *The Andes of Southern Peru.* New York: Henry Holt, 1916.

Kent, R. B. "Geographical Dimensions of the Shining Path Insurgency in Peru." *Geographical Review*, 83(4) (1993): 441.

Knapp, G. *Andean Ecology: Adaptive Dynamics in Ecuador.* Boulder, Colo.: Westview, 1991.

Parsons, J. J. "The Northern Andean Environment." *Mountain Research and Development* 2(1982): 253–262.

Stadel, C. "Altitudinal Belts in the Tropical Andes: Their Ecology and Human Utilization." *Benchmark 1990: Conference of Latin American Geographers*, 17/18 (1992): 45–60.

Weil, C. "Migration Among Landholdings by Bolivian Campesinos." *Geographical Review* 73(1983): 182–197.

Zimmerer, K. S. "Labor Shortage and Crop Diversity in the Southern Peruvian Sierra." *Geographical Review* 81(1991): 414–432.

CHAPTER 12

Brazil

C. GARY LOBB

INTRODUCTION

Brazil has the largest land area of any country in South America, and geologically it represents the core of the continent (Figure 2.17). The Guiana Highlands and the Brazilian Highlands or Plateau consist largely of ancient crystalline rocks that were formerly attached to the ancestral continent of Gowdwanaland before the South American plate broke from Africa and drifted westward, pushing up the Andes in the process.

THE PHYSICAL BACKGROUND

Physiographically, Brazil can be divided into the following regions (Figures 12.1a and b):

1. The Guiana Highlands
2. The Amazon Basin
3. The Brazilian Plateau
4. The Atlantic Coastal Plain

► *The Guiana Highlands*

The Guiana Highlands extend from Brazil into Venezuela, Guiana, Suriname, and French Guiana. In Brazil, the highlands drain into the Amazon system and do not rise to great elevations. The highest peak is Roraima (9,432 feet) in the Serra Pacaraima, which forms the border between Brazil, Venezuela, and Guyana. The country is rugged, having been sculpted into high level erosion surfaces, down which the rivers tumble in waterfalls and rapids. Transportation is difficult, and few resources have been exploited, although the igneous and metamorphic rocks of the region are rich in minerals.

► *The Amazon Basin*

The Amazon contains a greater volume of water than any other river in the world. The Amazon basin is filled with sediments, and the underlying rocks are deep beneath a mantle of alluvium that has been brought down from surrounding

Figure 12.1a Iguaçu Falls in Southern Brazil. One of the country's foremost tourist attractions.

highlands—the Andes, Guiana Highlands, and Brazilian Plateau. The river channels in the Amazon basin are flanked by natural depositional levees that provide higher ground and sites for settlement that usually, but not always, stand above the waters that can inundate the adjacent floodplain in the high-water season.

We think of the Amazon basin as being covered with rain forest, but the region is varied ecologically. Rainfall is not uniform throughout the basin, and in zones with a dry season the forest is replaced by tropical grasslands, except along the water courses. The rivers and adjoining areas contain many ecological niches including the main channels, the levees, flood plains, and old river courses that support lakes and marshes throughout the year. Plant and animal species and human usage vary in the ecological niches.

Mean monthly temperatures in Amazonia vary little throughout the year, although the cloud associated with months of heavy rain can reduce sunshine hours and lower temperatures slightly, as happens at Belém (1° S), near the mouth of the

Amazon in February and March (Figure 2.4). Even at stations close to the equator, rainfall is not evenly distributed throughout the year. At Manaus (3° S), with approximately 70 inches of rain a year, July, August, and September are relatively dry, each receiving around 2 inches of precipitation and having less than 10 rainy days in the month. By contrast, January, February, March, and April average just under 10 inches of rainfall per month, and there are few days without precipitation.

► *The Brazilian Highlands or Plateau*

Geologically the Brazilian Plateau is similar to the Guiana Highlands, being composed of a basement of ancient igneous and metamorphic rocks, overlain by deposits of limestone and sandstone. In places basaltic lava flows cover the surface, as in the Paraná basaltic area, which occupies the south of the plateau. When the basalts break down under weathering, they create the *terra roxa*—the red soils—of the states of Paraná and São Paolo. Spectacular rapids and waterfalls occur as the rivers

Figure 12.1b Brazil: Major physical regions.

drop off the edge of the erosion-resistant basalts. The best known falls are at Iguaçu, on the Brazilian–Argentine border, where the Iguaçu River flows through miles of rapids before plunging to join the Paraná River.

The Atlantic edge of the plateau is exposed in the Great Escarpment, which forms a barrier between the coast and the interior. Few rivers flow from the interior to the coast across the escarpment. A major exception is the São Francisco, which rises north of Rio de Janeiro in Minas Gerais and flows parallel to the escarpment before trending east-southeast and falling toward the Atlantic at the Paulo Afonso Falls, which are now used to generate hydroelectricity. Like a number of South American rivers, the São Francisco has changed direction of flow as the South American plate has shifted away from Africa.

Climates on the Brazilian Plateau are varied. The northern part of the region, abutting the Amazon basin, is wet enough to support rain forest. Southward in the zone of tropical wet-and-dry climates, the dry season is longer, and forest gives way to a mixture of tropical grassland. Rainfall along the Atlantic face of the Great Escarpment is heavy as moist maritime air, coming off the ocean, ascends, cools, forms clouds, and delivers precipitation. The port of Santos receives 80 inches of rain. Inland, on the other side of the escarpment, São Paulo gets 50 inches a year.

Moving westward away from the ocean to the interior of the plateau, rainfall decreases, and there is a marked dry season, as at Cuiabá (Figure 2.13). The city is in the tropics, 15° south of the equator, but there is a drop in temperature in the winter dry season, associated with the northward penetration of modified maritime polar air. This maritime polar air can move eastward in frontal disturbances into the state of São Paulo, where frost damage to coffee plants may result. South of Cuiabá is a region of savanna grassland (the Pantanál), poorly drained by tributaries of the Paraguay. The Pantanal is frequently flooded in the rainy season.

The contrast between coastal and interior rainfall is marked in northeastern Brazil. Recife, on the coast, 8° south of the equator, receives more than 60 inches per annum, although there is a dry season in October, November, and December. Inland in the sertão, or thorn bush region (Caatinga), rainfall averages less than 20 inches and is variable. In some years, only 10 to 15 inches fall, resulting in the loss of crops and livestock. Historically the sertão has suffered devastation by drought and outflows of refugees.

▶ *The Atlantic Coastal Plain*

The plain is not extensive but is best developed from Fortaleza to the coast northeast of Rio (Figure 12.1b). Much of the plain from Natal south to the Serra do Mar was once in rain forest, which has been cleared for sugar and other crops. At Rio de Janeiro the coastal plain is narrow and the old hard rocks of the Brazilian Plateau are spectacularly exposed as in the Sugar Loaf and other outcrops.

EARLY PORTUGUESE SETTLEMENT

Before the Portuguese explorer Pedro Alvares Cabral first sailed along the coast of Brazil, the limits of Portuguese settlement in the New World had been determined. The papal bull of Alexander VI and the subsequent Treaty of Tordesillas (1494) divided the Western Hemisphere between Portugal and Spain. What followed in the Portuguese sphere was a settlement stratagem that favored citizens of Portugal at the expense of Africans and Native Americans.

The Portuguese already had experience in tropical lands. The settlement of Gôa on the Indian subcontinent and especially on tropical Atlantic Islands off the coast of Africa brought the Portuguese into contact with products and production methods of tropical crops, the most important of which was sugar. On island groups such as the Cabo Verde, Canarias, Madeira, Fernando Pó, and Principe the Portuguese mastered the techniques and the social apparatus of sugarcane production. In Europe there was a large and growing market for sugar, which supported the sugar venture on the islands. The labor and race relations that developed on the Atlantic islands had a profound impact on the economy and society of Brazil.

The earliest Portuguese settlements, *feitorias*, were established between 1500 and 1530. They were modest settlements along the tropical coast of Brazil that supported themselves by trading with Amerindian groups. The main product of the coastal forests was *pau-brasil* (*Caesalpinia echinata*), a tree from which a purple dye could be made that was used in the European textile industry. It is from this tree that Brazil takes its name. The initial *feitorias* were unsuccessful, and the Portuguese crown considered abandoning Brazilian settlements to concentrate resources on more profitable ventures in Gôa and the Atlantic islands. However, in 1530 the crown consented to the establishment of a colony at São Vicente on the Brazilian coast east of modern São Paulo, and a new strategy was followed. Sugarcane and production techniques were imported from the Atlantic islands. The initial sugar colony was a success, and the *solução agricola* (agricultural solution) was applied to settlement ventures all along the Brazilian coast.

Of the many sugar settlements three developed into major centers. São Vicente remained an important production region, but two areas in the *Nordeste* (Northeast) developed into leading production and export centers. The Recôncavo, the flat accessible lands of the coastal plain surrounding Todos os Santos Bay, became the most important, with Salvador, the largest city in the region, serving as the colonial capital of the Viceroyalty of Brazil until 1763. The third sugar area was the Pernambuco region northeast of the Recôncavo (Figures 12.2 and 12.3).

From these sugar-producing areas thousands of tons of *rapadura* (semirefined sugar) were shipped to Portugal and northwestern Europe. Relying at first on Amerindians as a source of labor, the Portuguese brought death to tens of thousands of tribal people from European diseases and severe working conditions. Confronted by labor scarcity, the *fazenda* owners turned to Africa, buying enslaved Africans to work in the Brazilian sugar regions. The first ships carrying enslaved Africans arrived at São Vicente in 1530, and by the time slavery ended in Brazil in 1888, over 3.6 million had been imported (Table 10.3). These black immigrants are an important component of the modern Brazilian population with African culture, including cuisine, religion, and language, widely dispersed throughout the country, although the greatest concentrations of African Brazilians are in the colonial sugar areas of the tropical coast.

The society of colonial Brazil reflected the quasi-feudal structure of the *fazenda de açucar* (sugar plantation). At the top of the socioeconomic hierarchy were the *senhores do engenho*, the landowning families, who operated the sugar plantations and the *engenhos* (sugar mills). Their lands had been obtained by royal grants called *sesmarias*. None of the colonists, who had emigrated from Portugal to better themselves, intended to lead the humble life of a small peasant-proprietor in Brazil. *Sesmarias* had been granted in Portugal since the thirteenth century when they were used as a means of breaking up the holdings of the wealthy, landed families. It is ironic that this instrument of divestiture became the means of establishing a landed elite in Brazil known as *fidalgos*.

A land use system emerged based on the cultivation of commercial sugar on plantations. The three characteristics of the system—the large estate, monoculture, and slavery—complemented one another. Characteristically each estate employed between 150 and 200 slaves. A sixteenth-century account comments that a

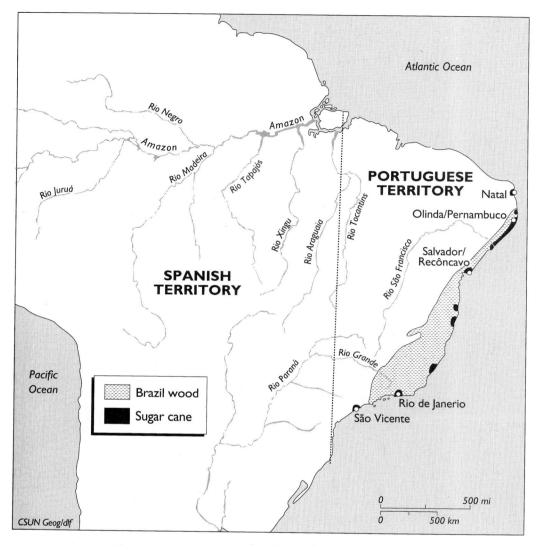

Figure 12.2   Brazil: Economic activity in the sixteenth century.

"plantation owner that employed fewer than eighty slaves is deemed a very poor *senhor do engenho*" (Vilhena, 1927). Those who could not qualify for *sesmaria* grants occupied less prestigious and less lucrative socioeconomic positions. *Lavradores* (sharecroppers) leased land from the *senhores do engenho*, kept slaves (rarely more than 15) and cultivated sugar. Few of the *lavradores* owned an *engenho,* and therefore they depended upon the plantation mill to crush cane. Lower in rank were the *moradores*, who maintained a small subsistence plot on the plantation and were of little economic value to the *senhor do engenho* except to reinforce his position as manor lord. At the bottom of the hierarchy were the slaves, who had few opportunities for upward mobility.

During the first 150 years of Brazil's history, settlement spread along the tropical Atlantic coast in a region known as the *zona da mata*. The rain forests were replaced with sugar plantations. Large amounts of firewood were needed to heat the cauldrons of the sugar mills, and coastal forests were cut for that purpose, too.

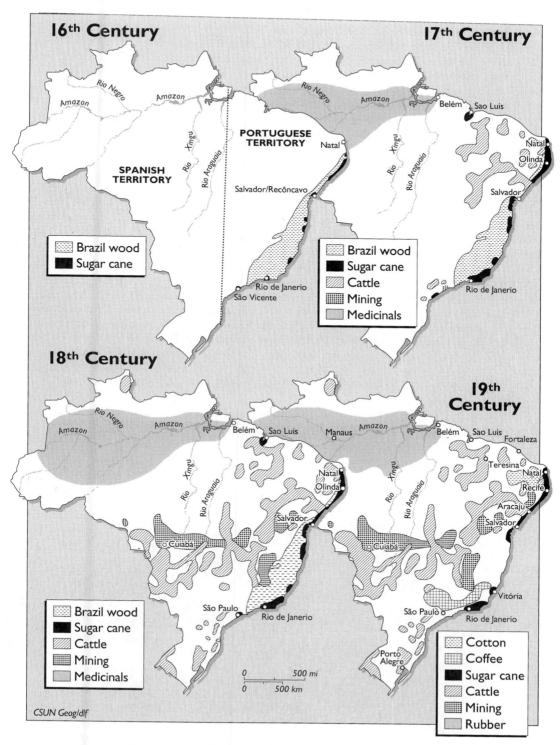

Figure 12.3 Brazil: Economic activity.

Emanating from population centers such as Pernambuco, the Recôncavo, and the original settlement of São Vincente, population occupied a coastal zone from Belém at the mouth of the Amazon River to São Vincente in the south (Figure 12.2).

Settlement of the interior consisted of occasional ranches in the São Francisco River basin and scattered mission settlements among the Tupi, Guaraní, and Amazonian Indians. One mission, São Paulo, founded 1552, became the base for expeditions into the interior during the seventeenth century. São Paulo became a large settlement of mixed Europeans and Amerindians, a rare combination in Brazil where most racial mixing involved Europeans and Africans. This mixed population, known as *mamelucos*, was responsible for the exploration and settlement of much of the interior of the country.

Organized into paramilitary bands called *bandeiras*, the *mamelucos* raided Jesuit mission communities in southern Brazil and Paraguay to capture Amerindians to sell to coastal planters. Some *bandeiras* penetrated deeply into the interior in search of Amerindians, gold, and precious stones. Gold was discovered in the 1690s in the Serra do Espinhaço in what is today the state of Minas Gerais. The gold frenzy that followed shifted population from the coast into the mining regions, and migration was accentuated by the decline of sugar production due to competition from the Caribbean islands.

Migration into the mining area came from the depressed coast and from Portugal. Associated with the mining boom was the development of the city of Rio de Janeiro, which the crown designated the chief export center for the "royal fifth," which was collected as a government tax. A highway was constructed from the mining region to Rio de Janeiro in 1705, and the modest town became a major port. Rio's location on Guanabara Bay was well suited for maritime commerce. The city became Brazil's largest port and, for a time, the major metropolis. The headquarters of the viceroyalty was moved here in 1763, by which time Rio's 100,000 population easily outnumbered Salvador's. Rio de Janeiro remained the capital of Brazil until relocation to Brasília in 1960.

Large bays, such as Guanabara Bay, are important features of the Brazilian coast. Historically (and today) they have contributed to the Brazilian economy. The early colonial capital and center of the sugar industry, Salvador, is located on the

Rio de Janeiro.

largest bay, called Todos os Santos. Southward along the coast is Vitória Bay, which has become one of the world's most efficient iron ore ports. South of Guanabara Bay is São Vicente Bay, the site of the first Portuguese colony. Most of São Paulo's industrial exports pass through the port of Santos on the bay. The southernmost bay, Paranaguá, is a leading center of coffee exports.

By the early eighteenth century the gold deposits of Minas Gerais had been largely mined out. *Mineiros*, migrants from the mining region, moved into nearby areas and began to farm. One important destination was the *terra roxa* area of northern São Paulo state, a region named for the red soils. The soils are not exceptionally fertile but could be converted into subsistence and small-scale commercial agricultural plots. These soils would support the early coffee industry, which developed in the nineteenth century. *Mineiros* also settled in the Rio Doce and Rio Paraíba valleys to the east and south of Minas Gerais, respectively. Both of these river valleys became important agricultural regions with the lower Rio Doce valley serving as one of modern Brazil's most important sugar-producing zones (Figure 12.3).

In the grasslands of southern Brazil, outside the tropics, land use was based on cattle ranching. Beginning in the 1730s large *sesmaria* land grants were given to Portuguese applicants for the purpose of raising cattle on the grasslands of the *campanha* (southern Rio Grande do Sul). In many cases the ranches could be started with feral cattle abandoned on the *campanha* by Jesuit Amerindian communities during the *bandeira* raids. The *estancias de gado* (ranches) produced hides, tallow, and dried salted beef (*charque*) for widespread markets, and there was a significant export of Brazilian *charque* to the United States during the late eighteenth century. Fresh meat sales were local before refrigeration.

INDEPENDENCE AND NINETEENTH-CENTURY LAND USE

Unlike most of Latin America, Brazil became an independent nation without a revolution. During the Napoleonic wars, with British assistance, the Portuguese crown was relocated to Brazil in 1808 to avoid capture by the French. With the headquarters of the Portuguese empire in Rio de Janeiro and the population and

Ouro Prêto, Brazil.

economic importance of Brazil having surpassed that of Portugal, the balance of power shifted to the west side of the Atlantic. The final break came in 1822 when Pedro I declared Brazil free and independent of Portugal. There was little resistance. Some of the garrisons were activated by the Portuguese *Cortes,* but most defected to the newly proclaimed Brazilian monarchy. (The monarchy persisted in Brazil until 1889.) At the time of *emancipacão* (independence) Brazil had a population of nearly 4 million people, twice that of Portugal. The agricultural economy was becoming more diversified. Sugar was cultivated in the Rio Doce valley, the old sugar lands of the *zona da mata,* and the *terra roxa* soils of São Paulo state. Cotton was an important export crop in the *meio norte* of the north coast, and in southern Brazil ranching dominated. Coffee emerged as an export crop in the Paraíba Valley, north of Rio de Janeiro, during the 1790s.

To appreciate the development of Brazil's economy and society during the nineteenth and twentieth centuries it is necessary to understand the legacy of the colonial period. The census of 1800 reveals that among the almost 4 million population was a relatively small number (300,000) of individuals of European descent (*criolhos*). This white or quasi-white group were the promoters and beneficiaries of the colonial system. The *criolhos* received landholdings (*latifundia*) through royal grants and had large numbers of slaves and poorly paid laborers working their fields. The great mass of the population operated small subsistence agricultural plots (*minifundia*). In fact the subsistence sector of the agricultural economy grew rapidly during the colonial period. Some subsistence farmers adopted Amerindian systems of production such as *roça,* emphasizing manioc, sweet potatoes, peanuts, and maize. Others, in nontropical regions, cultivated European crops such as wheat. These usually landless peasant farmers, known as *caboclos,* make up a large segment of the modern Brazilian population and can be found from the Amazon basin to the temperate lands of Rio Grande do Sul.

Colonial Brazil produced commodities for export, and the country continued this role after independence. There was no social revolution—indeed no revolution at all—to break up the commodity production monopolies. The colonial momentum was hard to reverse, and the problems of *latifundismo* and *minifundismo* were perpetuated as Brazil became independent.

Two Brazils emerged at the beginning of the nineteenth century. The southern third of the country enjoyed economic growth associated with the cultivation of coffee and the cattle industry, and the rest of the country, especially the very poor *Nordeste,* suffered depression. Many of the problems of the *Nordeste* were associated with droughts in the *sertão.* Just before the beginning of the nineteenth century the "great drought" (*seca grande*) drove population from the area to cities along the coast and to southern Brazil. It was during the *seca grande* that the term *flagelados,* meaning whipped or beaten ones, was coined for the victims of the drought. This term aptly describes the condition of the refugees from the droughts. With modern industrialization having come to southern Brazil and recurrent droughts causing problems in the *Nordeste,* the split between north and south is more evident than ever.

The nineteenth century was the beginning of an economic cycle involving coffee. Coffee had been drunk in the Arab world for centuries before it became common in Europe. Coffee houses, or Java houses as they were known, sprang up all over Europe after 1650. The main source of supply was Indonesia. In the 1790s some plants were introduced into Brazil from Africa, where it was known by its Arabic name *kahveh.* First planted in the Paraíba Valley, production spread to the plateau surface around the city of São Paulo. Here on the *terra roxa* soils cultivation spread rapidly and the small towns of São Carlos, Riberão Prêto, São José do Rio Prêto, and Campinas, established earlier by *mineiros,* became regional centers.

Campinas was Brazil's fastest growing city during the last half of the nineteenth century. São Paulo also grew rapidly as the center for the entire coffee region.

Coffee production improved over the years, but the methods of cultivation remained exploitative. Soils were cultivated until exhausted and then abandoned, usually after 30 years when the *arabica* shrubs came to the end of their life. Planters then moved westward into new production areas. The abandoned coffee lands of the "old zone" (*zona velha*) or the "hollow frontier" were eventually colonized by immigrant groups such as Italians and Japanese.

Coffee cultivation requires a large available labor force. While there is a certain season of harvesting the berries, they ripen throughout the year and must be picked at the proper time. Slave labor was used during the early period, but as production expanded, there were not sufficient slaves in the region to perform the work, and slavery was abolished in 1888. The imperial government, along with the state government of São Paulo, devised a program to recruit Europeans to work as contract laborers. The *falta de braços* (literally "lack of arms") program organized immigration companies that advertised for workers and subsidized their passage to Brazil. More than 1 million Europeans migrated to Brazil between 1887 and 1900. Most of the immigrants were Italian and settled in the state of São Paulo. Of the 140,000 who immigrated in the peak year of 1895, 133,000 were Italian. The large foreign-born population in the state of São Paulo made it the most cosmopolitan, progressive, and prosperous state, accentuating the gulf between northern and southern Brazil. By 1910, Rio de Janeiro, São Paulo, and Rio Grande do Sul accounted for 71 percent of Brazil's gross domestic product. With a large unemployed and underemployed population in the *Nordeste* the imperial government chose to recruit labor in Europe rather than bring workers from the north.

The infrastructure associated with the coffee industry in the *zona velha* of São Paulo state consisted of a railroad system, largely built with capital from the United Kingdom. The Paulista line, constructed in 1867, tied the coffee towns of São José do Rio Prêto, Riberão Prêto, and Campinas to São Paulo and the port of Santos. From Santos coffee was exported throughout the world. All the cities along the rail line, especially Santos, grew rapidly in population as the coffee frontier pushed westward. The Sorocabana line, built in 1900 to service the coffee land of western São Paulo state, connected with the Paulista line in São Paulo and provided access to the sea through Santos.

In 1925 the coffee frontier expanded into neighboring Paraná state. This *zona nova* rapidly became the fastest-growing coffee region in the country. The Ouro Verde rail line, built by the British Paraná Plantations, Ltd., claimed rights to lands adjacent to the railroad. These lands (*loteamentos*) were subdivided and sold as large coffee estates. The Ouro Verde line linked the new centers of the *zona nova*. From Maringá, founded in 1946, the line ran through Apucarana (founded in 1940) and Londrina (founded in 1936) to connect with the Sorocabana. All of these cities, especially Londrina, have grown to be large and important places. As the *zona nova* developed, the coffee frontier pushed toward the southwest and was exposed to outbreaks of Antarctic air associated with the polar front of the Southern Hemisphere. *Friagem*, blasts of cold air during the month of July, are a threat to the coffee crops of the *zona nova*. Crops can be damaged and coffee shrubs killed by prolonged exposure. As a result, coffee has been replaced by soybeans, a crop that is sown in the spring and harvested in the fall before the danger of frosts arrives. Coffee shrubs, of course, have to survive from year to year and cannot resist frost.

Immigration has always been important to the development of Brazilian culture and society. Brazil is a cornucopia of cultures and ethnicities. Portuguese immigration in particular has proceeded at a steady rate. In addition to the massive Portuguese, African, and Italian immigration, German and Japanese settlement has had a significant impact on the country.

German settlement during the first half of the nineteenth century was concentrated in Rio Grande do Sul and Santa Catarina. The imperial government gave German settlers free passage and free land to establish small family-sized farms on about 200 acres, called *datas da terra*. These were distributed within a region known even today as the *zona colonial*, lands in the northern portion of Rio Grande do Sul that had not been distributed to ranchers through *sesmaria* grants. Rich soils in the zone have developed from volcanic materials, and in many places the terrain is flat. São Leopoldo, established in 1824, was the first of the German settlements followed by Novo Hamburgo. Joinville and Blumenau in Santa Catarina state were settled in 1849 and 1850, respectively. Italian settlements, not associated with the *falta dabraços* program, began in the *zona colonial* after 1870, but Italian settlers found available lands only in the steeper and more remote areas and had to accept *datas da terra* that had been reduced in size to around 65 acres. The mystique of the *zona colonial* lives on in modern Brazil. Agricultural products and meat, especially pork, from the *zona colonial* are considered to be of a higher quality and bring a higher price than products from other areas. Precision machine tools and other industrial products made in the Joinville–Blumenau region of Santa Catarina are highly regarded.

Japanese immigration began in 1913 with a small group settling in Iguape Registro near São Paulo. Successful in the reclamation of hollow frontier land, the Japanese settlers prospered growing tea and producing silk. The reclamation of the worn-out hollow frontier lands impressed the state government, and in 1917 the state of São Paulo entered into an agreement with a Japanese emigration company, K.K.K.K. (Kaigai Kogyo Kabushiki Kaisha) to transport settlers to the state. Their expenses were paid, and they were encouraged to settle on the abandoned coffee land of the hollow frontier. More than 200,000 came in all with 25,000 during the peak year of 1933. Using compost and horticultural techniques, the Japanese produced crops for sale in urban centers such as São Paulo and Campinas, and Curitiba in the state of Paraná. Although discriminated against at first, the Japanese are now regarded as having made a significant contribution to Brazilian national life. The *nikkei*, third-generation Japanese-Brazilians, number more than 1 million. They are found in the the national congress and at all levels of business and industry. Brazilians of Asian origin have the highest incomes of all ethnic groups.

STRUCTURAL ASPECTS OF THE POLITICAL ECONOMY

Brazil ceased to be a monarchy in 1889 and established a republic based on the Constitution of 1891. The presidents of the new republic allowed the military to meddle, and the economy was sluggish. The economic stagnation of the late 1920s, linked to worldwide depression, created a crisis that changed the political order. Declining coffee prices, plus a loss of the rubber market to Southeast Asia, paved the way for military intervention that brought to power Getúlio Vargas, a powerful political figure from the state of Rio Grande do Sul. Vargas ruled Brazil as a dictator from 1930 until 1945. During that time the economy revived, and Vargas established Brazilian state capitalism. Under the slogan of *Estado Nôvo* (new state), state funds were invested in basic industries. In the state of Minas Gerais were large deposits of iron ore, which had been known since the gold-mining period. The deposits were extensive and of high quality, containing 69 percent iron. United States and British investors organized the Itabira Iron Ore Company in 1911 and began mining the rich ore deposits. The ore was exported via a railroad, which had been built in 1903 connecting the mines with the port city of Vitória. However, there was little successful exportation until the *Estado Nôvo* was declared in 1937. The Itabira Iron Ore Company became the Companhia Brasileira de Mineração e Siderúrgica. Reorganized again in 1942 with 80 percent government ownership, it became Companhia Vale do Rio Doce (C.V.R.D.), now Brazil's largest enterprise with investments throughout the country.

The 15 years of Vargas's tenure saw changes in his philosophy of government. Coming into power as a conservative "strong man," he gradually became more democratic, allowing elections in 1945. Not a candidate himself, he stayed out of politics but ran again for the presidency in 1950 on the Socialist-Labor ticket as an advocate for state ownership of the means of production and rights for the working classes. Elected, he encountered political problems that took a toll, and he committed suicide in 1954.

Free elections were held in 1955 and Juscelino Kubitschek was elected. Committed to the "structuralist" approach of Vargas, Kubitschek, a second-generation Polish-Brazilian, continued the policy of utilizing state capital to develop an industrial base. Structuralist theoreticians such as Raúl Prebisch of Argentina, Celso Furtado of Brazil, and Victór Urquides of Mexico convinced many Latin American heads of state that development required industrialization. Prebisch argued that as long as Latin American countries produced only the world's "desserts"—coffee, sugar, cacao, bananas—they could never diversify their economies. Kubitschek continued government funding of industrial enterprises and started Brazil toward large-scale industrialization. The policy resulted in the need to borrow money from international sources.

With a mandate from the 1891 Constitution, Kubitschek started construction on the new national capital of Brasília in 1957, planned by Brazilian architect Lucio

Brasília 1961, showing aspects of the original plan. Today the margins of the city are filled with shanty towns.

Costa. The city plan, from the air, resembled a great plane with swept-back wings. The wings comprise residential neighborhoods, and the monumental axis, the fuse-lage of the airplane, houses the government buildings. Planned for a population of 500,000, the city was to have no central business district. Each of the residential apartment blocks (*super quadras*) was to have a nucleus of small shops and markets to satisfy the needs of the residents. On April 21, 1960, the government officially transferred to Brasília. A university opened in 1962. A number of significant build-ings were designed by Oscar Niemeyer, an internationally known Brazilian archi-tect. The city has grown to more than 1 million people and has spread beyond Costa's original plan. Two large shopping centers in the central areas of the city are another departure from the original design.

> *Away from the center are the inevitable shanty towns. Brasilia, the purpose-built capital . . . seems gripped by something very much like apartheid. The Federal officials and bureaucrats who run the country—the vast majority of them white—live in apartment blocks within the city's modernistic central area. Other workers, forbidden to mar Brasilia's architectural purity, live miles away in ragged settlements that look almost like South African townships. (Robinson, 1995)*

Few observers of Brazil in the 1930s would have predicted that the country would be so economically diversified today. In addition to the state-controlled iron and steel industry are the durable consumer goods industries, another pillar of the industrialization program. The object was to promote import substitution, and high tariffs were adopted to protect locally manufactured goods. For example, the tariff on cars and appliances were placed at 85 percent of value to discourage imports and pro-tect local industry. Furthermore, the government required that automobiles, as well as appliances, be manufactured of 78 percent Brazilian parts. The policy kept prices arti-ficially high. For example, appliances are generally 50 percent to 275 percent more expensive than in the United States. Inflation was considered tolerable as long as wages and prices were indexed. The policy, known as the Propagation Mechanism, has gradually been abandoned as Brazil has moved toward a free market economy since 1990.

The automobile and truck industry, started during the Kubitschek years, is Brazil's largest. It is the biggest source of foreign investment (General Motors, Fiat, Ford-Volkswagen, Mercedes Benz, and Scania Vabis), as well as the largest source of manufacturing export income. Automobiles and auto parts represented 15 per-cent of total exports in 1988 and were valued at more than $4 billion dollars. The Volkswagen Fox and Passat are examples of Brazilian cars that have sold well over-seas. Industrial momentum began to build in the late 1950s. In the period from 1952 to1966 the city of São Paulo averaged a growth of over 17 percent per year in production of capital and durable goods. The gross domestic product of the state of São Paulo alone, Brazil's center of industry, is $75 billion—greater than that of any Latin American country except Mexico. While the automobile and truck industry is largely privately owned, four of Brazil's ten largest companies are state-owned.

The election of Jânio Quadros in 1960 by the largest plurality in history assured a continuation of the structuralist program. However, faced with the prob-lems of foreign debt and inflation, Quadros resigned after only seven months in office. He was succeeded by the leftist João Goulart. During his three years in of-fice he nationalized additional companies and reestablished ties with Castro's Cuba. He was overthrown by a military coup (*golpe*) on March 31, 1964. For 22 years the generals ruled Brazil. This was a time of massive violations of the civil rights of those citizens who opposed the regime. Houses were searched without warrants,

Brazilians were detained without appearing in court to face charges, and many simply disappeared. The media were censored.

There were three major political parties in Brazil at the time of the *golpe* in 1964: Social Democratic party (PDS), National Democratic Union (UDN), which had support from the growing middle classes, and the Brazilian Labor party, Vargas's old party and the party of Goulart. When they seized power in 1964, the generals outlawed all existing parties, and the National Renovating Alliance (ARENA) was created to represent the military leadership. The military created the Brazilian Democratic Movement (MDB), a puppet opposition. All other political dissent was repressed.

Economically the military let civilian professionals run business and industry and in a period of global expansion Brazil enjoyed a "Golden Age" of industrialization and economic growth (1964–1974). Throughout most of the period, the economy grew at around 9 percent per year. Expansion was brought to an end by the oil crisis of 1974 and rising foreign debt.

THE MODERN POLITICAL ECONOMY

Although the oil shock of 1974 slowed down its growth, the Brazilian economy has remained the largest in Latin America and the eighth largest in the world. Slow rates of growth and severe recession from 1981 to 1983 have been counterbalanced by rapid expansion of exports. In the period from 1983 to 1988 industrial output soared; steel production rose by 50 percent, coal 35 percent, aluminum 180 percent, and from a low base, oil production increased by 200 percent. The country has maintained a large export surplus since 1983. In 1988, the surplus was up 33 percent from 1986 and reached $11.2 billion, third highest after Japan and Germany. Brazil is now entering a difficult stage of industrialization and must manufacture intermediate industrial products and capital goods. To accomplish this, the government outlawed certain imports altogether, including computers, to protect local manufacturers. The policy was viewed with outrage by United States and Japanese computer manufacturers. State intervention has produced problems. The 2,200 state owned industries, most created under the generals, generate large deficits and employ over 8 million people. Hundreds of these enterprises are so debt ridden that they cannot survive without government subsidies. In a recent spate of privatization, these as well as profitable government-owned businesses have been sold to investors.

Automobile manufacturers are Brazil's largest private enterprises. With the exception of Fiat, located near Belo Horizonte, the automobile industry is concentrated in the southern suburbs of São Paulo. The so-called ABC towns of Santo Andrés, São Bernardo do Compo, and São Caetano are centers of the automotive and related industries. There are presently more than 18 million Brazilian-made cars on the road, one car for every fourteen Brazilians. The industry has the capacity to produce 1 million cars per year, a number that has been constant for the last several years. The largest manufacturer, Autolatina, is a Ford Motor–Volkswagen venture that holds over 50 percent of the market.

Ethanol-burning cars are an important facet of Brazilian life. One third of Brazil's cars run on ethanol with 80 percent of the 1990 models being ethanol-burning. The 4.5 million ethanol car owners are not entirely delighted. At first there were mechanical problems, but since 1988 the "second-generation" ethanol engines are said to be more reliable. Nevertheless, ethanol burning cars have a relatively low resale value. Operated by the government, the PROALCOOL program produces ethanol from sugarcane. Production and retailing have been erratic, resulting in long lines of cars at the pumps with no availability at all in some rural areas. When the program was initiated in 1975, the government agreed to keep the cost at 65 percent of that for gasoline. It is now common for ethanol to sell for 25 percent more than gasoline. The country is even importing ethanol now, because as subsidies were lowered for sugarcane cultivators, growers switched to soybeans or other more lucrative crops.

Brazil's mining and heavy industry are largely operated by the federal government. The largest enterprise in the country is the profitable Companhia Vale do Rio Doce, 80 percent owned by the Brazilian government. C.V.R.D. operates the large mines in the Itabira or southern zone. Reserves of iron ore in this zone are over 6 billion metric tons. Most of the production from Itabira is moved by rail to the Atlantic port of Tubarão, near Vitória. Over 70 percent of the production is sold overseas and transported on ships of the C.V.R.D. The ore facilities opened in 1966 and load 80 million metric tons a year. The iron is pelletized at a plant opened in 1970. The route along the Itabira–Tubarão railroad is heavily industrialized with steel mills such as Belgo-Mineira, Acesita, Açominas, Companhia Siderúrgica de Tubarão, and the new, profitable Usiminas plant. Brazil's largest steel facility is the Getúlio Vargas mill at Volta Redonda. The location was scientifically selected, midway between São Paulo and Rio de Janeiro in one direction and midway between the iron mines of Itabira and the port of Santos in the north–south axis. It is the largest integrated iron and steel mill in Latin America. Build in 1942 as part of Vargas's *Estado Nôvo*, capacity was doubled in 1970. Another concentration of heavy industry is located at Cubatão in the state of São Paulo. Here 23 large industries including COSIPA, a government steel mill, a PETROBRAS refinery, copper smelter, and others are lined up in a narrow valley between São Paulo and Santos. Residents call it the "Valley of Death," and it is said to be Brazil's most polluted environment.

A second, and potentially even larger, C.V.R.D. project is under development in the Carajás region of the eastern Amazon basin, where there are billions of tons of high-grade (66 percent iron) hematite. The overall development, called O *Projecto Grande Carajás*, comprising over 900,000 square kilometers (more than France and the United Kingdom combined), includes a railroad built in 1985 from the mines to the specialized iron ore port, Ponta da Madeira, near São Luis, Maranhão. Already more than 25 million metric tons of ore are being exported annually to Japan, the United States, Germany, and Italy. Associated with the project is the massive Tucuruí Dam, which will generate 8,000 megawatts in its first phase. The power is to be used to run the mines and to support industrial development in São Luis, Belém, and nearby Marabá. The controversial second facet of the project calls for the clearing of over 3 million *hectares* of tropical rain forest land to be used for agriculture and ranching. Squatters are able to obtain title to three times the amount of land they clear, up to 3,000 hectares. The third phase involves the maintenance of eucalyptus tree farms to provide charcoal for local small-scale industries. Those who support the Carajás projects are referring to it as the "fulcrum for a new Amazon."

In addition to the government-operated iron ore projects, there are other large mining developments involving manganese in Córregao do Azul, in the Carajás, and the Navio Mountains in the territory of Amapá. One of the world's largest tin mines is located in the Amazon 155 miles north of Manaus. The Companhia Vale do Rio Doce operates a bauxite (aluminum ore) mine at Trombetas east of Manaus. This mine is thought to have the third largest reserves in the world and supplies the Barcarena complex near Belém, a center of aluminum production (Figure 12.4).

With the exception of by-products from the pits of Itabira and another mine at Fazenda Brasileiro, gold is mined on a small scale very much the way it was during the colonial gold cycle. The pit at Serra Pelada near the Carajás is worked by as

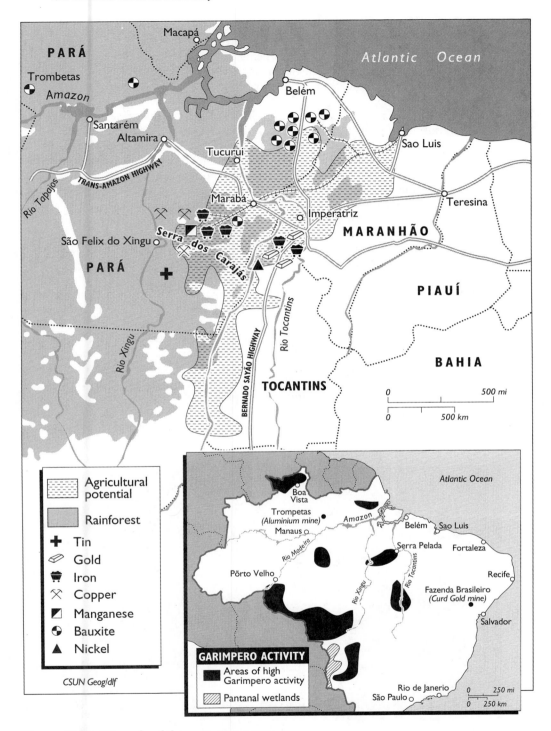

Figure 12.4 Minerals of the eastern Amazon.

many as 30,000 independent miners or *garimpeiros*. A discovery in the state of Roraima, near the Venezuela border, produced an influx of *garimpeiros*, which created conflict with local Amerindian groups whose lands were invaded and forest cleared. The problem has been most acute in the case of the Yanomani Indians. A number of Indians have been killed and the introduction of malaria is taking a toll. Mercury, used in mining operations, is being introduced into the streams, and fish are being killed in large numbers. According to John Hemming, "many Indian villages have seen their population drastically reduced, and some have been wiped out completely" (1990, p. 38).

The *Fundação Nacional do Indio* (FUNAI) is understaffed and some say undercommitted to protecting the rights and lands of Amerindians. On the urging of FUNAI hundreds of clandestine air strips in the Roraima region have been destroyed to prevent penetration by *garimpeiros*. Amerindian lands are supposed to be protected by FUNAI, but tribal lands are being encroached upon by farmers, ranchers, and miners. The typical attitude of the ranchers and farmers is that the Amerindians are lazy and have done little to improve their lands. There is always pressure to infiltrate Amerindian lands. Agencies such as FUNAI and CIM (*Conferencia Indigena Missional*), an arm of the Catholic Church, are having little success in isolating Amerindians from settlers and newly introduced diseases.

Until recently it seemed that Brazil had little petroleum. A national oil monopoly was created in 1953, mainly for the purpose of importing foreign oil and refining it in Brazil. PETROBRAS (*Petrolios Brasileiros*) was the government's energy arm and, at one time, a major oil importer. PETROBRAS's role as oil importer has changed since Brazil initiated the ethanol program for automobiles and discovered large oil reserves within the country. The first discoveries were in the Recôncavo and in adjacent offshore areas. Within recent years there has been another major discovery in the Urucu region of the western Amazon, 400 miles southwest of Manaus. Local oil now accounts for 60 percent of Brazil's consumption (Figure 12.5).

ELETROBRAS is the government enterprise responsible for the development and transmission of hydroelectric energy. Beginning with small projects, which are still in use, on the western slope of the southern plateau, the company expanded to

Brazil: Plant producing ethanol.

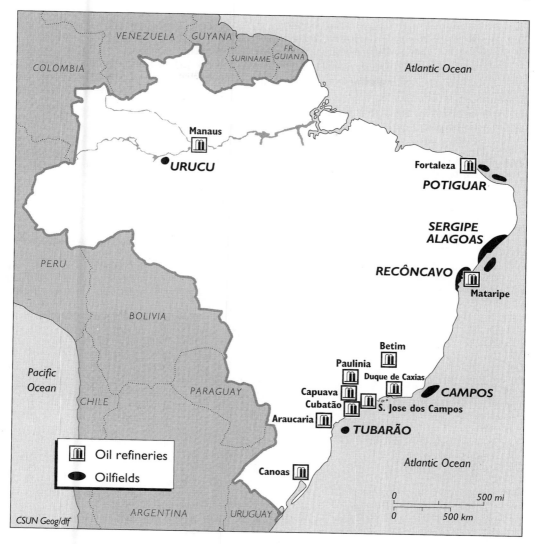

Figure 12.5 Brazil: Oil fields and refineries.

larger projects such as Paulo Afonso Dam on the São Francisco River in northeastern Brazil in the 1960s. The world's largest hydroelectrical project, Itaipu, came on line in 1983. Built on the Paraná River along the Paraguay–Brazil border, its electrical power is shared by the two countries. The project cost $16 billion to build and will supply Brazil with one fifth of its total electrical power. The Tucuruí Dam on the Tocantins River in the central Amazon, built by ELETRONORTE, an ELETROBRAS subsidiary, is linked to the massive Carajás project. Opened in 1985 with a 3.3 million kilowatt capacity, it is being expanded. A recent addition is the Balbinas project east of Manaus, completed in 1989. Throughout the country there has been an 11 percent per year increase in hydroelectric energy use since 1961 and the country has an impressive national grid. There are plans for an additional 25 dams, but the 6.3 million kilowatt Cararão Dam project on the Xingu River in the central Amazon was shelved. The huge foreign debt and sensitivity to Amerindian lands, which would have been flooded, were given as reasons. It may also be that the country has a surfeit of electrical energy and the hydroelectric facilities are having a difficult time selling cheap power and avoiding financial deficits (Figure 12.6).

In light of these surpluses, a plan to build 63 small nuclear power plants is in limbo. One plant, Angra I, came on line in 1985, seven years late and at a cost of $1.5 billion. There is a considerable discussion within and outside Brazil regarding the country's ability to build a nuclear weapon. NUCLEBRAS, the agency in charge of nuclear policy and power plants, has been scaled down.

The state-run manufacturer of airplanes, EMBRAER, has been successful in exporting products to a number of countries including the United States. The Xavante fighter bomber is the mainstay of the Brazilian Air Force and is exported to other Latin American nations. Brazil has become an important producer of munitions, including longrange missiles (Sonda IV) and armored personnel carriers. Even

Figure 12.6 Brazil: Hydroelectric power stations.

more successful are the passenger planes, the Bandeirante and the larger Brasília. Widely used in commuter fleets, they are in commission in the United States, Canada, and Europe.

While there are still millions of subsistence *caboclo* farmers in Brazil, there are also millions of hectares of commercially grown agricultural crops. In addition to sugarcane and coffee plantations there are many fields of traditional commercial crops such as cacao, cotton, wheat, and corn. More recent developments include orange orchards in the Paraíba Valley and elsewhere for orange juice concentrate, which is exported throughout the world. Brazil is a net exporter of food, depending on coffee for only 10 percent of exports.

Brazil is a major producer and exporter of soybeans, which now surpass coffee in acreage and value. Soybean production began in Rio Grande do Sul and diffused northwestward through western Santa Catarina, and Paraná, eventually spreading through the states of Mato Grosso do Sul and Mato Grosso, where soybeans are being cultivated in an area known as the Cerrado. The Cerrado was little used for agriculture before the 1970s, but there are 6 million hectares under cultivation producing 6 million metric tons of soybeans per year. Because this is a tropical region, there are problems with soil quality. The soils are poor; they are acidic, deficient in cations, devoid of phosphate, and exhibit aluminum toxicity. Yet, with careful management and the prudent use of fertilizer, phosphates, limestone, and legumes as green manure, these stunted tropical rain forest environments have become highly productive.

Brazilian soybean production in the Cerrado region averages 2 metric tons per hectare. In Iowa and Illinois, considered the most productive soybean area in the world, production averages 2.3 metric tons per hectare. The success of the Cerrado resulted from the application of agricultural research. There are now 15,000 agricultural agents and 30,000 salespersons of seeds and fertilizer who give advice to

Manaus on the Rio Negro near the confluence with the Amazon. Marked changes in river level require the use of floating quays for some vessels.

farmers. An influx of farmers from the south who were used to operating machinery and cultivating soybeans helped the program. The Brazilian government seems to be responding to the success in the Cerrado by de-emphasizing efforts in the Amazon. Climate, soils, and infrastructure, though not ideal, are better in the Cerrado that they are in the Amazon Basin.

In the heart of the Amazon Basin, at the confluence of the Rio Negro and the Amazon, the city of Manaus was the rubber capital of the country in the late nineteenth century. Today, Manaus is thriving in spite of its isolation from the rest of the country. (It is closer by road to Caracas, Venezuela, than it is to Brasília.) Manaus was designated a freetrade zone in 1966, and since that time it has been the fastest growing city in Brazil. It is now the second largest manufacturing center after São Paulo with more than 600 factories, most of them in a large industrial park on the eastern outskirts of the city. Honda, Kodak, Olivetti, Sanyo, Sony, Seiko, Toshiba, 3M, and Yamaha are a few of the international companies with plants there. Manaus has become an electronics bazaar attracting Brazilian tourists from all over the country to buy duty-free goods. Although only 5 percent of local production is sold duty-free to tourists, the city is sometimes overwhelmed by visitors. Special provision must be made at the airport and on board jetliners so that 21-inch television sets can be taken home to São Paulo. With 1.3 million inhabitants, Manaus is the twelfth largest city in Brazil, but it ranks eighth in generated sales tax.

POLITICAL AND SOCIAL GEOGRAPHY

Brazil, with a gross domestic product approaching $300 billion, has the biggest economy in Latin America. It also has the world's largest inequities of wealth distribution. Inequities are obvious in large cities such as São Paulo, Rio de Janeiro, Salvador, Belo Horizonte, and Recife. Huge squatter settlements or slums have developed around and within Brazilian cities, and millions of people live in substandard housing. The squatments, *mocambos* in the Northeast and *favelas* in the South, are home to hundreds of thousands of migrants from rural areas, especially from the *sertão* of the Northeast. In the case of the *favelas* of São Paulo and Rio de Janeiro they represent the final destination of migrants who have already spent time in northeastern cities. Migrants perceive cities of the South to offer opportunities for social and economic improvement.

For the past 30 years, urban population growth has been more than twice that of the natural national increase, reaching 6.1 percent a year. A population growing at this rate doubles itself every 12 years. While Brazil's overall growth rate is down from 2.8 percent in 1978 to 2.1 percent in 1990 and 1.7 percent in 1995, the population will double in 41 years at the 1995 percentage rate of increase. The Population Reference Bureau (1995) estimates that Brazil will have a population of 194 million in A.D. 2010.

Greater São Paulo has a population of 22 million and ranks among the world's five largest metropolitan areas. Of these 7.5 million live in *favelas*. In Rio de Janeiro, the inequalities in housing are conspicuous. *Favelas* stand across the street from luxury apartments and hotels. The difference between the northern neighborhoods and the more prosperous southern neighborhoods (Copacabana, Ipanema, and Leblon) is striking. It is estimated that there are 2 million *favelados* in the city distributed among 450 illegal *favelas*. There may be as many as 4 million on the outskirts of the city. Some are large; Rocinha has a population of 250,000. The world of the *favelado* (resident of a *favela*) is one of relative misery. Houses usually lack running water, sewer connections, and often electricity. The economic system can be described as a kind of hunting and gathering. Recyclable materials are gathered from throughout the city and sold for a small profit. Bands of juveniles just roam the streets in the hope of finding or stealing something valuable.

Most migrants do not find good jobs. There is always the story of the *flagelado* (a beaten one from the Northeast) who finds a well-paying union job on the Volkswagen assembly lines, but this is the exception. More common is domestic underemployment: shining shoes or selling lottery tickets or small amounts of goods purchased wholesale. Illicit sales of drugs have become a source of income. Some drug king-pins control *favelas* and extort protection payments from the residents. The police rarely enter the *favelas*, a world unto themselves.

Although the rate of population growth in Brazil had slowed, other patterns remain the same. Rural-to-urban migration continues. In 1960, Brazil was 22 percent urban; by 1995, 77 percent of the population lived in urban areas. One third of all Brazilian families live below the poverty line. Infant mortality rates are high: 58 per 1,000 live births in 1995. At the present rate, one child in 20 does not survive the first year of life. It is not uncommon for homeless children who live on the streets to be murdered, and child workers often live in conditions near slavery. Literacy has improved from 43 percent in 1940 to 75 percent in the mid-1990s, but school attendance after age 14 drops off considerably. This reflects the need of adolescents to work and the lack of public secondary schools. Those students who cannot afford the parochial and private schools generally lose out. There are other distressing trends. In 1960, 17 percent of the population was considered middle-class whereas in 1989 only 10 percent were so considered even though 72 percent had television sets.

The situation in the Northeast remains difficult. Here we find poverty on a monumental scale. The region is on a par with Haiti and some of the hemisphere's poorest countries. Accounting for 30 percent of Brazil's population, it generates only 15 percent of the gross national product. Infant mortality rates are well above the national average. Literacy rates are 30 percent below the overall rate for Brazil. The poverty cycle of the Northeast is deeply entrenched and seems to correspond to the *secas* or droughts that periodically smite the area. Yet, even during wet years poverty remains. There are approximately 42 million people living in the area affected by the droughts. Out-migration from the area has been increasing, but the total resident population has increased due to high local birth rates. Through the government agency SUDENE (*Superintendência de Desenvolvimento do Nordeste*) over 850,000 people are employed in public and private *Obras Contra a Seca* (works against the drought). The agency has been notoriously loose with its money, and corruption is said to abound. Northeasterners cynically talk about *a Industria da seca*, (the drought industry) or to *enricar na seca* (get rich off the drought). The reasons for the poverty are complex: the drought, land tenure, *minifundismo,* and more. The region has spawned some of Brazil's most radical movements, from the Marxist peasant leagues to the radical liberation theology of the Catholic Church. The bishops of the Northeast are the most powerful block of radical bishops in the Church.

Brazil is a Catholic country; however, formal Roman Catholicism is probably a minority religion. Missionary work by mainstream Protestant churches began in the nineteenth century. Protestant churches, especially noncentralized evangelical and pentecostal congregations, are ubiquitous. Many of these congregations meet in storefronts and rented homes or in shanties in the *favelas*. Some rapidly growing evangelical churches such as the Universal Church of the Kingdom of God have been exported to Portugal, Angola, and southern California. Syncretistic religions borrowing from the Catholic and African traditions are popular. Umbanda, the largest of them, was organized in the 1920s and is found throughout the country. There are now *umbandistas* from all walks of life. There are Japanese-Brazilian,

German-Brazilian, and Italian-Brazilian centers, and the number of adherents has reached 10 percent of the population. A purer form of African spiritism, Condomble, is prevalent in the Northeast.

The modern political structure of Brazil evolved out of the military dictatorship. Gradually, press censorship was relaxed, and there was a cutback in illegal detentions. In a celebrated incident in 1975, Vladimir Herzog, a journalist, claimed to have been beaten by the military police. This act infuriated the Brazilian intelligencia, and resulted in international embarrassment. The political process began to open up with the *abeirtura*, as the Brazilians called it. Opposition politicians rallied around the MDB, and the party won the majority of seats in the National Assembly in 1987. Presidential elections were finally held in 1985, but through an electoral college rather than popular vote. Tancredo Neves, governor of Minas Gerais and a staunch opponent of the military governments, was elected. Before he was inaugurated, however, he died from complications surrounding an abdominal infection. Cynical Brazilians say that his greatest mistake was to go for treatment to the military hospital in Brasília.

Vice President José Sarney assumed the presidency and served through a tumultuous time of rampant hyperinflation and growing debt problems but presided over the enactment of the liberal yet pro-structuralist Constitution of 1988. The Constitution implements many social and political reforms, not the least of which is the insistence that the military give up the role of maintaining domestic order and concentrate on defending the country from foreign invaders. Fernando Collor de Mello quit the MDB in 1988 and founded the National Reconstruction party (PRN). He and Luis Inácio da Silva, of the Labor party fought it out in a runoff election that offered a real political choice: the structuralist, the populist da Silva or a new approach involving austerity and free trade. In a close election, Collor won in December 1989. Regarding losses by the MDB, which had provided the opposition throughout the military dictatorship, O *Estado de São Paulo*, Brazil's leading newspaper, wrote in a headline "*MDB navagou durante vinte e trés anos para terminar encalhado*" ("MDB navigates for 23 years and dies on the beach").

Brazil has serious economic problems: maldistribution of wealth, land tenure problems, underemployment, unemployment, foreign debt, and inflation. Foreign debt reached $121 billion in 1990, and inflation averaged 260 percent per year during the 1980s, reaching 7,000 percent in the early 1990s. Investment collapsed, and the already wide gap between rich and poor grew larger. Debts had been incurred with the massive ELETROBRA public works projects, the drought remedies for the Northeast, and the capital expended to start and maintain so many state-owned industries. There are additional expensive government projects including the Trans-Amazonica Highway begun in the 1970s and still under construction, other road projects such as the BR 364 of the 1970s into the northwest, and the marginal highway around the western edge of the country scheduled for the 1990s, tax incentives to promote settlement in the Amazon basin, the *Calha Norte*, a system of military bases along the borders of Bolivia, Peru, and Colombia, the designated growth poles throughout the country to promote economic development, and the infrastructure to maintain them.

Inflation is an obstacle to sustained development because it stifles savings and productive investment. From 1964 until 1990, everything had been indexed to inflation—salaries, currency, and interest rates. The mechanisms took off on their own and drove inflation. Brazilians from all walks of life have suffered. The Brazilian middle class was hard hit by inflation. Those with capital could hedge against inflation by buying real estate. Durable consumer goods, when the rate of hyperinflation is running around 2,000 percent per year, appreciate in value: a TV set in a store will cost more next week than it does today. Purchases on credit are popular: a product purchased but not paid for until next month when the currency is worth less, has cost less. For many citizens the overnight markets (overnighters) are popular. The strategy is to deposit the weekly or monthly paycheck in a savings account where it collects a high rate of interest and then transfer small amounts into the checking account to cover checks that are about to be written. The secret is to keep the maximum amount of money in the interest-earning savings account and the minimum in the checking account. These techniques may reduce the impact of inflation, but most working Brazilians lose buying power and fail to keep up with inflation. José Sarney, Brazil's first civilian president since the military *golpe*, initiated the Cruzado Plan in 1986. To quell inflation, he froze prices, boosted wages, and devalued the currency, creating the cruzado to replace the overvalued cruzero. The Cruzado Plan was followed by the Bresser Plan which devalued the currency by another 10 percent. The plans curtailed inflation for a few months.

When President Collor de Mello came into office in 1990, another round of anti-inflationary policies began. A new currency (back to the cruzero) replaced the Bresser Plan's cruzado novo, the fourth currency in four years. Collor's emergency measures lowered the inflation rate by imposing price and wage freezes and abandoning the propagation mechanism, but inflation was up to 20 percent per year and rising in January 1991. As part of the Bresser Plan, Brazil stopped payments on its foreign debt. Collor has proposed eliminating up to 400,000 of the over 700,000 federal jobs, closing 23 of the state-run companies, privatizing others, and lowering government subsidies. The steel manufacturer Companhia Sidererúrgica Nacional was sold off in 1993.

Collor moved away from the import-substituting industrialization model. Under the banner of free trade, Collor initiated reforms in tariff policy like those undertaken by Mexico and Argentina. Tariffs were removed on imported products that had no Brazilian-made counterparts. For those products that were commonly produced in Brazil and abroad, tariffs were lowered from 85 to 20 percent by 1994. This includes most of the durable consumer goods such as automobiles and appliances. Products similar to those being produced by Brazilian "nascent industries," high technology manufactures incur a tariff of 40 percent. Computers and other electronic equipment remain protected by a ban on imports.

Collor played an important role in launching MERCOSUR in 1990, a free trade zone including Brazil, Argentina, Uruguay, and Paraguay. All tariffs affecting trade among the four countries will be eliminated by 1996. Collor's popular appeal cooled after his inauguration in March 1990, and he was impeached and found guilty of fraud and racketeering. Collor was succeeded by his vice president, Itamar Franco, who served until the elections of 1994.

The 1994 election was won by Fernando Enrique Cardoso, finance minister in the Collor-Franco government. Cardoso had created yet another currency, the *real*, which was tied to the U.S. dollar in value. Prices stabilized, and the *real* retained its value. Cardoso won the election by an impressive margin because the anti-inflationary measures he developed as finance minister were seen as likely to prevent the return of hyperinflation.

THE PROBLEMS AND PROMISE OF AMAZONIA

The Amazon basin incorporates an area representing 60 percent of Brazil's national territory. The drainage basin of the river forms an enormous region in northern Brazil. Amazonia is largely tropical rain forest, but includes savanna grasslands, wetlands, and shrublands. The Amazon is the largest expanse of tropical rain forest in the world. It contains one fifth of the earth's freshwater, 20 percent of all world's bird species, and 10 percent of all the mammals (*Science*, March 16, 1990). The destruction of the Amazon tropical rain forest would be irreversible and would be a global calamity of unknown proportions. Depending on point of view, Amazonia is being destroyed or developed. Of course, the region is being destroyed *while* being developed. Fires rage throughout the region as forests are cut and burned to make way for various human uses. Fires are seen from commercial airliners and from satellite photographs, some of which revealed, for example, that on August 24, 1987, 6,800 fires were burning in a relatively small area of Mato Grosso, southern Pará, and eastern Rôndonia.

Destruction of the rain forest has major ramifications. The central issue is global warming. It is argued that the burning rain forest contributes to the greenhouse effect and may cause global warming. Burning releases CO_2, and the destroyed forest can no longer absorb CO_2 and produce free oxygen. However, it can be argued that the secondary forest, called *caápoeira*, which replaces the primary rain forest, is also a CO_2 sink and that the problem is not acute. Estimates of rain forest being destroyed annually vary from 1.5 million hectares to 20 million hectares. A reasonable figure seems to be 8 million hectares per year (*Science*, September 30, 1988). It must be remembered that most tropical soils are poor and do not support intensive land use. They have been leached of trace minerals, are acidic, and exhibit aluminum toxicity. Few Amazonian soils can be cropped or used for pasture for more than a few years without intensive fertilization. Ranching is responsible for about 80 percent of the forest destruction, and is the leading cause of deforestation. Small-scale farming is increasing too, but farm plots are often sold or abandoned after a few years and converted to pasture for ranching.

Attitudes vary as to what the best policy is for Amazonia. Environmentalists condemn rain forest destruction. Many Brazilians see the taming of the wilderness as a sign of progress. "What is occurring in Brazil is no longer a Brazilian disaster, it is a real threat to the global climate" (Maxwell, 1991, p. 24). The author goes on to decry the fact that 7 percent of the Amazonian rain forest may have already been destroyed.

The native populations of Amazonia may have numbered as many as 5.0 million prior to the arrival of Europeans in 1541 when Francisco de Orellana, a Spanish explorer, crossed the Andes from Peru and sailed down the Amazon River in Indian boats (Newson, 1996). Today these populations have been reduced to less than 200,000. Belém at the mouth of the Amazon was settled in 1616, and cultivation of rice and sugar spread out around the city. Beginning in 1649 missionary groups of the Jesuit, Carmelite, and Mercedarian orders penetrated the basin to proselytize local Amerindians. Ironically, the Jesuits who cared so much about protecting the natives from European exploitation contributed to their demise by bringing them together into *reduções*, communities of Amerindians, where disease spread rapidly and took a heavy toll.

Portuguese hunters, including *bandeirantes*, entered the region in the 1680s and 1690s exploiting tropical animal products: turtles for eggs, meat, and leather; manatee for meat and leather; other animals for their furs and skins. There was little permanent settlement in the Amazon basin until Miles Goodyear invented rubber vulcanization in 1839 and created a market for the latex of the *Hevea brasiliensis* tree. The Brazilian rubber tree is native to the American tropics, and as the demand grew, an economic boom developed. Under the banner, *marcha para oeste* (trek to the west), the settlers poured into the region. In the period 1870–1890, 200,000 migrants from the Northeast arrived, many of them *flagelados* from the drought-stricken *sertão*. Rubber was a valuable commodity; a kilo was worth 40 days of farm labor and the average *seringero* (rubber tapper) could collect 5 to 10 kilos a day. The slogan was *enricar na seringa*—get rich off rubber. The best wild trees were in the upper waters, or *altos rios* of the tributaries, and ship owners, *aviadores*, made huge profits transporting *seringeros* and their rubber from the upper river to Belém.

The crest of the rubber boom lasted from 1890 to 1914. Belém and Manaus became prosperous cities as centers of the rubber trade. Manaus had an elegant opera house and residential streets lined by houses of rubber merchants. Eventually rubber trees were transported from Brazil to Southeast Asia via Kew Gardens in London, and by 1915, Brazil was outproduced by Southeast Asian countries. Although Brazilian rubber could not compete with the new plantations in Asia, large-scale rubber production was not abandoned in Brazil until additional schemes were tried. The Madeira–Mamoré railroad was built to provide better access to the *altos rios*, but the project was a financial disaster. It is said that a life was lost to tropical disease for every tie in the railroad, and the project opened for business amid the collapse of the rubber boom. There were failed attempts at industrialized plantations such as Fordlandia on the Tapajós River in the 1930s and the more recent massive woodpulp plant that was abandoned in the 1980s.

Before opening up Amazonia to settlement in the 1950s, economic activity in the region consisted of small-scale rubber tapping, small-scale lumbering, and Brazil-nut harvesting. The vast area was virtually empty. In a country where 70 percent of the rural population does not own land and less than 1 percent of the country's farms occupy 43 percent of the arable area, it is easy to see why the vacant lands of Amazonia held out hope for an equitable and viable alternative to Brazilian rural poverty. In the 1950s, settlement plots were granted by the government to those who

Near Manaus, Brazilians of Japanese extraction grow black peppers on a plantation. There is little plantation agriculture in the Amazon region but on fertile ground higher value crops can be produced and stand the cost of transport to distance markets.

chose to settle along the newly completed Belém–Bragança Railroad. In 1953 a development agency, SUDAM (*Superintendencia para o Desenvolvimento de Amazonia*), was created to manage the influx of people into the Amazon basin. Fiscal and tax incentives were provided to settlers. It can be argued that from its inception SUDAM favored the large property owners. The tax concessions favored those who practiced ranching over small-scale cultivators, and powerful, conservative organizations such as the Rural Democratic Union (UDR) successfully lobby against land reform. The 1988 Constitutional Assembly voted down expropriation of *latifundia* as a part of land reform but did back limited agrarian reform in principle.

In the 1960s, land grabbing began along the BR 14, the Belém–Brasília Highway, and the government expanded its Amazonia development program. The military takeover of the government in 1964 marked a watershed for development of Amazonia. Since infrastructure such as roads and railroads had played an important role in the early phase, it was decided that a highway network throughout the basin would enhance settlement and economic opportunity. The Trans-Amazon Highway was built in the early 1970s to connect the Northeast coast with the deep Amazon basin. Completed in 1978, the highway extends 3,100 miles from Rio Branco, near the Peruvian border, to the Atlantic coast, intersecting three major north–south routes along the way (Figures 12.7 and 12.8).

Settlement was to be organized under government control. INCRA (*Instituto Nacional de Colonização e Reforma Agraria*) was the agency organized to deal with land reform and small-scale settlement. The program began optimistically. Settlement was to take place along the route of the Trans-Amazon Highway where plots of 20 hectares were distributed; 50- to 100-hectare lots were offered after 1970. Along the route every 6 miles, small urban centers with schools, hospitals, commercial, and recreational facilities were to be located. Some 22 of these *agrovilas* were designated regional centers and called *agrópolises*. Much of this settlement has been transient. Abandonment of lots settled in the 1970s has reached 80 to 95 percent in sections along the Trans-Amazon Highway. The federal subsidy was paid to the settler for six months, and many left after that time. A large military project has also been undertaken in Amazonia. The *Calha Norte* (Northern Headwaters) program, a plan to create a string of military security posts along the borders with Colombia and Peru, will eventually extend all the way to the Atlantic Ocean.

Laws, subsidies, and tax incentives make the clearing of Amazonia profitable for large landholders involved in ranching. In an economy like Brazil's, land becomes a hedge against inflation and leads to speculation. Agricultural and ranching profits are not taxed, creating an incentive for large-scale commercial farms and ranches to proliferate. There are lower property taxes for "improved" or cleared land, and rural credit is subsidized. These legal and policy matters, regulated by SUDAM, encourage ranchers and large-scale farmers to clear the forest. Many ranch owners in Amazonia received millions of dollars in credit but never produced any beef and abandoned their ranches after receiving subsidies. Large scale mechanized soybean farming is moving into the region, and thousands of small farmers are being pushed on to new frontiers. *Latifundismo* is on the rise.

There are compelling reasons why clearing Amazonia should cease. Amazonian soils are infertile and cannot sustain productive agriculture or pasture. Conservationist groups have recommended debt-for-nature swaps. First tried in Costa Rica, such a plan has the country's creditors sell off their discounted debt to a conservation group, which retires it in exchange for the nation's setting aside and managing an area of pristine rain forest. The Brazilians have shown little interest. The myth of Amazonia as a vast, fertile, empty space capable of absorbing landless settlers from throughout the country is not tenable.

Figure 12.7 Brazil: Administrative divisions.

No area of Brazil exemplifies the impact of forest clearing more than the state of Rondônia in the western Amazon. The region was opened to settlement in the late 1970s, the population grew by 21 percent, and a third of its forest has been cut. The BR 364 highway was built and eventually paved between Cuiabá and Pôrto Velho with a grant of $250 million from the World Bank. Designated a growth pole, the Polonoroeste Plan encompasses some 100,000 square miles of tropical forest in the states of Rondônia and Mato Grosso, where the government has encouraged the settlement of thousands of landless peasants. Upon completion of the paving of the BR 364, a total of 5,000 migrants a month were arriving in the area. There is support in Rondônia and neighboring Acre for a proposed highway across the Andes to the Pacific coast of Peru. The new road would be used to export hardwood timber.

Those who favor development in the Amazon Basin see the present situation as an early stage of economic growth. Similar phases of development in countries

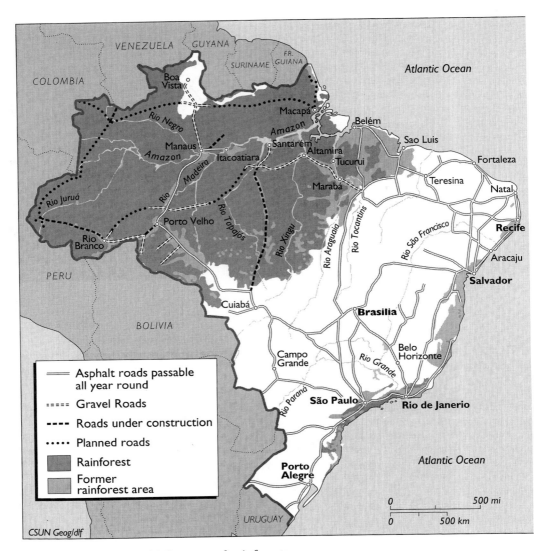

Figure 12.8 Brazilian highways and rainforests.

such as the United States brought waste and failure, common frontier phenomena. Greater labor, capital, and most of all appropriate land use systems will be required for stabilization. This position is rarely heard in the United States but is common in Brazil. Many Brazilian officials favor the development of restricted and appropriate sites by colonists who use appropriate techniques. The floodplains, *varzeas*, which represent 2 percent of the Amazonian region, have better soils that could be used for intensive cultivation. Examples of intensive cultivation in floodplain environments at Yurimaguas, Peru, indicate that continuous cropping for 50-year periods is sustainable.

Some political leaders have become defensive about the rain-forest problem. President Sarney, responding to criticism of Brazil's treatment of the Amazon, lashed out in a 1989 speech against the "campaign of scientific falsehood" and argued that only a tiny percentage of the rain forest has been cut since the Portuguese arrived in 1500, far less than the deforestation in Europe and North America.

Clearing the rainforest in Western Brazil to create low-quality grazing land.

He ended by saying "Brazil controls the Amazon, the Amazon is ours" (*New York Times*, April 7, 1989).

Many recent settlers in Amazonia, Brazil's fastest growing region, are resentful of government policies creating Native American and biological reserves that tie up land and prevent development. For example, they will point out that the Yanomani Indians, numbering fewer than 20,000 have been allocated a reserve of over 37,000 square miles, the size of Portugal, France, and Switzerland combined. Policies toward burning also trouble Amazonian settlers. With sarcasm, a Pôrto Velho newspaper editorialized, "fire in Kayapo (Indian) hands is a nurturing force for both humans and nature. In the hands of Amazon ranchers and developers or peasant colonizers, however, fire has become a major force of destruction with global implications."

Brazil has announced some important changes in its Amazon policy. A $100 million project financed in part by the Food and Agriculture Organization of the United Nations will study the condition of the Amazon over the next five years. Environmental protection courses will be taught in the public school system, and most importantly, tax breaks, subsidies, and credits are being temporarily suspended. Increasingly, land reform is being associated with ecological reform. A new agency IBAMA (*Instituto Brasileiro do Meio Ambeiente e dos Recursos Naturais Renováveis*) has been created for the purpose of enforcing environmental regulations. In theory, there is more protection, ranch sizes are being limited, permits are required for burning, and fewer permits will be granted. However, with a relatively small staff and vast territory to police, it is doubtful that IBAMA will be able to control burning. Nevertheless, burning has decreased since 1988. Forest destruction in 1990 equaled 1 million hectares, down from 3 million in 1988. The most recent development is a ban on all government financing for agriculture and ranching north of the thirteenth parallel.

While there was evidence that burning had been falling off since 1988, there are now indications that there has been a resurgence. As far as fires are concerned,

1995 may be the worst year ever according to satellite remote sensing by both Brazilian and U.S. observers. Alberto Setzer, who tracks burnings in the Amazon for the Brazilian National Institute for Space Research, said that in July 1995, "weather satellites detected 39,000 fires, a nearly fivefold increase over the 8,503 fires in July of 1994. In general, we notice a very strong increase in fires in the Amazon region, and in regions where we didn't expect to see the fires. It means that for sure we'll have a very high figure of new deforestation this year" (Schemo, 1995).

There has been violence in Amazonia since the area opened up in the 1950s. The most celebrated example was the murder of Francisco "Chico" Mendez in Zapurí, Acre, in December 1988. Mendez, an ecologically minded rubber tapper, was well known as a symbol of the poor, landless *caboclo*. Mendez stood up to the powerful ranching interests destroying the forest. A pair of ranchers, a father and son, were convicted of his murder. Some 1,566 Brazilians have been killed in land disputes since 1964, involving struggles between the landed and the landless. Few killers have been convicted (*New York Times*, February 5, 1991).

SUMMARY

A theme that has shaped Brazilian society since the colonial period is the tension created by land tenure. The *latifundia* and *minifundia* have historically differentiated between the rich and the poor. Tension can be seen in the modern period as industrialization has produced lucrative exports, wealthy business people, and a limited middle class who share the cities with destitute *favelados*. With industrialization have come foreign debt and hyperinflation. The world is watching Brazil as it deals with Amazonia. Can Brazilian development proceed without touching off an ecological time bomb? Can Brazilians, who comprise a blend of cultural and ethnic populations, confront problems of ecological concern with equity? Among the nations of Latin America, Brazil has the greatest potential for development. Impressive resources and a position in the international arena give Brazil a presence that is unsurpassed in the region.

FURTHER READINGS

Godfrey, B. J. "Boom Towns of the Amazon." *Geographical Review* 80(1990): 103–117.

————. "Modernizing the Brazilian City." *Geographical Review* 81(1991) 18–34.

Page, J. A. *The Brazilians*. Reading, Mass: Addison-Wesley, 1995.

CHAPTER 13

The Southern Cone

CHARLES S. SARGENT AND
SUSAN R. SARGENT

INTRODUCTION

The Southern Cone includes Argentina, Chile, Paraguay, and Uruguay, countries that share many physical, cultural, and economic characteristics. Argentina and Chile share the Andes, Tierra del Fuego, Patagonia, and claims to Antarctica. Paraguay and Argentina both covet the Chaco region, and Uruguay and Argentina share the Pampas (plains). All have a Spanish colonial heritage and are predominantly Roman Catholic. With urbanization rates of 85 percent, Argentina, Chile, and Uruguay are on a par with Europe's most urbanized nations. Argentina, Chile, and Uruguay have a shared history of large-scale European immigration, a pervasive Hispanic cultural imprint, and the Spanish language. By contrast, Paraguay and interior regions of Argentina and Chile are home to Amerindian populations, where indigenous languages and ways of life prevail.

In the Spanish colonial period—the years between Asunción's foundation (1537) and independence from Spain after 1810—socio-economic development in the Southern Cone was limited by the small population of the region, restrictive colonial economic policies, and distances from the South American political and economic core in Lima, Peru. The physical environment, with mountains, deserts, marshlands, and scrublands, as well as the "friction" of great distances presented significant obstacles to contact and interaction. Additionally, colonial economic policies together with the limitations of transportation and production technology restricted early growth. The Southern Cone was in every respect peripheral to the major Spanish Andean mining focus to the northwest and intentionally apart from the development of Brazil to the northeast.

After independence, the new countries of the region shared decades of economic and political transition. But progress, interaction, and development were impeded by political instability, military upheavals, and boundary disputes as the newly formed countries sought internal cohesiveness and well-defined borders. Much was resolved by the late nineteenth century, as the rapidly growing nations developed economic and social ties with the world economy and tentative but limited relationships with one another. During the early decades of the twentieth century, the effects of economic progress were uneven, both regionally and internally. Many immigrants from Europe, especially Spain and Italy, reshaped the character of Argentina and Uruguay.

Chile was also affected by migration from Germany and Austria. Technological development in manufacturing and transportation served to entrench early linkages, rather than facilitate contact and interaction between nations. Paraguay, with its interior location, remained beyond the major economic and migration flows that molded the European commercial quality of the Southern Cone.

Argentine school children learn that "the Mexicans descended from the Aztecs, the Peruvians descended from the Inca, and the Argentines descended from the boats." In fact, an estimated 8 percent of Argentines are mestizo (a mix of Indian and European). Historically, mestizos were numerically dominant and clustered in Argentina's Andean northwest, but since the 1930s a large number have migrated to the *littoral*, Atlantic or coastal areas, particularly into metropolitan Buenos Aires. Officially, the "pure European" is dominant in Argentina and Uruguay. Chile, by contrast, is 65 percent mestizo, while Paraguay is almost wholly (97 percent) mixed.

NATIONAL SIZE AND POPULATION: ARGENTINA, CHILE, PARAGUAY, AND URUGUAY

Argentina accounts for 67 percent of the land area and 61 percent of population in the Southern Cone (Table 13.1). It is South America's second largest and most populous nation after Brazil. Argentina is four times the size of Texas. Chile is one fourth the size of Argentina and has two fifths the population. Chile is ten times as long as it is wide, some 2,650 miles north–south and only 265 miles east–west at its widest. It is a classic example of an "elongated state." Chile is deceptively twice as large as California. Paraguay is almost as large as California, but has a population of only about 5.5 million, although at the present annual growth rate, the population will double in 25 years. Uruguay, in contrast, has a relatively slow population growth rate, and the 3.2 million population will double in over 100 years, if current rates continue. Each country has a single dominant metropolitan area—Buenos Aires (Argentina), Santiago (Chile), Montevideo (Uruguay), and Asunción (Paraguay).

At independence, Chile broke away from the Viceroyalty of Peru; Paraguay and Uruguay (as well as Bolivia) split from the Viceroyalty of la Plata. Paraguay (like Bolivia) is landlocked, and Uruguay, originally a buffer state, is one of South America's smallest nations.

The demographic characteristics of the Southern Cone nations since independence reflect changing political and socio-economic conditions together with combinations of natural increase, immigration, emigration, and the toll of wars. Between 1821 and 1932, 6 million immigrants entered Argentina, a world level surpassed only by the thirty million immigrants who entered the United States. In the same period, between 650,000 and one million European immigrants entered Uruguay, many of them via Argentina. Between 1881 and 1930, 200,000 migrants (including significant numbers from Germany, Switzerland, Austria, and Britain) entered Chile. Isolated Paraguay received fewer migrants and lost perhaps as much as 75 percent of its population, including almost all the adult males, in an 1860s war with Argentina, Uruguay, and Brazil. Paraguay's population remains indigenous in character, with Guaraní, the dominant Amerindian language, widely spoken, although Spanish is the primary official language. Paraguay is the poorest of the four countries of the Southern Cone.

After 1930 European immigration slowed; thereafter, natural increase accounted for most population growth, and rural-to-urban migration expanded the cities. Post-1930 population growth is shown in Table 13.2. Suggestive of this rural-to-urban movement is the fact that the Buenos Aires and Santiago metropolitan areas contain over a third of each nation's population. Over 50 percent of Uruguayans live in greater

TABLE 13.1 Countries of the Southern Cone: Basic Data

Country	Area Square Miles (000)	Square km. (000)	% of Cone	Pop. (millions)	% of Cone Pop.	% Urban	1995 Annual Growth Rate (%)	Yrs. to Double Pop.	Per Capita GNP ($)[1]	% of Labor Force in Agriculture	Pop. Under 15 Yrs. of Age (%)
Argentina	1,068	2,767	67	34.6	61	87	1.3	55	7,290	12	31
Chile	292	757	18	14.3	25	85	1.7	41	3,070	20	31
Paraguay	157	407	10	5.4	9	51	2.8	25	1,500	43	41
Uruguay	68	176	4	3.2	6	90	0.7	102	3,910	16	27

[1]South America average is $3020

Argentina is four times the size of Texas; Chile twice as large as California. Paraguay is almost as large as California; Uruguay somewhat smaller than Washington state.

TABLE 13.2   Population Estimates and Projections by Decade, 1930–2020

Country	(Population, 1000's)				
	1930	1960	1980	1990	2020
Argentina	11,896	20,616	28,237	32,880	45,564
Chile	4,424	7,609	11,127	12,987	17,724
Paraguay	880	1,778	3,168	4,231	7,930
Uruguay	1,704	2,538	2,908	3,128	3,782
Southern Cone Total	18,904	32,541	45,440	53,226	75,000
Latin America	102,889	209,311	351,435	439,624	719,169
United States	123,070	180,680	227,660	250,000	290,000
Cone as percent of Latin America	18.4	15.5	12.9	12.1	10.4

Source: CELADE, 1985

Montevideo. Even in relatively rural Paraguay, 51 percent of its population resides in cities, with about a quarter of the population in Asunción.

Although the Southern Cone contains the most European-derived populations of Latin America's many regions, the Amerindian presence is significant. For example, about 7 percent of Chile's population is Amerindian, consisting of various indigenous groups, including the Mapuches (called Araucanians by the Spanish). Reservation lands have been established for some half million Mapuches in Middle Chile.

People of African heritage are scarcely to be found in national censuses, even though during the colonial period enslaved Africans were introduced throughout the region via Argentina. The "disappearance" of the region's black population can be attributed to war losses, racial mixing, and statistical reclassification of census data.

THE PHYSICAL SETTING OF NATIONAL DEVELOPMENT

Southern Cone nations share the narrowing South American landmass south of tropical Brazil and Andean Peru and Bolivia. Although most of the area lies within the mid-latitude zone, there is considerable diversity of the physical environment with significant variations in topography, climate, and vegetation (Figures 13.1, 13.2, and 13.3). The most physically complex of Latin America's regions, the Southern Cone is influenced by ten major and distinct climatic regimes. The region contains the continent's warmest, coldest, highest, and lowest locations.

Within this diverse physical environment, there are four major physiographic regions, each with several subdivisions: (1) the highland Andean system and the west coast, (2) the plains and interior lowlands between the Andes and the Atlantic, including the Gran Chaco and the Pampas, (3) the interior plateau and hill districts, and (4) the valleys and tablelands of Patagonia and Tierra del Fuego in the extreme south (Figure 13.4).

▶ *The Andean System and the West Coast*

The dominant and most complex physical feature of the Southern Cone is the Andes. Structurally, the Andes are a triumvirate of building blocks: ancient crystalline rocks, geologically more recent volcanics, and sedimentary deposits. The Andes are reminders of plate tectonics and crustal instability that are expressed by frequent earthquakes, volcanic eruptions, and associated natural disasters. In Chile, severe earthquakes were recorded in 1575, 1647, 1730, 1822, 1906, and 1985.

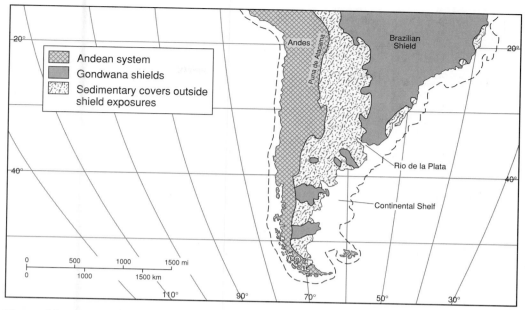

Figure 13.1   Southern Cone: Structural regions.

Seismologists notice a periodicity of 83 years (±9) for the tremors but cannot account for the timing. Disastrous tsunamis also strike the Pacific coast. For example, in May 1960, ports between Concepción and Puerto Aisén were inundated. Offshore is a subduction zone where the ocean floor plunges down to create the Chilean Trench.

The Peru (Humboldt) Current sweeps northward along the Chilean coast. It has two major segments: a slower coastal, cold, green-brown belt extending some 30 miles from the coastline; and a faster and less cold deep blue flow. The coastal

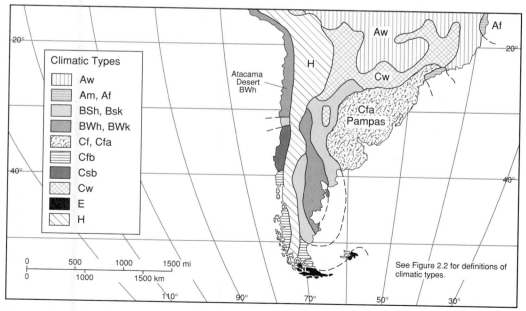

Figure 13.2   Southern Cone: Climate zones.

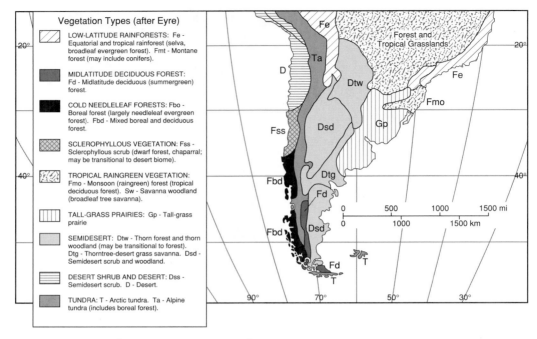

Figure 13.3 Southern Cone: Vegetational zones.

segment is an upwelling of cold waters, rich in nutrients, that sustains marine life along the coast. Fishing has always been important to Chile.

In the colonial period, the north-flowing current made for fast sailing from Valparaíso to Lima. The southbound voyage was correspondingly difficult and reduced interaction between Chile and Peru. The current contributed to the discovery of Chile's three Juan Fernandez Islands. Lying almost 400 miles off the Chilean mainland, the islands were named for their discoverer, a navigator on the Lima–Callao–Valparaíso run who tried to avoid the Peru Current by keeping away from the coast. The largest island became world famous as the locale of the stranding of Alexander Selkirk, popularized by Daniel Defoe as Robinson Crusoe.

Over 60 percent of the more than 4,000 mile extent of the Andes lies in the Southern Cone. The linear continuation of the peaks, *cordilleras*, and dry Altiplano of Peru and Bolivia narrows into a single, higher range which includes some of the continent's tallest peaks, many exceeding 19,000 feet. Beyond 48° south latitude the Andes curve toward the east, become lower in elevation, and are heavily glaciated. Here the mountains form islands; one of these, Cape Horn, marks South America's southernmost point. Beyond this, the range is submerged below the Drake Passage but resurfaces as the Antarctic Peninsula.

Chile's northern provinces, or *Norte Grande*, and Argentina's northwest are arid lands. The highlands of the Altiplano and Puna de Atacama (Figure 13.1) are a series of sedimentary basins, up to 300 miles wide and 11,500 to 14,700 feet above sea level, bordered by isolated mountain ranges and volcanic cones. Among these snow-capped peaks is Ojos del Salado (22,572 feet), the world's tallest active volcano. Within the basins are salt flats, or *salares*. In these cold, windy drylands vicunas and llamas feed on the scattered grasses and xerophytic vegetation.

At the eastern, Argentine side of the Puna, streams that ultimately flow into the Chaco cut deep valleys that lead down to fringing parallel ranges, rift troughs (*valles*), erosional valleys (*quebradas*), and hogback ridges (*cuestas*) that define the

Figure 13.4 Physical features of the Southern Cone.

margin between the Andean system and adjacent lowlands. Among the most striking of the moderately high ranges is the 17,000-feet-high Sierra de Aconquija, west of Tucumán. Farther to the east, facing the interior Pampean lowland, the Sierra de San Luis and Sierra de Córdoba rise up to 8,000 feet and stretch 300 miles north–south. The territory lying between Tucumán, Cordobá, and Mendoza is a dry land of broad basins with a number of large *salares*.

West of the Bolivian Altiplano and the Puna, the sierras of the Western Cordillera drop in elevation. Mountain spurs extending toward the Pacific form basins filled with Andean erosional debris. Lying at 6,500 feet above sea level, these sedimentary basins of *Norte Grande* make up the Atacama desert. These barren lands fill the northern portion of the structural trough between the Western Cordillera of the Andes and the 1,800 mile parallel Pacific coastal range (*Cordillera de la Costa*) that is Chile's longitudinal Central Valley. The coastal range runs much of the length of Chile and rises abruptly from the sea forming a cliff face 2,000 to 3,000 feet high. There are few natural harbors except where wave-cut platforms, the result of coastal emergence, provide sites for towns, notably Arica, Iquique, and Antofagasta. At Valparaíso, Chile's major port, a promontory creates a sheltered bay.

The hot, dry, barren Atacama desert (Figure 13.2) dominates the landscape of northern Chile. It extends north to south about 600 miles. A brown landscape of hills and plains, the Atacama is the world's most arid desert. At Antofagasta, rainfall averages a few millimeters per year (Table 13.3) Calama, near Chile's most important mining center, has never recorded rainfall (Figure 13.5); other locations record no rainfall for years at a time. This aridity is influenced by the offshore cold Peru Current and a persistent oceanic subtropical high-pressure cell of stable, subsiding air that prevents marine air masses from moving onshore. Along the coast, however, morning fogs (*camanchacas*) provide some moisture.

Only one Andean stream, and Chile's longest river—the Río Loa—crosses the Atacama to reach the Pacific Ocean. Roads and rail lines extend from the few ports inland across the coastal range into the high basins where rich copper ores, nitrate ("desert gold") beds, potassium and borax deposits, and the world's largest known

TABLE 13.3   Annual Precipitation at Comparable Latitudes: Andean Chile and Argentina

Chile	Elevation (ft.)	Max. Month	Annual Precipitation (inches) Chile	Annual Precipitation (inches) Argentina	Max. Month	Elevation (ft.)	Argentina
North							
Antofagasta	308	July	0.5	32	January	4,275	Jujuy
Taltal	16	July	1	27	January	3,865	Salta
Caldera	46	June	1	37	January	1,467	Tucumán
Coquimbo	88	June	5	13	February	1,673	La Rioja
Middle							
Santiago	1,706	June	15	8	February	2,477	Mendoza
Los Ángeles	427	June	52	10	June	2,792	Chos Malal
Temuco	371	July	53	7	October	869	Neuquén
Puerto Montt	328	May	83	42	June	2,799	Bariloche
South							
Melinka	16	July	130	19	June	1,864	Esquel
Puerto Aisén	16	July	114	7	May	1,089	Colonia Las Heras
Punta Arenas	33	April/ May	17	23	March	23	Ushuaia

Sources: *Chile: a Remote Corner on Earth*. Santiago, Chile: Empresas Cochrane, 1992. Grau, P.C. 1979. *Geografía de Chile*. Santiago, Chile: *Editorial Universitaria*. Roccatagliata, J.A. 1988. *La Argentina: Geografía general y los marcos regionales*. Buenos Aires, Argentina: *Planeta*.

Figure 13.5 Annual precipitation: West and east of Andes.

reserves of lithium—a high-tech mineral salt used in medicines, batteries, explosives, and drying agents—are commercially mined. Several interior oasis towns, most predating Spanish settlement, are found where streams or springs flow out of the mountains. Many settlements harbor archaeological remains of pre-Columbian cultures, including fortresses (*pukarás*), canals, terraces, ceramics, textiles, and mummies—all well preserved in the dry desert air.

Marginally less austere is *Norte Chico*, the lands between the Río Copiapó and the Río Aconcagua, where some valleys are under irrigated cultivation. South of the Puna de Atacama, the mountain ranges narrow, reaching their apex in the 22,834-feet Cerro Aconcagua ("stone watchtower"), the highest mountain in the Western Hemisphere. The Southern Cone claims 31 Andean peaks above 19,600 feet. Nearby, Uspallata Pass (12,650 feet) has linked Chile and Argentina since colonial days. The mountain passes become lower and more open as the cordillera stretches southward to Tierra del Fuego.

In Middle Chile, the Vale of Chile is the western approach to Uspallata Pass. On the southern frontier of the Inca Empire, this fertile valley of the Aconcagua River to the north of Santiago is the southernmost of the distinct basins before the longitudinal trough becomes continuous and progressively wider toward the south. This extensive Central Valley of Middle Chile, with Santiago its focus, is the economic and historic heart of the country. Andean streams bearing silt-laden meltwater formed alluvial fans in the valley before cutting through the coastal range to the ocean. The fine, fertile alluvial soils, enriched by volcanic ash and combined with sufficient water for irrigation and a Mediterranean climate, are the basis of a rich agricultural economy. In its southern districts, the Central Valley sustains commercial forests as well (see Figure 2.15).

Middle Chile's Mediterranean climate is characterized by dry, sunny summers and mild, moist winters. Yearly rainfall decreases toward the interior Central Valley and northward with proximity to the arid *Norte Grande*. Rainfall increases to the south where a marine west-coast climate prevails. At the latitude of Puerto Montt (comparable to the California–Oregon border), frontal storms occur throughout the year.

The native vegetation of Middle Chile includes drought-resistant Mediterranean-type sclerophytes such as deep-rooted chaparral, cypress, and laurel. Introduced varieties include poplar, willows, eucalyptus, and fruit trees and vines. Pasture lands have been improved with European grasses, alfalfa, and clover. South of the Bío-Bío River, in the *Sur Chico* transitional zone, are mixed forests of beech and cedars with dense undergrowth. Commercially grown, fast-growing radiata pines are replacing these forests. By contrast, in the eastern piedmont of the Andes (in the Argentine areas of Mendoza, Tucumán, and Santiago del Estero) the rain-shadow effect creates an arid climate.

South of Puerto Montt is Chile's *Sur Grande*, a cool, wet world of islands, channels, fjords, glaciers, and stunted vegetation. Charles Darwin described these sparsely populated, remote lands as a "green desert." Since 1976, a new highway, the Carretera Austral Longitudinal (now over 500 miles long), has been opened along the Andes south of Puerto Montt. Crustal subsidence and rising sea levels following the Ice Ages have flooded the Central Valley and glacial troughs at the edge of the Andes, creating fjords. The numerous islands, many forming archipelagos, are structurally the continuation southward of the Coast Range. The largest of these islands is Chiloé.

The Andes extend southward as a progressively lower range, where water erosion has broadened some valleys and glacial action has deepened and widened others. The Southern Icefield extends from 46° to 51° S. Many of its glaciers flow to sea level, the nearest to the equator in the world. North of the icefield, glacial troughs cut through the Andes, providing passage around the shores of glacial lakes onto the drier lands of Patagonia to the east. The westward drainage of many of these lakes has substantiated Chile's claim to territory east of the crest of the Andes. National parks and spectacular scenery abound. South of the icefield, submerged Andean peaks become the islands of Magallanes and Tierra del Fuego. Cape Horn, the island named by Dutch merchants after their port city of Hoorn, marks the Andes' southern extreme. In these far southern lands, driving westerly winds, the "Roaring Forties" and the "Screaming Fifties" (named for their latitudinal locations), still hinder transportation. The transit westward, especially through the Strait of Magellan or "around the Cape" (most difficult before the era of the steamship), remains a route that mariners respect.

► *The Plains and Interior Lowlands Between the Andes and the Atlantic*

In sharp contrast to the Andean system is the zone of plains and lowlands between the Andes and the Atlantic littoral, or coastal region. Major elements of this system are (1) the Gran Chaco, which slopes eastward from the Andes toward the fertile lands of the Argentine Pampas and eastern Paraguay and (2) the flat to gently rolling Pampas.

Geologically, the lowland is a combination of waterborne and windblown deposits that overlay the ancient crystalline rocks of the Brazilian (Gondwana) shield. In the structural depression defined by the Shield in the east and the Andes in the west is the silt-laden Paraguay–Paraná River system. Most of the system's tributaries descend from the shield and in places flow over resistant basalt flows as spectacular waterfalls. The most famous of these are the cataracts of the Iguaçú just before that river joins the Paraná. The confluence of the Uruguay and Paraná rivers is at the head of the Río del la Plata estuary. The Pilcomayo and Bermejo rivers intermittently flow southeastward across the Gran Chaco, transporting sediments from the flanks of the Andes. A number of smaller Andean streams never reach the Paraná, for their waters are absorbed or evaporated en route.

The Gran Chaco ("hunting ground") of Paraguay and northern Argentina lies to the west of the north–south flowing Río Paraguay ("water-big"). The Chaco is home to jaguar (*tigre*), wild boar, tapir, deer, armadillo, anteater, aquatic rodents, poisonous snakes, tropical birds, and a multitude of insects, including swarms of mosquitoes. A great interior lowland, the Chaco is a flat, semidesert of thorn and scrub forests, cacti, grass savanna, seasonally flooded marshlands, and hardwood forests. Of particular importance is the hard *quebracho colorado* ("red axe breaker"), the world's principal source of tannin. A white quebracho is also harvested for use as fencing, fuel, and building material.

The infertile soils of the Chaco are derived from Andean sediments, mostly sand and clays, deposited by the transitory streams that descend from the Bolivian highlands to the west. Principal among these rivers is the Pilcomayo, which forms the border with Argentina and is considered the southern limit of the forested (boreal) Chaco. The meandering, braided, sluggish, marshy, intermittent nature of these streams preclude their use for navigation or hydroelectric power. The highest temperatures in South America have been recorded in the Argentine Chaco.

A second major physical subregion is the Pampas, the vast, almost flat plains that lie east of the Andean system and south of the Chaco. These plains, which stretch west from Buenos Aires to Córdoba and northeast into the Paraná lowlands of Entre Ríos ("between rivers" Paraná and Uruguay) and western Uruguay are divided into the wetter, eastern parts known as the Humid Pampa (*Pampa Húmida*) and the progressively drier, western portions, the Dry Pampa (*Pampa Seca*). At the western extremity of the Pampas is the Sierra de Córdoba, related geologically to the Andes.

Along with the depositional sands, clays, and silts are large quantities of windblown soil (loess) carried from the drier west and south. The Pampas, the primary source of Argentina's agricultural wealth, is one of the world's major areas of loess deposition. These fine-grained soils (mollisols) are extremely fertile and advantageous for agriculture. There are no rocks or gravels in the Humid Pampa; however, neither are there mineral resources. Fuels, ores, building stone, and material for road making must be transported from afar. Only the Sierra de Tandil and the

396 The Southern Cone

Sierra de la Ventana (outcrops of the underlying Brazilian Shield in the southern Dry Pampa) are sources of building materials. Since the nineteenth century, coal has been imported, at first from Welsh and English coalfields.

The delta of the Paraná and Uruguay rivers forms an intricate series of islands, used for fruit and vegetable production. Below the delta lies the Río de la Plata ("river of silver") estuary. The name was given by explorers who hoped the waterway would lead to inland silver. About 100 miles long and some 60 miles wide between Montevideo and the Argentine shore, the estuary is shallow and silt-laden. The port of Buenos Aires requires constant dredging. When Atlantic winter storms from the southeast combine with high tides, flooding occurs within the delta and along the city's southern shore, where low-lying areas are inundated.

The weather of the Pampas is variable, responding to influences of the contrasting air masses and frontal storms that impact the region. Movements of cold polar air from the south bring the cool, invigorating *pampero* wind; warm, humid tropical air from the north produces the sultry *norte*. Violent thunderstorms occur along boundaries where the two air masses meet. Summers are warm and humid, modified along the southeast coast by the cold Falkland Current. Not far to the north, however, the Brazilian Current helps warm seaside resorts around Mar del Plata. Winters are mild, but cold spells and frosts are common. The growing season ranges from 300 days in the northeast to 140–150 days in the southeast. On the Pampas there is no true "dry season," but precipitation decreases to the west and south of Buenos Aires.

At the edge of the Pampa Húmida (roughly along the 16-inch per year rainfall isohyet) the native grasslands of tall bunch grasses (*pasto duro*) grade into the xerophytic deciduous trees, low scrub, and *monte* grasses of the Pampa Seca. Introduced trees (such as poplars and eucalyptus) serve as windbreaks and break the monotony of the landscape all across the Pampas.

► *The Interior Plateau and Hill Districts*

The hilly uplands that make up three fourths of Uruguay, the extreme northeast of Argentina, and southeastern Paraguay are transitional lands between the low-lying Argentine Pampas and the Brazilian Highlands. A conspicuous feature of Uruguay's rolling landscape is the *cuchillas* (ridges), gentle ranges of schists and granites that are extensions of the Brazilian Plateau. The *cuchillas* lack fossil fuels and metallic ores but are sources of marble and semiprecious stones. Two of the major ridges, *Cuchilla Grande* (980 feet) and *Cuchilla de Haedo* (1,378 feet) to the northwest divide drainage between the Atlantic to the east and the Río Uruguay to the west. Between them lies the drainage basin of the Río Negro, which originates in Brazil and empties into the Río Uruguay below Mercedes. Construction of a dam in the 1930s to control flooding and provide hydroelectric power created a 4,000 square mile lake, Rincon del Bonete. The Río Negro is navigable from the dam downstream to Mercedes. Pastures dominate in these hills and valleys, making them centers of livestock production.

To the north, the Paraná Plateau, a continuation of the Brazilian Highland, extends into eastern Paraguay (*Region Oriental*) and northeast Argentina. Rainfall is abundant, averaging over 60 inches per year. The dense, semitropical "Brazilian" woodlands, containing hundreds of species of hardwoods, varieties of palms, and medicinal plants, cover the eastern third of Paraguay and Misiones province of Argentina. Pumas, *tigres*, monkeys, deer, and more than 400 species of birds fill the forests. Native stands and plantations of yerba (*Ilex paraguayesis*) are harvested. The dried and ground leaves are soaked in hot water to make *yerba maté*, a tealike beverage. The woods, together with the quebracho of the Chaco, account for the strength of the forest products industry in Paraguay.

The Paraná Plateau is bisected by the Paraná River, which has deeply incised the successive layers of lavas and red sandstones that overlay the crystalline Brazilian Shield. The river forms Paraguay's eastern border with Brazil. Beyond the escarpment that delineates the plateau lies a belt of hills, the exposed and weathered shield. This fertile area includes Asunción and is Paraguay's major settlement zone.

► *The Valleys and Tablelands of Patagonia and Tierra del Fuego*

The tapering southern lowland of South America is a region of cool, arid, windswept steppes. The climate is relatively mild because increasing latitude is offset by decreasing distance from the moderating effects of the oceans. South of Buenos Aires, ranges of temperature diminish with increased latitude. Summers are warm rather than hot; winters are less cold than their latitude would suggest, although cooler than in similar latitudes west of the Andes in Chile. Patagonia becomes moister and its grasslands are more productive as one moves inland and south. This is the reverse of the Pampas region, which becomes drier farther inland. The best Patagonian grazing lands thus lie close to Tierra del Fuego and parallel to the Andes.

East of the Andes, much of Argentine Patagonia represents a rare east-coast desert, with rainfall under 10 inches per year. Two factors are responsible. The Andes are a barrier to westerly winds blowing off the Pacific, and the cold-water Falkland Current that cools Atlantic airmasses produces heavy sea fogs but little precipitation. What little moisture is received is derived mainly from frontal storms that originate out of the south Atlantic subtropical high pressure center that moves north in winter and allows moist air to penetrate the region.

These austral lands are composed of (1) Argentine Patagonia, (2) Chilean Patagonia and Atlantic Chile, the trans-Andean lands of *Sur Grande*, and (3) *Isla Grande* (Big Island) of Tierra del Fuego, shared by Chile and Argentina. Both lay claim to wedges of Antarctica as well.

Sheep farming in Patagonia.

The terraced uplands (*mesetas centrales*) of Argentine Patagonia are vast areas of plateaus that rise like stairsteps to the west. They are cut by a number of deep, east–west river valleys that provide transportation corridors and shelter from the ever-present westerly winds. Many of these broad valleys, cut by glacial-age melt-water from the Andes, contain no surface water. Others have intermittent streams or underground reservoirs. A few valleys have permanent watercourses, notably the Ríos Negro, Colorado, Chubut, and Santa Cruz, which flow across the desert to the Atlantic. The northernmost of these fertile river valleys sustain commercial irrigation agriculture. The short steppe grasses of the tablelands support sheep grazing (see Figure 2.7).

Along the Atlantic coast the plateaus form a wall facing the sea and reduce potential port sites. The Valdés Peninsula is a wildlife refuge, providing breeding grounds for marine mammals and penguins. The peninsula itself contains *salares* and has the lowest point on the South American continent, 132 feet below sea level. Lying 340 miles offshore on the continental shelf are the Falkland Islands (*Islas Malvinas*). The archipelago consists of approximately 100 small islands used predominantly for sheep farming. These cool, wet, treeless islands, named for early French contact out of St. Malo (Malvinas), are claimed by Argentina but held and settled by Great Britain.

Along the eastern piedmont of the Andes lies a series of glacially carved basins and marginal lakes that drain to the Pacific. Only Lakes Argentino and Viedma find an outlet to the Atlantic, via the Río Santa Cruz. The westward drainage has given Chile territorial ownership of lands east of the Andes, including a small section extending to the Atlantic along the Strait of Magellan (Atlantic Chile).

The Strait of Magellan was created at the end of the ice age as a glacially carved valley that was flooded. The strait is named after Ferdinand Magellan, the Spanish explorer who took 38 days to gain passage through the waters in 1520. He referred to the area of islands as "the land of smoke," later renamed the Tierra del Fuego ("land of fire"), after the smoke from the Indian campfires.

The *Isla Grande* of the Tierra del Fuego archipelago lies at the southern extreme of the continent, separated from the mainland by the Strait of Magellan. This "big island," roughly the size of Ireland, commonly called Tierra del Fuego, is shared by Chile (70 percent) and Argentina (30 percent). The subantarctic climate brings cool weather, but temperature extremes are moderated by marine influences. Prevailing strong southwesterly winds off cold southern Pacific waters bring high precipitation to the mountainous, forested western and southern portions of the island. These winds frequently develop into gale-force storms. The northeastern part of the island is treeless plains of sheep farms and oil fields. Paralleling the dramatic mountain peaks of the southern coast is the Beagle Channel.

THE DEVELOPMENT OF CULTURAL LANDSCAPES

Four distinct periods have shaped the cultural landscape of the Southern Cone. In each period, a revised and expanded set of agricultural, industrial, and cultural practices has modified the natural physical landscape into a sequence of culturally shaped landscapes.

1. Aboriginal, Pre-European Cultures. This era lasted from about 8000 B.C. to about A.D.1540.
2. Spanish Colonial Culture, which shaped the landscape from about 1540 to about 1810, when the independence movements began to release all of Latin America from Spanish political and economic control.

Strait of Magellan

3. Large-scale and extensive outside contact, particularly with non-Spanish Europe. These mercantile contacts were particularly active from the mid-nineteenth to the mid-twentieth century.

4. Incorporation of the Southern Cone into the broader world economy, in part due to improved communications and transportation links to the world. The era of jet aircraft arrived in 1960, followed by regional and international economic unions that forged new trade ties.

► *Aboriginal Cultures*

The native population ranged from sedentary agriculturalists to wandering hunters and gatherers. The sedentary Indians—resident in the lower Andean valleys of northwestern Argentina and north and central Chile—with small numbers and rudimentary technology produced few landscape changes. Within, but on the periphery of, the Inca world, they created agricultural terraces, irrigation systems, crude trails, and small villages. A handful of larger trading/agricultural centers served regional needs.

The grassland regions were home to nomadic people, who hunted animals and gathered plants. In the humid forests of the northeast, others practiced simple agriculture, while isolated groups harvested seafoods on the cool, wet coast of southern Chile and Argentina.

► *The Spanish Colonial Culture*

The first camps of Spanish explorers have left no trace, and some of the early attempts at town creation are gone. However, scores of colonial settlements survived and formed a system of cities and towns that by 1600 defined the economic potential of the region. Central Chile and northwestern Argentina were the centers of the Southern Cone's colonial world, with Asunción and Buenos Aires as relatively minor places at the margin of the system. Southern Chile and the Pampas of Argentina and Uruguay were essentially unsettled and only minimally part of an economic system that focused upon the political center of Lima and the silver mines of Potosí.

Towns and cities preceded agricultural settlement and were imperative to effective Spanish occupance, as indicated in chapters 3 and 6. Spanish farms focused on a network of towns that served as trading, manufacturing, and administrative centers. Towns, cities, and farmlands altered the landscape that had been little changed by indigenous peoples. Roads linked towns, and political jurisdictions tied them together or kept them apart. An increasing share of the colonial population became urban. Some towns became important provincial political, market, and production centers, their affluence evident in fine churches, homes, and public buildings.

In the colonial period, the major economic function of the Southern Cone was to provision the Potosí silver mines of upper Peru (Bolivia) with foodstuffs, mules, and horses. Valleys in northwestern Argentina and a portion of the Central Valley of Chile, which were home to sedentary Indians, came into the Potosí sphere. Elsewhere, fierce nomadic Indians denied access to the Pampas and to much of the Central Valley of Chile. Official directives forbade direct trade with Europe in favor of trade through Lima. Until late in the colonial period, Lima merchants suppressed what would have been natural trade routes to Europe from Atlantic ports of the Southern Cone.

Until the mid-1700s, Buenos Aires thus dealt more in contraband than in legal trade. Montevideo was established (1726) to block southward Portuguese expansion rather than as a trading focus. The small port of Asunción (1537) on the navigable Paraguay River was isolated from Potosí by the Chaco and blocked from direct outside ties by Buenos Aires. Politics in Lima and the difficult passage around Cape Horn precluded much contact with Europe for Santiago and Valparaíso.

► *Expanding European Contact, 1820–1950*

The nineteenth century altered the cultural landscape of the Pampas, the littoral (coastal zones) of Argentina, and southern Chile. Independence brought ties to the urban markets of the United States and Western Europe and led to a restructuring of the new national economies. Port cities of the Southern Cone grew in importance as they exported agricultural and mineral resources to world markets.

The evolution in the colonial period of distinct clusters of population, each cluster separated by great distances and/or physical obstacles, gave force to separate national aspirations in the nineteenth century. The heart of Chile lay within the bounds of the colonial Captaincy General of Chile, with Santiago its capital. The remainder of the Southern Cone was under the jurisdiction of the Viceroyalty of La Plata, created in 1776 with Buenos Aires as its capital.

The United Provinces of La Plata (modern Argentina, excluding Patagonia) sought to hold the colonial viceregal boundaries in place, but Bolivia, Paraguay, and Uruguay were independent by the late 1820s. Moreover, it was only later civil wars and economic disparities that ultimately bound the once-prosperous colonial oasis economies of the Argentine northwest to the economic ascendancy of the province of Buenos Aires and created modern Argentina.

The colonial city of Asunción gave an urban focus to Paraguay, still a distant and environmentally difficult transitional zone between Spanish and Portuguese interests. Montevideo was the nucleus of Argentina's *Banda Oriental* (land east of the Río Uruguay), but the Portuguese had territorial aspirations and there were proponents of an independent Uruguay. Britain sought commercial access and in 1828 sponsored a compromise resulting in Uruguayan independence. The country, consequently, became a buffer state between Argentina and Brazil with major economic ties to Britain.

The bounds of Argentine Patagonia were established in 1881, but lingering boundary disputes between Argentina and Chile were resolved only in 1902 after a British-mediated treaty averted war. The treaty led to the partition of Tierra del

Fuego, demilitarization of the Strait of Magellan, and its classification as an international waterway. As late as the 1980s, national jurisdiction for a handful of islands was still being settled.

Conflict between Argentina and Chile paled in comparison to the territorial disputes of Paraguay with Bolivia, Uruguay, Argentina, and Brazil. Two major wars proved costly to Paraguay. Particularly bloody was the War of the Triple Alliance (1865–1870), pitting the strong Paraguayan army against Brazil, Uruguay, and Argentina. Paraguay was devastated, its population drastically reduced. In the 1930s, the Chaco War resulted in Paraguay gaining territory from Bolivia.

As the new nations developed, they became more dependent upon the world economy. Economically and politically, Britain was the dominant outside influence until the worldwide 1930s depression. United States economic interests focused largely on Chile's nitrates and copper.

By the 1880s the crop and livestock potential of the Argentine and Uruguayan Pampas was developing. The grasslands, largely owned by powerful Argentine families, were cultivated by immigrant labor and provided wheat and beef for export to Europe. In Paraguay, the influence of immigration and world markets was less, and apart from the export of tannin from the rich quebracho wood, the country was not drawn into the global economy.

Paralleling the export of agricultural and mineral wealth was a high volume of European immigration that reshaped colonial social structures by reducing the influence of the colonial *criollo* interests. Immigrants from Spain, Italy, Germany, and eastern Europe arrived in the late nineteenth and early twentieth centuries. Many migrants came as agricultural workers but later moved to the cities, resulting in rapid growth of major ports and regional centers.

► Changing Global Impact, 1950–2000

The political and economic geography of the Southern Cone was transformed by the 1930s Depression and World War II (1939–1945). During the years 1930–1945, countries broadened their industrial base, developing import-substitution industries. After the war, Latin American manufacturers faced more competition, but nationalistic economic policies provided protection. At the end of the Korean War, world prices for industrial minerals and agricultural products declined, contributing to economic stagnation and inflation.

Economic problems were exacerbated by politics. Corrupt and incompetent civilian governments alternated with corrupt and inefficient military governments. The dictatorship of Juan Perón (1945–1955) and later military governments in Argentina, the dictatorship in Paraguay of Alfredo Stroessner (1954–1989), and the repressive government of Ernesto Pinochet in Chile (1973–1989) are the best known examples of quasi-fascist political influences shaping society and economy. Uruguay, which had established social programs by the 1920s, was spared the worst but suffered repression of human rights.

Rural-to-urban migration accelerated after 1930; Buenos Aires grew from 2 million inhabitants in 1930 to some 5 million in 1950 and 6.8 million by 1960. Today, the sprawling city contains more than 12 million inhabitants. Similar rural-to-urban forces were at work in Chile, and metropolitan Santiago grew from 600,000 in 1930 to more than 5 million today. Asunción grew from 110,000 to 310,000 between 1930 and 1960. Montevideo, as capital of a small and already urban land, grew more slowly, from 450,000 in 1930 to 1,200,000 in 1960, a level it has retained. In the last 60 years national and metropolitan population growth has come almost entirely from natural increase and rural-to-urban movement rather than international immigration.

Efforts are made to redirect internal migration away from primate cities. In Chile, growth is promoted in existing regional centers, and the national Congress was moved in 1989 to Valparaíso from Santiago. In Paraguay and Uruguay, efforts to improve the attractions of outlying areas are underway. In Argentina, plans were developed to move the capital of national government from Buenos Aires to Viedma, at the northern edge of Patagonia. The scheme was designed to promote growth on the Argentine frontier and to deflect criticism from the military for its failure to capture the Falkland Islands from Britain in 1982. The plan is moribund.

The four national economies depend on a somewhat different mix of agricultural and mineral exports and the more recent growth of domestic manufacturing and services (Tables 13.4 and 13.5).

TABLE 13.4   Southern Cone: Economic Structure

Employment by Major Economic Sectors				
	Percent of Employed Population			
	Argentina[a]	Chile[b]	Paraguay[b]	Uruguay[c]
Agriculture, hunting, forestry, and fishing	12	18	38	15
Manufacturing	20	16	11	18
Construction	10	7	8	5
Wholesale and retail trade, restaurants and hotels	17	16	13[d]	12
Transport, storage, and communication	5	6	3	5
Finance, insurance, real estate (FIRE)	4	5	n.a.	4
Government, social, and personal services	24	25	13	31
Mining and quarrying	e	2	e	e
Other	8	5	16	10
Total labor force	9,989,000	4,794,000	1,584,000	1,177,000

a = 1980 data; b = 1991 data; c = 1985 data; d = includes FIRE; e = less than 1 percent; n.a. = not available.
Source: *Europa Year Book*, 1995

	Gross Domestic Product by Economic Activity (percent)			
	Argentina	Chile	Paraguay	Uruguay
Agriculture, forestry, and fishing	7	9	27	10
Mining and quarrying	2	9	a	b
Manufacturing	24	18	17	24
Construction	5	5	6	4
Electricity, gas, and water	2	3	3	3
Transport and communications	5	7	6	6
FIRE	15	n.a.	a	25
Trade, finance, restaurants, and hotels	16	18	39	12
Government, social, and personal ser vices	25	n.a.	4	18

The data is more suggestive than strictly comparable due to different national accounting systems. Totals do not always equal 100 percent.
a = subsumed under another category; b = less than 1 percent.
Source: *Europa Year Book*, 1995

TABLE 13.5   **Southern Cone: Agriculture**

	Production (metric tons) 1980 and 1991						
	Ranking of Agricultural Production					10^3 metric tons	
Country	1	2	3	4	5	Wheat	Corn
1991 Argentina	Sugarcane	Soybeans	Corn	Wheat	Grapes	9,867	7,768
1980	Sugarcane	Corn	Wheat	Sorghum	Grapes	7,780	12,900
1991 Chile	Sugar beets	Wheat	Potatoes	Corn	Apples	1,589	836
1980	Wheat	Grapes[a]	Potatoes	Sugar beets	Corn	966	405
1991 Paraguay	Sugarcane	Cassava	Soybeans	Cotton	Corn	260	401
1980	Sugarcane	Cassava	Soybeans	Corn	Cotton	44	506
1991 Uruguay	Sugarcane	Rice	Potatoes	Wheat	Sugar beets	188	124
1980	Sugar beets	Sugarcane	Wheat	Rice	Potatoes	300	119

a = #6 in 1991.

	Livestock Inventory (000 head) 1981 and 1991			
Livestock	Argentina	Chile	Paraguay	Uruguay
1981 Cattle	54,200	3,800	5,400	10,971
1991	50,580	3,461	7,627	8,889
1981 Sheep	30,700	6,800	430	20,429
1991	26,500	4,689	257	25,986
1981 Goats	5,100	1,100	135	n.a.
1991	3,320	600	102	n.a.
1981 Horses	2,900	450	330	530
1991	3,400	339	320	470
1981 Pigs	4,649	1,000	1,310	450
1991	2,600	1,226	2,580	215

n.a. = not available.
Source: *Europa Year Book*, 1990 & 1995.

PRINCIPAL ECONOMIC-CULTURAL
REGIONS OF THE SOUTHERN CONE

We have already described the major physical regions of the Southern Cone and their subregions and suggested the manner in which four major time periods created distinct cultural landscapes. It is now useful to identify three major regions as they exist today. These are (1) the Chilean, Argentine, and Uruguayan heartlands, (2) a set of long-existing and distinctive peripheral areas, including Paraguay, in the northern reaches of the Southern Cone, and (3) new southern frontiers that have been developing particularly in the past several decades (Figure 13.6).

Figure 13.6 Generalized regions of the Southern Cone.

▶ *Chilean, Argentine, and Uruguayan Heartlands*
The Chilean Heartland

The economic and political heartland of Chile is known as Middle Chile, the area between Copiapó, at the southern fringe of the Atacama Desert, and Puerto Montt, at the southern end of the Central Valley (Figures 13.7 and 13.8). In turn, distinctions are made between Middle Chile's core and its transitional zones. The

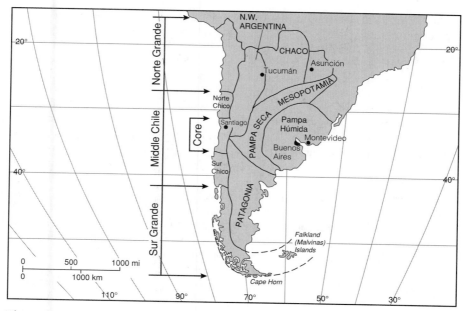

Figure 13.7   Southern Cone: Physical regions.

core lies between the ports of Valparaíso and Concepción, the second and third largest cities. Santiago is the focus of the core.

Within Middle Chile, the transitional zones are *Norte Chico* and *Sur Chico*. Beyond the transitional zones are *Norte Grande*, harsh, arid mountains and basins of the Atacama Desert and Andean highlands, and *Sur Grande*, a distant southern frontier of humid, cold, isolated forests and mountains.

Spanish conquistadors arrived in Middle Chile from Lima, moving south through the dry highlands of Bolivia, northwestern Argentina, and the arid Atacama.

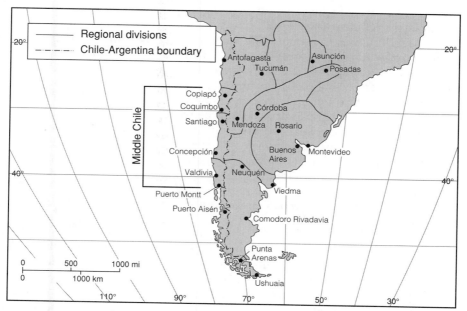

Figure 13.8   Southern Cone: Major economic regions and urban places.

In the Central Valley between the Andes and the coastal range were a Mediterranean-type climate, fertile lands, and Andean streams to provide water for agriculture.

The hostile Araucanian Indians initially made Spanish settlement in the Central Valley difficult. The 1541 settlement of Santiago del Nueva Extrema ("St. James on the New Frontier") and its 1543 port at Valparaíso ("Paradise Valley") survived. Farther south Concepción was destroyed (for the first of three times) within seven years of its 1552 foundation, while Valdivia (1552), Osorno (1559), and lesser settlements were laid waste by Araucanian Indians at the end of the century. In the Chilean district of Cuyo, across the Andes, the settlements at Mendoza (1561), San Juan (1562), and San Luis (1594) endured, even if Mendoza had to be relocated within a year because of an earthquake. Earthquakes are a major hazard to cities on both sides of the Andes.

Under the terms of a post-1600 treaty, the Indians retained the lands south of the Bío-Bío, and Spanish settlement was checked at the river for centuries (Figure 13.9). North of the river, settlement was initiated late in the 1730s and included the reestablishment of towns earlier destroyed by the Indians. North of Santiago, towns were established in Norte Chico. Some of them, like San Felipe (1740), were tied to gold and copper mining as well as agriculture.

Settlement in the forests south of the Bío-Bío coincided with German immigration in the mid-nineteenth century. The influx led to the growth of Valdivia, the port of Corral, Puerto Montt (1853), and Puerto Varas at the southern end of the Central Valley. A penal colony and coaling station were established at Punta Arenas (1849) in the extreme south. The ultimate submission of the Indians to European pressures occurred between 1880 and 1883, the same period as the last campaigns against the Indians of the Pampas and the western United States.

The growth of the Chilean urban system since the mid-nineteenth century is indicated in Table 13.6. Since its founding, Santiago has been the major city and the ports at Valparaíso and Concepción became the largest outlets for the agricultural exports of the Central Valley. Viña del Mar (1556) is now part of Valparaíso, just as Talcahuano (1730) is an integral part of the Concepción metro area.

Central to the urban system is Santiago (1541). Santiago was the capital of the Captaincy General of Chile, an area that encompassed present-day Chile, as well as the Cuyo district east of the Andes across the Uspallata Pass. The Cuyo district, in which Mendoza, San Juan, and San Luis were settled from Chile before 1600, became part of Argentina (the Viceroyalty of Río de la Plata) in 1776.

At independence in 1818, Santiago became the capital of Chile. On the eve of modern commercial and industrial growth, in 1865, the city had a population of 116,000. Today, the metropolitan area has over 5 million inhabitants (Table 13.6).

The port of Valparaíso was the link between colonial Lima and the Chilean "new frontier." Its role expanded modestly after trade was allowed between Lima and Spain via Cape Horn late in the colonial period. Rapid development paralleled the nineteenth-century growth of Santiago and the Central Valley and increased traffic around Cape Horn. With the 1914 opening of the Panama Canal, Valparaíso and Concepción lost trade and ship repair work.

Valparaíso is Chile's second largest city, its principal port, and a manufacturing center. Adjacent Viña del Mar evolved as a beach resort and fashionable suburb of Valparaíso. The economic base has broadened as the cities have grown together. Today the Valparaíso area has a population of more than 600,000. Beyond Viña del Mar is Concon, where oil from Tierra del Fuego is refined. A four-lane highway links Valparaíso to Santiago, 80 miles inland.

A promontory protects the Valparaíso harbor from the southerly winds that impel the offshore current. The downtown is built on reclaimed land, with major residential areas on hillsides and atop low hills, several of which are reached by

(A) 1541–1730

Arica

Taltal

Copiapó

San Juan

Valparaíso/
Viña del Mar
Santiago

Mendoza

San Luis

(to Argentina, 1773)

Concepción

Bio-Bio River

Valdivia

● 1541–1600 (20)
○ 1600–1730 (4)

(B) 1730–1850

Pisagua

Huasco

San Felipe

Constitucíon

Lebu

Bio-Bio River

Punta Arenas

● 1730–1850

(C) 1850–1990

Chuquicamata

Caldera

Coquimbo

Bio-Bio River
Temoco

Puerto Montt

Puerto Aysen

● 1850–1990

```
0          400          800 mi.
|-----------|-----------|
0          650          650 km
```

Figure 13.9　Town foundings in Chile.

now-antique inclined elevators. Like most Chilean cities, Valparaíso is subject to earth
quakes and was destroyed by one in 1906.

The metropolitan district of Concepción/Talcahuano is the third largest in Chil
and marks the southern end of central, Mediterranean Chile. Concepción lies 6 mile
up the Bío-Bío River from the Bay of Concepción; a deep-water port at Talcahuan
shelters a large naval base and fishing center. Under Indian attack Concepción wa

TABLE 13.6   Chile: Population of Selected Cities, 1865–1990 (in thousands)

	1865	1895	1920	1940	1960	1975	1990[a]
Santiago	116	257	508	953	1,908	3,232	4,386
Viña del Mar*	—	11	36	66	116	235	281
Valparaíso*	71	123	183	210	253	249	277
Talcahuano**	3	11	23	36	84	187	247
Concepción**	14	40	65	86	149	171	307
Antofagasta	—	14	52	50	88	153	219
Temuco	—	8	29	43	73	141	212
Rancagua	6	7	18	32	54	111	191
Talca	18	34	37	51	69	118	165
Chillán	15	29	31	43	66	104	146
Arica	—	3	10	15	44	118	178
Iquique	—	33	38	39	51	74	149
Valdivia	4	9	27	35	62	99	114
Puerto Montt	—	4	10	22	42	79	107
Punta Arenas	1	4	21	30	50	72	120
Los Ángeles	4	8	14	21	36	61	94
Calama	—	—	4	5	27	61	96
Coquimbo	8	8	16	19	34	61	81
Chile	1,820	2,688	3,715	5,024	7,375	10,254	13,348
% Urban	29	46	47	53	69	81	85

*Valparaíso and Viña del Mar constitute a single metro area, 584,000 total in 1985.
**Concepción and Talcahuano constitute a single metro area, at least 439,000 total in 1985.
a = 1990 figures are census estimates. Chilean total is for 1992.

abandoned several times after foundation; the present town is the fourth. Earthquakes have taken a toll. Of three major quakes, the 1939 event proved particularly severe.

The metropolitan area has development potential. The broad valley of the Bío-Bío provides access to the rich agricultural and forested hinterland. There is hydroelectric power from the Andes. Since 1950, Huachipato (within the metro area) has been the site of a state-run iron and steel plant, built when import-substitution industries and a domestic iron and steel capacity were considered elements of economic independence. Local and imported coal are mixed for coke production. Iron ore is shipped from Coquimbo province in the north and limestone from the southern archipelago. Concepción is the location of Chile's second largest oil refinery and a growing petrochemical industry.

In the colonial period, the Central Valley was divided into haciendas (*fundos*), to which workers were bound by custom, not law. Agriculture was extensive, not intensive, oriented to the production of livestock, cattle feed, and a few cash crops. After Independence, the pattern persisted. In the 1960s agrarian reform began to reduce the dominance of large holdings. With the shift came diversification. Pastoralism gave way to fruits and vegetables for export.

During the rule of Salvador Allende's Marxist government (1970–1973), agricultural production dropped due to unrealistically low farm prices, rising production costs, badly administered agricultural reform, inefficient collective farms, and the unwillingness of private farmers to produce. In the Allende period, most farms over 200 irrigated acres were expropriated and even smaller farms were seized.

Policies of the post-1973 military government contributed to a rebound in agriculture but increased the flow of unskilled rural workers to the cities.

The primary activities of farming, fishing, forestry, mining, and quarrying are important in contemporary Chile and provide most export income. Chilean manufacturers, protected by tariffs and distance from world competition, produce for domestic markets. With the exception of mining, Middle Chile is the major focus of all sectors of the economy, although one of Chile's three largest copper operations, El Teniente, is southeast of Santiago.

In the mid-1970s, copper accounted for 80 percent of the value of Chilean exports. As a result of economic diversification, by the late 1980s it represented 44 percent. Today, copper still accounts for more than 40 percent of the value of exports. Other exports include apples, grapes, wine, fish (including fish-farmed varieties of Northern Hemisphere salmon), lumber and forest products, and diversified mineral resources, including gold. Fruit accounts for 12 percent of the value of exports. Table grapes account for over half the total, but wine exports are increasing. The government began to encourage grape exports in 1960, and today hundreds of million pounds reach northern markets annually. Vineyards are concentrated in the Central Valley. With drip irrigation technology, however, vineyards have been brought into production at the northern and southern extremes of Middle Chile extendiing the vineyard zone over 700 miles with the most productive areas being south of Santiago.

Agricultural exports benefit from the fact that the markets for Chilean farm products lie in the Northern Hemisphere, where seasons are the reverse of those of the Southern Cone. Apples, apricots, cherries, citrus fruits, pears, peaches, plums,

In the Central Valley of Chile grapes are a major crop. Exports of wine are growing rapidly.

Lake Todos los Santos in the Lake District of Southern Chile–Argentina.

almonds, and walnuts are other Central Valley crops exported to Northern Hemisphere markets.

Sur Chico lies between the Bío-Bío River and Puerto Montt. It is a transition zone between Mediterranean Chile and the cold, wet fjord regions of *Sur Grande*. A land of beech and pine forests, lakes and volcanic peaks, *Sur Chico* was until the 1880s the last redoubt of the Araucanian Indians. The area around Temuco is called Araucania and contains the last native population. Forest remains but land has been cleared for agriculture, principally by German settlers after 1850. The region is Chile's major wheat district; other crops include apples, potatoes, and peas.

Valdivia processes agricultural and forest products and is the focus of *Sur Chico*. At the southern end of *Sur Chico* is Puerto Montt, which is a center of lumbering, livestock raising, fishing and salmon farming, and tourism. Tourist appeal is based on trout-rich lakes that stretch south from Lago Colico near Temuco, past Lago Osorno to Lago Llanquihue, where a German colony was established at Puerto Varas (founded 1853). Above the lakes are volcanic cones, including Villarica and Osorno, and Andean peaks. Earthquakes are common. A road connects Puerto Varas and Bariloche, linking the Chilean and Argentine system of morainal lakes that attract tourists from the Southern Cone and Brazil.

Wheat is the leading grain crop, but forests dominate the landscape between Concepción and Valdivia. The native tree cover is characterized by mixed beech and cedar forests. Commercial forests primarily grow radiata pine, introduced from California's Monterey Peninsula. The radiata reaches harvest size in about 25 years, more quickly than in the Pacific Northwest or California. The mid-1970s surge in plantings is now being harvested and contributes to Chilean timber exports. The largest market is Japan. In the 1980s the first eucalyptus plantations were established to provide raw material for wood pulp plants. The trees are harvested in 8 to 10 years, and labor costs are low.

Heartland of Argentina: The Pampas

The forces that shaped the urban and agricultural landscape of the Argentine Pampas after about 1860 are similar to those that molded the American West. Like the grasslands of North America, the Pampas were scarcely settled by Europeans before the mid-nineteenth century. Colonial policies and mercantile constraints combined with fierce nomadic Indians and the absence of adequate transportation to preclude mercantile growth.

At the beginning of the nineteenth century, European settlement on the Argentine Pampas was confined to the shores of the Paraná and large *estancias* (ranches) between the Río de la Plata and a line of forts on the Salado River. The frontier area was home to the gaucho (a mixed-breed cowboy) and supplied cattle, horses, and mules to Buenos Aires. Wheat was grown close to the city. Cattle were slaughtered, their hides treated, and the beef dried in the city's *saladeros* for export to foreign markets.

The first railroad onto the Pampas opened in 1857, and the first to link the littoral with the colonial core in the Argentine northwest opened in 1870 between the river port of Rosario and Córdoba. Since colonial times, Córdoba has been a major transition point between the towns and economy of the Argentine northwest and the Pampas. Rosario grew as the outlet for the agricultural produce of the northwest and the nearby Pampas lands being developed by European agricultural colonies.

By 1870, the Pampas was being transformed into a major agricultural zone by industrialization in Europe, the opening of the Pampas by the railroad, technological advances in agriculture and transportation, and large-scale immigration from Europe onto the unsettled Pampas lands, especially after the Indian threat was removed. Cattle and sheep raising in the south and in Patagonia remained the dominant activities, but arable farming expanded rapidly, with wheat and corn being the major crops.

From the 1870s, railroads linked the ports of Buenos Aires, La Plata, Rosario, and Bahía Blanca to the increasingly settled and farmed Pampas lands (Figure 13.10). Agriculture, it was said, arrived on the train. Initially, the rail lines served existing agricultural areas close to Buenos Aires but then opened up more distant lands. Older military settlements became agricultural marketing centers, and scores of small farm towns grew up along railroads. By the 1890s, Argentina had an extensive railroad system, the largest in South America.

European industrialization and urbanization created demand for farm products. European prosperity created surplus capital, particularly in Britain. British capital, technology, and business methods were at the forefront in the development of the Pampas. Technological advances, notably the windmill and barbed wire, allowed areas to be fenced and farmed for the first time. Nowhere else in Latin America was there such a large, mid-latitude region suitable for mechanized agriculture. The Pampas continues to be the continent's principal producer of grains. Meat, wool, and hides are still of major importance.

After 1877, the meat-packing plant (*frigorífico*) replaced the beef-salting plants (*saladeros*). Larger and faster ships with iron hulls, screw propellers, and shipboard refrigeration carried foodstuffs across the Atlantic and accelerated agricultural development (Chapter 4).

The ships off-loaded a flood of immigrants, particularly from Italy and Spain, into Argentina. Smaller numbers of eastern Europeans came later. Many

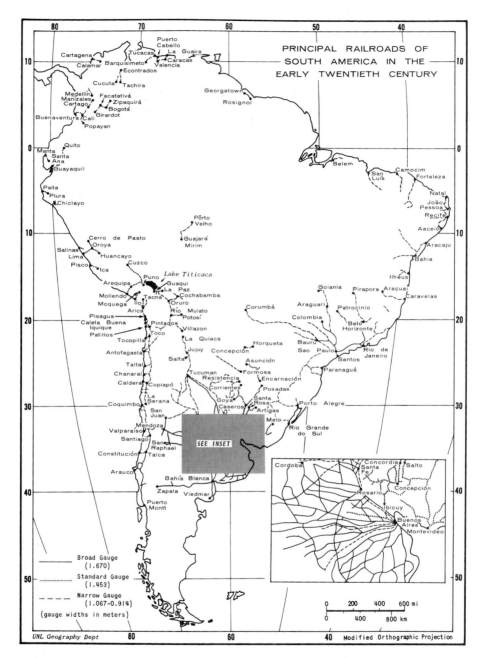

Figure 13.10   Principal railroads of South America in the early twentieth century.

immigrants went to the Pampas as laborers under a system that favored the planting of grains and alfalfa before the workers were forced, by terms of their land leases, to move on to new lands or return to Buenos Aires. Many chose to do the latter given the few opportunities for landownership. On the Pampas, Argentine society became European in character with little Indian admixture.

Within a few decades, between 1880 and 1910, the Pampa Húmida was transformed from Indian lands and cattle country into an agricultural zone producing wheat, corn, flax, and meat for export. Livestock farming was dominant, and meat, wool, and hides were the major exports. The adoption of imported breeds of cattle and sheep was combined with advances in agricultural machinery. Rich native grasses (*pasto duro*) were replaced by European grasses (*pasto terno*), with alfalfa being the principal fodder crop.

Wheat growing was characteristic on the frontier of the Pampas while corn was important in the northern areas around Rosario. Cattle were raised throughout the Pampas, but sheep raising was more important in the drier regions toward the *Pampa Seca*.

Buenos Aires: A Case Study Buenos Aires has been transformed since the 1870s from a small port city surrounded by villages into a metropolitan area of 12 million inhabitants. In 1850 Buenos Aires was a city of 90,000. At the edge of the city, along the roads that led onto the Pampas, were small colonial towns and farming districts. By 1870, there were 170,000 people in the city and 50,000 in the outlying areas connected by rail to the port at Buenos Aires (Table 13.7).

TABLE 13.7 Argentina: Population of Selected Cities, 1869–1991 (in thousands)

	1869	1895	1914	1947	1980	1990
Greater Buenos Aires	223	774	2,019	4,722	9,928	10,887
Buenos Aires (city)	171	656	1,561	2,981	2,923	2,961
Córdoba*	29	48	122	370	983	1,198
Rosario	24	93	227	468	955	1,096
Mendoza*	9	29	59	98	597	774
La Plata	—	46	102	208	561	641
Tucumán*	18	35	94	195	497	623
Mar del Plata	—	6	28	115	416	520
Santa Fe	11	23	60	169	288	395
Salta*	12	17	29	68	261	371
San Juan*	9	11	22	83	291	354
Bahía Blanca	—	10	50	113	224	256
Resistencia	—	2	9	53	219	292
Corientes	12	17	29	59	180	258
Paraná	11	25	37	85	160	212
Santiago del Estero*	8	10	24	61	149	265
Jujuy*	4	5	8	32	125	183
San Nicolás	6	13	20	26	99	115
Catamarca*	6	8	14	32	89	134
La Rioja*	5	6	9	24	67	105
San Luis*	4	10	16	26	71	111
Argentina	1,831	4,045	7,904	15,894	27,950	32,900
% Urban	29	37	53	63	83	87

*Northwest Argentina.

Note: 1980 and 1991 data, excluding Buenos Aires city, are for metropolitan areas and not strictly comparable with earlier periods. In 1991, the census added additional areas to Greater Buenos Aires, resulting in a population of 11,255,618.

Within Buenos Aires there was a single commercial and political core, oriented north–south in response to the port and east–west along the major road out to the countryside. Until late in the nineteenth century the population was concentrated in this core. The population grew rapidly, to 656,000 in 1895, mostly by immigration. Italians and Spaniards, in a ratio of 5 to 3, accounted for much growth.

Access to the expanding urban fringe before 1900 was by suburban trains and horse-drawn trams (Chapter 6), but the residents of Buenos Aires (*porteños*) remained clustered in the heart of the city. Few had the money to buy land on the suburban fringe, and the slow horse trams offered only an expensive and long journey to work. The horse tram effectively served the dense core, not the far reaches of the city.

Figure 13.11 Buenos Aires CBD and trolley network.

The population within Buenos Aires rose to 2,287,000 in 1930, resulting in central-city crowding and the development of the periphery. *Porteños* could use the faster electric trolley, introduced in 1898, to escape the crowded core (Figure 13.11 and Chapter 6). Demand for land, trolley transportation, and new forms of credit resulted in urban development along the trolley lines. The introduction of the bus freed developers from the trolley lines in the 1920s, and other areas were developed.

Access to the countryside improved as suburban trains encouraged movement into the railroad corridors, creating new towns and turning summer resorts along the estuary into commuter communities. By 1930 the population of Greater (metropolitan) Buenos Aires was more than 3 million.

The poor could not afford to move to the periphery, and most of the wealthy did not want to leave affluent inner-city *barrios*. The middle-income office worker and the skilled industrial worker built up most of the peripheral *barrios* that slowly acquired urban amenities. Initially, entire *barrios* were subdivided without regard to sanitation standards, and rarely did the seller put in a paved street or other facilities. There was no zoning, and land uses were mixed, with residences, stores, and industrial facilities often being close together. By 1930, the metropolitan area had a crowded core—composed of the poor in tenement slums ("conventillos"), the wealthy in mansions and apartments—and a sprawling periphery.

From 1930 to 1990, the population of Greater Buenos Aires rose from just over 3 million to 12 million, and a large part of the growth went to the periphery. By 1960 more people were living outside the Federal District than within it, many in squalid squatter settlements (*villas miseria*).

The growth of Buenos Aires after 1930 exhibited distinctive patterns: (1) in-filling of the periphery; (2) high central-city population densities; (3) increasing population density in the upper-income sectors as high-rise apartment blocks were built, in part replacing older mansions; (4) population stagnation in the older industrial lowlands; and (5) large-scale growth of peripheral *villas miseria* as migration from the provinces added to a housing shortage. The central city continued to grow in importance after 1930 as subway lines built in the 1930s and early 1940s focused on the Central Business District (CBD) as did the trolley and train lines. In the 1950s and 1960s, outlying commercial centers grew, but the core of Buenos Aires is still dominant in commerce, culture, and entertainment. Buenos Aires is the primate city of Argentina, although some of the interior centers are recently becoming increasingly economically viable.

On the Pampas, Rosario, with over a million inhabitants, is the chief city and river port of Santa Fe province. At San Lorenzo, upstream from Rosario, is the terminus of an oil pipeline from Campo Duran and a large petrochemical complex.

Along the river between Rosario and Buenos Aires is San Nicolás, a small colonial city with a minor port, which became an important industrial center after the opening of the large state-owned steel plant in the 1950s. Metal processing is important to Villa Constitución, between San Nicolás and Rosario. The nearby industrial center of La Emilia gained a reputation for fine woolen textiles and yarn. In the modern period, Rosario has vied with Córdoba to be the second largest city, but Córdoba is now second and is likely to remain so.

After the selection of Buenos Aires as national capital in 1880, the provincial government was moved in 1882 to La Plata, a wholly new, planned town and port 35 miles downstream from Buenos Aires. The new city was laid out in a gridiron of streets with an overlay of diagonal avenues. Tied to Buenos Aires by commuter bus and rail service, it is a commercial outlier of downtown Buenos Aires. South along the Atlantic coast is Mar del Plata, Argentina's principal seaside resort, fishing port, and center of woolen textile manufacturing.

Bahía Blanca evolved as the major wheat port for the southern Pampas after the development of railroads, but its proximity to drier and less fertile portions of the Pampas makes it less important than its location would suggest. A natural gas pipeline from the Neuquén field in Patagonia to the port has led to the development of a petrochemicals industry.

The Uruguayan Heartland

The grasslands of Uruguay, like the Pampas of Argentina, were effectively settled only after Independence. As with Argentina, it was foreign investment in railroads and the introduction of European livestock and agricultural technology that converted simple grasslands to commercial agriculture.

In the colonial period, Uruguay was disadvantaged by distance from Potosí, lack of minerals, and the ferocity of the nomadic Charrua Indians. The area began to emerge in 1680 when the Portuguese fortified Colonia do Sacramento on the north shore of the Río de la Plata, opposite Buenos Aires. In 1726, the Spanish governor at Buenos Aires ordered the creation of Montevideo as an outlying military base to deter Portuguese advances. From 1776 to 1810 Uruguay was part of the Viceroyalty of la Plata; until Independence in 1828, it was coveted by Argentina and Brazil.

Physically, Uruguay is both a continuation of the Argentine Pampas and an extension of the Brazilian Shield. The dominant ground cover in the colonial period was tall grasses, and more than 70 percent of Uruguay remains in pasture. Pasture use by Europeans began with the introduction of cattle from Paraguay onto the east bank of the Uruguay River—*la banda oriental del Uruguay*—in 1603. For decades cattle roamed wild and were slaughtered for meat and hides by mixed-blooded gauchos who fought the Indians and resisted European encroachment. Large-scale trade in hides and tallow was initiated by buyers from Buenos Aires, and by 1800 most of Uruguay was subdivided into *estancias*, or cattle ranches. *Saladeros* near the Plata estuary and along the Río Uruguay produced salted and dried beef, cured hides, and tallow. Commercial farming was limited to the area adjacent to Montevideo and small towns on the Río Uruguay. By 1850, the indigenous population had been replaced by a mestizo majority.

The political creation of Uruguay as La República Oriental del Uruguay was complex. Between 1807 and 1828 the territory was alternately occupied by British, Spanish, Argentine, and Portuguese forces. Eventually, British mediation led to independence in 1828 and opened the country to trade and investments in land, railroads, and banking. Until the 1930s, British investments were the major force that shaped the economies of both Argentina and Uruguay.

By the 1890s, Hereford and Shorthorn cattle began to upgrade the quality of Uruguayan beef, and the first *frigorífico*, or beef-freezing plant, was opened in 1904. Earlier, the lower-quality meat had largely been converted into meat extract, using the Leibig process introduced at Fray Bentos in 1864. In the mid-nineteenth century, English sheep were introduced into Uruguay to upgrade the wool clip. No other Latin American country is so dependent on exports of cattle and sheep products.

Today, on the grasslands of Uruguay are 26 million sheep and almost 9 million cattle (Table 13.5). About half the grassland is in *estancias* of more than 2,000 acres. The bulk of the arable farmlands, 12 percent of Uruguay's land area, is in the lowlands adjacent to the Río Uruguay and Río de la Plata. The preference for pasture over cropland is due to the excellence of the grasslands and the variable rainfall that makes grain production unreliable. The ratio between sheep and cattle production shifts with demand.

Climate, relief, and soils vary and influence the pattern of land use. Wheat, flax (for vegetable oil), corn (as a feed grain), oats, and barley are grown primarily for domestic consumption. Exports vary because levels of production are closely related to frost damage and drought. Sugarcane and citrus are characteristic of the Salto area, and dairying, truck farming, and fruit growing—particularly apples, pears, and peaches—are common in zones facing the Río de la Plata. Grapes for wine are grown near Montevideo. Sugar beets are common to the east, near Maldonado. The major crops are indicated in Table 13.5. Extensive areas of rice production are found west of Mirim Lagoon and along the northern tributaries of the Río Negro. Exports take 80 percent of production.

The coastal zone is the major industrial district of Uruguay. The industries reflect the nation's agricultural base, with much production tied to livestock and food processing. The economic Depression of the 1930s and World War II favored the creation of import-substitution industries. Until the 1930s, Britain was the major export market for beef, wool, and lamb and the major source of manufactured imports.

As in Argentina, British investment in railroads (nationalized in 1948), meatpacking plants (*frigoríficos*), and other industries spurred industrial development. Like the road system, the railroads radiate from Montevideo. Almost 90 percent of the nation's import/export business passes through Montevideo, whose port has been deepened and channels protected by breakwaters. Most exports move to port by truck rather than rail, and rail passenger service has been wholly replaced by buses. In Argentina, the luxury motorcoach has also largely replaced the railroad, providing faster, more frequent, more comfortable, and less expensive passenger service.

Fishing is of modest importance, but efforts are being made to expand fishing as a source of foreign currency.

The sandy beaches that lie between Montevideo and Punta del Este, and beyond to the Brazilian border, constitute a major focus of the tourism industry that caters to Argentines and Brazilians. After 1965, development along the 43-mile strip between Piriapolis and Punta del Este was helped by installation of a water supply system to serve beach districts. Punta del Este, a whaling and fishing center by the 1830s, opened its first hotel in 1890 and saw its first property boom in 1905. Today it is the major seaside resort of Uruguay and draws a large number of Argentines to it and other beaches along the coast east of Montevideo.

Founded in 1726, the city of Montevideo had a population of some 15,000 by 1800, but was politically and economically subservient to Buenos Aires (population 40,000), and both were smaller than Rio de Janeiro (100,000). Between 1836 and 1926, 700,000 Europeans migrated to Uruguay; by 1840 Montevideo had more foreign- than native-born inhabitants. By 1875 the population of Uruguay had risen to 450,000 from 74,000 in 1830; it would reach 1 million by 1900, with one third foreign-born. In addition to Spanish and Italian immigrants there were Germans, eastern Europeans, and British. European Jews arrived in the 1890s, with subsequent waves in the 1920s and the 1930s. The last law to promote immigration was passed in 1890; the first to curtail the flow in 1911.

Immigrants caused Montevideo to expand beyond its colonial framework, with residential accretions at the edge and the poorer immigrants crowded into inner-city slums (*conventillos*). The expansion of electric trolley service after 1900 led to more distant suburbs, and by 1920 the city's population had grown to 300,000, up from 90,000 in 1880. By 1940 that had risen to 770,000, and there was major residential growth to the north and east along the beaches at Pocitos, Buceo, Malvin, and as far as the Garden City of Carrasco. Lower-income groups tended to move northwest of the city. Since 1945 pockets of squatter settlements (*cantegriles*) have appeared within established neighborhoods.

Today, the city has grown to more than 1.5 million, more than 50 percent of the national population. By and large, the population is highly literate, middle class, and protected by one of the broadest social welfare systems in the Americas.

► *Peripheral Regions of the North*

The Argentine Northwest

The role of towns and cities as loci of regional wealth, points of production and exchange, is evident in the northwestern provinces of Argentina where urban areas evolved to supply the Potosí silver industry. Oldest of the towns oriented to Potosí was Santiago del Estero (1553). Later came Tucumán (1565) and Salta (1582), which lay along the supply route to Potosí. La Rioja (1591) was an extension south along the Andean foothills, while Jujuy (1593) filled in the route north. Catamarca was not established until 1683.

Farther south settlement advanced from Chile via the Uspallata Pass into the Cuyo region. Mendoza was founded in 1561 and rebuilt in 1562 after an earthquake. San Juan (1562) was a second oasis, while San Luis was established in 1594 as a link to Buenos Aires, founded for the second time in 1580.

For many decades, the towns enjoyed a prosperity that came from Potosí's wealth, but as the silver trade declined, so did the towns of the northwest. Ultimately, independence separated the northwest from Potosí, now in Bolivia. The northwest was landlocked and peripheral in the nineteenth-century Argentine economy that focused on the Pampas and Buenos Aires. In the colonial period, the Pampas had been at the economic margin; now it was the core of modern Argentina. Tucumán grew slowly, and Córdoba lost its number two position to the river port of Rosario and would not regain it until the 1970s.

The economic contrast between the poorer interior and the affluent littoral was magnified by large-scale rural-to-urban migration after the 1930s from the northwest to the Pampas and Greater Buenos Aires and the decline of industry in the northwest (Table 13.7).

Although Santiago del Estero was the oldest town in the northwest, Tucumán, a former Inca center, became the economic focus and most important town along the Andes. Platted by the Spanish in 1565, Tucumán benefited from an abundant water supply from the Sierra de Aconquija, a frost-free, subtropical climate, and available Indian labor. The principal sugarcane district of Argentina, with both large plantations and numerous small farms, the area still has Argentina's highest rural population density. Corn and rice are also important.

Extension of the rail system to Tucumán in the 1880s introduced heavy industry, textiles, and additions to food processing and small-scale manufacturing. The city is Argentina's sixth largest, surpassed in the interior by Córdoba and Mendoza.

Since its foundation in 1573, Córdoba has linked the towns of the northwest with the littoral. The city of Santa Fe, on the Paraná River, was founded the same year and for the same purpose, but Córdoba's centrality gave it the edge in trade and manufacturing. In the twentieth century, heavy industry, armaments, automobiles, aircraft parts, tractors, glass, ceramics, leather, and food processing became part of the industrial mix. The economic base also includes universities and tourism, based on the adjacent Sierra de Córdoba. The range stands apart from the Andes and runs 300 miles north–south. There are some forests along the eastern slopes of the mountains, but most is in scrub and grass. Man-made lakes are stocked with fish, and dams provide electric power. Livestock ranching and irrigated agriculture have been important since the colonial period.

Mendoza was founded from Santiago, Chile, but now looks to Buenos Aires. Irrigation is essential here, but the city's location at the foot of the Andes assures adequate water. The smaller oasis town of San Juan, founded contemporaneously with Mendoza, lies to the north and has a similar agricultural base.

The Mendoza region is famous for wines. Large-scale production dates to Italian immigration in the late nineteenth century. A city of tree-lined broad avenues, large plazas, and a vibrant commercial center, Mendoza is a tourist gateway to Andean summer and winter recreation areas. Oil from the Andean piedmont is refined at the city. Freight still moves over the mountains via the Uspallata Pass, but rail passenger service has ceased and the highway is closed in winter. Airline service has become important for Mendoza and other interior cities.

Founded in 1582 along the route to Potosí, Salta lies north of Tucumán in an intermontane basin. In the colonial period, the town was an agricultural center and mule market providing animals for the mines of Potosí and for the transportation of goods to and from the mining district.

A relatively unimportant railroad crosses the Andes from Salta to Antofagasta. The surrounding lands produce corn, wheat, sugarcane, tobacco, and wine. Small fields at Campo Duran and Tartagal yield oil and gas, which are piped to the Atlantic littoral. Sulphur and borax are exported, but mineral deposits in northwestern Argentina are small. There is no copper production to speak of, unlike the Chilean flank of the Andes. Lead, zinc, and sulphur are more common to the Argentine side. Small industries process local agricultural products.

In an intermontane basin to the north of Salta is Jujuy. Agriculture is based on wheat, sugarcane, and crops such as beans, potatoes, and quinoa. Corn is marginal because the cool weather impedes maturation. Llama skins, sheepskins, and wool are produced locally. Salt is extracted from the Salinas *Grande Salar*, and some lead and zinc are produced. Local iron ores at Zapla, south of Jujuy, supply an iron and steel industry, which uses local charcoal as fuel.

Catamarca and La Rioja are small regional centers, located away from both the Mendoza–Buenos Aires route and the Buenos Aires–Córdoba–Tucumán–Salta corridor.

Northern Chile

The hot, dry, barren Atacama Desert dominates the landscape of *Norte Grande* where only one major stream, Río Loa, crosses from the Andes to the Pacific. At Arica, where the Andes press closer to the sea, several streams provide water for irrigation. A number of ports on the coast have to pipe water from the Andes.

As we have seen, agriculture shaped the Chilean heartland. It was, however, mineral wealth that transformed the landscape and economy of *Norte Grande*. The major mineral resources—nitrates and copper—came into production in the nineteenth century with the growth of foreign demand. As a consequence, the importance of *Norte Grande* to Chile is disproportionate to the area's population. For many decades, nitrates and copper were the economic mainstay of the government. Between 1890 and 1920, for example, over half of government revenues came from taxation of nitrate exports. Until the 1970s, copper accounted for some 80 percent of the value of all exports and still represents over 40 percent of the total.

The Nitrate Industry of Northern Chile: 1830–1930

In *Norte Grande*, the Atacama begins at the 3,500-foot crest of the barren coast range and stretches eastward in a series of tablelands (known in Quechua as "pampas") until the highest pampas merge with the Andes. Particularly on the western side of this pampas highland, in the Tarapaca and Antofagasta areas of the Atacama, lie large beds of *caliche* deposited when lakes occupied the upland basins. The nitrate-rich *caliche* beds contain deposits of sodium nitrate, potassium nitrate, and iodine salts.

Deposits of potassium nitrate (saltpeter) were exploited in the colonial period to produce explosives for mining operations. By 1830 European farmers used it as a fertilizer. Chilean and British firms developed the Tarapaca nitrate deposits along the western edge of the *salares*, exporting through Iquique, Antofagasta, and Pisagua.

Before 1883 the nitrates lay in Peru and Bolivia. In the War of the Pacific (1879–1883) Chile, backed by British mining interests, gained the nitrate districts and the ports of Arica and Antofagasta.

At the turn of the twentieth century, there were 200 nitrate *oficinas* mining, refining, and shipping by rail to the coast. Approximately 80 percent of the capital invested was foreign, particularly British. The industry peaked during World War I, and declined when Germans developed synthetic nitrates. Today a few *oficinas* export through Iquique, Tocopilla, Antofagasta, and Taltal.

In 1888, Chile produced a quarter of the world's copper, but by 1904, it was producing only 4 percent. The mines had depleted the richest ores and used outdated machinery. The industry revived in the next decade with United States investment and the development of lower-cost open-pit mining that made lower-grade ores exploitable.

For example, Chuquicamata was developed by Anaconda Copper. One of the world's earliest open-pit mines (1915) and its largest, it still has large reserves of low-grade ore. The operation smelts and refines copper, which is exported via Antofagasta.

Another mine, El Teniente, was developed in 1904 by Kennecott Copper. It is the largest underground mine in the world, producing over 300,000 tons of copper a year. The modernized mine lies in the Andes 50 miles to the southeast of Santiago within the Chilean heartland. Hydroelectric power is used to refine copper, which moves by rail to the port of San Antonio. Smaller mines closer to Santiago supply a smelter at Las Ventanas and export via Quintero.

Today, Chile is responsible for many advances in copper smelting and hydrometallurgy. The state-owned Corporación Nacional del Cobre de Chile (Codelco) now owns four of the major mines—including Chuquicamata and El Teniente, plus El Salvador and Andina—and supplies about 75 percent of production. Since the 1950s, part of Chile's copper income has gone, by law, to the military. In the 1980s, the military government drew up a new "foreigner-friendly" mining code, which has resulted in major outside investments and the development of new mines. Limited environmental restrictions and a labor force that works 12-hour shifts also helped attract investment, especially from the United States, Japan, Australia, and South Africa.

Newest of the large, modern mines is the foreign-owned (Australia, Japan) La Escondida open pit near the Llullaillaco volcano. The richest copper mine in the world (2.8 percent copper ore) initiated production in 1990. Electric power is conducted from Tocopilla, and copper concentrate slurry is shipped through a 170-mile pipeline that descends to port facilities at Colosa, south of Antofagasta.

By the year 2000, Chile is expected to produce over 4 million tons of fine copper per year, compared with 2.5 million tons in 1995. This will increase its share of the world market to 40 percent from the present 31 percent. World copper surpluses are expected by 1999 that will force inefficient operations worldwide to close.

Today, the port cities of *Norte Grande* handle nitrates, copper, and other mineral exports. Diversification has broadened the economic base, and they have grown in recent decades. Antofagasta is the largest city of *Norte Grande* (Table 13.6), a port for iron ore and copper exports and a regional trading center. The city has a rail link to La Paz, Bolivia, and Salta, Argentina.

Arica gained early fame as the port of Potosí. Here, the Andes approach the coast and streams allow irrigation. Arica is the major port for Bolivia. The city was declared a free-trade zone in the 1960s and is a regional minerals processing (copper and sulphur) and manufacturing center.

Iquique, the third largest city of *Norte Grande*, has a well-protected harbor and is Chile's major fishing port and canning center. The city has grown industrially as a result of being made a free-trade zone in 1975. Companies can import goods tariff free and are exempt from local taxes. Coca-Cola is shipped, for instance, from Iquique to southern Peru and Bolivia, and woolen goods are manufactured for the United States market.

Paraguay

The Río Paraguay divides the country into two distinct regions: the lightly settled *Chaco Boreal* to the west of the north–south flowing Río Paraguay, and the *Región Oriental*, home to the majority of the population and the economic focus of the country.

Until several Mennonite colonies, with Filadelfia as their settlement focus, were reestablished in the 1920s and 1930s, the forest and scrub lands of the Chaco ("hunting ground") were largely home to nomadic Indians. The Chaco War (1932–1935) created a large number of small military camps (*fortins*) and established the western boundaries of Paraguay but led to little economic growth or agricultural settlement.

In sharp contrast to the economic and physical marginality of the Chaco is the multifaceted *Región Oriental* that lies east of the Paraguay River. The forests contain hundreds of species of hardwoods, varieties of palms, and medicinal plants. The woods, the native stands and plantations of yerba, plus the quebracho of the Chaco, account for the strength of the forest products industry in Paraguay.

Commercial agriculture is the major economic activity and largest employer of Paraguay. The major food crops are corn, cassava, rice, and wheat. Industrial crops include cotton, vegetable oils, coffee, and tobacco. Cattle are raised in the southern districts and in the Concepción area, with much of the production destined for export.

The Paraguayan mining industry is little developed. Hopes of finding oil deposits in the Chaco have been disappointing. Hydroelectric power potential is huge. The Acaracy hydro plant (1968) released Paraguay from wood- and oil-fired generating plants. The 1982 completion of the Itaipu hydroelectric project on the Alto Paraná River allows Paraguay to export electricity to Brazil and Argentina. A recently completed hydroelectric project is Yacyretá Dam, reputedly the world's largest. On the Paraná River along the Paraguay/Argentina border, the megaproject is destined to stimulate economic growth in the region as well as in northeast Argentina.

Industries in Paraguay include tannin extraction, vegetable oil production, soybean meal, ginned cotton, and yerba. The textile, leather products, soap, furniture, and sugar industries are oriented to domestic markets. The principal outside market was traditionally Argentina, but Brazil is now dominant. Major bridges link Asunción and Encarnación with Clorinda and Posadas in Argentina, while improved paved roads link the *Región Oriental* with the Brazilian port of Paranaguá. Increasing levels of Brazilian immigration, particularly into eastern Paraguay, are strengthening the ties between the countries.

There is a fragile rail service to Argentina that runs, via Villarica, for 274 miles from Asunción to Encarnación and a link at Posadas. Passenger service is available once a week and is slower than the frequent luxury bus service between Asunción and Buenos Aires. Throughout the *Región Oriental*, truck and bus are of major importance. International air service is provided by Lineas Aereas Paraguayas (LAP), formed in 1962. The domestic service has expanded, and the modern airport at Asunción is the hub of the system.

Asunción is the oldest and largest city of Paraguay. The first European settlements came in the period 1536–1556 as Domingo Martínez de Irala peacefully settled colonists among the Guaraní Indians. The fragility of Buenos Aires led to the founding of Asunción in 1537 and the abandonment of Buenos Aires by 1541.

Asunción became the focus of Spanish power along the Paraguay River and downstream to the Atlantic. Attempts to link up with new mining areas of the Andes led to excursions to the northwest and ultimately the founding of Santa Cruz, Bolivia. Small, racially mixed villages grew adjacent to Asunción. A military frontier was established to the east to repress Portuguese expansion. New settlements were founded south from Asunción along the Paraná River, notably at Santa Fe (1573), Corrientes (1588), and Buenos Aires, resettled in 1580.

Many of the forest Indians were brought into Jesuit *reducciones*, or missions. Between 1609 and the expulsion of the Jesuits in 1767, the missions were centers of religion, agriculture, manufacturing, and trade and contained over 100,000 Guaraní. Among the economic products of the 32 Jesuit *reducciones* was yerba for export. Brazilian and Spanish landowners, unable to exploit the Indian labor and envious of the success of the missions, helped expel the Jesuits in 1767. Today, some missions in Paraguay and Misiones province in Argentina (lost by Paraguay in the 1865–1870 War of the Triple Alliance), have been restored and are tourist attractions.

The importance of Asunción was further reduced in 1776 when Buenos Aires became capital of the Viceroyalty of la Plata and Paraguay one of its regions. Paraguay refused, however, to acknowledge that the 1810 Declaration of Independence by Argentina also meant continued Argentine control of Paraguay. In 1811, Paraguay declared independence.

► *The Southern Frontiers of Argentina and Chile*

The tapering southern portion of South America is composed of (1) Argentine Patagonia and (2) Chilean *Sur Grande*, which includes Chilean Patagonia, Atlantic Chile and the large island of Tierra del Fuego, shared by Chile and Argentina. Both countries lay claim to wedges of Antarctica as well.

Argentine Patagonia

The interior tablelands and valleys of Patagonia remained largely unsettled until the nomadic Indians were subdued in the early 1880s and Patagonia was legally included in Argentina in 1881. Thereafter, Argentines, English, and Scots moved into northern Patagonia from the Pampas and from Mendoza, establishing enormous sheep ranches across the broad, hilly plateaus. In the far south, Chileans moved eastward from Punta Arenas to establish other *estancias*. Even earlier, in the 1860s, Welsh settlers had established irrigation agriculture and wheat production on the lower Río Chubut.

After the southern Argentine–Chilean boundary was adjudicated in 1902, many Chileans withdrew into Chile from Patagonia and Tierra del Fuego. In 1984 a dispute over Beagle Channel islands was defused by Vatican intervention. At issue was not so much the islands as the principle that "Chile stays in the Pacific, Argentina in the Atlantic."

Sheep *estancias*—some covering hundreds of thousands of acres and raising equal numbers of sheep—still extend over much of Patagonia, utilizing the bunch grasses and small shrubs common to the area. Cattle are unimportant. Sheep are processed in the small *frigoríficos* in the southern ports from which the wool is sent via Buenos Aires to world markets. Tierra del Fuego and the Falklands are sheep-raising islands.

In the valleys of the Negro and Chubut, protected from incessant winds that sweep over the plateaus, irrigated agriculture produces orchard fruit crops. Vineyards are important as is the growing dairy industry, particularly in the older Welsh region of the Chubut. "Chubut" cheese is nationally distributed.

Other primary economic activities include low-grade coal mined at Río Turbio and shipped to Buenos Aires from Río Gallegos (1897). Hydroelectricity is generated at the El Chocon–Cerros Colorados power and irrigation project whose dams on the Limay and Neuquén rivers are sources of electric power conducted to Buenos Aires.

Oil was first exploited after 1907 and led to the creation of Comodoro Rivadavia from which crude is shipped by tanker and gas by pipeline to Buenos Aires. Production and refining are controlled by a national corporation, Yacimientos Petróleos Fiscales (YPF), which is actively privatizing its operations.

Patagonian districts closest to the Pampas are the most developed, with railroads and highways linking the valleys to Bahía Blanca. The major city within the region is Neuquén in the upper valley of the Río Negro. The linear metropolitan area includes dozens of small towns that together make the Neuquén urbanized area the tenth largest in contemporary Argentina.

Southwest of Neuquén is the Lake District, a major center of tourism since the 1920s, which blends pine forests, mountains, and glacial lakes. Most notable is Lago Nahuel Huapi, on whose shore the resort town of San Carlos de Bariloche was founded in 1895.

Offshore, on the continental shelf, the Falkland Islands (*Islas Malvinas*) are devoted to sheep production. Argentina established Puerto de Soledad (Port Stanley) in 1789 but subsequently lost the islands to Britain in the 1830s. Ownership is in dispute, its most recent manifestation the Falklands War (1982). The less than 2,000 inhabitants are British.

Chilean *Sur Grande*

South of Puerto Montt, the southern edge of Middle Chile, lies *Sur Grande*, a world of islands and the southern reaches of the Andes that extends to Tierra del Fuego and Cape Horn. Here the archipelagos and Andean system become increasingly cool and moist, with low clouds and driving westerly winds for most of the year. Along the western margins evergreen forests dominate the vegetation and represent one of the world's two major Pacific Coast forest ecosystems, comparable to North America's Pacific Northwest. The cool climate and long winters lead to slow growth and slow recovery when forest is harvested or destroyed.

Until the 1980s, *Sur Grande*'s forests had been partly protected from destruction by remoteness and limited markets for their wood, although woodland was destroyed to create pasturelands. In the 1940s, for example, forest fires were started to clear land for cattle grazing. The impact was dramatic. Heavy rainfall and the slow regeneration of plant cover in the cool climate led to erosion. Puerto Aisén, a fishing and trading port for the region, was silted and ship traffic forced to use Puerto Chacabuco downstream.

The dominance of Chilean commercial forestry in the Southern Cone is evident in Table 13.8. In contrast to the other Cone countries, very little of the total harvest is consumed as fuelwood. The first mechanized timber mill in South America went into operation on the island of Chiloé in 1828.

In the last decade, scores of lumber and paper-pulp companies from the United States, Europe, and Japan have rushed into southern Chile to clear-cut thousands of acres of old-growth trees, a practice usually forbidden at home. About 19 mil-

TABLE 13.8 Southern Cone: Forest Products Harvested, 1991 (000 cubic meters)

Country	Logs	Pulpwood	Fuelwood	Total
Argentina	2,563	3,584	4,764	11,251
Chile	8,355	8,557	163	17,075
Paraguay	2,692	n.a.	5,360	8,466
Uruguay	600	214	3,021	3,878

Source: Europe Year Book, 1995

lion acres (79 percent) of Chile's woodlands are privately held. The conservation ethic is weak; in the 1980s, the military dictatorship sold forests for as little as 50 cents an acre.

Although forest products are Chile's second largest earner of foreign exchange (1994, after copper), Chileans are typically more aware of copper in the north and agriculture in central Chile than of their southern forest resources, especially those in distant Tierra del Fuego. In 1990, however, a prohibition on the harvesting of Araucania trees (which need 500 years to reach maturity) eased the danger of their eradication.

At the tip of the continent is Atlantic Chile, one part Chilean Patagonia, one part a share of the island of Tierra del Fuego. Cool, wet, and windy, its grasslands are devoted to sheep and cattle. Tierra del Fuego has been Chile's most important oil region, and crude is shipped north from Caleta Clarencia. Some 10,000 Chileans live on the island's western side and some 80,000 Argentineans on the eastern side. The last of Tierra del Fuego's indigenous population died out in the 1930s.

The most important Chilean center is Punta Arenas (1849), originally a penal colony on the Strait of Magellan that also served passing ships. Punta Arenas had its heyday in the steamship period before the opening of the Panama Canal. The port exported wool, hides, and frozen meat from Chilean Patagonia and Tierra del Fuego. Today the town functions as a launching point for Chilean Antarctica, a growing fishing port (notably king crab), and the site of chemical plants based on local natural gas and a new coal mine. The port suffered with the opening of the Panama Canal in 1914, but ships too large for the canal still stop here. It is a tax-free port, as is Ushuaia in Argentina, which has become an increasingly important manufacturing, tourism, and retail trade center thanks to legislative incentives to growth.

SUMMARY

At the beginning of the twentieth century, Argentina was about to join the economically developed countries of the world. Life-styles and living standards were comparable to a number of West European countries. We can debate why the promise only partly materialized. Is it that Argentina occupied a peripheral place in an economic system that was badly disrupted by two world wars? Has economic dependency, a "colonial-like" tie to European nations, impeded growth? Has progress been retarded by political instability and the manner in which the military takes over in times of supposed turmoil? Did social inequality and a maldistribution of wealth retard the growth of demand for goods and services that the country did produce? Whatever the causes of delay, Argentina now has a more stable and regionally integrated economic and political system and a stable currency (tied to the U.S. dollar) with potential to check the inflationary forces that have been another cause of economic stagnation.

We can ask many of these same questions of the other nations of the Southern Cone. Uruguay, too, was poised for economic diversification in the early decades of the century and is now stable once more after a long period of uncertainty. Paraguay is still isolated and backward economically. But in 1996 its economic bonds to Mercosur (The Southern Cone Common Market) and the legal mechanism of the OAS (Organization of American States) were important to the failure of a military coup d'etat that would have hindered the pace of democratic and economic advancement. Chile is emerging from a period of military rule. It has an elected government, but the military dictators still retain some veto powers. The country is thriving economically with mineral, timber, fish, and agricultural exports. Chile, with one of Latin America's strongest economies, wants to join NAFTA. It has the economic strength to compete in the organization, but to become a member, the Chilean military politicos will have to march off into complete retirement.

FURTHER READINGS

Keeling, D. *Contemporary Argentina: A Geographical Perspective*. Boulder, CO: Westview Press, 1996.

Moorhead, A. *Darwin and the Beagle*. New York: Harper and Row, 1969.

Sargent, C. S. *The Spatial Evolution of Greater Buenos Aires, Argentina, 1870–1930*. Tempe: Arizona State University Press, 1974.

Skidmore, T.E., and P.H. Smith. *Modern Latin America*, 3rd ed. New York: Oxford University Press, 1992.

South American Handbook. Bath, U.K.: Trade and Travel Publications, 1996.

Winn, P. *Americas: The Changing Face of Latin America and the Caribbean*. New York: Pantheon Books, 1992.

CHAPTER 14

Latin America and
the World Scene

BRIAN W. BLOUET AND
OLWYN M. BLOUET

INTRODUCTION

Since 1492, a major theme in the history of Latin America has been the creation of transatlantic and transpacific linkages. It is a complex theme involving the establishment of routeways and trade connections, the migration of peoples, and contact between cultures. The intrusion of the Iberians brought Western European, Native American, and African cultures into contact, conflict, and creation. Several indigenous cultures, including the Incan and the Aztec, were destroyed or damaged in the process of exploration, conquest, and settlement. However, innumerable pre-Columbian cultural traits survived and contributed to the evolution of a distinctively Latin American culture realm.

The tales of relatively small armies of *conquistadors* subduing and exploiting the indigenous inhabitants have many chroniclers. But the Iberians were not only destroyers. They, too, contributed traits to the new culture realm that was being created in Middle and South America. The introduction of Spanish eventually gave the region a *lingua franca*. Europeans transferred plants and animals to the region: wheat, vines, coffee, sugar, citrus fruits, cattle, sheep, pigs, and horses. In turn, the indigenous cultures contributed traits to the present-day life of the region, together with plants that have diffused into many parts of the globe. Maize (corn), tobacco, the tomato, cacao, the potato, and species of beans and squash are examples of domesticated plants that are grown by farmers in regions far removed from Latin America. Africans provided rich cultural traits, and most importantly their labor was used to develop the resources of Latin America. Racial and cultural diversity characterizes much of Latin America.

The mines of Mexico and Peru provided circulatable wealth at a critical period in the economic growth of the North Atlantic region. The additional coinage, which came into use in the sixteenth century, proved to be inflationary, but it stimulated trade and investment in the Atlantic world and contributed to the emergence of capitalized, industrial societies in Europe and eventually in the Americas. As the thirteen colonies moved towards independence Spanish silver, pieces of eight were common currency on the Atlantic seaboard.

FROM EMPIRE TO INDEPENDENCE

How did the colonial American empires of Spain and Portugal, discussed by Robert West in Chapter 3, evolve along the road to independence? How did control from Spain and Portugal cease and independent successor states emerge? Unlike the United States' experience, where revolution and independence led to the creation of one federal state from 13 diverse colonies, the seemingly centralized Spanish empire fragmented into numerous republics. Instead of integration, there was disintegration. Conversely, the large Brazilian empire emerged as one unitary state—initially as a monarchy, then as a republic. First, let us consider the last century of the Spanish American empire, and the transition to independence.

► *The Spanish American Empire*

In 1700, King Charles II of Spain, the last of the Spanish Hapsburg line, died without a direct male heir. There were several claimants to the throne, and Europe was plunged into the War of the Spanish Succession that ended in 1713 with the Treaty of Utrecht. Under the treaty, the Spanish crown passed to the French House of Bourbon—to Philip of Anjou, grandson of Louis XIV—on condition that the Spanish and French thrones were never to be united. Spain retained control of the Indies but relinquished its European possessions in Flanders (Belgium) and parts of Italy to rival claimants. Spanish territorial power was simplified,[1] allowing the Bourbon monarchs to concentrate on rule in Spain and the Americas.

The eighteenth century was a period of colonial rivalry and international warfare, principally between France, Great Britain, and Spain. Territory and trade of the Americas were frequently contested, and the costs of war rose significantly

Ancient and modern methods. A tractor pulls traditional plows formerly powered by human energy, in Otavala, in the Andes of Ecuador.

---

[1]This was temporary. Later in the eighteenth century, Spain gained territory in the Italian peninsula.

throughout the century. During this time, the population and trade of Spanish America grew, but Spanish power declined. The colonial economy of Spanish America was dependent on an underdeveloped Iberian core. Economically, Spain needed the Americas more than the Americas needed Spain.

The new Bourbon rulers designed reforms to tighten control of the empire and reap more economic benefits. Administratively, Spanish America had been divided into two viceroyalties since the sixteenth century: New Spain (1535) and Peru (1542) with their centers in Mexico City and Lima, respectively. But in the eighteenth century, the Viceroyalty of New Granada, with its center in Bogotá, was established in 1717 and refounded in 1739. It linked Venezuela, taken from the Viceroyalty of New Spain, with New Granada (Colombia, Panama, Ecuador) (Figure 14.1). This represented an attempt to administer the region more efficiently and provide protection against British penetration in the region. Venezuela, with its prosperous center of Caracas and expanding cacao and tobacco cultivation, developed more rapidly than the region to the west in the Andes, centered on land-locked Bogotá.

The Bourbon reform effort came to a peak in the reign of Charles III (1759–1788) after the weakness of Spanish power was exposed during the Seven Years War/French and Indian War (1754–1763), when the British briefly occupied Havana. This war was a great turning point in the history of the Americas and had important geographical and geopolitical implications that indirectly spawned the era of revolutions, including the American Revolution, French Revolution, and Haitian Revolution. The Seven Years War led to the end of European control in North and South America, although independence would arrive sooner for the 13 small seaboard colonies of British North America than the much more extensive Spanish American lands.

The Peace of Paris that concluded the French and Indian War in 1763 brought major geographical changes to the Americas. First, French control of and claim to American mainland territory was ended. The French lost Canada and the routeway from the Great Lakes, via the Mississippi to New Orleans. Second, British imperial claims to lands as far west as the Mississippi River were acknowledged. Third, the Spanish North American empire expanded to include lands west of the Mississippi (the future Louisiana Territory),[2] even though Spain temporarily lost Florida to Britain. In 1763, the Spanish American empire included South America (except Brazil); Mexico and Middle America; the Caribbean colonies of Cuba, Santo Domingo, and Puerto Rico; and large sections of North America. The Spanish empire theoretically bordered the British empire at the Mississippi River.

Meanwhile, in addition to territorial adjustments, the expense incurred during wartime encouraged administrative and financial reform in both the British and Spanish empires, as monarchs and ministers sought to centralize empires, raise more revenue, and fashion colonial commerce to better serve the interests of the imperial state. Attempts by the British to rationalize colonial government and tax colonists to pay for defense and administration ultimately led to the American Revolution and independence for the United States. At the close of the American Revolutionary War (in which France and Spain sided with the American colonists), a further Treaty of Paris was negotiated (1783), recognizing the independence of the 13 colonies, with boundaries as far west as the Mississippi. Florida was returned to the Spanish American empire, which then stood at its most extensive

---

[2]Spain, in 1800, ceded the Louisiana Territory to France, and Napoleon sold it to the United States three years later, when Thomas Jefferson was president. This gave the fledgling United States an outlet on the Gulf of Mexico.

(Figure 14.1). The United States later acquired Spanish Florida in the Transcontinental Treaty of 1819.

The Spanish similarly tried to reform colonial government, with the objectives of raising more taxes, asserting metropolitan authority, and developing the

Figure 14.1 Seats of Viceroyalties and Audiencias in 1799.

economies of regions bordering the Atlantic, to stimulate the lethargic Spanish economy. The Spanish reform plan included replacing creole bureaucrats with officials (intendants) sent out from Spain, increasing rates of taxation, and attempting to prevent production in the colonies of commodities such as wheat, wine, and olive oil that competed with Iberian products. State monopolies were expanded to include tobacco and spirits. However, the mercantilistic system of trade was liberalized to some extent under a new system described as *un comercio libre y protegido* (free trade under the protection of the state), introduced in 1778. The monopoly of Mexico City and Lima, of Seville and Cádiz, was broken. Colonial ports, such as Havana, Cartagena, and Buenos Aires were allowed to trade directly with several metropolitan ports. Agricultural exports rose, and trade boomed for a while. But trade was still exclusively with Spain. The colonies desired new markets for agricultural and other exports and resented reliance on Spanish-controlled imports at a time when British manufactured goods looked increasingly attractive. International smuggling was, of course, widespread.

In administrative terms, the Spanish reform program included the creation of the Viceroyalty of Río de la Plata in 1776, split from the Viceroyalty of Peru. Its focal point was the small port town of Buenos Aires. Sparsely populated areas east of the Andes were included in the new viceroyalty, as was Upper Peru (modern Bolivia) with its rich silver mining industry. The objective was to halt the expansion of Brazil and cut out smuggling at the coast. The impact of the administrative reform was to reorientate trade from Lima and the Pacific to Buenos Aires and the Atlantic. This might be seen as continuing the process of developing points of economic attachment in the Indies to Spanish centers.

A further area of Bourbon reform was in religious affairs. The Jesuit order was expelled from the colonies in 1767 as part of the program to increase state power. At that time, there were more than 200 Jesuit missions with over 250,000 Indians in the Spanish empire. All the missions eventually declined, even though some were put into the hands of secular clergy, others were transferred to various religious orders (such as the Franciscans), and still others collapsed rapidly. The crown, however, continued to support missionary expansion in frontier regions such as Alta California that were vulnerable to foreign threat from the English or Russians (Merino and Newson, 1995).

Widespread colonial dissatisfaction greeted some of these reform measures introduced by an interventionist state. Serious resistance first occurred in Peru, Upper Peru, and New Granada. Most notably, in 1780, Tupac Amaru, an Indian *cacique*, organized an Indian uprising that protested high tribute payments, forced purchase of Spanish trade goods (which led to debt), and the work requirements of the *mita*. Tupac Amaru promised freedom to black slaves who joined the movement. Indians took control of La Paz for a time, and the rebellion, which took 5 years to suppress, is estimated to have cost 100,000 lives (Winn, 1995). Other disturbances hit New Granada and Venezuela and made the colonial elite nervous about social control.

Remember that this was the Age of Revolution when the American Revolution (1776–1783), French Revolution (1789–1799), and Haitian Revolution (1791–1804) rocked the world scene. The United States emerged as the first American republic. Haitian slaves, led for a time by Toussaint L'Ouverture, threw off French colonial rule and withstood British and French military expeditions to become the second republic in the Americas. Spanish colonies gained independence in the first quarter of the nineteenth century, aided by events in Europe.

Initially in the revolutionary era, the ruling elite in the Spanish empire remained loyal to Spain, frightened by the excesses of the French Revolution, the horrors of race war that occurred in Haiti, and a belief that Spanish authority was necessary for internal order and security, especially in areas where there were large concentrations of Indians.

In the period of the Napoleonic Wars, however, colonial commerce with Spain was dislocated after the British blockaded Spanish ports in 1796 in response to Spain's entry into the war as an ally of France. As a result, the Spanish colonies became accustomed to trading with other countries (even Britain), and the economic link with Spain was severed. Then, in 1807–1808, Napoleon invaded Spain, imprisoned King Ferdinand VII, and placed his own brother Joseph Bonaparte on the Spanish throne. The political link between crown and colonies was broken. The Treaty of Utrecht, which had forbidden the union of Spain and France, was overturned. The Spanish American colonies were possessions of France. Thus, by virtue of external events in Europe, the Spanish American colonies were pushed to make decisions about the colonial relationship. Some members of the elite opted for independence, and wars ensued. It is generally believed that without Napoleon's escapades, the empire might have lasted many more years.

Wars of Independence

There were three major centers of independence in Spanish America:

1. In Mexico, initiated by Father Hidalgo
2. In northern South America, led by Simón Bolívar
3. In southern South America, led by José de San Martín

Mexico Father Miguel Hidalgo, a creole priest, organized an uprising of mestizos and Indians on September 16, 1810 (September 16 is now Independence Day) who marched on Mexico City from Querétaro and Guanajuato, killing creoles and peninsulars on the way. Hidalgo's aims included the abolition of Indian tribute, the return of Indian lands, and the end of slavery. Hidalgo was captured and executed. His mass revolution failed, but the dangers of revolution from below were made clear to the elite.

When Mexican independence eventually came in 1821, it was orchestrated by a conservative creole army officer, Augustín de Iturbide. Mexican independence was proclaimed to prevent liberal reform emanating from Spain. Iturbide became Emperor Augustín I of Mexico. The Yucatán and the Central American provinces (Chiapas, Guatemala, Honduras, El Salvador, Nicaragua, and Costa Rica) joined the empire. But within two years a military coup established a Mexican republic, and the Central American provinces (with the exception of Chiapas) seceded from the union, forming the United Provinces of Central America, until it fragmented into five separate republics in 1838 (Table 14.1).

Northern South America After involvement in a failed rebellion in Caracas in 1810–1811, Simón Bolívar (1783–1830), a wealthy, well-educated creole born in Caracas, was a leader in the independence struggles of Colombia, Venezuela, and Ecuador. These three territories briefly formed the Republic of Gran Colombia between 1822 and 1830 (Table 14.1 and Figure 14.2) before splintering into three independent republics, largely to protect their own diverse economic interests. Bolívar's ambition of pan-American unity and a league of Latin American states did not materialize, although he convened a continental congress at Panama in 1826.

Southern South America As early as 1810, creoles in Buenos Aires declared independence for their region. It was felt that, in order for independence to be secured, the center of Spanish power in Peru had to be eliminated. To that end, José de San Martín (1778–1850), an Argentine military officer, organized the Army of the Andes, defeated Spanish troops in Chile in 1817–1818, and established an independent government in Chile under Bernardo O'Higgins.

But it took the combined strength of San Martín's and Bolívar's forces to erode Spanish power in Peru. Eventually, in 1824, at the Battle of Ayacucho (Figure 14.2), in the Andes, Spanish forces were overcome, and South America was finally liberated in 1826 when the remaining Spanish troops at Callao surrendered. The Battle of Ayacucho was the South American equivalent of the U.S. Battle of Yorktown. By the end of the first quarter of the nineteenth century, the Spanish American empire was reduced to Cuba and Puerto Rico. Santo Domingo was embroiled in the aftermath of the Haitian Revolution. It is usually suggested that Cuba remained loyal to Spain partly from fear of slave revolt.

TABLE 14.1   Independence and Emancipation

Country	Independence from Colonial Power		Independence as a Separate State	Abolition of Slavery
I. Mexico and Central America				
Mexico	[1821–1823	1821	1823	1829
Costa Rica	Mexican	1821	[1823–1838	1838	1823/4
El Salvador	Empire	1821	United	1838	1823/4
Guatemala		1821	Provinces	1838	1823/4
Honduras		1821	of Central	1838	1823/4
Nicaragua	]	1821	America]	1838	1823/4
Panama		1821	[1821–1903 part of Colombia]	1903	1851
Belize	1981		1981	1833/4
II. South America				
Argentina	1816		1816	1853
Bolivia	1825		1825	1826
Brazil	1822		1822	1888/89
Chile	1810		1818	1823
Colombia	1811	[1822–1830	1830	1851
Ecuador	1822	part of	1830	1852/53
Venezuela	1811	Gran Colombia]	1830	1854
Guyana	1966		1966	1833/34
Guyane	Not yet		Not yet	1848
Paraguay	1811		1814	1869
Peru	1821		1821	1854
Suriname	1975		1975	1863
Uruguay	1814		1828	1853

Figure 14.2 Latin America after independence, 1829.

▶ *Brazil*

During the eighteenth century, Brazil faced similar "reform" measures as those encountered in Spanish America. The Portuguese crown tried to intervene directly in Brazil to centralize administration, monopolize trade, and realize more revenue.

Reform peaked under the Marquis of Pombal, first minister in Portugal from 1750 to 1777. He expelled the Jesuits from the empire in 1759, moved the viceregal capital from Salvador in the Northeast to Rio de Janiero (Figure 14.1), and attempted to eliminate smuggling. His imperial plans included policies to diversify exports from Brazil, lower the trade deficit with Britain, and stimulate the Portuguese and Brazilian economies. The Brazilian economy thrived in the 1780s and 1790s, as new export staples such as coffee, cotton, rice, and indigo joined the traditional exports of sugar, hides, and tobacco. Portugal relied on Brazilian production.

As in the case of Spanish America, events in Europe precipitated Brazilian independence. When Napoleon invaded Portugal in 1807–1808, the Portuguese royal family fled to Brazil where they set up court. King John VI opened Brazil's ports to free trade, and Rio de Janiero developed as a prosperous commercial center. In contrast to Spanish America, independence for Brazil evolved peacefully in 1822, after King John VI returned to assume rule in Portugal, leaving his heir Dom Pedro as prince regent in Brazil. Fearing increased control from Portugal, the prince regent declared Brazilian independence in 1822, and was crowned emperor. There was no violent break with the colonial past and no civil war. Brazil had self-government and retained legitimate royal authority until becoming a republic in 1889, after a military coup deposed the emperor.

CHALLENGES FACING NEW STATES

In mainland Latin America, direct rule from Spain and Portugal had ended by 1826. Nevertheless, European influence continued as parts of Latin America were drawn more and more into the world economy. Political independence was gained, but economic dependence continued. The new states faced severe political, economic, and social challenges. In addition, they had to establish territorial integrity and gain international recognition.

The Spanish American empire eventually splintered into 15 independent states: Mexico, Guatemala, Honduras, El Salvador, Nicaragua, Costa Rica, Colombia, Venezuela, Ecuador, Bolivia, Peru, Chile, Paraguay, Argentina, and Uruguay. *Audiencias* (territorial jurisdictions of royal law courts), not viceroyalties, acted as the bases for new states. J. L. Phelan suggested that a major reason for this development was that "The Spanish concept of the state was essentially medieval in that the administration of justice, not legislative or executive authority, was regarded as the highest attribute of sovereignty" (Phelan, 1967, p. 38). In addition, economic regionalism and political parochialism played a role in fragmentation.

Audiencias had been centered in Santo Domingo, Mexico City, Panamá, Lima, Guatemala, Guadalajara, Bogotá, Sucre, Quito, Santiago, Buenos Aires, Caracas, and Cuzco (Figure 14.1). Not all these *audiencias* became cores of states. Guadalajara, Sucre, and Cuzco did not give issue to new states. The territory of the *audiencia* of Guatemala eventually split into five republics. In the cases of Paraguay and Uruguay, *audiencias* were lacking, but the "buffer" nature of these regions, between Spanish and Portuguese spheres of influence, explains their existence.

Most new states embraced the principle of *uti possidetis, ita possideatis*—as you possess, so you may posses (Cukwaruh, 1967). Under this concept, new states succeeded to the territory of the preexisting administrative division on which the emergent nation was based. "Critical dates" were recognized, which meant that in South America administrative boundaries operative in 1810 were the bases of new states, while in Central America, the date was fixed at 1821. In practice, it has often been impossible to establish boundaries at the critical dates. The maps or descriptions on which divisions had been based were frequently inaccurate or ambiguous, while many areas had never been effectively occupied by the Spanish or Portuguese empires. A considerable amount of land appeared to be available for possession.

With the exception of Brazil, most of the new states lacked internal political stability, and numerous civil wars ensued. Politics was characterized by the development of *caudillismo*—the rule of a local strong man, frequently a military dictator. Sometimes rural-based *caudillos* gained national power, as was the case with José Páez, ruler in Venezuela between 1830 and 1848, and Manuel de Rosas of Argentina between 1829 and 1852. *Caudillos* were frequently popular with the masses, but their power reflected the lack of central authority. Some commentators talk of a ruralization and militarization of power in the new republics.

The independence wars in Spanish America were longer and more violent than North American experience and led to economic dislocation. Trade, agriculture, and mining were severely disrupted. During the wars, herds of livestock were slaughtered, mines flooded, and equipment destroyed (Burns, 1994). There was flight of capital and labor. Once the new governments were in place, they lacked effective power over fiscal or labor policy. The new states lost the uniform monetary union and credit policies of the Spanish empire. Independence meant the creation of new national tariff structures, and former inter-colonial trade networks (such as that between Mexico and Cuba) were adversely affected.

External commerce, however, was generally liberalized, as trade no longer had to go through Spain or Portugal. Producers of agricultural exports, for example coffee, cacao, and hides, benefited. Formerly marginal areas such as Chile, the Río de la Plata countries, northern Mexico, Venezuela, and the São Paulo province of Brazil would eventually become dynamic export regions. The new states tended to encourage the development of the export sectors of their economies because this was relatively cheap. Lack of investment in roads and infrastructure hindered the development of domestic markets. National unity faced obstacles of difficult geography and slow communication.

Initially, Great Britain (as the leading trading force in the world at the time) acted as the major supplier of investment capital, manufactured goods, markets, and entrepreneurial skills. Economic development had to face the great problems of distance and poor transportation, which were ameliorated by the construction of railroads, telegraph lines, steamboats, and port facilities after mid-century.

Socially, the revolutionary era did not lead to great transformations. An elite —now creole rather than peninsular—controlled political, economic, and social power in the new states. The elite acquired land from the state, Indians, and the church. Indians lost lands (often communal lands) to expanding estates. In some places, the legal differences between races was lifted, and slavery was abolished, relatively early in some central American states, later in countries heavily reliant on slave labor, such as Brazil (Table 14.1).

A further challenge that the new states faced was establishing stable borders. There have been many boundary disputes and wars between neighbors in Latin America, due to ambiguously defined or poorly demarcated boundaries, unoccupied lands, strategic considerations, and desire for valuable resources. Examples of warfare in the nineteenth century include an early contest between Argentina and Brazil (1825–1828) that resulted in the creation of Uruguay as a buffer state and the War of the Triple Alliance (1864–1870) when Argentina, Brazil, and Uruguay defeated Paraguay, at an estimated cost of 300,000 lives. One outcome of the war was that the la Plata River system was opened to international commerce. A further conflict, the War of the Pacific (1879–1883), ended when Chile won nitrate lands claimed by Peru and Bolivia. Chile took over the major nitrate port of Antofagasta and deprived Bolivia of direct access to the Pacific.

One of the best known conflicts in the nineteenth century was the U.S.–Mexican War of 1846–1848, during which United States troops, under General Winfield Scott, occupied Mexico City. By the Treaty of Guadalupe Hidalgo (1848), Mexico lost between 40 and 50 percent of its territory—lands that would become the states of California, Arizona, New Mexico, and parts of Utah and Colorado. Texas had previously broken away from Mexico and after a short period as an independent republic (1836–1845), joined the expansionist United States.

The other vast territorial change in Latin America involved the gradual expansion of Brazil from a narrow coastal strip to control half the continent. At the outset, the line of demarcation separating Spanish and Portuguese territory (Treaty of Tordesillas, 1494) was open to dispute (Owens, 1993). The general trend was Brazilian westward expansion, whether in search of Indian slaves, in hopes of annexing Potosí silver mines, or for exploitation of gold or timber resources. In the colonial period, several agreements, including the Treaty of Madrid (1750)[3] and the Treaty of Ildefonso (1777) modified the boundary. In essence, Spain recognized Brazilian penetration into the heartland of South America. Today, Brazil extends far to the west of any possible Tordessillas line, and Brazil is the most active country in Amazonia.

DEVELOPMENT OF EXPORT ECONOMIES AFTER THE MID-NINETEENTH CENTURY

In the second half of the nineteenth century, the export economies of many Latin American countries expanded. Latin America was a supplier of raw materials and foodstuffs consumed by the industries and industrial workers of Western Europe. Exports were largely agricultural products and minerals. Sugar was joined by bananas, coffee, wheat, meat, and wool; silver was joined by nitrates, copper, lead, and tin. By 1900, Brazil supplied almost all the world's rubber. The annual economic growth rate in Argentina averaged 5 percent between 1880 and 1914; it averaged 8 percent per year in Mexico between 1876 and 1910, when Porfirio Díaz was in power (Williamson, 1992).

Successful export economies stimulated foreign investment in Latin America, and increased profits led to a rise in manufactured imports entering Latin America. Cities thrived, and modernization, such as railroads, roads, and schools, developed. The case of Argentina illustrates this trend. The development of the Argentine Pampas, from the construction of railroads, ports, and shipping facilities, to the introduction of breeds of sheep and beef suitable for the British market, were financed through London. Britain adopted policies of free trade in the mid-nineteenth century and opened its markets to wheat and meat produced outside the British Isles (see Chapter 4). By 1914, Argentina had relatively high per capita incomes, a rapid rate of economic growth, and the ability to expand its economy, absorb migrants from Europe, and sell commodities into European markets. Buenos Aires was an elegant city with over 1 million inhabitants (see Chapter 6) and a G.N.P. per capita similar to that of Germany, the Netherlands, and Belgium (Rock, 1986). Development was fueled by an expanding export economy.

Some commentators have described the Latin American export economies as "neocolonial" and suggested that Latin America faced severe disadvantages in the global arena. Latin American economies relied on a few exports, most often primary

---

[3]Spain recognized Brazilian expansion west of the Tordesillas line, and Portugal recognized Spain's control of the Philippines.

products, that were vulnerable to price fluctuations. Manufactured goods were usually imported and were frequently purchased with foreign loans. In addition, export sectors were controlled by external business interests. It is argued that these factors contributed to a cycle of underdevelopment and dependency.

Recently, some scholars have suggested that it was logical for Latin American states to develop export economies. They lacked capital, education, and personnel for internal development and had to rely on external finance for modernization and industrial growth. By developing export economies, Latin America was tied into the global industrial economy, and the late nineteenth century saw an acceleration of modernization and urbanization in many regions of Latin America. Some of the more prosperous countries, such as Brazil, Argentina, Chile, and Mexico, were later able to diversify and industrialize their economies.

THE UNITED STATES AND LATIN AMERICA

By the end of the nineteenth century, the United States was emerging as a major economic force in the region. Politically, militarily, and strategically, the United States has imposed control on Latin America. The Monroe Doctrine of 1823 declared that Latin America was not to be the object of new colonial activity by European states. The immediate impact of this declaration was not great, but by the end of the century, America exerted a major influence on the foreign affairs of the region. During the 1890s, the United States initiated arbitration between Venezuela and Britain concerning the Venezuela-Guiana boundary dispute. In the Spanish-American War (1898) Spain was deprived of all remaining possessions in the Western Hemisphere, and the United States became the predominant power in the Caribbean. Puerto Rico was acquired, and the right to interfere in the internal affairs of Cuba and the Dominican Republic was assumed.

In 1903 the British reduced their naval strength in the Caribbean, and in 1904 the United States leased the Canal Zone from Panama, successfully opening the Panama Canal ten years later in 1914. Admiral Alfred Thayer Mahan (1840–1914) had predicted in *The Influence of Seapower upon History* (1890) that the canal would change the Caribbean from a terminus into a major trade route linking the Pacific and the Atlantic, and he was correct. The enhanced strategic value of the Caribbean required the United States to acquire bases in the region and develop naval stations on the Gulf coast. A naval base was developed in Cuba and the

Many crops from Latin America enter into world trade. Here coffee berries can be seen on the bush. The beans are extracted after the berries have been dried.

United States Virgin Islands acquired from Denmark (1917). Under the Roosevelt Corollary (1904) to the Monroe Doctrine, the United States assumed the power to supervise the internal affairs of states in the Caribbean and Latin America to insure security. Above all, large military facilities were built in the Panama Canal Zone to protect U.S. interests in the region. The 1979 treaty that negotiated future Panamanian control of the canal has not stopped the United States from supervising the affairs of that country. In 1989, U.S. troops invaded Panama, and in 1992 the Panamanian head of state, General Manuel Noriega, was tried in Miami on criminal charges and given a long prison term in a U.S. jail.

United States political pressure and direct interference are asserted mainly in the lands around the Caribbean, but economic activity knows no such bounds. U.S. investment has touched every country in the hemisphere. Well-known examples include the development of sugar estates in Cuba, banana growing in Central America, oil extraction in Venezuela, Peru, and Colombia, copper in Chile, and iron ore in Venezuela. Investment in the production of industrial raw materials and agricultural commodities is complemented by the control of banks, import-export companies, shipping lines, railroads, manufacturing plants, and retail stores. The region is an outlet for the export of U.S. capital and goods and a source of tropical commodities, industrial raw materials, and manufactured goods consumed in the United States.

The United States established economic predominance in Latin America during the twentieth century largely as a result of two world wars. At the beginning of the twentieth century, the United States was the most important trading partner of Mexico, Central America, Colombia, Venezuela, Brazil, Ecuador, and Peru. Britain was the leader in trade with Chile, Argentina, and Uruguay. Argentina had by far the most dynamic economy in Latin America.

In terms of foreign-owned capital investment in Latin America, the United States was an important—but secondary—player in 1914. In that year, just over half of all long-term foreign capital investments in the region were owned by British financial interests. United States investors owned 15 percent of foreign-owned assets, French interests had 10 percent, and the German contribution was less than 5 percent, although increasing.

► Impact of World War I

The outbreak of World War I in August 1914 for a time brought the export economies of South America to a halt. German ships were trapped in port, and British ships awaited instructions. Banks ceased to advance credit, and orders for commodities were on hold. There was no money to pay laborers and no certainty that goods produced would find a market (Albert, 1988). Mexico and Central America, with more dependence upon North American than European markets, suffered less, but the sailing schedules of the French, German, and British shipping lines that served the Caribbean and its shores were disrupted. As a region, Latin America came to realize that exports were dependent upon overseas markets and that goods were carried to those markets in ships owned almost exclusively by non–Latin American shipping lines.

World War I altered trade and investment relationships. The war removed Germany as a trade partner and as a source of new investment funds. Britain and France could still trade with and invest in Latin America, but their energies were diverted to the war effort. European overseas investment was sharply decreased. Exports of machinery and consumer goods from Europe were reduced, as industry switched to the production of armaments.

At the end of World War I, prewar economic relationships were not restored. France, Germany, and Britain had little capital to invest overseas. Europe had been

Cut sugar cane brought from fields on
two-wheeled carts awaits crushing.

badly damaged—financially and industrially—by the war. The French geographer, Albert Demangeon, declared in *The Decline of Europe* that European productive capacity had been weakened and Europe's losses would be the United States' gain. This proved to be the case, and Latin America came increasingly into the economic sphere of the United States, the same power that exercised the greatest political and strategic interest in the Western Hemisphere. Occasionally, the United States combines economic, military, and strategic interests. In 1954, for example, when a left-wing government in Guatemala began to take land from the United Fruit Company in exchange for government bonds, the United States used the C.I.A. to engineer a right-wing takeover and protect the assets of the American company.

The emergence of the United States as the predominant economic power in the region was not in the best all-around interests of most Latin American countries. The United States provided valuable equipment exports to Latin America, but many Latin American commodities—wheat, beef, and cotton—could not easily be sold in United States markets. These and other commodities were produced in the United States, and there were tariffs or other restrictions on products entering the United States. For Latin American countries, more trade with the United States usually meant balance-of-payment problems, since the United States exported more to countries of the region than it took in imports. As a result, the countries of Latin America became heavily indebted to the United States, and their currencies were weak against the dollar.

World War I had other impacts. First, wartime demand stimulated the production of primary products everywhere, but when hostilities ceased, postwar markets were glutted with surplus production that depressed prices. Second, wartime shortages prompted technological innovations. For example, before the war, Chilean nitrates had been a major component in explosives and fertilizers. When Germany was cut off from Chile, German scientists synthesized nitrates. Postwar demand for the natural Chilean product declined. Third, because most Latin American countries were at an early stage of development and had little manufacturing capacity, the war did not do much to stimulate industrialization, particularly as imports of manufacturing equipment were reduced. Overall, the export commodity booms of World War I did not lead to diversified economic development. When the war ended, the demand and the price for many commodities declined.

A pronounced side effect of the war was the weakening of the finance and trade system that had focused on London. The system was organized largely on the basis of free trade, and it involved British shipping lines and import-export merchants trading goods around the world, using brokerage, insurance, financial services, and sterling credits provided by institutions in London. Britain had an adverse trade balance with most other countries, including those in Latin America. The banking, shipping, and insurance services, together with dividends from overseas investments, provided an income that helped balance Britain's international payments.

Because the British industrial, trade, and financial system was weakened by World War I, the peripheral regions in the London-centered system were also adversely affected. Just as the British shipbuilders on the Clyde, Tyne, and Wear suffered after World War I, so too did the Río de la Plata exporters of beef, wheat, hides, and wool. When Britain declined economically as a result of World War I, so did the prospects of Argentina and Uruguay, because they were linked into the capital sources, shipping system, and consumer markets of Britain.

► *The Interwar Years*

The 1920s were marked by periods of expansion when Latin American exports improved and recessions when prices fell. The recessions of the 1920s were short-term downturns, but then came the Great Depression of the 1930s, triggered by the Wall Street crash of October 1929. Following the stock market collapse, banks failed, pushing depositors and businesses into bankruptcy. World trade contracted; the price of agricultural commodities and industrial raw materials declined. Regions like Latin America that had been primarily involved in world trade as suppliers of meat, grains, fruits, coffee, oils, and ores were badly hit. Between 1929 and 1932, world trade shrank by 60 percent. In the Depression years, Latin American countries sold less coffee and copper, for example, and received a lower price for what they did sell. Countries in the region defaulted on the loans they owed to banks in North America and Europe. Imports of all goods from luxury items to spare parts and essential machinery were drastically cut because Latin American countries were not exporting enough to earn the dollars and pounds needed to pay for manufactured imports from North America and western Europe.

Harvesting cocoa in the tropical lowlands of Bolivia. The cocoa beans are in the pods on the tree.

The results of the Depression were mixed. On the down side, producers of primary commodities—the products of farming, fishing, forestry, mining, and quarrying—lost business and laid off workers. All aspects of economic life in the cities concerned with imports and exports were depressed. Ports received fewer ships, dockworkers loaded and unloaded fewer cargoes, warehouses handled fewer goods, retailers did less business, and banks cleared fewer payments. On the other hand, if foreign currency to pay for imports was scarce and the price of imported goods was high due to the devaluation of Latin American currencies, then manufacturers of foodstuffs, footwear, clothing, household goods, and agricultural machinery in Mexico, Colombia, Brazil, Argentina, and Chile had less competition and were able to increase market share in the region and expand manufacturing activities.

By the mid-1930s world trade was recovering, and Latin America began to see exports of primary products increase. The rise of expansionist powers in Europe and Asia was, in the second half of the 1930s, an economic advantage to Latin America. The German, Japanese, and Italian armaments industries drove up demand for ores and oil. Furthermore, Germany began to use trade as a political weapon in its attempt to become the hegemonic power in Europe and the Atlantic world. After Hitler came to power in 1933, Germany cut trade with the Soviet Union, reduced trade with western Europe, and worked to exclude the United States from European markets (Hearden, 1987). Using bilateral trade agreements, economic interaction was promoted with eastern Europe, the Middle East, and Latin America. In 1932, 9.6 percent of German imports came from Latin America. By 1938 the figure was 15 percent. In 1932, 4.1 percent of German exports went to the region, and by 1938 slightly less than 12 percent of German exports were sold to Latin America (Barkai, 1990). Brazil, Argentina, Chile, and Uruguay signed bilateral trade agreements with Germany in which they agreed to be paid for their products in so-called Aski marks—money that could only be used to buy goods from Germany (Hearden, 1987). In 1933, Argentina and the United Kingdom signed a trade treaty that contained many elements of a bilateral deal.

In 1936, German territorial ambitions in Europe caused alarm. Britain and France tried to appease Hitler in 1936, 1937, and 1938. But Britain began a rearmament program: the keels of new warships were laid, and military aircraft like the *Hurricane* fighter were ordered. The demand for strategic raw materials such as aluminum, copper, tin, and oil was stimulated. Rearmament in Britain was particularly important for Latin America, as the United Kingdom was a large importer of raw materials, and historically Britain had been a major trade partner with the region since the independence era of the early nineteenth century.

The rise of aggressive and expansionist powers had some political advantages for Latin America. For example, when Mexico nationalized the oil industry in 1938, the U.S. fear of driving Mexico into oil trade with Germany and Japan contributed to a cautious response to the act of nationalization.

For Latin America, the second half of the 1930s represented the start of a new era. Economies began to grow, not only in the primary sector of farms and forests, oil wells and ore pits. The manufacture of consumer goods increased, and as cities grew, the service sector developed, too. Living standards began to improve. The region entered the second phase of the demographic transition in which death rates fall and birth rates remain high or rise—in Latin America they rose (see Chapter 5). During the second phase of the transition, population numbers grew and—in the phrase of Wilber Zelinksy—a great shaking loose of inhabitants of the countryside began (Zelinsky, 1971). Rural-to-urban migration increased among young adults, who moved to the larger cities that started to grow rapidly (Skeldon, 1990).

The 1930s also saw the implementation of government policies that promoted national economic expansion in Latin America. Governments established development banks and government-owned corporations to promote activity in key sectors. The rules on foreign corporations operating within Latin American countries were tightened, and enterprises that were owned and operated by nationals were promoted. As we shall see, this movement toward promoting diversified economic activity within states was given a boost by World War II, even though some traditional, agricultural, export activities were hurt during the war years.

As a postscript to this section, it is worth noting that during the 1930s and 1940s in a number of Latin American countries, expansionist geopolitical views were common. For example, many members of the Argentine military admired Germany. It was suggested that just as Germany was striving for hegemony in Europe, so Argentina should become the dominant power in South America. In part, Argentine geopolitics was a reaction to a Brazilian version of Manifest Destiny that envisioned Brazil as projecting power "to the Pacific, to the Amazon Basin and the Caribbean, to the River Plate Basin, to the South Atlantic and Antarctica" as a prelude to achieving great power status (Child, 1985, pp. 34–35). A Brazilian geopolitician developed the notion of *living frontiers*, which is an extension of the German geographer Friedrich Ratzel's idea of *Lebensraum*. Under *Lebensraum*, powerful states expand at the expense of weaker states to gain living space. The concept of living frontiers provides a rationalization of a strong nation's right to "move its borders into the territory of the weaker neighbor" (Child, 1985, p. 39). Most of the Amazon road network was conceived and laid out by Brazilian military governments. The network was not wholly an economic enterprise, and the roads secured Brazilian influence in the heart of the continent. This type of thinking and action has engendered suspicion and antagonism, promoting conflict rather than cooperation.

► Impact of World War II

Although the rearmament of the late 1930s benefited producers of primary products in Latin America, the world war that broke out in Europe on September 1, 1939, reshaped the pattern of world trade and altered the demand for industrial raw materials and agricultural commodities. The changes in trading patterns outlasted the war.

When Germany invaded Poland, Britain and France declared war and blockaded German ports. The German war economy was closed as a market for raw material producers in Latin America. Then in the spring of 1940, German forces overran all the neutral countries of Western Europe and France. Only Britain and the Iberian Peninsula remained as a market for Latin American commodities. To reach the British Isles, cargo ships had to evade German U-boats and pocket battleships. German ships were initially successful. For example, the *Admiral Graf Spee* sank nine vessels before being damaged by British cruisers, forced into Montevideo, and later scuttled in Uruguayan territorial waters by the German crew to avoid capture.

The Pacific also underwent a radical change in strategic terms. Japan attacked China in the 1930s, but at the outbreak of World War II weighed the advantages of attacking either the Soviet Union, on its Pacific flank, or the colonies of Britain, France, the Netherlands, and the United States (the Philippines) in Southeast Asia. In the meantime, Japan stockpiled strategic materials and increased trade with Latin America.

In 1940, as France was taken over by Germany, the United States created mechanisms to purchase strategic materials and build additional armament plants. Federal funds were used to establish the Rubber Reserve Company and the Metals Reserve Company. These organizations were soon placing orders for rubber and strategic metals like tin and manganese in Latin America and helping to push prices up (Eckes, 1979).

In June 1941, Germany attacked the Soviet Union. The Japanese bombed Pearl Harbor on December 7 of the same year. At the same time, Japanese attacks were launched on Hong Kong, Malaya, Singapore, Burma, the Philippines, and a little later, the Dutch East Indies (Indonesia) to gain a tropical empire in Southeast Asia, rich in timber, tin, oil, and rubber.

The Japanese assault brought the United States into the war, and Germany also declared war on the United States on December 11, 1941. The Japanese attack on Southeast Asia was successful at first. The United States and the United Kingdom were deprived of sources of oil, tin, rubber, and timber from Indonesia (oil, tin, rubber), Malaya (tin, rubber), and Burma (teakwood). Latin American producers of these commodities, such as the tin mines of Bolivia and the Maracaibo oil fields, benefited. Even American demand for wild Amazonian rubber, supposedly eliminated by the development of plantations in Southeast Asia, recovered as Japan took over the plantations. To retake Southeast Asia, Britain, Australia, and the United States had to fight in the tropics, and chincona bark from Bolivia and Peru became valuable as a source of quinine used to combat malaria.

By the beginning of 1942, the world had been divided into trade blocs by force. Most of Europe (with the exception of Spain, Portugal, and the United Kingdom) had been overrun by Germany and could only trade with Germany, in a European New Economic Order. The seeds of the European Economic Community were sown in World War II when most of Europe had to trade with Germany. What we now call the Pacific Rim had been conquered by Japan and could only trade in a Japanese-dominated Greater Co-Prosperity Sphere. The Soviet Union traded with no one but accepted American Lend-Lease and aid from Britain.

Latin America, the Caribbean, and North America were isolated from markets outside the hemisphere. Latin America was forced to trade increasingly with the United States and became economically dependent upon the United States for markets, technology, and financial services. The forces leading to the creation of NAFTA emerged in World War II when the economies of Mexico, the United States, and Canada were brought closer together.

Submarine warfare affected Latin America and brought some countries into the war. In 1941, German submarines sank 3.3 million tons of shipping. In 1942, with the United States in the war and U.S. east coast shipping a legitimate target, sinkings by U-boats increased to 6.3 million tons. The sinkings included the Mexican vessels *Potrero del Llano* (May 13, 1942) and the *Faja de Oro* (May 21, 1942), which brought Mexico into the war, although without unanimous sentiment for the declaration. Sinkings of Brazilian vessels resulted in Brazil entering the war against Germany in August 1942. On the coast of Venezuela, tankers moving oil from the mainland to the Dutch islands of Aruba and Curaçao were stopped for a time in the spring of 1942, as submarines had become a menace (Humphries, 1982). Venezuelan oil production declined.

Thus, the war created a new trade situation for Latin American countries. Competitors in Southeast Asia were removed as sources of supply. Markets in mainland Europe were lost. Materials needed in the war economies of Britain, Canada—which had entered the war against Germany in September 1939—and, later, the United States enjoyed increased prices. Nonessential commodities could not be shipped, since cargo space in convoys was used for high-priority goods.

Most of the sizable countries of Latin America enjoyed growth in the total value of exports. But the composition of exports changed, and the annual rate of export growth varied from country to country. Brazilian exports grew at an average annual rate of over 10 percent in the period 1940–1945. Argentina attained a rate of 4 percent, and Chile less than 2 percent (Rock, 1994, p. 46, Table 3). Before World War II Argentina had produced 6.8 million metric tons of wheat, much of it going to Britain. By 1945, production was down to approximately 4 million metric tons (Mitchell, 1993) as British farmers increased arable land and nearly doubled wheat acreage. Before the war, Britain had been the major market for Argentine beef. During the war, exports continued but did not increase much because in wartime, many foodstuffs, including meat, were rationed in Britain. Even so, 40 percent of British beef came from Argentina in 1943, when a military coup brought a group of officers to power. Among the coup leaders was Colonel Juan D. Perón, who wished to make Argentina the hegemonic power in South America. Perón was an admirer of fascism, having served as a military attaché in Italy and Germany (Pendle, 1963). The United States, alarmed by the military takeover, wanted a trade embargo against Argentina, but Britain resisted because food imports from Argentina were essential to the war effort. The meat trade between the Río de la Plata and Britain continued, even though Argentina did not declare war on Germany until hostilities were virtually over in 1945.

Exports of nonessential foodstuffs from Latin America were disrupted, and the production of many fruits was reduced. For example, Jamaica had been the first country to get seriously into the banana shipping business, and the United Fruit Company had its origins there. In 1938, Jamaica produced 345,000 metric tons of bananas, with Britain as a major market. But in 1943, only 4,000 metric tons of bananas were produced (Mitchell, 1993). In general, the agricultural sector in Latin American wartime economies was stagnant, although there were exceptions. For example, Mexico exported fruit, vegetables, and edible oils to the U.S. market.

Exports of strategic raw materials rose to offset the decreases of many nonessential foodstuffs. For example, exports of Bolivian tin, Chilean copper, and iron ore from several countries increased. Exports of molybdenum, used to make steel for high-speed tools, increased from Mexico, Chile, and Peru. Vanadium production grew in Mexico and Peru. Vanadium was used in steels made to cut ordinary steel. Manganese mining increased in Mexico, Cuba, and Brazil. Alloyed with steel, manganese produced a light-weight metal used in aircraft engines.

The major producers of antimony available to the Western Allies were in the Americas by early 1942 when China (with 60 percent of prewar production), Yugoslavia, Czechoslovakia, and Algeria were under Axis control. Mexico, Bolivia, and Peru benefited. Antimony was used in cosmetics, camouflage paint, and lead (as a hardener). The hardened lead was used in truck batteries, lead bullets, and shrapnel.

Tungsten production in Latin America also rose. Tungsten was used in electrical equipment, radar sets, and other military capacities. China and Burma had been major producers before the war. Now they were in Japanese hands, and production in Latin America grew. There were some spectacular increases. At the beginning of the war, Brazil produced the equivalent of 8 metric tons of tungsten; by 1944 it produced 1,492 metric tons (Mitchell, 1993). In the years before World War II, the Philippines was a major source of chrome for the United States, but when the Philippines fell into Japanese hands, production in Cuba was rapidly increased.

Similarly with tin. In 1940, just over a third of world tin production came from Malaya, the Dutch East Indies produced about 20 percent, and Bolivia was the third largest producer with 16 percent of total world output. When the Japanese invaded Southeast Asia, they occupied countries that had produced 62 percent of the world's tin (Jones, 1950). As the war progressed, Bolivia became the major source of tin available to Britain and the United States. In 1938, Bolivia produced 25,894 metric tons of tin; by 1944 production stood at over 39,341 metric tons (Mitchell, 1993).

During World War II, Latin American countries found that imports from North America and Europe were no longer available. Canada, Britain, and the

United States were so busy making armaments that little industrial capacity existed to produce textile machinery, motor vehicles, or consumer goods for export. The lack of machinery imports was an impediment to industrialization, but existing producers of consumer goods in Latin America had less competition, and domestic production increased. In most cities of the region, small-scale production units sprang up to make spare parts and consumer items that had formerly been imported. In Argentina for example, during the war years, the number of manufacturing companies increased by approximately one third, but most were small and collectively only accounted for 11.4 percent of manufacturing output when the war ended (Bulmer-Thomas, 1995).

Some basic heavy industries, requiring substantial capital outlays, were expanded. For example, in 1940, Mexico produced 147,000 metric tons of steel, rising in 1945 to 230,000 metric tons as capacity was increased at Monterrey, and a new plant began production at Monclova in 1944. Steel production started in Argentina in 1939 when 18,000 metric tons were produced. By 1944, production stood at 150,000 metric tons. Brazil produced 114,000 metric tons of steel in 1939 and over 200,000 metric tons in 1945, as the Volta Redonda plant, began in 1942 and run by the Companhia Siderúrgica Nacional, came into full production. Before the war, the German company Krupps discussed building the plant, but the war resulted in the United States playing a part in the venture (Humphries, 1981). Numerous other basic industries including glass, chemicals, paper, cellulose, and textiles were expanded in Latin America during the war years (Hughlett, 1946). The manufacturing sectors of most countries in the region grew in excess of 5 percent per annum, but the starting base was low and many of the new manufacturing activities were small-scale and vulnerable to postwar competition.

The major countries of Latin America developed positive trade balances as exports grew and imports were scarce. Large balances of foreign currency were built up in dollars and pounds. However, all countries suffered inflation (Thorp in Rock, 1994), and wholesale prices approximately doubled in most countries between 1938 and 1945 (Mitchell, 1993).

Mexico illustrates the advantages and disadvantages of wartime economies in Latin America dependent upon the United States for markets, finance, and technological assistance. Mexico and the United States share a border, and their economies became more interactive during World War II. There were positive developments. The Rockefeller Foundation supplied the technical assistance that resulted in the development of improved varieties of corn and wheat. Mexico enjoyed the first modern green revolution. On the other hand, Mexican agriculture, responding to market opportunities in the United States, exported produce to the north and left the Mexican market short of corn.

In some areas, labor was scarce in Mexico as a result of the Bracero program that allowed Mexican farm workers to enter the United States. The program stimulated both legal and undocumented migration across the border and set patterns that continue to the present day, for once migration chains and channels are established, they are difficult to break.

Mexico attracted U.S. investment as efforts were made to increase production of goods that could complement the war effort that Mexico had joined in the spring of 1942. The iron and steel company at Monterrey had been in existence since 1900, but during World War II, U.S. capital financed the installation of more blast furnaces and steel-making capacity.

Throughout Latin America, urbanization had quickened in the 1930s. The war accelerated the trend. Big cities grew at faster rates than smaller urban places. Ports that exported strategic materials were busy. The machinery of government grew, and more jobs were created in the bureaucracy. As urban areas enlarged, so did the service sector of the economy as more retail stores, transportation facilities, and places of entertainment emerged to meet the needs of the population living in cities. An important political by-product accompanied urbanization. With a larger percentage of the population living in cities and often concentrated in primate cities, it became possible to organize mass movements that had an impact on political development and government policies. For example, in 1945, when Juan Perón was arrested by political rivals, mass demonstrations in Buenos Aires forced his release. In Brazil, political pressure for cheap food in urban areas led to a disadvantageous price structure for farmers.

► Government Control of Economies

Would World War II be just another Latin American export boom? When the war ended, would demand decrease, old competitors reemerge, prices fall, and unemployment rise? In response to these questions, many Latin American governments after World War II tried to establish policies that protected against the swings in the world economy. Although not widely recognized, the move toward more government control of economies in Latin America had started well before World War II. For example, in the 1920s, Argentina established *Yacimientos Petrolíferos Fiscales*, a state corporation to run the oil industry. Bolivia nationalized the oil industry in 1937, and Mexico did the same in 1938, establishing Pemex to produce, refine, and distribute petroleum products with the objective of promoting industrialization and agricultural improvement in Mexico. In Brazil during the *Estado Nôvo* (new state) era of 1937 to 1945, attempts were made by the central government to improve everything from architecture (Pendle, 1963) to steel making, eventually resulting in the creation of a large steel industry, using indigenous resources and capable of competing in international markets. The origins of the new national capital of Brasília, inaugurated in 1960 and located 600 miles into the interior, can be found in the *Estado Nôvo* era when Oscar Niemeyer, an architect who worked at the new city, came into prominence as a result of his designs for government buildings. The notion of creating an interior capital can be traced back into the nineteenth century.

By the 1930s, the economic ideas of totalitarian regimes were understood in Latin America. Politicians on the left knew about Soviet five-year economic plans and those on the right understood that fascist governments, although allowing large corporations to operate, controlled economic policy. Mediterranean Europe, to which Latin America was connected by history and culture, was under fascist control by the end of the 1930s. Italy under Benito Mussolini and Portugal under Antonio Salazar (a professional economist) had fascist systems in place. In 1939, at the end of the Spanish Civil War, Francisco Franco tried to organize the Spanish economy on a fascist model. In the fascist economic model state-directed industry controlled imports and exports, promoted public works, set food prices, and developed agricultural self-sufficiency. Although the apparatus of state control was abhorrent to some, the economic theory was attractive, at least as a reaction to the Great Depression, in which economic failure in one country had been transmitted to others in the international system.

The economic ideas of totalitarian regimes shared a common characteristic. They wanted to create autarkic economic systems: economies that were self-contained and produced most requirements within their own borders. When imports were needed, they were obtained from bilateral trade partners on carefully negotiated terms. Latin Americans understood the concept of bilateral trade deals because Germany negotiated several in the region in the 1930s.

At first sight, economic autarky for Latin America was impossible because the countries of the region needed to export commodities to earn hard currency to buy the technology of industrialization. However, parts of totalitarian economic theory were adaptable to Latin America. The state could control imports and encourage manufacturing within the national territory to supply the home market. The policy was termed import substitution. Most countries in the region adopted a formal import substitution policy after World War II, but import substitution had already emerged when goods were in short supply in Latin American markets in the 1930s and during World War II. One country, Cuba, was later to adopt a Soviet model of autarkic economic development complete with bilateral trade deals that ensured that all Cuban sugar went to trade partners in the Soviet bloc in exchange for goods produced there.

Much of the economic analysis used to justify import substitution for Latin America was published after World War II by Raúl Prebisch at the Economic Commission for Latin America (ECLA), established by the U.N. in 1948. Prebisch thought:

> In Latin America, reality is undermining the outdated schema of the international division of labor . . . Under that schema, the specific task that fell to Latin America, as part of the periphery of the world economic system, was that of producing food and raw materials for the great industrial countries. There was no place within it for the industrialization of the new countries. It is, nevertheless, being forced upon them by events. Two world wars in a single generation and a great economic crisis between them have shown the Latin American countries their opportunities, clearly pointing the way to industrial activity. (Hirschman, 1961, p. 14)

The formal implementation of import substitution policies with protective tariffs and government licensing of protected manufacturing activities was a product of the post–World War II era. Analysis has shown that the theory of import substitution was worked out in eastern Europe and in Germany in the 1920s and fine-tuned in Stalin's Soviet Union and Hitler's Nazi Germany (Fitzgerald in Rock, 1994).

Whatever the origins of import-substitution policies, they became the basis of industrialization in the postwar decades in Latin America. Eventually, import-substitution became discredited as protected producers turned out high-cost, low-quality goods in an economic environment regulated by bureaucrats rather than by competition in the marketplace. Alan Gilbert in Chapter 7 provides more on import substitution which was abandoned by many Latin American governments in the 1980s and 1990s. Some of the state-owned corporations were sold to investors—privatized—but several, including Pemex, continue to play a leading role and enjoy a monopoly in Mexico.

POST–WORLD WAR II

After World War II, the price of strategic raw materials did not decline because demand remained high. Some former competitors for Latin American primary producers were removed from world markets because they were part of the Soviet autarkic system.

The Western European economies had been badly damaged by the war, and even after the end of hostilities in May 1945, living standards declined in many areas of Europe. There were fears that economic and political instability would result in more countries going under Communist control. In 1947, the United States launched the Marshall Plan to help reconstruct the economies of Western Europe. Latin American countries were disappointed not to be included in the plan, but world demand for commodities increased, prices improved, and Latin American primary exports benefited.

The Korean War (1950–1953) started when Communist North Korea invaded South Korea. Demand for raw materials surged as the United States enlarged stockpiles of strategic materials and sent forces under the U.N. flag to repel the invaders. The United States used bases in Japan to support the Korean War and procured many supplies in Japan. The Korean War stimulated the Japanese economy, which had not recovered after World War II. Japan has few mineral resources, and the reemergence of Japanese industrial combines increased world demand for ores and oil.

Exports of primary products, from most Latin American countries, increased in the postwar decade. There were adjustments. Brazilian and Peruvian rubber exports declined as the liberated rubber plantations of Southeast Asia regained lost markets. Against that decline can be set the reemergence of banana and fruit exports, especially from Central America and the Caribbean, as postwar European living standards rose and shipping was again available to transport consumer commodities.

► *Latin American Manufacturing in the Postwar Years*

During World War II, trade in manufactured goods between the countries of Latin America increased. Brazil and Argentina also exported manufactured items to markets beyond the region. After World War II, as North American and Western European manufacturers reestablished themselves in overseas markets, Latin American industries lost export markets and had difficulty competing at home against imported products.

Given that much new manufacturing activity in the war years had been improvised in a workshop environment, it was not surprising that many small units were uncompetitive. In the normal course of events, after numerous new producers enter the marketplace, a period of competition results in the survival of a few. Companies created during World War II competed against each other and faced competition from North American and European manufacturers reentering markets after World War II. Many companies failed.

To reduce competition from nonnational companies, Latin American governments adopted import-substitution policies, with protective tariffs and licensing of imports. Some new industries emerged. For example, in the postwar era, Mexico, Brazil, and Venezuela began to produce aluminum. During World War II, only Mexico, Argentina, and Brazil were making steel, but 20 years later, Chile, Cuba, Colombia, Peru, and Uruguay also produced steel.

In the case of Brazil, the World War II investment in steel did yield results. Brazil eventually was able to compete in world export markets for some types of steel. In the case of Venezuela, it can be argued that, although the investment in steel and aluminum making at Ciudad Guayana on the Orinoco River is unlikely to show a profit, putting government funds into the economic development of the Orinoco region and the northern margin of the Guiana Highlands is a worthwhile venture. Other countries, like Chile, Colombia, and Peru with the Pacific coast Chimbote steel plant saddled themselves with high-cost steel when they could have imported the metal for less.

In the 1950s, Mexico, Argentina, and Brazil started manufacturing motor vehicles. Later, Chile, Colombia, Peru, and Venezuela did the same. At first, the auto industry consisted of plants that assembled cars from imported parts. Tariffs ensured that imported vehicles cost more than the expensive, locally assembled models. The results were mixed. Some countries achieved little more than higher-priced transportation. Brazil did get into export markets. Argentina developed a viable agricultural equipment industry. In Mexico, the assembly of motor vehicles became a major export, as U.S., Japanese, and German car makers put together cars and trucks for the North American market. Assembly activity was established in Mexico before NAFTA was enacted, but the agreement gave vehicle assembly a boost.

Import substitution policies had their greatest impact in the area of consumer goods. By the 1960s, many countries in Latin America were producing a significant proportion of consumer goods at plants in the national territory. Products included textiles, clothing, footwear, furniture, radios, television sets, and electronic equipment. Usually, products were relatively expensive, and quality control was lacking.

Sometimes the politics of regional development resulted in ignoring the normal rules of industrial location. Chile managed to locate motor car assembly activities in the north, far from the main markets in the cities of Santiago and Valparaíso. Brazil established a Foreign Trade Enterprise Zone at Manaus to promote development in that region. Manufacturers were given tax and tariff advantages to locate in the heart of Amazonia. Common sense suggests that we view with skepticism a plan that has Brazilians flying from Rio, Recife, Santos, and São Paulo to buy televisions and electronic equipment because they are cheaper on the Amazon. It cannot be efficient to move consumers thousands of miles on round trips from the main centers of population in the east to shopping malls on the banks of the Amazon.

▶ Postwar Trade Patterns

As we have seen, World War II altered trade patterns. For example, between 1939 and 1945, trade between Argentina and Brazil increased several-fold. The pace of growth was not maintained in the postwar era, but a pattern of increased trade among the countries of the region was in place. In 1961, a Latin American Free Trade Association (LAFTA) was created; but never established itself, because policies of import substitution and protectionism cannot promote free trade.

World War II laid the foundation of global trade zones. During World War II the United States became the dominant economic power in the Western Hemisphere, with Latin America in its orbit. World War II created the trade patterns, focused on Germany, that characterize the European Union (E.U.). World War II further disrupted the free-trade system that had centered on Britain, and World War II saw the creation of a Pacific Rim group under Japanese tutelage.

The E.U., established in 1957,[4] adopted a Common Agricultural Policy that paid E.U. farmers more for crops than world prices. This had an impact on Latin America. With the price incentives, European farmers produced more grain, meat, vegetable oils, wine, butter, and cheese, reducing the market for Latin Americans. Latin American farmers could still sell to E.U. markets, but a tariff was imposed to bring the price of the import up to the artificially high price that European farmers received. Britain joined the E.U. in 1974. The country that, by the repeal of its Corn Laws in 1846, had ushered in a long period of free trade in which Latin America had easy access to British markets, was behind a high protective tariff for imported foodstuffs.

Most countries, including the United States and Japan, have their "import-substitution" and "export-enhancement" programs, especially in the agricultural sector. We do not need to feel sympathy for Latin American farmers disadvantaged

---

[4]The Treaty of Rome (1957) establishing the European Economic Community (E.E.C.) was signed by France, West Germany, Italy, Belgium, Netherlands, and Luxembourg. The object was to promote economic growth by allowing the free movement of goods, labor, and capital between the member states. Britain, Denmark, Ireland, Greece, Spain, Portugal, Austria, Finland, and Sweden have since joined and the name of the organization has evolved from the E.E.C., or Common Market, to the European Community (E.C.) to the European Union (E.U.). Many economic planners in Latin America like the idea of a Latin American Common Market, but it is not an easy concept to implement.

by an E.U. common external tariff that protected European farmers. The governments of Latin America were doing exactly the same thing when they protected home industries against the import of manufactured goods.

▶ *The Alliance for Progress*

In the postwar era, most Latin American countries found that the United States dominated trade, with the trade balance strongly in favor of the North American partner. Other economic trends moved against the region in the 1950s. Raw material prices dropped after the Korean War. The terms of trade shifted against commodity producers, who were receiving low prices for their exports, and in favor of the countries that exported manufactured goods at increasing prices. Within Latin America, population growth rates were high, rural-to-urban migration was rapid, cities grew quickly, and living standards rose slowly, particularly in urban squatter settlements around the major cities. Inflation made life difficult for the urban poor.

In the late 1950s, the vice president of the United States, Richard Nixon, visited Latin America. He was forced out of Lima and nearly assaulted in Caracas. The tour was canceled. In September 1959, Fidel Castro gained control of the Cuban Revolution and converted the island to a socialist system.

These events led President John Kennedy to propose an Alliance for Progress (1961), a ten-year plan to hasten economic and social development in Latin America. The plan sought to promote democracy, accelerate development, sponsor agrarian reform, and improve housing, working conditions, education, public health, and taxation policies. Further aims were to reduce inflation, stimulate private enterprise, smooth commodity price fluctuations, and promote economic integration. The Alliance for Progress had little impact on any of the problems. The countries of Latin America were not ready to cooperate and implement the region-wide goals of the plan. The Latin American Free Trade Association (1961), which might have been a vehicle for regional economic integration, never functioned as a free-trade association and came to consist of bilateral agreements between countries, rather than a region-wide commitment to reduce tariffs.

The United States provided over $10 billion to the aims of the plan. However, in the 1960s, as military governments took power in many countries of the region, most of the aid helped to support the old regimes rather than promote democracy and progress. The social and economic problems identified in the Alliance for Progress were not effectively addressed—because individual countries lacked either resources or political will or both. No country had the financial, technical, and urban planning resources to bring about the controlled growth of cities. Few countries tackled the issue of agricultural reform.

▶ *The 1960s*

In general, Latin American economies grew at a faster percentage rate than developed economies after World War II until around 1970. However, population also grew faster than the world average and much faster than in the countries of the economically developed world. By the 1960s, the major Latin American issues being discussed were the high rate of economic growth, the rapid increase in population numbers, and the need for land reform, meaning the breakup of large estates and the creation of small farms to satisfy the aspirations of the rural peasantry.

Land reform in Latin America is a political, economic, social, and emotional issue. The problem is grounded in the emergence of the great estates—*haciendas*—in the colonial era and the foreign-owned plantations that were developed in the nineteenth and twentieth centuries. In an agrarian society, which Latin America was until a few decades ago, there were stark divisions between the great farms *(latifundia)*, peasant landholdings *(minifundia)*, and landless laborers. Social justice seemed to demand that governments pursue policies of land reform, buy up large estates, subdivide them, and make land available to the rural poor. For a variety of reasons, land reform was beyond the reach of most countries. Ruling elites were connected to large, landowning families. When populist movements demanded land reform, it was likely to lead to a military coup, as happened in Brazil in 1964.

In order to have effective land reform, agricultural extension services, credit for farmers, and roads to link areas of agricultural production to growing, urban markets, was necessary. In practice, there was plenty of political rhetoric but little bureaucratic and technical support for the problems of rural development.

Partially because rural prospects were so poor and land so difficult to acquire for small family farmers, the 1960s was marked by high rates of rural-to-urban migration. The young, the ambitious, the better educated left the countryside and moved to urban areas. Urban majorities emerged in most Latin American countries, making rural issues less pressing. Land reform was hardly addressed.

The fast pace of rural-to-urban migration resulted in the creation of sprawling, disorganized primate cities that dominated the urban hierarchies of most countries in the region. In the 1960s, the population of Latin America grew around 3 percent per annum, and at that rate a population will double in about 23 years. Rural populations grew at less than the average because of out-migration. The population of urban areas increased at nearly 5 percent and in the big, primate places, it grew at 7 percent. Big cities doubled their size in 10 years and were unable to plan for orderly expansion. The old, colonial cores of many cities were overwhelmed by traffic congestion. The prosperous middle- and upper class housing areas made sure that services were provided to them. In the squatter settlements sprouting at the margins of rapidly growing urban areas, dwellings were made from packing cases, scrap wood, and galvanized iron. The squatter settlements—*ranchos* to Venezuelans and *callampas* (mushrooms) to Chileans—lacked piped water, electricity, sewers, and paved roads. Eventually, the municipal authorities tried to provide services, and the original, temporary dwellings were replaced by brick or cinder-block homes. Living conditions in squatter settlements seemed awful to outside observers, but a journey into the rural hinterlands of growing cities revealed that life was worse there.

Postwar economic trends tended to favor urban rather than rural areas. As governments expanded, they attracted workers to capital cities. New manufacturing activities usually located in cities, very often at ports. As cities grew, so did the service sector of the economy, outpacing employment in manufacturing and in primary production. The service sector includes wholesaling, retailing, finance, transportation, administration, and education. The service sector produced some high-quality white-collar jobs and many low-level positions, besides spawning informal economic activities, as armies of vendors, hucksters, shoeshine boys, and lottery ticket sellers lined the streets to collect tiny profits on innumerable small transactions.

Once people moved to the cities, the birth rate dropped and the rate of population growth decreased. By the end of the 1970s, most countries in Latin America had urban majorities, and the rate of population increase began to decline although, of course, the total population of the region continued to grow.

INTERNATIONAL ECONOMIC GROUPINGS WITHIN LATIN AMERICA

In Latin America, there were attempts in the 1960s and 1970s to create economic groupings. Within Middle and South America, the Central American Common Market (1960), the Latin American Free Trade Association (1961), the Caribbean Free Trade Association (1968), and the Caribbean Common Market (1974) came into operation. These organizations had few successes.

Three major types of international cooperative economic groupings exist: the free-trade area, the customs union, and the common market. A free-trade area consists of two or more countries that agree to remove tariffs and other restrictions on trade between them. The free-trade partners can trade with nonmember states on their own terms. A customs union involves a group of states removing restrictions on trade between them and establishing common external tariffs, which are applied equally to all imports entering the territory of member states. A common market provides for the removal of restrictions on trade between member states, the establishment of common external tariffs, and the creation of economic policies and business laws that apply equally to all the member states. In short, with a common market an attempt is made to create a multinational region with common business conditions. The European Community is the most advanced example of a common market, and it is now moving toward political integration in a European Union. The European Community allows the free movement of goods, labor, and capital between its 15 European members. NAFTA (the North American Free Trade Agreement), as a free-trade association, does not determine trade arrangements with nonmembers and does not provide for the free movement of labor between members. Mercosur (Common Market of the South) reduces tariffs on trade between Argentina, Brazil, Paraguay, and Uruguay. As in a common market, it imposes tariffs on imports from nonmember states.

The aim of a free-trade association, a customs union, or a common market is the creation of larger market area in which investment is stimulated, competition promotes efficiency, economies of scale are gained, and the pace of economic growth accelerated. Politically, there is an increasing degree of commitment from a free-trade association to a common market, and in the case of a common market, a central authority makes some decisions formerly arrived at by national governments.

Unlike the European Community, Latin America lacks a regional transportation network, and it is difficult to envisage the large-scale movement of goods over the region. Some countries have good transportation links with neighboring states; others do not. The different levels of economic development present another difficulty. Countries like Chile, Brazil, Argentina, and Mexico have manufacturing sectors and goods to sell in adjoining markets. Poorer states like Bolivia, Paraguay, and Ecuador have limited manufacturing capacity and few goods to sell in the markets of neighboring states.

► *The Latin American Free Trade Association (LAFTA)*

LAFTA was established by the Treaty of Montevideo in 1960 with the aim of liberalizing trade among the member states of Argentina, Bolivia (acceded February 1967), Brazil, Chile, Colombia (acceded October 1961), Ecuador (acceded October 1961), Mexico, Paraguay, Peru, Uruguay, and Venezuela (acceded September 1966).

The treaty stated that within a period of 12 years the member states would eliminate duties and trade restrictions on goods originating in the territory of any member state. Elimination of the trade barriers was not achieved on a regional basis although some countries succeeded in removing tariffs on certain goods moving between them. The failure of LAFTA was not surprising. The organization was launched when most countries were promoting import-substitution manufacturing and protecting home markets with high tariffs. A free-trade agreement requires that member states open markets to each other. Free trade and import substitution are opposite policies, and LAFTA was killed by the conflict.

► *The Andean Group*

LAFTA's inability to achieve aims strengthened the desire of some countries to form subregional organizations. The Andean countries wanted to coordinate development, and in 1969 Colombia, Ecuador, Peru, Bolivia, and Chile agreed to establish a customs union, coordinate economic policies, and build a regional communications system. Venezuela joined the organization in 1973. Chile has since left. The Andean Group has had minimal impact on economic development in the member states.

► *The Central American Common Market*

The Central American Common Market was established by the Treaty of Managua in 1960. The member states were Costa Rica (acceded 1962), El Salvador, Guatemala, Honduras, and Nicaragua. For a time the common market was a success because (1) the member states were at about the same level of economic development; (2) they formed a relatively compact unit; (3) the regional transportation network, though not fully developed, had established the initial linkages in a system that was capable of growth; and (4) the members were reaching the point where the development of import-substitution industries might be profitable and the profits would be greater within an enlarged market area.

Political problems undermined the organization. In 1969 El Salvador and Honduras went to war with each other, and subsequently Nicaragua and El Salvador experienced civil wars. The organization ceased to have an impact on the economic life of Central America.

THE OIL CRISIS OF THE 1970s

In the 1970s, there were world shortages of oil and other raw materials. Oil prices went up, and producers in Latin America, such as Mexico, Venezuela, Colombia, and Trinidad, benefited. In the early months of 1973, the price of oil was less than $3 a barrel. When the year closed, oil was bringing over $10 a barrel. The price increase continued, and in 1979 there was another huge jump. Oil traded at around $18 per barrel in the first part of the year and, as scarcities developed, soon rose to $36. In the season of panic it fetched $40 a barrel. As with most internationally traded commodities, worldwide price levels are established for oil by traders in the big commodity exchanges in London, New York, and Chicago.

In general, commodity exports from the Latin American region were buoyant for much of the 1970s, and with scarcities developing, prices were good. Many countries in the region wanted to increase the pace of economic development and found the banks of North America, Europe, and Japan willing to lend to finance economic growth. International banks in Europe and North America were full of money, principally from Middle Eastern countries. Latin American countries received many loans in dollars, pounds, francs, yen, and deutsche marks, at moderate rates of interest. Countries like Venezuela and Mexico, which exported oil, found it particularly easy to borrow on international markets, but their ability to repay and service the debt was to be greatly reduced by the oil price slump of the early 1980s and increasing interest rates.

THE DEPRESSION OF THE 1980s

By the early 1980s, the developed economies were in recession and imported fewer commodities, including oil, from Latin America. The economies of North America and Europe experienced recession and inflation. The cure adopted for that was high interest rates. As rates went up, payments due on loans made to Latin American countries increased at a time when Latin American commodity exports and the income from them were falling. As Latin American economies came under pressure, the value of their currencies declined. More pesos, for example, were required to buy the dollars to pay the interest on the loans from U.S. banks.

In 1982, Mexico defaulted on loans, and other countries followed. To help pay the debts, Latin American governments decreased expenditures, and imports were reduced. In Latin America during the 1980s, there was economic contraction. Few countries avoided the trend. Most countries "suffered an absolute decline in manufacturing" (Bulmer-Thomas, 1994, p. 400, and Table 11.6, p. 401). Wages fell and living standards sank. In 1993, the average per capita income in Latin America was 5 percent less than it had been in 1980 (Naím, 1995). In some countries, Peru for example, declines in infant mortality were reversed, and the percentage of newborn infants that failed to survive the first year of life increased in the mid-1980s. Decreases in nutritional standards were common, and caloric intake went down. Everywhere governments had less revenue, resulting in lower expenditure on disease prevention and public health. There were epidemics. Polio broke out in Bolivia. Cholera appeared in the coastal towns of Peru, as ineffective sewage disposal systems allowed the cholera bacillus to invade drinking water. Death rates from cholera were low, and less than 1 percent of those infected died, as opposed to a 30 to 40 percent death rate in European and North American cities in the nineteenth century, when the disease was frequent in rapidly growing cities.

The population trends of the 1970s continued into the 1980s. The rate of population growth slowed, birth rates declined, and the average number of births per woman dropped. In 1963, the average woman in Mexico gave birth to 6.7 children. By 1993, the number of births was down to 3.2, having fallen by just over 50 percent in 30 years. The trend was the same in Brazil, Peru, Venezuela, and Costa Rica. In Colombia, the percentage drop in the number of births per woman, in the same 1963–1993 period was 60 percent, from 6.7 to 2.7 children. For the United States, the figures were 3.2 (1963) and 2.1 (1993).

LOWER TRADE BARRIERS AND NAFTA

The mid-1930s to the late 1980s was a period in which Latin America continued to sell raw materials into the world economic system and attempted to diversify economically, using protective devices to shelter home-grown manufacturing. It was an era of economic nationalism that was successful in creating new consumer goods industries that supplied national markets and did create some trade in manufactured goods between countries in the region. However, economic nationalism, with its import-substitution policies, created a developmental cul-de-sac. High-priced manufacturers within Latin America were protected from competition and innovation by tariffs, import controls, the licensing of producers, and restrictions on investments by nonnationals wanting to start economic activity in Latin America. The protected producers that were profitable within national markets were seldom competitive in world export markets.

The perceived way out of the developmental cul-de-sac and the cure for economic stagnation, high unemployment, rapid inflation, weak national currencies, and shortage of hard currency to pay off debts was to open national economies to international competition. Foreign investment was encouraged to bring in hard currency. Government-owned corporations were sold to generate revenue. Foreign investors were allowed to buy or create companies to bring in capital. Tariffs were reduced on imports to lower prices to consumers, with the aim of stimulating demand for goods and promoting economic growth.

The new policies were illustrated by the experience of Mexico in the 1990s. Restrictions on imports into Mexico were reduced, the banking sector was privatized, and a number of state-owned corporations, including the national telephone company, were sold to private investors.

In 1989, Canada and the United States signed a free trade agreement. The two countries were already major trade partners with strong linkages in several manufacturing industries, including the motor car industry. In the mid-eighties Mexico sought a closer relationship with the European Community. The president of Mexico at the time, Miguel de la Madrid, went on a grand tour of Europe. For obvious reasons, de la Madrid was well received in Spain, and at every capital there were splendid pageants that recalled the age of discovery and the European contribution to the creation of the Latin American culture region.

Behind the scenes, the Mexican president was courteously told that, because the Soviet grip on Eastern Europe was slipping, the first claim on E.C. financial resources would be investment in East Germany, Poland, Czechoslovakia (as it was), and Hungary. Mexico and Latin America could expect no special treatment from the European Community.

Mexico had to face the reality that after striving, like Canada, to avoid becoming a farm, a forest, and a quarry supplying U.S., raw material needs, an era of trade regions had arrived, and Mexico needed to sign a North American Free Trade Agreement. Canada, the United States, and Mexico agreed to form a free-trade zone. The U.S. Senate ratified NAFTA in 1993. The strongest economic power in NAFTA, the United States, gave little away, and when American financial institutions and agricultural interests begin to penetrate Mexican markets more fully, problems will emerge for Mexico. The issue of NAFTA allowing Mexico to attract manufacturers out of the United States, at the expense of the U.S. blue-collar workers is spurious. As the *maquiladoras* illustrate, industry has been migrating to Mexican border towns for decades. If not to Mexico, the manufacturing and assembly operations would go, in one form or another, to South Korea, Taiwan, China, Singapore, Hong Kong, Malaysia, or Indonesia. Having the manufacturing plants just south of the border, to feed into the wholesaling and transportation systems of California and Texas is more profitable than having Japanese *zaibatsu* (money cliques) organizing production in South Korea or Malaysia and exporting the goods via their own shipping lines to West Coast ports. In a few years, if NAFTA is fully implemented, U.S. trucking companies, will be running cargo between Mexico City, St. Louis, and Minneapolis. U.S. banks will be financing the exports and the trucking companies and clearing the payments between Mexico City and banks in San Francisco, New York, Chicago, and Charlotte.

U.S. exports to Mexico were given a boost by NAFTA, as formerly significant tariffs on imports from the United States into Mexico were reduced. Mexican exports were not given the same lift because import duties on Mexican goods entering the United States were low before the agreement was signed. NAFTA did enhance investment interest in Mexico. Existing production facilities in motor car assembly were enlarged, and European and Japanese companies made plans to build plants. Such manufacturers have to meet the "rules of origin" laid down in the NAFTA agreement. The greatest part of a car, or any other good, will have to originate in North America. It will not be possible, for example, to ship Japanese motor bike parts to Veracruz and then assemble the pieces for sale in the United States and Canada.

U.S. retailers have extended operations into Mexico, often in partnership with Mexican companies. Mutual funds offer investments that specialize in Latin American and Mexican stocks. The economic integration of the U.S., Canadian, and Mexican markets is proceeding on course with all countries gaining benefits from the interchanges, although specific industries in all countries are exposed to more competition.

The weakest currency in NAFTA is the Mexican peso. However, the currency was stable in the runup to NAFTA, and it strengthened when the agreement was signed, as outside investors bought pesos to pay for the construction of new factories in Mexico or to buy shares on the Mexico City stock exchange, where prices are quoted in pesos.

In the fall of 1994, the holders of Mexican pesos perceived that the currency was vulnerable and sold them. The value of the peso fell sharply against other currencies. Henceforth, more pesos were required to buy industrial equipment from the United States. Conversely, due to the devaluation of the peso, Mexican goods were cheaper on entering U.S. and Canadian markets. The export of goods from Mexico surged, and the country had a large trade surplus in 1995. Currency devaluation has disadvantages (imports cost more) and advantages (exports are more competitive).

In 1994 and 1995, Mexico experienced political instability with the uprising in the southern state of Chiapas. There were battle scenes on international television of fighting in San Cristóbal de las Casas. Foreign investors feared the stability of the Mexican currency and the political system. The stock market and the value of the peso fell further, and the selloff of Latin American stocks by international investors spread to Argentina and Brazil.

Mexico learned that free trade, a freely convertible currency, and attachment to the international trade and currency speculation system could come at a high price. With a loan from the United States, the peso and the stock market stabilized. The devaluation and stock market crash scenario is likely to be repeated in the future.

Recently, there has been a revival of interest in Latin American economic integration. Perhaps it will lead to the creation of a Western Hemisphere Free Trade Association. In 1991, Argentina, Brazil, Paraguay, and Uruguay signed an agreement to create the Mercosur (Mercado Común del Sur, or Common Market of the South). The agreement has promoted rapid increases in trade between members. In 1994 the countries of the Caribbean established an Association of Caribbean States to promote regional integration.

The creation of more trade between the countries of Latin America and the gradual opening up of national economies to global competitive forces will not necessarily result in a sustained improvement in economic conditions. Latin America is still on the periphery of the world economy. While Mexico, via NAFTA, may succeed in attaching itself to the core, the region as a whole is still far from stable economically. The governments of the region have decided that progress can be made by exposure to international economic pressures. It remains to be seen whether more competition and interactions will produce economic growth in Latin America, or whether the region will return to the era of marked swings in economic fortunes brought on by exogenous forces generated in Europe, North America, and Japan.

SUMMARY

Latin America's history of the last 500 years may have been a story of increasing contact with the wider world, but the region is still spatially peripheral to the major centers of economic activity. As a late-developing region, it tends to be more dependent on external economic forces than is the case for older, established regions, such as Western Europe and North America. From the colonial past through independence and into the modern era there has been a significant interplay between economic and political developments within Latin America and events in the international arena.

Between 1914 and 1945, policymakers in Latin America saw a large number of economic lessons played out in their region. Latin America absorbed the lessons of globalism. From World War I through the economic swings of the 1920s, the Depression of the 1930s, the strategic materials surge in the prewar years, the loss of markets as World War II broke out, and the reorientation of economies in wartime, Latin American observers learned that faraway quakes in the global system reached the region in a magnified form.

In the post–World War II era, Latin American countries attempted to seal themselves off from global economic forces by the adoption of import-substitution policies. Recent attempts to open economies to competition have led to some export industry successes and the attraction of foreign investment. International investors, however, tend to move capital in and out of the region, leading to currency and stock market fluctuations.

► *Sustainable Development*

Latin American nations have been striving for economic development and modernization throughout the twentieth century. Recently there has been interest in what has been termed "sustainable development" (Wilbanks, 1994). This term has two meanings. In the narrow sense, it refers to management of the environment, including the prevention of pollution and maintaining biodiversity, so that resources will be sustained. In other words, economic development should be pursued only with the interests of environmental health firmly in the forefront of policy. To give a simple example, it is realized that continued cutting and burning of the Amazon forest cannot go on forever without dire consequences. In the broad sense, "sustainable development" has come to mean development that takes human rights and economic justice into account. It means working against poverty and pushing for education and prosperity for all. It means striking a balance between protection and production, between tradition and modernization, between the past and the future. It means governments, nongovernment organizations, and private enterprise working together to spread the benefits of development widely among citizens. The message is that development must be in the interest of all sections of Latin American society for it to be considered sustainable.

FURTHER READINGS

Bulmer-Thomas, V. *The Economic History of Latin America Since Independence.* Cambridge: Cambridge University Press, 1994.

Dodds, K-J. "Geopolitics, Cartography, and the State in South America." *Political Geography* 12(1993): 361–381.

Grayson, G. W. *The North American Free Trade Agreement.* Lanham, Md.: University Press of America, 1995.

Knox, P. and Agnew J. *The Geography of the World Economy,* London: Arnold, 1989.

Mitchell, B. R. *International Historical Statistics: The Americas, 1750–1988,* 2d ed. New York: Stockton, 1992.

Williamson, E. *The Penguin History of Latin America.* New York: Penguin Books, 1992.

Glossary

Bold type within a definition indicates a term referred to elsewhere in the glossary.

Abattis   A **shifting cultivation** plot cleared in the forest in French Guiana (Guyane).

Aboriginal (aborigine)   Original inhabitant of a region; native American; native to America. Original characteristics, especially when contrasted with invaders or colonizers. See **Indigenous**.

ACS   See **Association of Caribbean States**.

Adelantado   A man given permission by the Spanish Crown to explore and claim territory in colonial Spanish America.

Aguardiente de caña   Rum; hard liquor distilled from the fermented juice of crushed sugar cane.

Altiplano   Literally, high plain. Specifically, the high basin adjacent to Lake Titicaca, in northwestern Bolivia and southeastern Peru.

Alpaca   Animal of the camel family, as are vicuna, guanaco and llama.

Altitude   Height above sea level, usually expressed in feet or meters. One meter is equivalent to 3.281 feet, or 39 inches.

Altitudinal zonation   The atmosphere cools, on average, 3.7°F every thousand feet ascended. Even in the tropics cooling with increasing altitude produces changes in natural vegetation and influences crop selection. The limits of temperature zones vary with local circumstances. In general, the hot land (*tierra caliente*) extend up to around 3,000 feet. The *tierra templada* (temperate land) lies between 3,000 and 6,000 feet; the *tierra fria* (cold land) is in the range of 6,000 to 12,000 feet. *Tierra helada* (ice land), where frosts are common and snow or ice is encountered, are found at the highest elevations.

Amazonia   Lowland drainage area of the Amazon River and many of its tributaries. Mainly in Brazil, eastern Ecuador and Peru, and northeast Bolivia.

Amerindian   Indigenous inhabitant of the Americas.

Animism   The belief that features in the natural environment are sacred and have power.

Antilles; Greater and Lesser   *Greater Antilles:* The large islands in the Caribbean Sea: Hispaniola, Cuba, Jamaica, Puerto Rico. *Lesser Antilles:* The Leeward and Windward islands stretching from the Virgin Islands to the coast of South America.

Araucanians   Aboriginal people of central Chile.

Arawaks   Aboriginal people, originally located on the upper Amazon, who migrated into the Greater Antilles, taking their manioc-sweet potato based farming with them.

Association of Caribbean States (ACS)   Established in 1994 to promote cooperation. Includes most countries of the Caribbean, Central America and Mexico.

Atacama Desert   Desert on the coast of northern Chile and Peru. The desert is primarily a result of the cold, off-shore **Peru (Humboldt) Current**.

Audiencia   A tribunal or court, in the Spanish colonial system, composed of a president and usually four judges, with jurisdiction over a geographical area—also known as an *audiencia*. Only the viceroy and Council of the Indies were superior in administrative and judicial affairs. The *audiencia* also acted as a court of appeal.

Autarky   Economic self sufficiency.

Aztec realm   With its center at Tenochtitlán, in the Valley of Mexico, the Aztec domain extended from the Pánuco River in the northeast to the modern frontier of Guatemala. It was a tribute state in which conquered towns paid dues to the center.

Bagasse   Woody remnants of crushed sugar cane stalks used to fuel furnaces in sugar boiling houses.

Baja California   California peninsula, lower California. Present Mexican state, as opposed to Alta (upper) California (U.S. state).

Banda Oriental   Uruguay. The eastern bank of the Uruguay River.

Bandeirantes People of mixed Portuguese and Indian ancestry who banded together to explore the interior of Brazil and Paraguay in search of Indian slaves for coastal plantations, and for deposits of precious metals, particularly gold and diamonds. Also known as *Paulistas*.

Barrio Neighborhood.

Bauxite A hydrated oxide of aluminum from which aluminum is refined. Bauxite is widely found in the tropics where igneous rocks, rich in aluminum silicates, are weathered in humid conditions.

Birth rate, crude The number of live births, per thousand of a population, that occur each year.

Border Industries Program (BIP) (1965) A policy established by the Mexican government after the conclusion of the U.S. **Bracero Program** in 1964. Assembly plants built under the program could be foreign-owned and employ foreign managers and technicians. No Mexican duties were charged on imported equipment, material, and components. The United States charged duty on the value added during assembly in Mexico. BIP created many assembly plants—**maquiladoras**—in Mexican cities bordering the United States.

Bola or Bolas A weapon used by Indians and gauchos consisting of two or more balls (usually stone) attached to the end of a cord for throwing at animals in order to entangle their legs.

Bororó A tribe of mainly nonagricultural aborigines in east-central Brazil, today greatly reduced in numbers and area occupied.

Bracero Program A U.S. immigration policy, initiated in 1942, that allowed Mexican workers access to the United States when labor was short during World War II. The program ended in 1964.

Brazilwood A tree—pau-brazil (*Caesalpinia echinata*)—native to Brazil that gave the country its name. Before synthetic dyes, brazilwood was used as a dyestuff.

British Honduras (Belize) Established as a wood-cutting settlement in the seventeenth century, administratively recognized as a British colony in 1862, independent as Belize since 1962.

Buffer state An independent country separating more powerful states. For example, Uruguay was created in 1828 to divide Brazilian and Argentinean interests in the Rio de la Plata region.

Caatinga A deciduous thorn scrub region in northeast Brazil subject to periodic droughts. The northeast of Brazil is also referred to as *Sertão*. Both *Caatinga* and *Sertão* have a regional connotation and a general meaning. *Caatinga* can be used to describe a vegetation type, and *Sertão* is used to describe sparsely populated frontier zones.

Cabildo Town council.

Caboclos Landless peasants in Brazil.

Cacao (*Theobroma cacao*) A small tree native to the American tropics, producing pods containing cocoa beans. Cocoa is consumed as a drink or as chocolate.

Camino real Main road or trail. Literally, royal road.

Campesino Peasant; small farmer; inhabitant of a rural area.

Capitânia The Portuguese Crown divided the American territory into 12 large proprietary grants called capitanias, each fronting onto between 100 and 400 miles of coast and extending inland.

Caribbean Basin Initiative (CBI) U.S. program that came into effect January 1, 1984 to promote economic growth in the Caribbean and Central America.

Caribbean Community and Common Market Abbreviated as **CARICOM**. Established in 1973 by Barbados, Guiana, Jamaica, Trinidad and Tobago to promote economic integration. Other former British West Indian areas have joined.

Caribs Aboriginal people originally from the Guianas and Venezuela who migrated into the Lesser Antilles. Gave their name to the Caribbean.

CARICOM See **Caribbean Community and Common Market.**

Carreta Two-wheeled cart.

Casa de Contratación The Board of Trade established in Seville in 1503 to oversee Spanish trade with the Americas.

Casa-grande In Brazil the mansion which constitutes the headquarters of a large landed estate.

Cassava See **Manioc.**

Casco The cluster of buildings at the heart of a **hacienda** or farm.

Caudillo/caudilho (Portuguese)/caudillismo A "strong man" who ruled powerfully, though sometimes popularly. The name also carries the meaning "little big head."

CBD Central business district of a city or town.

CBI See **Caribbean Basin Initiative.**

Cédula Official document for identification or authorization.

Central American Common Market (CACM) **Common Market** established by the Treaty of Managua, 1960, signed by El Salvador, Guatemala, Honduras, and Nicaragua. Costa Rica acceded to the treaty in 1962. Trade was disrupted by the 1969 **Soccer War** between El Salvador and Honduras, and the civil wars in El Salvador and Nicaragua.

Centrale   Plant for crushing sugar cane in Cuba, Puerto Rico, and the Dominican Republic.

Chaco War   Fought between Bolivia and Paraguay, 1932–1935, over possession of the Chaco Boreal, a semi-arid scrub region that Bolivia believed contained oil and might offer a passageway to the Atlantic. Paraguay won the war and retained the territory.

Charque   Beef that has been cut, salted and dried in the sun.

Chicha   Beer fermented from corn.

Chichimecs   Nomadic native inhabitants of northern Mexico who resisted Spanish penetration.

Chinampa   Raised cultivation plots along the margins of shallow, freshwater lakes. Used in the Valley of Mexico and in other parts of east-central Mexico. Chinampas produce several crops a year due to water availability and soils rich in organic matter.

Chicle (*Manilkara zapota*)   Tree from which gum is harvested to make chewing gum. Today the gum in chewing gum is produced from petrochemicals.

Cholera   Disease of the intestine caused by the bacterium *Vibro cholerae* that is spread by infected drinking water or foodstuffs.

Chuñu   Preserved tubers, including the potato, in Andean highland areas.

Ciboney   Shellfish-gathering people who once occupied the Greater Antilles.

Ciudad   City.

Coca (*Erythoxylum coca*)   Plant native to the tropical forests of South America. Dried leaves are chewed as a stimulant or prepared as a tea. Cocaine can be derived from the plant.

Cocal   A plantation of coca.

Cochineal   A crimson dye extracted from insects raised on prickly pear; exported to Europe for use in the cloth industry.

Colono   In Brazil a small freeholder, especially in southernmost region.

Common Market   A group of countries that promotes the free movement of goods, capital, and labor between their national territories; establishes a common external tariff on goods entering the market from non-member states; and works toward the creation of common business conditions.

Congregación   The Spanish policy known as congregación involved congregating/concentrating the native Americans/Indians into villages.

Conquistadores   Conquerors.

Conuco   Permanent raised beds used for planting crops in the **Greater Antilles** and parts of the mainland.

Continental Tropical (cT)   An air mass acquiring characteristics over a hot desert.

Convective rainfall   Precipitation resulting from the lifting of air from the earth's surface, often in a thunderstorm system.

Cordillera   Mountain range.

Council of the Indies/*Consejo de las Indias*   Established in 1524, the Council, seated in Seville, advised the Spanish monarch on policy in the Spanish American empire.

Creole/ criollo (Sp.)/criolho (Port.)   A person of Iberian heritage/extraction born in the Americas. In the Caribbean, black creoles are African-Americans born in the Caribbean, and white creoles are persons of European stock born in the Western Hemisphere. Probably derived from two Spanish words: *criar* (to create) and *colono* (a colonist).

Cyclonic rainfall   Precipitation that results from the passage of a depression.

Death rate, crude   The number of deaths, per thousand of a population, that occur each year.

Debt crisis   Period beginning in 1982 when large numbers of poor countries found their external debt obligations rising to unsustainable levels relative to their exports.

Demography/demographic   The study of human populations, especially size, density, distribution, and vital statistics.

Demographic Transition Model   A series of changes in **birth and death rates** that result in high birth rates and high death rates being replaced by low birth rates and low death rates. As death rates fall sooner in the transition than birth rates, there is a surge in population numbers before populations assume slow rates of growth.

Dependency   A territory or state under the political and economic control of another country although not formally annexed by it.

Direct migrant   A person moving directly from place of origin to destination without residing at an intervening place.

Donatario   A Portuguese noble (proprietor) who was given use of a large land grant (*capitânia*) in the Americas and was expected to develop it at his own expense.

Doubling time   The number of years it will take for a population to double in size. It is calculated by dividing the percentage rate of increase into 70. A population consistently growing at 2% annually would double in 35 years.

Eco-tourism   Tourist activities that feature contact with the natural environment.

Ejido   Communal lands (Mexico).

Ejidatarios   Occupiers of communal lands.

El Niño   A warm current accompanied by warm, moist air that occasionally displaces the cooler waters of the **Peru (Humboldt) Current** off the coast of Peru and Northern Chile.

Encomendero   The person who gained an *encomienda*.

Encomienda   Literally, an entrustment. A social and economic system used by the Spaniards, principally in New Spain and Peru, where the native population was large. Groups of Indians, usually in villages, were "commended" to a Spaniard who could exact tribute in the form of economic products or labor from the native Americans he was entrusted with. The Spaniard did not own the land or villages. In return, the Spaniard was to provide instruction in the Spanish language and Catholic religion. In 1549, labor service by *encomienda* Indians was abolished, but tribute payments continued.

Engenhos de açucar   Portuguese term for a mill to crush sugar cane to extract the juice.

Environmental determinism   The view that the physical environment controls, or determines, human activity.

Epiphyte   A plant that grows on another plant. In the tropical rainforest, many epiphytes including orchids, ferns, and lianas, perch on trees.

Era   Permanent, ridged fields on hillslopes, made for cultivation of potatoes or other tubers in the Andes.

Estado Nôvo   Literally, new state. A period in Brazilian history from the mid-1930s to 1945, under President Vargas, in which the apparatus of government was centralized and the state promoted industrialization.

Estancia   Farm, or cattle ranch, particularly in southern South America.

Estância   Cattle ranch in southern Brazil.

European Union (EU)   The European Economic Community (E.E.C.) or **Common Market** was established in 1957. The European Union was created by the Maastricht Treaty of 1991.

Export-orientated industrialization (EOI)   A policy designed to encourage manufacturing industry to produce for foreign markets.

Export Processing Zones (EPZs)   Demarcated areas where companies are permitted to import materials free of tax providing that the majority of the final assembled product is re-exported.

Favela   Brazilian squatter settlement especially in Rio de Janeiro, São Paulo, and surrounding areas.

Favelado   Resident of a favela.

Fazenda   A large landed estate or plantation in Brazil, usually based on black enslaved labor and the export of sugar. A cattle ranch in **Amazonia**.

Fazenda de açucar   Sugar plantation, Brazil.

Feral   Wild, especially domesticated animals that have reverted to a wild state.

Fertility rate, total   The number of children a woman is likely to bear during the childbearing years based on the average number of births per female in a country.

Finca   Farm; estate; frequently a farm growing coffee.

Flota   Literally, fleet; convoy of transoceanic vessels plying between Spanish America and Spain.

'Footloose' industry   Light industry which is theoretically able to locate anywhere because it does not use heavy raw materials.

'Formal' sector   A term used to describe large-scale manufacturing, commerce, finance, and services (often plants with more than 10 employees), where employees are covered by social security arrangements and have formal contracts. Also used to describe housing designed by trained architects and built by construction companies.

Free Trade Area   Created when two or more states agree to remove tariffs and other restrictions on trade between them. Members can trade with non-members on terms of their own choosing. See **NAFTA**.

Fuegian people   Stone Age shellfish gatherers who inhabited the southern tip of South America.

Garifuna   Black Carib.

Garimpeiros   Small-scale miners exploiting alluvial gold or diamond deposits in Brazil.

Garúa   Mist and drizzle experienced on the coast of Peru and Northern Chile.

Gaucho   A cowboy on the **pampas**.

Gê-speaking   Aboriginal language family of central Brazil.

Gran Chaco   Forested lowlands in northwestern Paraguay, and northern Argentina.

Gross Domestic Product   The total value of goods and services produced, and recorded in monetary transactions, within the national territory of a country. **Gross National Product (GNP)** includes monies earned outside the country such as repatriated profits. Both GDP and GNP are often expressed on a per capita basis. Neither GDP nor GNP includes the value of subsistence activities, and both tend to exaggerate wealth differences between cash economies and subsistence economies.

Guanaco   An American camel related to the domesticated llama of the Andes.

Guano   Mineral-rich droppings of seabirds on Peruvian offshore islands, used as a fertilizer.

Guaraní   Aboriginal people of eastern Paraguay. In the seventeenth century, **Jesuits** Christianized and settled many Guaraní in mission towns, but folkways were preserved. The Guaraní language is spoken widely in Paraguay.

Hacienda   Large landed estate, with a great house, owned by an elite family or the church in Spanish America.

Hispaniola   Largest island in the Greater Antilles. The Spaniards established the city of Santo Domingo in 1496 and used the island to conquer and settle the mainland. By the Treaty of Ryswick (1697), France gained control of the western third of the island, which was known as St. Domingue (it was renamed Haiti in 1804 after a slave rebellion released the area from French control). Now Hispaniola is host to two independent countries, Haiti in the west and the Dominican Republic in the east.

Hurricane (tropical cyclone)   A storm, formed over tropical seas, with winds in excess of 74 m.p.h.

Hydrolic cycle   The cycle through which water passes on evaporation from water bodies, condensation into clouds, and precipitation as rain or snow.

Hydrophytes   Plants adapted to an environment in which water is abundant.

Iberia/Iberian Peninsula   The peninsula occupied by Spain and Portugal.

IMF   See **International Monetary Fund**.

Import substitution/Import-substituting industrialization (ISI)   A nationalistic economic policy that discourages imports and encourages the manufacture of goods within the territory of a country to replace imports with national production.

Inca Empire   The largest aboriginal state in the Americas with its major center in Cuzco. The empire included the central Andean highlands and adjacent Pacific coastal areas.

Indigenous   Having originated in a particular environment or region.

Indigo   Tropical plant usually cultivated on estates that used enslaved Africans. Indigo produced a dark blue dye used in the cloth industry in Europe. Today the dye is synthesized chemically.

Infant mortality   The number of children, per thousand live births, who fail to survive the first year of life.

'Informal' sector   The opposite of the 'formal' sector. Includes self-employed, low-productivity jobs such as shoe polishing and small-scale manufacturing. The informal sector is also used to describe 'self-help' housing.

Insolation   Solar radiation received at the earth's surface.

Interfluve   The higher ground lying between the flood plains of adjoining streams or rivers. In tropical environments, interfluves are often infertile, having experienced the leaching out of minerals by rainfall. Being above the flood plain, interfluves rarely receive nutrients when streams rise.

International Monetary Fund (IMF)   The formation of the IMF was agreed upon at Bretton Woods in 1944. It began operations in 1947.

Intertropical Convergence Zone (ITCZ)   A zone of low pressure, high humidity, cloudiness, and rainfall found close to the equator that migrates northward in the northern hemisphere summer and southward in the southern hemisphere summer. The migration of the ITCZ brings the wet season to **tropical wet and dry climates**.

Jaracho   An inhabitant of Veracruz.

Jesuits   The Society of Jesus, begun by Ignatius Loyola in 1535. The Jesuits were active in the mission field, especially in Paraguay, Brazil, and northwestern Mexico. Spain expelled the Order in 1767. Portugal expelled the Jesuits from the empire in 1759. The Pope, under pressure from European rulers, agreed to widespread ejection from other territories in 1773.

Ladinos   People of mixed Indian and Hispanic stock who do not practice traditional Indian lifestyles.

Latifundia   Large landholding/estate.

Leaching   The washing out of soluble minerals from the upper soil layers and their deposition lower down. Common in tropical rainforest soils which are, as a result, mineral deficient.

Life expectancy   The number of years a newborn child is expected to live.

Lignum-vitae (*Guaiacum sanctum*)   The "wood of life" tree is hard and durable. Wood from it was used to make pulleys and shafts before the widespread use of steel.

Llanos   Literally, plain. In Venezuela and eastern Colombia, the tropical grasslands of the Orinoco drainage basin.

Llanero   Plainsman; cowboy; inhabitant of the llanos of Venezuela.

Mahogony (*Swietenia mahagoni*)   A hardwood tree native to the American tropics used in the manufacture of furniture, panelling, and floors.

Maize (*Zea mays*)   Aboriginal, New World, annual crop, called corn in the United States and Canada.

Malaria   A disease transmitted by the Anopheles mosquito which introduces a protozoan parasite into the blood of humans. The parasite causes an intermittent fever. Malaria probably came to the New World from Europe and Africa shortly after contact.

Mamelucos   Miscegenated population in Brazil involving Europeans and native Americans.

Manioc/Cassava (*Manihot utilissima*)   Originally a native plant of America that grows six to seven feet or more. The roots are an important food crop in the tropics, consisting mostly of starch, and must be carefully prepared by boiling because they are poisonous if eaten raw. In Asia and the United States, the plant is known as tapioca.

Maritime Polar (mP)   An air mass acquiring characteristics over Arctic, Antarctic, or circumpolar seas.

Maritime Tropical (mT)   An air mass acquiring characteristics over tropical seas.

Maquiladora   Assembly plant in Mexico that produces for export markets primarily in the United States. On entry into the United States, products are charged duty on value added by assembly in Mexico, not on the total value of the product. Literally, *maquila* is the grain taken by a miller in payment for grinding flour. See **Border Industries Program**.

Mediterranean climate   A climate of warm, sunny, dry summers and mild, moist winters found in the Central Valley of Chile; however, both winters and summers are cooler than the Mediterranean climates of southern Europe.

Mercantilism   An economic theory that said a nation's wealth was determined by the amount of precious metals (usually gold and silver) in the treasury. One way to achieve wealth was to possess mines. Another was through trade, by exporting more than was imported and maintaining a favorable balance of trade. Colonies were important in mercantilism because they provided raw materials that most likely would have to be purchased from a foreign nation. Colonial raw materials usually enjoyed protected markets in the territory of the controlling power.

Mercosur   Southern **Common Market** established by the treaty of Ascunción signed by Argentina, Brazil, Paraguay, and Uruguay on March 26, 1991. Chile became an associate member in 1996. The agreement calls for the establishment of free trade among the member states and the creation of a common external tariff on goods entering from non-members.

Mesoamerica   Middle America, including Mexico and Central America.

Mestizo   A person of mixed European and native American heritage.

Métayage   Share cropping.

Metric System   Decimal system of measurement. The U.S. currency is metric, but otherwise the United States retains miles, feet, inches, fahrenheit, and pints. Some common approximate conversions:

2 inches	=	5 centimeters
10 feet	=	3 meters
5 miles	=	8 kilometers
1 meter	=	3.281 feet
70°F	=	21.1°C
80°F	=	26.7°C
90°F	=	32.2°C
1 hectare	=	2.471 acres
1 acre	=	.405 of a hectare
1 square mile	=	2.5899 square kilometers

Migration chain   The social mechanisms by which migration patterns are established and maintained along migration channels.

Migration channel   The movement of many migrants along a common routeway between a source region and a destination.

Migration field   The area from which a city, town, or village draws migrants. The larger the place, the wider the migration field.

Milpa   Literally, a field of **maize**; in Mexico and Central America, also used to mean any small cultivated field.

Minifundia   Small landholding.

Miscegenation   Racial inter-mixing.

Mita   A labor system in which able-bodied adult Indian males had to serve a few weeks of each year in the army or on public works. Used by the Inca and continued by the Spaniards, especially in the mining industry in Peru. Abolished in 1812.

Mocambo   Squatter settlement in northeast Brazil.

Montaña   Literally, mountain. Another usage is mountain forest. The eastern region of Peru (*oriente*) is referred to as *montaña*.

Mulatto   A person of mixed European and African heritage.

NAFTA   See **North American Free Trade Agreement**.

Nationalization   The takeover of domestic or foreign private companies by a nation/state.

Natural increase   The difference between the **birth rate** and the **death rate** expressed as a percentage. A population with a **crude birth rate** of 20 per thousand and a **crude death rate** of 10 per thousand has a 1 percent rate of natural increase each year.

Natural vegetation   The vegetation of a region prior to modification by human activities such as burning, logging, or agriculture.

New Granada   Viceroyalty established by the Spanish crown in 1717 and refounded in 1739. Present-day Colombia, Panama, and Ecuador.

Nortes   North winds experienced during the Northern Hemisphere winter in Mexico, Central America, and the Greater Antilles that bring colder air from North America into regions normally exposed to tropical air. *Nortes* can damage tropical crops. In Argentina, a *norte* is a warm, humid wind arriving from tropical lands to the north.

North American Free Trade Agreement (NAFTA)   Canada, Mexico, and the United States signed the agreement in 1993, which entered into force on January 1, 1994. Canada and the United States had a previous free trade agreement signed in 1989.

Obrajes   Small, Spanish-controlled factories involved in textile making.

Organization of American States (OAS)   Founded, 1948. Most Latin American and Caribbean states are members. Cuba is excluded.

Organization of Oil Producing Countries (OPEC)   Founded in 1960 by Iran, Iraq, Kuwait, Saudi Arabia, and Venezuela. Venezuela remains the only member from Latin America.

Oriente   Literally, east. The eastern regions of Bolivia, Ecuador, and Peru are referred to as the *Oriente*, as is the eastern part of Cuba.

Orographic rainfall   Precipitation resulting when moisture-bearing air rises over higher ground.

Pampas   A temperate grassland lying to the east of the Andes in the Southern Cone of South America. In the west, in the lee of the Andes, the region is in a rainshadow and precipitation totals are lower in the Pampa seca (dry) than the Pampa húmida (moist) which is closer to the Atlantic Ocean.

Pampeans   Pampean Indians; bands of hunters on the **Pampas**.

Pamperos   A cold wind that invades the pampas from the south, accompanied by squalls and rapid drops in temperature as skies clear with the passage of the cold front.

Panama Canal   A canal opened in 1914, that linked the Caribbean with the Pacific Ocean. The project was initially funded by a French company and then by the U.S. government. The labor to dig the canal came primarily from Jamaica and Barbados.

Pantanal   A swampy or marshy area in southwestern Brazil along upper Paraná River drainage.

Páramo   Bleak, damp lands found above the tree line from about 10,000 feet to the snow line.

Patagonia   Plateau and adjacent eastern slope of Andes in southern Argentina.

Patria chica   Literally, "little homeland," the expression is used in Latin America for one's home area and the area to which one attaches both culture and political loyalty.

Paulista   Native of São Paulo, Brazil.

Pemex   Acronym for *Petróleos Mexicanos*, the nationalized petroleum industry of Mexico, established in 1938.

Peninsulares   In Spanish America, persons born in Spain (the Iberian peninsula) who migrated to the Americas.

Peon   Unskilled, agricultural worker, employed on a daily basis.

Peru (Humboldt) Current   An ocean current of cool water that flows from south to north, along the coasts of Chile and Peru.

Petrobras   Acronym for *Petrolios Brasileiros*, Brazil's nationalized petroleum industry.

Placer   Erosional materials deposited by a stream that contain minerals of economic value such as gold or tin.

Plantains   Varieties of bananas that are frequently used in cooking.

Population density   The number of persons living in a square mile or square kilometer of earth space.

Population pyramid   A diagram used to illustrate the age and gender distribution of a population. The population is divided into 5-year age cohorts 0–4, 5–9, 10–14, 15–19, etc. If the lower age cohorts are larger than those above, the population is increasing. If the 0–4 age group is smaller than the 5–9 cohort, the rate of population growth is slowing.

Porteño Native of Buenos Aires, Argentina.

Potosí Large silver mining town in **Upper Peru** (modern Bolivia). Spaniards discovered the rich silver mines in 1545. In the mid-seventeenth century it produced two-thirds of Peruvian silver and was considered the leading town in South America.

Presidio A military post or fortified settlement in Spanish America.

Primary sector of the economy Economic activities that directly exploit natural resources: hunting, gathering, forestry, fishing, farming, mining, and quarrying.

Privatization The selling off of state companies to the private sector.

Primate city The major city in a country that has a population many times larger than any other urban place. In Latin America primate cities are frequently capital cities; examples include Mexico City, Montevideo, Buenos Aires, and Lima.

Pronatal policies Policies designed to maintain or increase the **birth rate**.

Pueblo Town or village.

Quechua The language of the Incas still in widespread use in Ecuador, Peru, and Bolivia.

Quinoa (*Chenopodium quinoa*) An annual herb with edible leaves that produces many small seeds that can be boiled or ground up to make meal. Grows well at high altitudes in the Andes.

Rain shadow Area of relatively low rainfall on the lee side of uplands over which rain-bearing winds have passed.

Rank-size rule A distribution of city size populations in which the largest city is twice the size of the second largest place, three times as big as the third-ranked city, and four times the population of the fourth largest urban place.

Rapadura Brown sugar from which none of the molasses has been extracted.

Repartimiento A labor system used in Spanish America whereby royal authorities designated Indian workers to labor on a given task (mining or agriculture) for a specific period. Repartimientos replaced the **encomienda** as the major Indian labor system. By the end of the colonial era, the system was being used to require Indians to purchase goods imported from Spain.

Rhea American bird like an ostrich.

Roza (Roça in Portuguese) A cultivation plot. In a system of **shifting cultivation**, a roza plot is cultivated for a few years before being left fallow for more years than it is in cultivation.

St. Domingue The western third of the island of **Hispaniola**, transferred from Spain to France in the Treaty of Ryswick in 1697. Modern-day Haiti.

Saladero A factory that salts and dries beef in the sun to produce **tasajo**.

Sandinista Member of a Nicaraguan guerrilla movement. The Sandinista movement is named after César Augusto Sandino (1893–1934), a guerrilla leader of the 1920s and 1930s, until he was murdered while discussing peace talks.

Santo Domingo The eastern two-thirds of the island of **Hispaniola** that was under Spanish colonial rule. Modern-day Dominican Republic.

Saprophyte An organism that lives on dead plants (or animals). Most fungi live on dead matter.

Secondary sector of the economy Manufacturing and construction activity.

Selva Tropical rainforest. A name for the eastern region (*oriente*) of Peru that is covered with tropical forest.

Seringueiro Rubber tapper in the Amazon.

Serranía Mountain chain.

Sertão Sparsely populated backlands in Brazil beyond areas of permanent settlement. The term is used with particular reference to northeastern Brazil inland from the coast.

Sesmaria A grant of land made by the crown in colonial Brazil.

Sesmarias Along the Brazilian coast *sesmarias* (large plantations) were established, each averaging 12 square miles in size.

Shifting cultivation/Slash-and-burn farming Small plots are cleared in the forest and the brush burned prior to planting seeds or tubers. After 2 or 3 years, due to declining fertility and weed infestation, the plot is abandoned and the farmer clears new ground.

Sierra Madre Occidental Mountain chain in western Mexico composed mainly of volcanic material. The western slope is characterized by deep canyons formed by rivers flowing into the Pacific Ocean.

Sierra Madre Oriental Mountain chain in eastern Mexico. The eastern slope can receive heavy rainfall brought in from the Caribbean by the North East Trade winds.

Soccer War A war that broke out, on July 14, 1969, between El Salvador and Honduras after a soccer match. The war disrupted the **Central American Common Market**.

Solum   The A and B horizons of a soil, lying above the parent material, which forms the C horizon.
Step-wise migration   Movement in which migrants move from a source area to the final destination by residences at intervening places.
Structural adjustment   The restructuring of the economy as ordained by the **IMF** and the **World Bank** as a condition for lending to help deal with the **debt crisis**. The typical structural adjustment package required cuts in government budgets and in consumer demand, opening up the economy to external competition and giving greater encouragement to exporters.
Subsistence farming   Agricultural activity in which farming families consume most of the produce.
Sustainable development   The use of resources in a manner that passes to future generations an environment undegraded by the extinction of species, soil erosion, pollution, or any other means.
Swidden   Another term for **shifting cultivation**.

Taclla   Footplow used by the Incas in the Andes; similar implements are used today.
Tallow   Animal fat used for making candles, soap, and lubricants for machinery.
Tarascan state   A military state that existed in the western highlands of central Mexico at the time of Spanish contact. Roughly corresponded to the present state of Michoacán.
Tasajo   Jerked beef (sun-dried and/or salt-cured) produced in a **saladero**.
Tectonic forces   Pressures in the earth's crust that produce faulting, earthquakes, and mountain-building forces.
Tenant farmer   A farmer who rents land and pays with cash, a share of the crop, or labor services to the landlord; *inquilino*; in Chile, *inquilino* refers to any small farmer and to farm workers.
Tenochtitlán   Aztec capital in the Valley of Mexico. The Spaniards built Mexico City on the site.
Terra roxa soils   Dark red, clay soils, derived from basic lavas in southern Brazil. Noted for initial fertility but easily degraded.
Tertiary sector of the economy   The service sector of the economy includes wholesaling, retailing, transportation, financial services, education, recreation, administration, and medical services.
Tordesillas, Treaty of   In 1493, the Pope awarded Spain all lands lying 100 leagues (about 300 miles) west of the Azores and Cape Verde islands. In 1494, the treaty signed at Tordesillas set the line at 370 leagues west and gave Portugal all the lands east of a north–south line extending from approximately the mouth of the Amazon southward.
Trade winds   Winds that blow from sub-tropical, high-pressure cells, north and south of the equator, toward the **intertropical convergence zone** (ITCZ).
Tropical lands   In terms of latitude, that part of the earth lying between 23 1/2°N (the Tropic of Cancer) and 23 1/2°S (the Tropic of Capricorn) of the equator. Climatically, it is a region free of frosts except where altitude reduces temperatures. Köppen defined the tropics as lands in which the mean temperature for any month did not drop below 64.4°F.
Tropical wet and dry climates   A climate region, north and south of the equator, with a marked dry season in the "winter" months. The seasonal rainfall regime is associated with the movement north and south of the equator of the **ITCZ**. See Köppen's Aw climatic regions in Figure 2.2.
Tupi   Aboriginal people of the lower Amazon Basin.

Upper Peru   Modern day Bolivia.
Urban hierarchy   The ranking of urban places on the basis of population numbers and functions.
Urban majority   A country has an urban majority when more than 50% of the population lives in towns and cities.
Usina   Modern sugar refinery.

Vanilla (*Vanilla planifolia*)   A climbing orchid native to Central America. The fruit is the source of vanilla flavoring.
Várzeas   Wide belts of marsh and swamp along either side of lower Amazon River.
Viceroy   Deputy of the monarch governing a Viceroyalty. Literally, vice-king.
Viceroyalty   Administrative division used by the Spaniards in the Americas. Several Viceroyalties were established: New Spain (1535); Peru (1542); New Granada (1717 and 1739); La Plata (1776). Brazil was the one Viceroyalty of the Portuguese crown.
Villa   Town. In Spanish America, villas were established mainly to administer surrounding areas, including control of aborigine population.

War of the Pacific   A conflict (1879–1884) in which Chile defeated Bolivia and Peru. Bolivia lost the nitrate deposits of the northern Atacama desert and the port of Antofagasta, landlocking the country. Chile also took Pacific coast territory, which contained rich copper deposits, from Peru.

World Bank Its formal title is the International Bank for Reconstruction and Development. Started operations in June, 1946 as the world's development bank.

Xerophyte A plant adapted to living in desert or semi-arid conditions.

Yellow Fever A disease native to the Americas transmitted by the mosquito.

Yucatán Peninsula Low area with limestone surface in southeastern Mexico, home of northern Maya civilization.

Zacatecas In 1546, discovery of silver deposits at Zacatecas, in present-day Mexico, pushed the frontier of settlement northward in New Spain.

Bibliography

Chapter 2

Aceituno, P. "On the Functioning of the Southern Oscillation in the South American Sector." *Monthly Weather Review* 116 (1988): 505–524.

Agassiz, L., and E. Agassiz. *A Journey in Brazil*. Boston: Ticknow and Fields, 1868.

Alho, C. J. R., T. E. Lacher, Jr., and H. C. Concalvey. "Environmental Degradation in the Pantanal Ecosystem." *Bioscience* 38 (1988): 164–171.

Allen, J. C., and D. F. Barnes. "The Causes of Deforestation in Developing Countries." *Annals of the Association of American Geographers* 75 (1985): 163–184.

Armesto, J. J., and J. Figueroa. "Stand Structure and Dynamics in the Temperate Rain Forests of Chiloe Archipelago, Chile." *Journal of Biogeography* 14 (1987): 367–376.

Blume, H. *The Caribbean Islands*. London: Longman, 1974.

Brooks, M. Y. "Drought and Adjustment Dynamics in Northeastern Peru." *GeoJournal* 6 (1982): 121–128.

Brooks, R. H. "The Adversity of Brazilian Drought." *GeoJournal* 6 (1982): 121–128.

Brush, S. B. "The Natural and Human Environment of the Central Andes." *Mountain Research and Development* 2 (1982): 19–38.

Burns, E. B. *Latin America: Conflict and Creation*. Englewood Cliffs, N.J.: Prentice Hall, 1993.

Burton, R. *Exploration of the Highlands of Brazil*. London: Tinsley, 1868.

Caufield, C. *In the Rainforest*. Chicago: University of Chicago Press, 1986.

Caviedes, C. N. "El Niño 1982–83." *Geographical Review* 74 (1984): 267–290.

_____. "Natural Hazards in South America: In Search of a Method and a Theory." *GeoJournal* 6 (1982): 101–109.

Chung, J. C. "Correlations Between the Tropical Atlantic Trade Winds and Precipitation in Northeastern Brazil." *Journal of Climatology* 2 (1982): 35–46.

Cole, M. "Cerrado, Caatinga and Pantanal: The Distribution and Origin of the Savanna Vegetation of Brazil." *Geographical Journal* 126 (1980): 168–179.

Correa, J., and K. Reichardt. "The Spatial Variability of Amazonian Soils Under Natural Forest and Pasture." *GeoJournal* 19 (1989): 423–427.

da Cunha, E. *Rebellion in the Backlands*, trans. by Samuel Putnam. Chicago: University of Chicago Press, 1944.

Dickenson, J. "The Naturalist on the River Amazons and a Wider World: Reflections on the Century of Henry Walter Bates." *Geographical Journal* 158 (1992): 207–215.

Dickinson, R. E., ed. *The Geophysiology of Amazonia: Vegetation and Climate Interactions*. New York: John Wiley & Sons, 1987.

Dixon, C. "Yucatan After the Wind: Human and Environmental Impact of Hurricane Gilbert in the Central and Eastern Yucatan Peninsula." *GeoJournal* 23 (1991): 337–345.

Fittkau, E. J., et al. *Biogeography and Ecology of South America*. The Hague: Junk, 1968.

Gardner, G. *Travels in the Interior of Brazil*, 2d ed. London: Reeve, Benham, and Reeve, 1849.

Gentry, A. J., and J. López-Parodi. "Deforestation and Increased Flooding of the Upper Amazon (Peru, Ecuador)." *Science* 10 (1980): 1354–1356.

Gerbi, A. *The Dispute of the New World: The History of a Polemic.* Pittsburgh: University of Pittsburgh Press, 1973. Originally published in Italy, 1955.

Glacken, C. J. *Traces on the Rhodian Shore: Nature and Culture in Western Thought from Ancient Times to the End of the Eighteenth Century.* Berkeley: University of California Press, 1973.

Herring, H. *A History of Latin America from the Beginnings to the Present.* New York: Knopf, 1961.

Hill, A. D., ed. *Latin American Development Issues.* East Lansing, Mich.: CLAG Publications, 1973.

Hudson, W. H. *Green Mansions: A Romance of the Tropical Forest.* New York: Bantam, 1962. Originally published 1904.

Hughes, R. *A High Wind in Jamaica.* New York: Time, Inc., 1963. Originally published 1929.

Humboldt, A. von. *Personal Narrative of Travels to the Equinoctial Regions of the New World*, English Translation. Philadelphia: M. Carey, 1829.

James, P. E. *Latin America,* 3d ed. New York: Odyssey Press, 1959.

_____. *Latin America,* 4th ed. New York: Odyssey Press, 1969.

_____, and C. W. Minkel. *Latin America,* 5th ed. New York: John Wiley & Sons, 1986.

Jordan, C. F., ed. *Amazonian Rain Forests: Ecosystem Disturbance and Recovery.* New York: Springer–Verlag, 1987.

Lentnek, B., R. L. Carmin, and T. L. Martinson, eds. *Geographic Research on Latin America: Benchmark 1970.* Muncie, Ind.: Conference of Latin Americanist Geographers, 1971.

Liu, Han-Shou. "Mantle Convection Pattern and Subcrustal Stress Field Under South America." *Modern Geology* 7 (1988): 231–248.

Lowenthal, D. "The Caribbean Region." In *Geographers Abroad: Essays on the Problems and Prospects of Research in Foreign Areas*, edited by M.W. Mikesell. Chicago: University of Chicago Press, 1973.

Major, R. H. ed. *Christopher Columbus: Four Voyages to the New World.* New York: Corinth, 1961.

Martinson, T. L., and G. S. Elbow, eds. *Geographic Research on Latin America: Benchmark 1980.* Muncie, Ind.: Ball State University, 1981.

Maximillian, Prince of Wied-Neuwied. *Travels in Brazil in the Years 1815, 1816, 1817.* London: Henry Colburn, 1920.

Momsen, R. P., Jr., ed. *Geographical Analysis for Development in Latin America and the Caribbean.* Chapel Hill, N.C.: CLAG Publications, 1975.

Moorehead, A. *Darwin and the Beagle.* New York: Harper and Row, 1969.

Nelson, R. *Popul Vuh.* Boston: Houghton Mifflin, 1977.

Nortcliff, S. "A Review of Soil and Soil-Related Constraints to Development in Amazonia." *Applied Geography* 9 (1989): 147–160.

Parsons, J. J. "The Contribution of Geography to Latin American Studies." In *Social Science Research on Latin America,* edited by C. Wagley. New York: Columbia University Press, 1964.

_____. "Latin America." In *Geographers Abroad: Essays on the Problems and Prospects of Research in Foreign Areas,* edited by M. W. Mikesell. Chicago: University of Chicago Press, 1973.

_____. "The Scourge of Cows." *Whole Earth Review,* Spring 1988, pp. 40–47.

Picon-Salas, M. *A Cultural History of Spanish America, From Conquest to Independence.* Berkeley: University of California Press, 1963.

Platt, R. S. *Latin America: Countrysides and United Regions.* New York: McGraw-Hill, 1942.

Proctor, J. "Tropical Rain Forests." *Progress in Physical Geography* 13 (1989): 409–430.

Ramos, G. *Barren Lives.* Austin: University of Texas Press, 1992.

Sauer, C. O. *The Early Spanish Main.* Berkeley: University of California Press, 1966.

Simpson, L. B. *Many Mexicos.* Berkeley: University of California Press, 1961.

Sioli, H. "The Effects of Deforestation in Amazonia." *Geographical Journal* 151 (1985): 197–203.

Smith, A. *Explorers of the Amazon.* Chicago: University of Chicago Press, 1994.

Veblen, T. T., and D. C. Lorenz. "Recent Vegetation Change Along the Forest/Steppe Ecotone of Northern Patagonia." *Annals of the Association of American Geographers* 78 (1988): 93–111.

Wallace, A. R. *A Narrative of Travels on the Amazon and Rio Negro.* New York: Haskell House, 1964. (Originally published 1853.)

Waterton, C. *Wanderings in South America.* New York: Dutton, 1925. (Originally published 1825.)

Watts, D. *The West Indies: Patterns of Development, Culture and Environmental Change Since 1492.* Cambridge: Cambridge University Press, 1987.

Waylen, P., and C. Caviedes. "El Niño and Annual Floods on the North Peruvian Littoral." *Journal of Hydrology* 89 (1986): 141–156.

Chapter 3

Alchon, S. *Native Society and Disease in Colonial Ecuador.* Cambridge: Cambridge University Press, 1991.

Armillas, P. "Gardens on Swamps." *Science* 174 (1971): 653–661.

Bakewell, P. J. *Silver Mining and Society in Colonial Mexico: Zacatecas, 1546–1700.* Cambridge: Cambridge University Press, 1971.

———. "Technological Change in Potosí: The Silver Boom of the 1570s." *Jahrbuch für Geschichte von Staat, Wirtschaft und Gesellschaft Lateinamerikas* 14 (1977): 57–77.

Bethell, L. *The Cambridge History of Latin America.* 8 vols. Cambridge: Cambridge University Press, 1984–1995.

Borah, W. "New Spain's Century of Depression." *Ibero-Americana,* No. 35. Berkeley and Los Angeles: University of California Press, 1951.

———. "Early Colonial Trade and Navigation Between Mexico and Peru." *Ibero-Americana,* No. 38. Berkeley and Los Angeles: University of California Press, 1954.

———, and S. F. Cook. "The Aboriginal Population of Central Mexico and the Eve of Spanish Conquest." *Ibero-Americana,* No. 45. Berkeley and Los Angeles: University of California Press, 1963.

Boxer, C. R. *The Golden Age of Brazil, 1695–1750.* Berkeley and Los Angeles: University of California Press, 1962.

Brading, D. A., and H. E. Cross. "Colonial Silver Mining: Mexico and Peru." *Hispanic American Historical Review* 52 (1972): 545–579.

Brading, D. A. *Miners and Merchants in Bourbon Mexico, 1763–1810.* Cambridge: Cambridge University Press, 1971.

Brooks, F. J. "Revising the Conquest of Mexico: Smallpox, Sources, and Populations." *Journal of Interdisciplinary History* 24(1) (1993): 1–29.

Caraman, P. *The Lost Paradise, the Jesuit Republic in South America.* New York: The Seabury Press, 1976.

Chaunu, H., and P. Chaunu. *Seville et l'Atlantique, 1504–1650.* 8 vols. Paris: Colin, 1955.

Chevalier, F. *Land and Society in Colonial Mexico: The Great Hacienda.* Berkeley and Los Angeles: University of California Press, 1963.

Cobb, G. B. "Potosí, A South American Mining Frontier." In *Greater America: Essays in Honor of Herbert E. Bolton.* Berkeley and Los Angeles: University of California Press, 1945, pp. 39–58.

———. "Supply and Transportation for the Potosí Mines, 1545–1640." *Hispanic American Historical Review* 29 (1949): 25–45.

Coni, E. A. *Historia de las Vaquerías del Río de la Plata, 1555–1750.* Buenos Aires: Editorial Devenier, 1956.

Cook, N. D. *Demographic Collapse: Indian Peru, 1520–1620.* New York: Cambridge University Press, 1981.

Cook, S. F., and W. Borah. *Essays in Population History, Mexico and the Caribbean.* 3 vols. Berkeley and Los Angeles: University of California Press, 1971–1979.

Cortesão, J. *Brazil, de los Comienzos a 1799.* Barcelona: Savat, 1956.

Crosby, A. W. "Conquistador y Pestilencia: The First New World Pandemic and the Fall of the Great Indian Empires." *Hispanic American Historical Review* 47 (1967): 322–337.

_____. *Ecological Imperialism: The Biological Expansion of Europe 900–1900*. Cambridge and New York: Cambridge University Press, 1986.

Curtin, P. D. *The Atlantic Slave Trade, A Census*. Madison: University of Wisconsin Press, 1969.

Dean, W. *With Broadax and Firebrand: The Destruction of the Brazilian Atlantic Forest*. Berkeley: University of California Press, 1995.

Deerr, N. *The History of Sugar*. 2 vols. London: Chapman & Hall, 1949.

Denevan, W. M. "A Cultural-Ecological View of the Former Aboriginal Settlement in the Amazon Basin." *Professional Geographer* 18 (1966a): 346–351.

_____. "The Aboriginal Cultural Geography of the Llanos de Mojos of Bolivia." *Ibero-Americana*, No. 48. Berkeley and Los Angeles: University of California Press, 1966b.

_____. "Aboriginal Drained-Field Cultivation in the Americas." *Science* 69 (1970): 647–654.

_____. "The Pristine Myth: The Landscape of the Americas in 1492." *Annals of the Association of American Geographers*, 82(1992b): 369–385.

_____. "The Native Population of Amazonia Reconsidered." In *Latin American Geography: Past and Present*, edited by D. R. Stoddart and P. Starrs. Oxford: Blackwell, forthcoming.

DeVorsey, L. *Keys to the Encounter: A Library of Congress Resource Guide for the Study of the Age of Discovery*. Washington, D.C.: Library of Congress, 1992.

Dobyns, H. F. "Estimating American Population." *Current Anthropology* 7 (1966): 395–416.

Doolittle, W. E. "Agriculture in North America on the Eve of Contact: A Reassessment." *Annals of the Association of American Geographers* 82(1992): 368–401.

_____. *Canal Irrigation in Prehistoric Mexico: The Sequence of Technological Change*. Austin: University of Texas Press, 1990.

Dunn, R. S. *Sugar and Slaves, The Rise of the Planter Class in the English West Indies, 1624–1713*. Chapel Hill: University of North Carolina Press, 1972.

Eidt, R. C. "Aboriginal Chibcha Settlement in Colombia." *Annals of the Association of American Geographers* 49 (1959): 374–392.

Evans, B. "Migration Processes in Upper Peru in the Seventeenth Century." In *Migration in Colonial Spanish America*, edited by D. J. Robinson. New York: Cambridge University Press, 1990.

Fisher, J. "Silver Production in the Viceroyalty of Peru, 1778–1824." *Hispanic American Historical Review* 55 (1975): 25–43.

Freide, J. *Los Quimbaya bajo la Dominación Española: Estudio documental (1539–1810)*. Bogotá: Banco de la República, 1963.

Gade, D. W. "Landscape, System, and Identity in the Post-Conquest Andes." *Annals of the Association of American Geographers*, 82(1992): 461–477.

_____, and R. Rios. "Chaquitaclla, the Native Footplough and Its Persistence in Central Andean Agriculture." *Tools and Tillage* 2 (1972): 3–15.

Gerhard, P. *The Northern Frontier of New Spain*, Rev. Ed. Norman: University of Oklahoma Press, 1993.

_____. *The Southern Frontier of New Spain*, Rev. Ed. Norman: University of Oklahoma Press, 1993.

_____. *A Guide to the Historical Geography of New Spain*. London: Cambridge University Press, 1972.

Gibson, C. *The Aztecs Under Spanish Rule*. Stanford, Calif.: Stanford University Press, 1964.

Greenleaf, R. E. "The Obraje in the Late Mexican Colony." *The Americas* 22 (1967): 227–250.

Greenlee, W. B. "The First Half Century of Brazilian History." *Mid-America* 25 (1943): 91–120.

Helmer, M. "Commerce et Industrie au Pérou à la Fin du XVIIIᵉ Siècle," *Revista de Indias* 10 (1950): 519–526.

Himmerich y Valencia, R. *The Encomenderos of New Spain 1521–1555*. Austin: University of Texas Press, 1991.

Hyslop, J. *Inca Settlement Planning.* Austin: University of Texas Press, 1990.

Jiménez de la Espada, M., ed. "Relaciones Geográficas de Indias: Peru." *Biblioteca de Autores Españoles.* Madrid: Ediciones Atlas, 1965, pp. 183–185.

Knapp, G. *Andean Ecology: Adaptive Dynamics in Ecuador.* Boulder, Colo.: Westview Press, 1991.

Lasserre, G. "Le Nord-Est du Brazil." *Les Cahiers d'Outre-Mer* 1 (1948): 40–67.

Levenson, J. E., ed. *Circa 1492.* New Haven, Conn.: Yale University Press, 1991.

Lévi-Strauss, C. "Saudades do Brasil." *The New York Review of Books* 42(20) (December 21, 1995): 19–26.

Lipschutz, A. "La Despoblación de las Indias después de la Conquista." *América Indígena* 26 (1966): 229–247.

Lockhart, J., and S. B. Schwartz. *Early Latin America: A History of Colonial Spanish America and Brazil.* Cambridge: Cambridge University Press, 1983.

Lovell, W. G. "Heavy Shadows and Black Night: Disease and Depopulation in Colonial Spanish America." *Annals of the Association of American Geographers* 82(1992): 426–443.

_____, and C. H. Lutz. "The Historical Demography of Colonial Central America." *Benchmark 1990, Conference of Latin Americanist Geographers* 17/18(1992): 127–138.

MacLeod, M. J. *Spanish Central America: A Socioeconomic History, 1520–1720.* Berkeley and Los Angeles: University of California Press, 1973.

MacNeish, R. S., et al. *Second Annual Report of the Ayacucho Archaeological Botanical Project.* Andover, Mass.: Robert S. Peabody Foundation for Archaeology, Phillips Academy, 1970.

Mauro, F. *Le Portugal et l'Atlantique au XVIIe Siècle (1570–1670).* Paris: Études Économique, 1960.

_____. "México y Brasil: Dos Economías Coloniales Comparadas," *Historia Mexicana* 40 (1961): 571–587.

McFarlane, E. *Colombia Before Independence; Economy, Society, and Politics Under Bourbon Rule.* New York: Cambridge University Press, 1993.

Mellaffe, R. *Negro Slavery in Latin America.* Berkeley and Los Angeles: University of California Press, 1975.

Melville, E. G. K. *A Plague of Sheep: Environmental Consequences of the Conquest of Mexico.* New York: Cambridge University Press, 1994.

Miller, J. C. *Way of Death.* Madison: University of Wisconsin Press, 1990.

Minchom, M. *The People of Quito, 1690–1810: Change and Unrest in the Underclass.* Boulder, Colo.: Westview, 1994.

Morgan, P. D. "Slaves and Livestock in Eighteenth-Century Jamaica Vineyard Pen, 1750–1751." *The William and Mary Quarterly,* 3d ser., vol. 52, no. 1 (January 1995): 47–76.

Mörner, M. *Region and State in Latin America's Past.* Baltimore, Md.: Johns Hopkins University Press, 1993.

_____. "The Spanish American Hacienda: A Survey of Recent Research and Debate." *Hispanic American Historical Review* 53 (1973): 183–216.

Morse, R. M. "Some Characteristics of Latin American Urban History." *American Historical Review* 67 (1962): 317–338.

_____. *The Bandeirantes: The Historical Role of the Brazilian Pathfinders.* New York: Knopf, 1965.

Newson, L. A. *Life and Death in Early Colonial Ecuador.* Norman: University of Oklahoma Press, 1995.

_____. "The Population of the Amazon Basin in 1492." *Transactions of the Institute of British Geographers.* 21 (1996): 5–27.

Palmer, C. A. *Slaves of the White God: Blacks in Mexico, 1570–1650.* Cambridge, Mass.: Harvard University Press, 1976.

Parry, J. H., and R. G. Keith, eds. *New Iberian World: A Documentary History of the Discovery and Settlement of Latin America to the Early 17th Century,* 5 vols. New York: Times Books, 1984.

Patch, R. *Modern Indians and the Inca Empire,* vol. 5, no. 8. Washington, D.C.: American University Field Staff Reports, West Coast South America Series, 1958.

Phelan, J. L. *The Kingdom of Quito in the Seventeenth Century.* Madison: University of Wisconsin Press, 1967.

Poppino, R. "Cattle Industry in Colonial Brazil." *Mid-America* 31 (1949): 219–247.

Rippy, J. F., and J. T. Nelson. *Crusaders of the Jungle.* Chapel Hill: University of North Carolina Press, 1936.

Robinson, D. J. *Social Fabric and Spatial Structure in Colonial Latin America.* Ann Arbor, Mich.: University Microfilms International, 1979.

Rosenblat, A. *La Población Indígena y el Mestizaje en América,* 2 vols. Buenos Aires: Editorial Nova, 1954.

_____. "The Population of Hispaniola at the Time of Columbus." In *The Native Population of the Americas in 1492,* 2d ed, edited by W. M. Denevan. Madison: University of Wisconsin Press, 1992.

Rowe, J. H. "Inca Culture at the Time of the Spanish Conquest." In *Handbook of South American Indians,* vol. 2, edited by J. H. Steward. Washington, D.C.: Smithsonian Institution, Bureau of American Ethnology, Bulletin 143, 1946, pp. 183–330.

Sánchez-Albornoz, N. "The Population of Colonial Spanish America." In *The Cambridge History of Latin America,* vol. 2, edited by L. Bethell. Cambridge: Cambridge University Press, 1984, pp. 3–35.

Schmieder, O. "The Historical Geography of Tucuman." *University of California Publications in Geography* 2 (1928): 359–386.

_____. "The Brazilian Culture Hearth." *University of California Publications in Geography* 3 (1929): 159–198.

Simpson, L. B. *The Encomienda in New Spain.* Berkeley and Los Angeles: University of California Press, 1950.

Smith, C. T. "Depopulation of the Central Andes in the Sixteenth Century." *Current Anthropology* 11 (1970): 453–464.

Stanislawski, D. "Tarascan Political Geography." *American Anthropologist* 49 (1947): 46–55.

Super, J. C. "Queretaro Obrajes: Industry and Society in Provincial Mexico, 1600–1810." *Hispanic American Historical Review* 56 (1976): 197–216.

Tenenbaum, B. A. ed. *Encyclopedia of Latin American History.* New York: Scribners, 1995.

Thomas H. *Conquest: Montezuma, Cortés and the Fall of Old Mexico.* New York: Simon and Schuster, 1993.

Vargas, J. M. *La Economía Política del Ecuador Durante la Colonia; Esquema Histórico.* Quito: Editorial Universitaria, 1957.

Vásquez de Espinosa, A. *Compendium and Description of the West Indies, Miscellaneous Collection,* No. 102. Washington, D.C.: Smithsonian Institution, 1942.

Warren, R. *Coercion and Market: Silver Mining in Colonial Potosí, 1692–1826.* Albuquerque: University of New Mexico Press, 1993.

Watts, D. *The West Indies: Patterns of Development, Culture, and Environmental Change Since 1492.* Cambridge: Cambridge University Press, 1987.

Webb, K. E. "Origins and Development of a Food Economy in Central Minas Gerais." *Annals of the Association of American Geographers* 49 (1959): 409–419.

Weber, D. J. *The Spanish Frontier in North America.* New Haven: Yale University Press, 1992.

West, R. C. "The Mining Community in Northern New Spain: The Parral Mining District." *Ibero-Americana,* No. 30. Berkeley and Los Angeles: University of California Press, 1949.

_____. *Colonial Placer Mining in Colombia.* Baton Rouge: Louisiana State University Press, 1952.

_____. "Ridge or 'Era' Agriculture in the Colombian Andes." *Proceedings,* 33rd International Congress of Americanists, San José, Costa Rica, 1 (1959): 279–282.

_____. "Population Densities and Agricultural Practices in Pre-Columbian Mexico, with Emphasis on Semi-terracing." *Proceedings,* 38th International Congress of Americanists, Stuttgart-Munich, 1968. 2 (1970): 362–369.

Whitmore, T. M. "A Simulation of the Sixteenth-Century Population Collapse in the Basin of Mexico." *Annals of the Association of American Geographers* 81(1991): 464–487.

_____, and B. L. Turner, II. "Landscapes of Cultivation in Mesoamerica on the Eve of Conquest." *Annals of the Association of American Geographers* 82(1992): 402–425.

Wilken, G. C. *Good Farmers: Traditional Agricultural Resource Management in Mexico and Central America.* Berkeley: University of California Press, 1987.

_____. "Drained-Field Agriculture: An Intensive Farming System in Tlaxcala, Mexico." *Geographical Review* 59 (1969): 215–241.

Wrigley, E. A., and R. S. Schofield. *The Population of England, 1541–1871: A Reconstruction.* Cambridge, Mass.: Harvard University Press, 1981.

Wrigley, G. M. "Salta, An Early Commercial Center of Argentina." *Geographical Review* 2 (1916): 116–133.

Young, E. V. *Hacienda and Market in Eighteenth-Century Mexico: The Rural Economy of the Guadalajara Region, 1675–1820.* Berkeley: University of California Press, 1981.

Zavala, S. *The Spanish Colonization of America.* Philadelphia: University of Pennsylvania Press, 1943.

Chapter 4

Austin, J. P. *Agribusiness in Latin America.* New York: Praeger, 1974.

Baleé, W. *Footprints in the Forest.* New York: Columbia University Press, 1994.

Barraclough, S., and J. C. Collarte. *Agrarian Structure in Latin America.* Lexington: D. C. Heath, 1973.

Brown, J. C. *A Socioeconomic History of Argentina 1776–1860.* Cambridge: Cambridge University Press, 1979.

Bray, W. "Crop plants and Cannibals: Early European Impressions of the New World." In *The Meeting of Two Worlds: Europe and the Americas, 1492–1650.* Oxford: The British Academy, 1994.

Butt, A. J. "Land-use and Social Organization of Tropical Forest Peoples of Guianas." In *Human Ecology in the Tropics,* edited by J. P. Garlick and R. W. Keay. New York: Pergamon, 1970, pp. 26–42.

Carneiro, R. L. "Slash and Burn Cultivation among the Kuikuru and its Implications for Cultural Development in the Amazon Basin." In *The Evolution of Horticultural Systems in Native South America: Causes and Consequences, A Symposium,* edited by J. Wilbert. Caracas: Editorial Sucre, 1961.

Carter, W. E. *New Lands and Old Traditions: Ketchi Cultivators in the Guatemala Lowlands.* Gainesville: University of Florida Press, 1969.

Chang, J. H. "The Agricultural Potential of the Humid Tropics." *Geographical Review* 58 (1968): 333–361.

Chonochol, J. "Land Tenure and Development in Latin America." In *Obstacles to Change in Latin America,* edited by C. Valez. New York: Oxford University Press, 1965.

Crossley, J. C. "The Location of Beef Processing." *Annals of the Association of American Geographers* 66 (1976): 60–75.

deJanvry, A. *The Agrarian Question and Reformism in Latin America.* Baltimore: Johns Hopkins University Press, 1981.

Denevan, W. M. "Aboriginal Drained-Field Cultivation in the Americas." *Science* 169 (1970): 647–654.

Duncan, K., and I. Rutledge. *Land and Labor in Latin America: Essays on the Development of Agrarian Capitalism in the Nineteenth and Twentieth Centuries.* Cambridge: Cambridge University Press, 1977.

Freyre, G. *The Masters and the Slaves.* New York: Knopf, 1946.

Furtado, C. *The Economic Growth of Brazil.* Berkeley: University of California Press, 1965.

_____. *Economic Development of Latin America.* Cambridge: Cambridge University Press, 1970.

Gallo, E. "The Cereal Boom and Changes in the Social and Political Structure of Santa Fé, Argentina, 1870–95." In *Land and Labor in Latin America: Essays on the Development of Agrarian Capitalism in the Nineteenth and Twentieth Centuries,* edited by K. Duncan and I. Rutledge. Cambridge: Cambridge University Press, 1977.

Grigg, D. *The World Food Problem,* 2d ed. Oxford: Blackwell, 1993.

Grindle, M. S. *State and Countryside: Development Policy and Agricultural Politics in Latin America.* Baltimore: Johns Hopkins University Press, 1986.

Grunwald, J., and P. Musgrove. *Natural Resources in Latin American Development.* Baltimore: Johns Hopkins University Press, 1970.

Gudeman, S. *The Demise of a Rural Economy: From Subsistence to Capitalism in a Latin American Village.* London: Routledge and Kegan Paul, 1978.

Harlan, J. R. "Agricultural Origins: Centers and Noncenters." *Science* 174 (1971).

_____. *Crops and Man.* Madison, Wisc.: American Society of Agronomy, 1975.

Harris, D. R. "Swidden Systems and Settlement." In *Man, Settlement and Urbanism,* edited by P. J. Ucko, R. Tringham, and G. W. Dimbleby. London: Gerald Duckwork, 1972.

Harrison, P. D., and B. L. Turner, eds. *Pre-Hispanic Maya Agriculture.* Albuquerque: University of New Mexico Press, 1978.

Inter-American Development Bank. *Economic and Social Progress in Latin America,* Annual Report. Washington, D.C.: Author, 1983.

Inter-American Development Bank. *Economic and Social Progress in Latin America,* Annual Report. Washington, D.C.: Author, 1990.

Jefferson, M. *Peopling of the Argentina Pampa.* New York: American Geographical Society, 1926.

Knapp, G. *Andean Ecology: Adaptive Dynamics in Ecuador.* Boulder, Colo.: Westview Press, 1991.

Lobb, C. G. "The Viability and Ecological Stability of Re-settlement in the Interior Lowlands of Guatemala. "In *Geographical Analysis for Development in Latin America and the Caribbean,* edited by R. P. Momsen, Jr. Chapel Hill, N.C.: CLAG Publication, 1975, pp. 140–153.

_____. "The Human Geography of Agricultural Origins." In *Proceedings of the Sino-American Symposium on Human Geography,* edited by G. Laixi, R. Hoffpauir, and E. McIntire. Beijing, China: Science Press, 1988.

Lucier, R. L. *The International Political Economy of Coffee.* New York: Praeger, 1988.

MacArthur, R. H. *Geographical Ecology, Patterns in the Distribution of Species.* New York: Harper & Row, 1972.

Nelson, M. *The Development of Tropical Lands.* Baltimore: Johns Hopkins University Press, 1973.

Parish, W. *Buenos Ayres and the Provinces of the Rio de la Plata.* London: John Murray, 1839.

Piñeiro, M., and E. Trigo. *Technical Change and Social Conflict in Agriculture: Latin American Perspectives.* Boulder, Colo.: Westview Press, 1983.

Rappaport, R. A. "The Flow of Energy in an Agricultural Society." *Scientific American* 225 (1971): 117–132.

Ruddle, K., and D. Odermann, eds. *Statistical Abstract of Latin America.* Los Angeles: Latin American Center, UCLA, 1972.

Sargent, C. S. *The Spatial Evolution of Greater Buenos Aires, Argentina, 1870–1930.* Tempe: Arizona State University, 1974.

Sauer, C. O. "Cultivated Plants of South and Central America." In *Handbook of South American Indians.* Washington, D.C.: Smithsonian Institution Bureau of American Ethnology, Bulletin 143, 1956.

_____. *The Early Spanish Main.* Berkeley and Los Angeles: University of California Press, 1966.

Schery, R. W., *Plants for Man.* Englewood Cliffs: Prentice-Hall 1963.

Scobie, J. R. *Argentina: A City and a Nation* 2d ed. New York: Oxford University Press, 1971.

Seavoy, R. E. "The Shading Cycle in Shifting Cultivation." *Annals of the Association of American Geographers* 63(4): 522–528 (1973).

Shoumatoff, A. *The World Is Burning.* New York: Little, Brown, 1990.

Smith, T. L. *The Race Between Population and Food Supply in Latin America.* Albuquerque: University of New Mexico Press, 1976.

Staff of the Land Tenure Center Library. *Agrarian Reform in Latin America: An Annotated Bibliography,* 2 Vols. Madison: University of Wisconsin-Madison: Land Tenure Center, 1974.

Viola, H. J., and C. Margolis. *Seeds of Change: A Quincentennial Commemoration.* Washington, D.C.: Smithsonian Institution Press, 1991.

Wilkie, J. W. ed. *Statistical Abstract of Latin America,* Vol. 28. Los Angeles: UCLA Latin American Center Publications, 1990.

Wolf, E. R. ed. *The Valley of Mexico: Studies in Pre-Hispanic Ecology and Society.* Albuquerque: University of New Mexico Press, 1976.

Chapter 5

Arreola, D. D., and J. R. Curtis. *The Mexican Border Cities: Landscape Anatomy and Place Personality.* Tucson: University of Arizona Press, 1993.

Bryce, J. *South America: Observations and Impressions.* New York: Macmillan, 1914.

Bulmer-Thomas, V. *The Economic History of Latin America Since Independence.* New York: Cambridge University Press, 1994.

Caldwell, J. C. *A Theory of Fertility Decline.* London and New York: Academic Press, 1982.

Collins, E. J. T. *Unseasonal Migrations: The Effects of Rural Labor Scarcity in Peru.* Princeton, N.J.: Princeton University Press, 1988.

Collver, O. A. *Birth Rates in Latin America: New Estimates of Historical Trends and Fluctuations.* Berkeley: University of California, Institute of International Studies, 1965.

Dickenson, J. P. "Brazil's Census Surprises." *Geographical Magazine* 44 (1972): 60.

Durand, J., and D. S. Massey. "Mexican Migration to the United States: A Critical Review." *Latin American Research Review* 27 (1992): 3–42.

Evans, B. "Migration Processes in Upper Peru in the Seventeenth Century." In *Migration in Colonial Spanish America,* edited by D. J. Robinson. New York: Cambridge University Press, 1990.

Gilbert, A. *Latin American Development: A Geographical Perspective.* Baltimore, Md.: Penguin, 1974.

_____. *Latin America.* London and New York: Routledge, 1990.

_____. *The Latin American City.* London: Latin America Bureau, 1994.

Gonzalez, A. "Population Growth and Socio-Economic Development: The Latin American Experience." *Journal of Geography* 70 (1971): 36–46.

Grigg, D. B. "E. G. Ravenstein and the 'Laws of Migration'." *Journal of Historical Geography* 3 (1977): 41–54.

McNeill, W. H., and R. S. Adams, eds. *Human Migration: Patterns and Policies.* Bloomington: Indiana University Press, 1978.

Merrick, T. W., and D. H. Graham. *Population and Economic Development in Brazil: 1800 to the Present.* Baltimore and London: Johns Hopkins University Press, 1979.

Merrick, T.W. *Population Pressures in Latin America.* Washington, D.C.: Population Reference Bureau, 1986.

Moreno, L. "The Linkage Between Populations and Economic Growth in Mexico: A New Policy Proposal." *Latin American Research Review* 26 (1991): 159–170.

Norris, R. E. "Migration as Spatial Interaction." *Journal of Geography* 71 (1972): 294–301.

Ravenstein, E. G. "The Laws of Migration." *Journal of the Statistical Society of London* 48 (1885): 167–227; and "The Laws of Migration." *Journal of the Statistical Society of London* 52 (1889): 214–301.

Sánchez-Albornoz, N. "The Population of Latin America, 1850–1930." In *The Cambridge History of Latin America,* vol. 4, edited by L. Bethel. New York: Cambridge University Press, 1986,

Smith, C. T. "Problems of Regional Development in Peru." *Geography* 58 (1968): 260–281.

Swann, M. "Migration, Mobility, and Mining Towns of Colonial Northern Mexico." In *Migration in Colonial Spanish America,* edited by D. J. Robinson. New York: Cambridge University Press, 1990.

Vernez, G., and D. Ronfeldt. "The Current Situation in Mexican Immigration." *Science* 251 (1991): 1189–1193.

Viel, V. B. *The Demographic Explosion: The Latin American Experience,* translated by J. Walls. New York: Halstead Press, 1976.

Wall, D. L. "Spatial Inequalities in Sandinista Nicaragua." *Geographical Review* 83 (1993): 1–13.

White, P., and R. Woods. *The Geographical Impact of Migration.* New York: Longman, 1980.

World Development Report 1990. New York: Oxford University Press, 1990.

Wrigley, E. A. "The Growth of Population in Eighteenth-Century England: A Conundrum Resolved." *Past and Present* 98 (1983): 121–150.

Chapter 6

Andrews, G. F. *Maya Cities: Placemaking and Urbanization.* Norman: University of Oklahoma Press, 1975.

Beyer, G., ed. *The Urban Explosion in Latin America.* New York: Cornell University Press, 1967.

Butterworth, D., and J. K. Chance. *Latin American Urbanization.* Cambridge: Cambridge University Press, 1981.

Collier, S., H. Blakemore, and T. E. Skidmore, eds. *The Cambridge Encyclopedia of Latin America and the Caribbean.* Cambridge: Cambridge University Press, 1985.

Cornelius, W. A., and R. V. Kemper, eds. "Metropolitan Latin America: The Challenge and the Response." *Latin American Urban Research* 6. Beverly Hills, Calif.: Sage, 1977.

Cross, M. *Urbanization and Urban Growth in the Caribbean.* Cambridge: Cambridge University Press, 1979.

Davis, K. "Colonial Expansion and Urban Diffusion in the Americas." *International Journal of Comparative Sociology* 1 (1960): 43–66.

Deffontaines, P. "The Origin and Growth of the Brazilian Network of Towns." *Geographical Review* 28 (1938): 379–99.

Epstein, D. *Brasilia: Plan and Reality. A Study of Planned and Spontaneous Settlement.* Berkeley: University of California Press, 1973.

Gade, D. W. "Landscape, System, and Identity in the Post-Conquest Andes." *Annals of the Association of American Geographers* 82 (1992): 460–477.

Geisse, G., and J. E. Hardoy, eds. *Regional and Urban Development Policies: A Latin American Perspective.* Beverly Hills: Sage, 1972.

Greenfield, G. M., ed. *Latin American Urbanization: Historical Profiles of Major Cities.* Westport, Conn.: Greenwood Press, 1994.

Griffin, E., and L. Ford. "A Model of Latin American City Structure." *Geographical Review* 70 (1980): 397–442.

Gwynne, R. N. *Industrialization and Urbanization in Latin America.* Baltimore, Md.: Johns Hopkins University Press, 1986.

Hardoy, J. E. *Urbanization in Latin America: Approaches and Issues.* New York: Doubleday, 1975.

_____. *Pre-Columbian Cities.* New York: Walker, 1964.

Harris, W. D., Jr., and H. Rodriguez-Camilloni. *The Growth of Latin American Cities.* Athens: Ohio University Press, 1971.

Hauser, P. M. *Urbanization in Latin America.* New York: Columbia University Press, 1961.

Herzog, L. A. *Where North Meets South: Cities, Space and Politics on the United States–Mexico Border.* Austin: University of Texas Press, 1990.

Hoberman, L. S., and S. M. Socolow, eds. *Cities and Society in Colonial Latin America.* Albuquerque: University of New Mexico Press, 1986.

Houston, J. M. "The Foundation of Colonial Towns in Hispanic America." In *Urbanisation and its Problems,* edited by R. P. Beckinsale and J. M. Houston. Oxford: Blackwell, 1968, pp. 352–390.

Katzman, M. T. *Cities and Frontiers in Brazil.* Cambridge, Mass.: Harvard University Press, 1977.

Lloyd, P. *The Young Towns of Lima: Aspects of Urbanization in Peru.* Cambridge: Cambridge University Press, 1980.

Mangin, W. "Latin American Squatter Settlements: A Problem and a Solution." *Latin American Research Review* 2 (1967): 65–98.

Morse, R. M., and J. E. Hardoy, eds. *Rethinking the Latin American City.* Baltimore, Md.: Johns Hopkins University Press, 1993.

Morse, R. E., ed. *The Urban Development of Latin America, 1750–1920.* Stanford, Calif.: Stanford University, Center for Latin American Studies, 1971.

Portes, A., et al. "Urbanization in the Caribbean Basin." *Latin American Research Review* 29 (1994):117–139.

_____, and J. Walton. *Urban Latin America: The Political Conditions from Above and Below.* Austin: University of Texas Press, 1976.

Sánchez-Albornoz, N. *The Population of Latin America: A History.* Berkeley: University of California Press, 1974.

Sauer, C. O. *The Early Spanish Main.* Berkeley: University of California Press, 1966.

Schnore, L. F. "On the spatial structure of cities in the two Americas." In *The Study of Urbanization,* edited by P. M. Hauser and L. F. Schnore. New York: John Wiley & Sons, 1965, pp. 347–399.

Stanislawski, D. *The Transformation of Nicaragua: 1519–1548.* Berkeley: University of California Press, 1983.

Travis, C. *A Guide to Latin American and Caribbean Census Material.* Boston: G. K. Hall, 1990.

Twinam, A. *Miners, Merchants and Farmers in Colonial Colombia.* Austin: University of Texas Press, 1982.

Vance, J. E. *The Merchant's World.* Englewood Cliffs, N.J.: Prentice-Hall, 1970.

Violich, F. *Cities of Latin America.* New York: Reinhold, 1994.

_____. *Urban Planning for Latin America: The Challenge of Metropolitan Growth.* Boston: Oelgeschlager, Gunn & Hain, 1987.

Ward, P. M. *Mexico City: The Production and Reproduction of an Urban Environment.* Boston: G. K. Hall, 1990.

Weinstein, B. *The Amazon Rubber Boom 1850–1920.* Stanford, Calif.: Stanford University Press, 1983.

Wilkie, R. W. *Latin American Population and Urbanization Analysis: Maps and Statistics, 1950–1982.* Los Angeles: UCLA Latin American Center Publications, 1984.

_____, and E. Ochoa. *Statistical Abstract of Latin America.* Los Angeles: UCLA Latin American Center Publications, 1978.

Wolf, E. *Europe and the People Without History.* Berkeley: University of California Press, 1982.

Wong, P. *Planning and the Unplanned Reality: Brasilia.* Berkeley: University of California, Institute of Urban and Regional Planning, 1989.

Chapter 7

Blakemore, H., and C. T. Smith, eds. *Latin America: Geographical Perspectives,* 2d ed. London: Methuen, 1983.

Deere, C. D. *In the Shadows of the Sun: Caribbean Development Alternatives and U.S. Policy, Policy Alternatives for the Caribbean and Central America.* Boulder, Colo.: Westview, 1990.

Dell, S. *Latin American Common Market?* London: Oxford University Press, 1966.

Diniz, C. C. "Polygonized Development in Brazil: Neither Decentralization Nor Continued Polarization." *International Journal of Urban and Regional Research* 18 (1994): 293–314.

Ewell, J. *Venezuela: A Century of Change.* Stanford, Calif.: Stanford University Press, 1984.

Friedmann, J. *Regional Development Policy: A Case Study of Venezuela.* Cambridge, Mass.: M.I.T. Press, 1966.

Garza, G., ed. *Atlas de la Cuidad de México.* Mexico City: DDR and El Colegio de México, 1987.

Germani, G., ed. *Modernization, Urbanization and the Urban Crisis.* New York: Little, Brown & Co., 1973.

Gilbert, A. G. *Latin American Development: Geographical Perspective.* Harmondsworth, England; Penguin Books, 1974.

_____, and D. E. Goodman. "Regional Income Disparities and Economic Development: A Critique." In A. G. Gilbert ed., *Development Planning and Spatial Structure.* London: John Wiley, 1976, pp. 113–141.

Goodman, D. E. "Industrial Development in the Brazilian Northeast: An Interim Assessment of the Tax Credit Scheme of Article 34/18." In R. J. A. Roett, ed., *Brazil in the Sixties.* Nashville, Tenn.: Vanderbilt University Press, 1972, pp. 231–274.

Gwynne, R. N. *New Horizons? Third World Industrialization in an International Framework.* Burnt Mill, England: Longman, 1990.

Hirschman, A. O. *Journeys Toward Progress.* New York: Anchor Books, 1965.

Klak, T., and J. Rulli. "Regimes of Accumulation, the Caribbean Basin Initiative, and Export Processing Zones." In E. G. Goetz and S. E. Clarke, eds., *The New Localism: Comparative Urban Politics in a Global Era.* Beverly Hills, Calif.: Sage, 1993, pp. 117–150.

Klein, H. S. "The Creation of the Patiño Tin Empire." *Inter-American Economic Affairs* 19 (1965): 3–24.

Kleinpenning, J. M. G. *The Integration and Colonization of the Brazilian Portion of the Amazon Basin.* Nijmegan, Netherlands: Katholick Universiteit, 1975.

Lieuwen, E. *Petroleum in Venezuela: A History.* Berkeley: University of California Press, 1954.

_____. *Venezuela,* 2d ed. London: Oxford University Press, 1965.

Maldonaldo, S. "La Rama Automovilística y Los Corredores Industriales en el Noroeste de México." *Comercio Exterior* 44 (1995): 487–497.

Osborne, H. *Bolivia.* London: Royal Institute of International Affairs, 1964.

Randall, L. *The Political Economy of Venezuelan Oil.* New York: Praeger, 1987.

Rodríguez, M., and P. Alvarez. "La Economía Mexicana, el Turismo y el Tratado de Libre Comercio." *Comercio Exterior* 44 (1995): 525–534.

Rodwin, L., et al. *Planning Urban Growth and Regional Development: The Experience of the Guayana Program of Venezuela.* Cambridge, Mass.: M.I.T. Press, 1969.

Sklair, L. *Assembling for Development.* Hemel Hempstead, England: Unwin Hyman, 1989.

Storper, M. "Who Benefits from Industrial Decentralization? Social Power in the Labor Market, Income Distribution and Spatial Policy in Brazil." *Regional Studies* 18 (1984): 143–164.

Townroe, P. M., and D. Keen. "Polarization Reversal in the State of São Paulo, Brazil." *Regional Studies* 18 (1984): 45–54.

United Nations. *Statistical Yearbook 1992.* New York: United Nations, 1994.

Villa, M., and J. Rodríguez. "Demographic Trends in Latin America's Metropoli, 1950–1991." In A. G. Gilbert, ed., *Megacities in Latin America.* Tokyo: United Nations University Press, 1996.

Waller, P. O. "The Spread Effects of a Growth Pole—A Case Study of Arequipa (Peru)." In F. M. Helleiner and W. Stöhr, eds., *Spatial Aspects of Development,* Proceedings of the I.G.U. Commission on Regional Aspects of Development, Vol. 2. Toronto: Allister, 1974.

Wilkie, J. W., C. A. Contreras, and C. Komisaruk, eds. *Statistical Abstract of Latin America* 31. Los Angeles: UCLA Latin American Center Publications, 1994.

Wilkie, J. W., and E. Ochoa, eds. *Statistical Abstract for Latin America*. Los Angeles: University of California—Los Angeles, 1989.

Chapter 8

Arreola, D. D. "Nineteenth Century Townscapes of Eastern Mexico." *Geographical Review* 72 (1982): 1–19.

Anuario Estadístico de los Estados Unidos Mexicanos. Mexico: Secretaría de la Economía Nacional, Dirección General de Estadística, 1994.

Butzer, K. W. "Cattle and Sheep from Old to New Spain: Historical Antecedents." *Annals of the Association of American Geographers* 78 (1988): 29–56.

Casteneda, J. "Ferocious Differences." *Atlantic Monthly*, (July 1995): 68–76.

Collins, C. O., and S. L. Scott. "Air Pollution in the Valley of Mexico." *Geographical Review* 83 (1993): 119–133.

Cumberland, C. C. *Mexico: The Struggle for Modernity*. New York: Oxford University Press, 1968.

De Janvry, A., E. Sandoulet, and G. G. de Anda. "NAFTA and Mexico's Maize Producers." *World Development* 23 (1995): 1349–1362.

Denevan, W. M., ed. *The Native Population of the Americas in 1492*, 2d ed. Madison: University of Wisconsin Press, 1992.

Economist. "A Survey of Mexico." (October 28, 1995): 1–18.

Garza, G., and Programa de Intercambio Científico. *Atlas de la Cuidad de México*. Mexico: Departamento del Distrito Federal y El Colegio de México, 1986.

Gerhard, P. *A Guide to the Historical Geography of New Spain*. Cambridge: Cambridge University Press, 1972.

_____. *The Northern Frontier of New Spain*. Norman: University of Oklahoma Press, 1993.

Grayson, G. W. *The North American Free Trade Agreement: Regional Community and the New World Order*. Lanham, Md.: University Press of America, 1995.

Hall, P. *The World Cities*, 3d ed. London: Weidenfeld and Nicholson, 1984.

Hassig, R. *Trade, Tribute, and Transportation. The Sixteenth Century Political Economy of the Valley of Mexico*. Norman: University of Oklahoma Press, 1985.

ISLA, Information Services on Latin America. I.S.L.A.: Oakland, Calif.: Oakland, CA, 1991–1995.

Jones, R. C. *Ambivalent Journey: U.S. Migration and Economic Mobility in North-Central Mexico*. Tucson: University of Arizona Press.

_____. "Immigration Reform and Migrant Flows: Compositional and Spatial Changes in Mexican Migration after the Immigration Reform Act of 1986." *Annals of the Association of American Geographers*, 85 (1995):715–730.

Kopinak, K. *Desert Capitalism: Maquiladoras in North America's Western Industrial Corridor*. Tucson: University of Arizona Press, 1996.

Malmström, V. H. "Geographical Origins of the Tarascans." *Geographical Review*, 85 (1995): 31–40.

Martinson, T. L., ed. *Benchmark 1990: Conference of Latin Americanist Geographers* 17/18 (1992).

Melville, G. K. *A Plague of Sheep: Environmental Consequences of the Conquest of Mexico*. New York: Cambridge University Press, 1994.

"Mexican Connection Grows as Cocaine Supplier to the U.S." *New York Times*, July 30, 1995.

"Mexico's Road Programme's Heavy Toll." *Financial Times (London)*, July 5, 1995.

"Mexican Tobbaco Firm Is Changing the Way Small Farmers Work." *Wall Street Journal*, July 26, 1995.

Philip, G. D. E. "Mexico," Rev. Ed. *World Bibliographical Series*, Vol. 48. Oxford: Clio Press, 1993.

Pick, J. B. *Atlas of Mexico*. Boulder: Westview Press, 1989.

Proffitt, T. D. *Tijuana: The History of a Mexican Metropolis.* San Diego, Calif.: San Diego State University Press, 1994.

Rees, P. W. *Transportes y Comercio entre México y Veracruz, 1519–1918.* Mexico, D.F.: Sepsentenas, 1976.

Siemans, A. H. "Wetland Agriculture in Pre-Hispanic Mesoamerica." *Geographical Review* 73 (1983): 166–181.

Simonian, L. *Defending the Land of the Jaguar: A History of Conservation in Mexico.* Austin: University of Texas Press.

Slayter, A. "Wetland Agriculture in Mesoamerica: Space, Time, and Form." *Annals of the Association of American Geographers,* 84 (1994): 557–584.

South America, Central America, and the Caribbean, 1988, 2d ed. London: Europa Publications, 1987.

Special Issue on the Privitization of Mexican Agriculture. *Urban Anthropology,* 23(3–4): 1994.

Statistical Yearbook for Latin America and the Caribbean, 1989. New York: Economic Commission for Latin America and the Caribbean, United Nations, 1990.

Townsend, J., and J. B. de Corcuera. "Feminists in the Rainforest." *Geoforum* 24 (1993): 45–54.

Ward, P. M. *Mexico City.* Boston: G. K. Hall, 1990.

Whitmore, T. M. "A Simulation of the Sixteenth-Century Population Collapse in the Basin of Mexico." *Annals of the Association of American Geographers* 81 (1991): 464–487.

Wilken, G. C. *Good Farmers. Traditional Agricultural Resource Management in Mexico and Central America.* Berkeley and Los Angeles: University of California Press, 1987.

Wilkie, J. W. ed. *Society and Economy in Mexico.* Los Angeles: UCLA Latin Ameircan Center Publications, University of California Press, 1990.

_____, and E. Ochoa. *Statistical Abstract of Latin America,* Vols 27–31. Los Angeles: UCLA Latin American Center Publications, University of California, 1990–1995.

Zea, L. *The Latin American Mind,* trans. by J. H. Abbott and L. Dunham. Norman: University of Oklahoma Press, 1963.

Chapter 9

Ameringer, C. D. *The Caribbean Legion: Patriots, Politicians, Soldiers of Fortune, 1946–1950.* University Park, Pa.: Penn State Press, 1996.

Arbingast, S., et al. *Atlas of Central America.* Austin: Bureau of Business Research, University of Texas, 1979.

Belt, T. *The Naturalist in Nicaragua.* New York: E. P. Dutton, 1911.

Browning, D. *El Salvador: Landscape and Society.* Oxford: Clarendon Press, 1971.

_____. *Conflicts in El Salvador.* London: Institute for the Study of Conflict, 1984.

Bethell, L., ed. *Central America Since Independence.* New York: Columbia University Press, 1991.

Camille, M. A. *Government Initiative and Resource Exploitation in Belize.* Unpublished Doctoral Dissertation. College Station: Texas A&M University, 1994.

Chomsky, A. *West Indian Workers and the United Fruit Company in Costa Rica, 1870–1940.* Baton Rouge: Louisiana State University Press, 1996.

Clayton, L. A. "The Nicaragua Canal in the Nineteenth Century: Prelude to American Empire." *Journal of Latin American Studies,* 19 (1987): 323–352.

Collins, C. O. "Refugee Resettlement in Belize." *Geographical Review* 85(1995): 20–30.

Danner, M. *The Massacre at El Mozote.* New York: Vintage, 1994.

Davidson, W. V. *Historical Geography of the Bay Islands, Honduras: Anglo-Hispanic Conflict in the Western Caribbean.* Birmingham, Ala.: Southern University Press, 1974.

Denevan, W. M., ed. *The Native Population of the America of 1492.* Madison: University of Wisconsin Press, 1992.

Dobyns, H. "Estimating Aboriginal American Population: An Appraisal of Techniques with a New Hemispheric Estimate." *Current Anthropology* 7 (1966): 395–416.

Faber, D. J. *Environment Under Fire: Imperialism and the Ecological Crisis in Central America*. New York: Monthly Review Press, 1993.

Gjording, C. N. *Conditions Not of Their Choosing: The Guaymí Indians and Mining Multinationals in Panama*. Washington, D.C.: Smithsonian Institution Press, 1991.

Gudmundson, L. *Costa Rica Before Coffee: Society and Economy on the Eve of the Export Boom*. Baton Rouge: Louisiana State University Press, 1986.

Hall, C. "Regional Inequalities in Well-Being in Costa Rica." *Geographical Review* 74 (1984): 48–62.

Herlihy, P. H. "Opening Panama's Darien Gap." *Journal of Cultural Geography* 9 (1989): 41–69.

ISLA, Information Services on Latin America. Oakland, Calif.: I.S.L.A. 1991–1995.

Jones, J. *Colonization and Environment: Land Settlement Projects in Central America*. Tokyo: United Nations University Press, 1990.

Jones, R. C. "Causes of Salvadoran Migration to the United States." *Geographical Review* 39 (1989): 13–19.

Kramer, W. *Encomienda Politics in Early Colonial Guatemala, 1524–1544: Dividing the Spoils*. Dellplain Latin American Studies, Report No. 31. Boulder, Colo.: Westview, 1994.

Leschine, T. M. "The Panamanian Sea-Level Canal." *Oceanus*, 24 (1981): 20–30.

Lovell, G. *Conquest and Survival in Colonial Guatemala: A Historical Geography of the Cuchumatán Highlands, 1500–1821*, Rev. Ed. Montreal: McGill-Queen's University Press, 1992.

Lutz, C. H. *Santiago de Guatemala, 1541–1773: City, Cast, and the Colonial Experience*. Norman: University of Oklahoma Press, 1994.

McCleod, M. J. *Spanish Central America: A Socioeconomic History, 1520–1720*. Berkeley and Los Angeles: University of California Press, 1973.

McCreery, D. *Rural Guatemala 1760–1940*. Stanford, Calif.: Stanford University Press, 1994.

Meggers, B. J. *Amazonia: Man and Culture in a Counterfeit Paradise*. Chicago: Aldine Press, 1971.

Melmed-Sanjak, J., C. E. Santiago, and A. Magrid, eds. *Recovery or Relapse in the Global-Economy: Comparative Perspectives on Restructuring in Central America*. Westport, Conn.: Praeger, 1993.

Nash, J., ed. *Crafts in the World Market: The Impact of Global Exchange on Middle American Artisans*. Albany: State University of New York Press, 1993.

Naylor, R. A. *Penny Ante Imperialism: The Mosquito Shore and the Bay of Honduras, 1600–1914: A Case Study of British Informal Empire*. Rutherford, N.J.: Farleigh Dickinson University Press, 1989.

Newson, L. A. *Indian Survival in Colonial Nicaragua*. Norman: University of Oklahoma Press, 1987.

Nietschmann, B. *The Unknown War: The Miskito Nation, Nicaragua, and the United States*. New York: Freedom House, 1989.

Pearce, D. G. "Cotton and Cattle in the Pacific Lowlands of Central America." *Journal of Inter-American Studies* 7 (1962): 149–159.

_____. "Planning Tourism in Belize." *Geographical Review*, 74 (1984): 291–303.

Raddell, D. "Realejo, a Forgotten Colonial Port and Ship-building Center in Nicaragua." *Hispanic American Historical Review* 51:2 (1971): 295–312.

Rohter, L. "Where Countless Died in '81, Horror Lives on in Salvador." *New York Times*, February 12, 1996.

Seligson, M. A. "Thirty Years of Transformation in the Agrarian Structure of El Salvador." *Latin American Research Review* 30 (1995): 43–74.

Schwartz, N. B. *Forest Society: A Social History of Petén, Guatemala*. Philadelphia: University of Pennsylvania Press, 1990.

South America, Central America, and the Caribbean, 5th ed. London: Europa, 1995.

The Diagram Group. *The Atlas of Central America and the Caribbean*. New York: Macmillan, 1985.

Vilas, C. M. *Between Earthquakes and Volcanoes: Market, State, and Revolutions in Central America*, trans. by T. Kusler. New York: Monthly Review Press, 1995.

Weaver, F. *Inside the Volcano: The History and Political Economy of Central America*. Boulder, Colo.: Westview Press, 1994.

Weinberg, B. *War on the Land: Ecology and Politics in Central America*. London: Zed, 1991.

West, R. C., and J. P. Augelli. *Middle America: Its Lands and Peoples*, 3d ed. Englewood Cliffs, N.J.: Prentice Hall, 1989.

Wilken, G. C. *Good Farmers: Traditional Agricultural Resource Management in Mexico and Central America*. Berkeley and Los Angeles: University of California Press, 1987.

World Bibliographical Series, *Belize*, 2d ed., vol. 21, 1993; *Guatemala*, Rev. Ed., vol. 9, 1992; *El Salvador*, vol. 98, 1988; *Honduras*, vol. 139, 1992; *Nicaragua*, Rev. Ed., vol. 44, 1992; *Costa Rica*, vol. 126, 1991; *Panama*, vol. 14, 1982. Oxford: Clio Press.

Chapter 10

Beckles, H. M. *A History of Barbados: From Amerindian Settlement to Nation-state*. Cambridge: Cambridge University Press, 1990.

Benedict, B. *Problems of Smaller Territories*. London: Athlone Press, 1967.

Berleant-Schiller, R., and L. Pulsipher. "Subsistence Cultivation in the Caribbean." *Nieuwe West Indische Gids* 60 (1986): 1–40.

Besson, J., and J. Momsen, eds. *Land and Development in the Caribbean*. London: Macmillan, 1987.

Boswell, T. D., and J. Curtis. *The Cuban-American Experience: Culture, Images, and Perspectives*. Totowa, N.J.: Rowman & Allanheld, 1984.

_____, and D. Conway. *The Caribbean Islands: Endless Geographical Diversity*. New Brunswick, N.J.: Rutgers University Press, 1992.

Bourne, C., et al. *Caribbean Development to the Year 2000: Challenges, Prospects, and Policies*. London: Commonwealth Secretariat, 1988.

Brittain, A. W. "Anticipated Child Loss to Migration and Sustained High Fertility in an East Caribbean Population." *Social Biology* 38 (1991).

Browning, C. E. "Urban Primacy in Latin America." In *Yearbook 1989, Conference of Latin Americanist Geographers*, vol. 15, edited by R. B. Kent and V. R. Harnapp. Baton Rouge: Louisiana State University Department of Geography and Anthropology, 1989.

Burton, R. D. E., and F. Reno, eds. *French and West Indian: Martinique, Guadeloupe, and French Guiana Today*. Charlottesville: University Press of Virginia, 1995.

Clark, C., and T. Payne. *Politics, Security and Development in Small States*. London: Allen & Unwin, 1987.

Collier, S., H. Blakemore, and T. E. Skidmore, eds. *The Cambridge Encyclopedia of Latin America and the Caribbean*. Cambridge: Cambridge University Press, 1985.

Conniff, M. L. and T. J. Davis. *Africans in the Americas: A History of the Black Diaspora*. New York: St. Martin's Press, 1994.

_____. "Caribbean International Mobility Traditions." *Boletín de Estudios Latinamericanos y del Caribe* 46 (1989): 17–47.

Curtin, P. *The Atlantic Slave Trade: A Census*. Madison: University of Wisconsin Press, 1969.

_____. *The Rise and Fall of the Plantation Complex: Essays in Atlantic History*. Cambridge: Cambridge University Press, 1990.

Davis, D. J., ed. *Slavery and Beyond: The African Impact on Latin America and the Caribbean*. Wilmington, Delaware: Scholarly Resources, 1995.

Deerr, N. *The History of Sugar*, 2 vols. London: Chapman and Hall, 1949–1950.

Delson, R. M., ed. *Readings in Caribbean History and Economics: An Introduction to the Region*. New York: Gordon and Breach, 1981.

Dunn, R. S. *Sugar and Slaves: The Rise of the Planter Class in the English West Indies, 1624–1713*. Chapel Hill: University of North Carolina Press, 1972.

Elbow, G. S. "Caribbean Regional Identity" in *Abstracts: The Association of American Geographers, 92nd Annual Meeting*. Washington: Association of American Geographers, 1996.

Foreign Broadcast Information Service. *Latin America Daily Report.*

Graham, N. A., and K. L. Edwards. *The Caribbean Basin to the Year 2000: Demographic, Economic and Resource-Use Trends in Seventeen Countries.* Boulder, Colo., and London: Westview Press, 1984.

Griffith, W. H. "CARICOM Countries and the Caribbean Basin Initiative." *Latin American Perspectives* 17 (1990): 33–54.

Grugel, J. *Politics and Development in the Caribbean Basin: Central America and the Caribbean in the New World Order.* Bloomington and Indianapolis: Indiana University Press, 1995s.

Higman, B. "The Spatial Economy of Jamaican Sugar Plantations: Cartographic Evidence from the Eighteenth and Nineteenth Centuries." *Journal of Historical Geography* 13 (1987): 17–39.

Hope, K. R. *Urbanization in the Commonwealth Caribbean.* Boulder, Colo.: Westview Press, 1986.

James, P. E., and C.W. Minkel. *Latin America,* 5th ed. New York: John Wiley and Sons, 1986.

Kent, R. B., and V. R. Harnapp, eds. *Year Book 1989, Conference of Latin Americanist Geographers,* vol. 15. Baton Rouge: Louisiana State University Geography and Anthropology Department, 1989.

Kimber, C. "Aboriginal and Peasant Cultures of the Caribbean." In *Benchmark 1990: Conference of Latin Americanist Geographers,* vol. 17/18, edited by T. L. Martinson, 1992, pp. 153–165. Auburn, AL: Conference of Latin American Geographers.

Knight, F. W., and C. Palmer, eds. *The Modern Caribbean.* Chapel Hill: University of North Carolina Press, 1989.

_____. *Slave Society in Cuba During the Nineteenth Century.* Madison: University of Wisconsin Press, 1970.

Lockhart, J., and S. B. Schwartz. *Early Latin America: A History of Colonial Spanish America and Brazil.* Cambridge: Cambridge University Press, 1983.

Lovejoy, P. E. "The Volume of the Atlantic Slave Trade: A Synthesis." *Journal of African History* 23 (1982): 473–501.

Lowenthal, A. F., ed. *Latin America and Caribbean Contemporary Record,* vol. 6. New York: Holmes and Meier, 1989.

Lowenthal, D. *West Indian Societies.* London: Oxford University Press, 1972.

Maingot, A. P. *The United States and the Caribbean: Challenges of an Asymmetrical Relationship.* Boulder, Colo. and San Francisco: Westview Press, 1994.

Martin, D. M., ed. *Handbook of Latin American Studies,* No. 53. Austin: University of Texas Press, 1994.

Martínez, S. *Peripheral Migrants: Haitians and Dominican Republic Sugar Plantations.* Knoxville: University of Tennessee Press, 1995.

Meditz, S. M., and D. M. Hanratty. *Islands of the Commonwealth Caribbean: A Regional Study.* Washington, D.C.: U.S. Government Printing Office, 1989.

Mintz, S. *Caribbean Transformations.* Chicago: Adline, 1974.

Momsen, J. H. "Linkages Between Tourism and Agriculture: Problems for the Smaller Caribbean Economies," Seminar Paper #45. Newcastle-upon-Tyne, England: University of Newcastle-upon-Tyne, Department of Geography, 1986a.

_____. "Migration and Rural Development in the Caribbean." *Tijdschrift voor Economische en Sociale Geografie* 77 (1986b): 50–59.

_____, and J. Townsend, eds. *Geography of Gender in the Third World.* London: Hutchinson, 1987.

1995 World Population Data Sheet. Washington, D. C.: Population Reference Bureau, 1995.

Olwig, K. F. *Global Culture, and Island Identity: Continuity and Change in the Afro-Caribbean Community of Nevis.* Switzerland: Harwood Academic Publishers, 1993.

Palm, R., and M. E. Hodgson. "Natural Hazards in Puerto Rico." *Geographical Review* 83(1993): 280–289.

Palmié, S. ed. *Slave Cultures and the Cultures of Slavery.* Knoxville: University of Tennessee Press, 1995.

Peach, C. *West Indian Migration to Britain: A Social Geography.* London: Oxford University Press, 1968.

Potter, R. B. "Urbanization and Development in the Caribbean." Paper presented at the Institute of British Geographers Seminar, Royal Holloway College, London University, 1995.

_____, ed. *Urbanization, Planning and Development in the Caribbean.* London: Mansell, 1989.

Richardson, B. C. "Caribbean Migration, 1838–1985." In *The Modern Caribbean* edited by F. Knight and C. Palmer. Chapel Hill: University of North Carolina Press, 1989.

_____. *Panama Money in Barbados, 1900–1920.* Knoxville: University of Tennessee Press, 1985.

_____. *Caribbean Migrants: Environment and Human Survival on St. Kitts and Nevis.* Knoxville: University of Tennessee Press, 1983.

Ritter, A. R. M. "The Cuban Economy in the 1990s." *Journal of Interamerican Studies and World Affairs* 32 (1990): 117–149.

Scarano, F. A. *Sugar and Slavery in Puerto Rico: The Plantation Economy of Ponce, 1800–1850.* Madison: University of Wisconsin Press, 1984.

Schomburgk, R. H. *A Description of British Guiana.* London: Simpkin, Marshall, 1840.

Serbín, Andrés. *Caribbean Geopolitics: Toward Security Through Peace?* Boulder, Colo. and London: Lynne Rienner, 1990.

Smith, R. T. *British Guiana.* London: Oxford University Press, 1962.

Smith, R. F. *The Caribbean World and the United States.* New York: Twayne Publishers, 1994.

Solow, B., and S. Engerman, eds. *British Capitalism and Caribbean Slavery: The Legacy of Eric Williams.* Cambridge: Cambridge University Press, 1987.

Statesman's Year Book. London: Macmillan, 1905, 1990, 1994–1995, and 1995–1996.

Szulc, T., ed. *The United States and the Caribbean.* Englewood Cliffs, N.J.: Prentice-Hall, 1971.

Viola, H. J., and C. Margolis. *Seeds of Change: A Quincentennial Commemoration.* Washington, D.C.: Smithsonian Institution Press, 1991.

Ward, J. R. *British West Indian Slavery, 1750–1834: The Process of Amelioration.* Oxford: Clarendon Press, 1988.

Watson, H. A. *The Caribbean in the Global Political Economy.* Boulder, Colo., and London: Lynne Rienner, 1994.

Weaver, D. "Ecotourism in the Caribbean." In *Ecotourism: A Sustainable Option?* edited by E. Cater and G. Lowman. New York: John Wiley, 1994.

West, R., and J. P. Augelli. *Middle America: Its Lands and People*, 3d ed. Englewood Cliffs, N.J.: Prentice Hall, 1989.

Williams, E. *Capitalism and Slavery.* Chapel Hill: University of North Carolina Press, 1944.

Wood, D. *Trinidad in Transition: The Years After Slavery.* London: Oxford University Press, 1968.

World Development Report, 1990. Oxford: Oxford University Press, 1990.

Chapter 11

Anderson, T. D. "The Gulf of Venezuela Sea Boundary: A Problem Between Friends." *Yearbook, Conference of Latin Americanist Geographers* 15 (1989): 97–105.

Basile, D. G. *Tillers of the Andes. Farmers and Farming in the Quito Basin.* Studies in Geography, No. 8. Chapel Hill: University of North Carolina at Chapel Hill, Department of Geography, 1974.

Bates, H. W. *The Naturalist on the River Amazons.* London: Murray, 1892 (originally published, 1863).

Bergquist, C. W. *Coffee and Conflict in Colombia, 1886–1910.* Durham, N.C.: Duke University Press, 1978.

Bromley, R. D. F. "The Functions and Development of 'Colonial' Towns: Urban Change in the Central Highlands of Ecuador, 1690–1940." *Transactions of the Institute of British Geographers* 4 (1970): 30–43.

Blanchard, P. *Markham in Peru: The Travels of Clements R. Markham, 1852–1853.* Austin: University of Texas Press, 1991.

Brush, S. B. *Mountain, Field, and Family: The Economy and Human Ecology of an Andean Valley.* Philadelphia: University of Pennsylvania Press, 1977.

_____. "The Natural and Human Environment of the Central Andes." *Mountain Research and Development* 2(1982): 19–38.

Darwin, C. *Journal of Researches into the Natural History and Geology of the Countries Visited during the Voyage of HMS Beagle Round the World.* London: 1890.

Eder, P. J. *Colombia.* London: Fisher Unwin. 1913.

Escobar, G. "Keeping Coca a Cash Crop: Bolivian Peasants Fight U.S. Campaign to End Cultivation of their 'Sacred Leaf'." *Washington Post*, September 29, 1995.

Ewell, J. *Venezuela: A Century of Change.* Stanford, Calif.: Stanford University Press, 1984.

Farah, D. "Colombians Boost Drug Production: Cocaine Barons Shift Operations." *Washington Post*, May 13, 1990, p. A23.

Fifer, J. V. *Bolivia: Land, Location and Politics since 1825.* Cambridge: Cambridge University Press, 1972.

Gade, D. W. *Plants, Man and the Land in the Vilcanota Valley of Peru.* The Hague: W. Junk, 1975.

_____. "Cultural Geography as a Research Agenda for Peru." *Conference of Latin Americanist Geographers* 14 (1988): 31–37.

Gilbert, A. *Latin America.* New York: Routledge, 1990.

Guillemoprieto, A. "Letter from Bolivia." *The New Yorker*, March 16, 1992, pp. 95–107.

Henkel, R. "The Bolivian Cocaine Industry." *Studies in Third World Societies* 37 (1986): 53–80.

"Inside Intelligence on the Operation of the Colombian Narcotics Cartels in America." *Latin American Times* 10 (1990).

Kane, J. *Savages.* New York: Knopf, 1995.

Kolata, A. *The Tiwanaku: Portrait of an Andean Civilization.* Cambridge, MA: Blackwell, 1993.

Markham, C. R. *Travels in Peru and India.* London: Murray, 1862.

McFarlane, A. *Colombia Before Independence: Economy, Society, and Politics under Bourbon Rule.* New York: Cambridge University Press, 1993.

McGlade, M. S., R. Henkel, and R. S. Cerveny. "The Impact of Rainfall Frequency on Coca (Etythrozylam coca) Production in the Chapare Region of Bolivia." *Yearbook, Conference of Latin Americanist Geographers* 20 (1994): 97–104.

McGreevey, W. P. *An Economic History of Colombia 1845–1930.* Cambridge: Cambridge University Press, 1971.

Mitchell, B. R. *International Historical Statistics: The Americas 1750–1988.* New York: Stockton Press, 1993.

Morales, E. "The Political Economy of Cocaine Production: An Analysis of the Peruvian Case." *Latin American Perspectives* 17(1990): 91–109.

Mörner, M. *The Andean Past: Land, Societies, and Conflicts.* New York: Columbia University Press, 1985.

Murra, J. B. "High Altitude Andean Societies and Their Economies." In *Geographic Perspectives in History*, edited by E. D. Genovese and L. Hochberg. New York: Basil Blackwell, 1989.

Ogilvie, A. *The Geography of the Central Andes.* New York: American Geographical Society, 1927.

Parsons, J. J. *Antioqueño Colonization in Western Colombia*, 2d rev. ed. Berkeley: University of California Press, 1968.

"Peru and Ecuador Are Hit Hard by Exceptionally Severe Effects of El Niño." *Latin American Weekly Report*, April 2, 1992.

Rausch, D. M. *A Tropical Plains Frontier: The Llanos of Colombia 1531–1831.* Albuquerque: University of New Mexico Press, 1984.

Robinson, E. "U.S. Tempers Expectations of Quick Antidrug Victory in Bolivia." *The Washington Post*, November 30, 1990a.

_____. "In Bolivia, Great Expectations." *Washington Post*, December 11, 1990b.

Sauer, C. O. *The Early Spanish Main*. Berkeley: University of California Press, 1966.

Sims, C. "Defying U.S. Threat, Bolivians Plant More Coca." *The New York Times*, July 11, 1995.

Skeldon, R. *Population Mobility in Developing Countries: A Reinterpretation*. New York: Belhaven, 1990.

South American Handbook, 67th ed. New York: Prentice Hall, 1990.

Starn, O. "New Literature on Peru's Sendero Luminoso." *Latin American Research Review* 27(1992): 212–226.

Stadel, C., and M. Luz del Alba. "Plazas and Ferias of Ambato, Ecuador." *Yearbook: Proceedings of the Conference of Latin Americanist Geographers* 11 (1988): 3–10.

Thorp, R. *Economic Management and Economic Development in Peru and Colombia*. Pittsburgh: University of Pittsburgh Press, 1991.

Townsend, J. G. *Women's Voices from the Rainforest*. London: Routledge, 1995.

Trewartha, G. T. *The Earth's Problem Climates*. Madison: University of Wisconsin Press, 1966.

Trombold, C. D. *Ancient Road Networks and Settlement Hierarchies in the New World*. Cambridge: Cambridge University Press, 1991.

Vander, J. C. "Culture, Place, and School: Improving Primary Education in Rural Ecuador." *Yearbook 1994: Conference of Latin Americanist Geographers* 20 (1994): 107–119.

Wesche, R. "Ecotourism and Indigenous Peoples in the Resource Frontier of the Ecuadorian Amazon." *Yearbook 1993: Conference of Latin Americanist Geographers* 19 (1993): 35–45.

Wilke, J. W., ed. *Statistical Abstract of Latin America*. Los Angeles: University of California at Los Angeles Press, Latin American Center Publications, 1995.

Chapter 12

Albas, L. A. *Dinamica Especial do Desenvolvimento Brasileiro*. São Paulo: Universidade de São Paulo, Instituto de Pesquisas Economicas, 1985.

Abranches, S. H. *Os Desposscido: Crescimento e Pobreza no Pais do Milagre*, edited by J. Zahar. Rio de Janeiro, 1985.

"Amazon Deforestation and Climate Change." *Science* 247 (1990): 1322.

Baer, W. *The Brazilian Economy*. New York: Praeger, 1989.

Barzelay, M. *The Political Market Economy: Alcohol in Brazil's Energy Strategy*. Berkeley: University of California Press, 1986.

Bruneau, T. C., and P. Faucher, eds. *Authoritarian Capitalism: Brazil's Contemporary Economic and Political Development*. Boulder, Colo.: Westview Press, 1981.

Bruneau, T. C. *The Church in Brazil: The Politics of Religion*. Austin: University of Texas Press, 1982.

Bunker, S. G. *Developing the Amazon: Extraction, Unequal Exchange and the Failure of the Modern State*. Urbana: University of Illinois Press, 1980.

Buschbacher, R., C. Uhl, and E. A. S. Serfao. "Abandoned Pastures in Eastern Amazônia: II. Nutrient Stocks in the Soil and Vegetation." *Journal of Ecology* 76 (1988): 682–699.

Colchester, M. "The Successful Fight of the Amazon Indians to Ban Brazilian Dam Projects." *Geographical Magazine*. June 1989, Vol. LXI, No. 6, pp. 16–22.

Davis, S. H. *Victims of the Miracle*. Cambridge: Cambridge University Press, 1977.

De Jesus, C. M. *Child of the Dark*. New York: Dutton, 1962.

Dean, W. *The Industrialization of São Paulo, 1880–1945*. Austin: University of Texas Press, 1969.

Demetrius, J. *Technology and Development in an Authoritarian Regime*. New York: Praeger, 1990.

Dickinson, R. E., ed. *The Geophysiology of Amazônia*. New York: John Wiley and Sons, 1987.

Dinsmoor, J. *Brazil: Responses to the Debt Crisis*. Baltimore, Md.: Johns Hopkins University Press, 1990.

Epstein, D. G. *Brasilia, Plan and Reality: A Study of Planned and Spontaneous Urban Development*. Berkeley: University of California Press, 1973.

Evans, P. *Dependent Development: The Alliance of Multinational, State and Capital in Brazil.* Princeton, N.J.: Princeton University Press, 1979.

Evanson, N. *Two Brazilian Capitals: Architecture and Urbanism in Rio de Janeiro and Brasilia.* New Haven, Conn.: Yale University Press, 1973.

Fearnside, P. M. "Agricultural Plans from Brazil's Grande Carajas Program: Lost Opportunity for Sustainable Local Development?" *World Development* 14 (1986a): 385–409.

_____. *Human Carrying Capacity of the Brazilian Rainforest.* New York: Columbia University Press, 1986b.

_____. "Brazil's Amazon Forest and the Global Carbon Problem." *Interciencia* 10 (1985): 179–186.

Forman, S. *The Brazilian Peasantry.* New York: Columbia University Press, 1975.

Furtado, C. *Economic Development of Latin America.* London: Cambridge University Press, 1970.

_____. *The Economic Growth of Brazil.* Berkeley: University of California Press, 1963.

Gereffi, G., and D. L. Wyman, eds. *Manufacturing Miracles: Paths of Industrialization in Latin America and East Asia.* Princeton, N.J.: Princeton University Press, 1990.

Goodland, R. J. A., and H. S. Irwin. *Amazon Jungle: Green Hell or Red Desert? An Ecological Discussion of the Environmental Impact of the Highway Construction Program in the Amazon Basin.* New York: Elsevier, 1975.

Graham, L., and R. H. Wilson. *The Political Economy of Brazil.* Austin: University of Texas Press, 1990.

Hall, A. L. *Developing Amazônia.* Manchester, England: Manchester University Press, 1989.

Hecht, S., and A. Cockburn. *The Fate of the Forest.* New York: Harper, 1990.

Hemming, J. "Invaded by Gold-Diggers." *Geographical Magazine,* May 1990, pp. 26–30.

Hewlett, S. A., and R. S. Weinert. *Brazil and Mexico: Patterns in Late Development.* Philadelphia: Institute for the Study of Human Issues, 1982.

International Herald Tribune. January 30–31, 1989, p. 4.

Kitching, G., ed. *Development and Underdevelopment in Historical Perspective.* New York: Routledge, 1990.

Kuczynski, F. *Latin American Debt.* Baltimore, Md.: Johns Hopkins University Press, 1988.

Lévi-Strauss, C. "Saudades do Brasil." *New York Review of Books,* December 21, 1995, pp. 19–21.

Lopes, F. *O Cheque Heterodoxo: Camnate a Inflação e Reforma Monetaria.* Rio de Janeiro: Campus, 1986.

Lucier, R. L. *The International Political Economy of Coffee.* New York: Praeger, 1988.

Mahar, D. J. *Frontier Development Policy in Brazil: A Study of Amazônia.* New York: Praeger, 1979.

_____. *Government Policies and Deforestation in Brazil's Amazon Region.* Washington, D.C.: The World Bank, 1989.

Mangin, W. "Latin America's Squatter Settlements." *Latin American Research Review* 2 (1967): 65–98.

Maxwell, K. "The Tragedy of the Amazon." *The New York Review of Books,* March 7, 1991.

McDonough, P. *Power and Ideology in Brazil.* Princeton, N.J.: Princeton University Press, 1981.

Moog, V. *Bandeirantes e Pioneiros.* Rio de Janeiro: Civilização Brasileira, 1966.

Newson, L. A. "The Population of the Amazon Basin in 1492: A View from the Ecuadorian Headwaters." *Transactions of the Institute of British Geographers* 21(1996): 5–26 .

Page, J. A. *The Brazilians.* Reading, Mass.: Addison-Wesley, 1995.

Perlman, J. E. *The Myth of Marginality: Urban Poverty in Rio de Janeiro.* Berkeley: University of California Press, 1976.

Prado, C., Jr. *The Colonial Background of Modern Brazil.* Berkeley: University of California Press, 1971.

Prince, G. T., and T. E. Lovejoy. *Amazônia.* New York: Pergamon, 1984.

Population Reference Bureau. *World Population Data Sheet.* Washington, D.C.: Author, 1995.

Ribeiro, U. F., and P. R. Leopoldo. "Colonizção ao Longo do Transamizonica: Trecho Km 930–1035." *Interciencia* 14 (1989): 311–316.

Roberts, J. T. "Squatters and Urban Growth in Amazonia." *Geographical Review* 82(1992): 441–457.

Robinson, E. "Over the Brazilian Rainbow." *Washington Post*, December 10, 1995.

Saffioti, H.I.B. *Women in Class Society*. New York and London: Monthly Review Press, 1978.

Schemo, D. J. "Amazon is Burning Again, as Furiously as Ever." *The New York Times*, October 12, 1995.

_____. "Hard choices ahead on biodiversity." *Science* 241 (1988): 1759–1761.

Shoumatoff, A. *The World Is Burning*. New York: Little, Brown, 1990.

Skidmore, T. *Politics in Brazil 1930–1964*. New York: Oxford University Press, 1967.

Tavares, M. C. *Da Substituica de Importações ao Capitalismo Financeiro*. Rio de Janeiro: Zahar, 1974.

_____. "Ecologists make friends with economists," *The Economist*, October 15, 1988, pp. 25–29. *The New York Times*, April 7, 1989, p. A6.

Vilhena, L. S. *Recopilação de Noticias Soteropolitanas e Brasilicas em XX Cartas*. Bahia: Bras do Amaral, 1927.

Wagner, E. *Policy of Occupation and Utilization of the Cerrados, Empresa Brasileira de Pesquisa Agropecuaria*. Brasilia: Departimento de Informação e Documentação, 1981.

Williams, J. R. *Manaus, Amazonas: A Focal Point for Development in Amazônia: Publication #4*. Washington, D.C.: Pan American Institute of Geography and History, 1971.

Wirth, J. D. *The Politics of Brazilian Development 1930–1954*. Stanford, Calif.: Stanford University Press, 1970.

World Resources Institute. *World Resources, 1990–91: A Guide to the Global Environment*. London: Oxford University Press, 1991.

Chapter 13

Blakemore, H. *Chile*. Oxford: Clio Press, 1988.

Caviedes, C. L. *The Southern Cone: Realities of the Authoritarian State in South America*. Totowa, N.J.: Rowman and Allenheld, 1984.

Drake, P. W., and I. Jaksić, eds. *The Struggle for Democracy in Chile, 1982–1990*. Lincoln: University of Nebraska Press, 1991.

Finch, M. H. J. *Uruguay*. Oxford: Clio Press, 1989.

Fifer, V. J. "Old Tracks and New Opportunities at the Uspallata Pass." *Yearbook 1994: Conference of Latin Americanist Geographers* 20 (1995): 35–48.

Gillespie, F. "Comprehending the Slow Pace of Urbanization in Paraguay Between 1950 and 1972." *Economic Development and Cultural Change* 31 (1983): 355–374.

Griffin, E. "Causal Factors Influencing Agricultural Land Use Patterns in Uruguay." *Revista Geográfica* 80 (1974): 13–33.

Gubetich, H. F. *Geografía del Paraguay*. Asunción: Orbis, 1985.

Keeling, D. "Transport and Regional Development in Argentina: Structural Deficiencies and Patterns of Network Evolution." *Yearbook 1993: Conference of Latin Americanist Geographers* 19 (1994): 25–34.

Kerr, J. G. *A Naturalist in the Gran Chaco*. Cambridge: Cambridge University Press, 1950.

Klacko, J., and J. R. Roade. *Uruguay: El País Urbano*. Montevideo: Ediciones de la Banda Oriental, 1981.

Miller, E. E. "The Frontier and the Development of Argentine Culture." *Revista Geográfica* 90 (1979): 183–198.

Nickson, R. Andrew. *Paraguay*. Oxford: Clio Press, 1987.

Porteous, J. D. "Urban Transplantation in Chile." *Geographical Review* 62 (1972): 455–478.

Revista Geográfica 1982, Special edition on Argentina.

Rock, D. *Argentina, 1516–1982*. Berkeley and Los Angeles: University of California Press, 1985.

Ross, S. R., and T. F. McGann, eds. *Buenos Aires: Four Hundred Years*. Austin: University of Texas Press, 1982.

Ruggiero, K. H. *And Here the World Ends: The Life of an Argentine Village*. Stanford, Calif.: Stanford University Press, 1988.

Schwerdtfeger, W., ed. *Climates of Central and South America*. Amsterdam: Elsevier Scientific Publishing, 1976.

Slatta, R. W. *Gauchos and the Vanishing Frontier*. Lincoln: University of Nebraska Press, 1983.

Solberg, C. E. "Peopling the Prairies and the Pampas" *Journal of Inter-American Studies* 24 (1982): 131–161.

Subercaseaux, B. *Chile: A Geographic Extravaganza*, translated by Angel Flores. New York: Hafner, 1971.

Whitaker, A. P. *The United States and the Southern Cone: Argentina, Chile, and Uruguay*. Cambridge, Mass.: Harvard University Press, 1976.

Young, G. "Paraguay." *National Geographic Magazine* 162 (1982): 240–269.

Chapter 14

Albert, B. *South America and the First World War: The Impact of the War on Brazil, Argentina, Peru, and Chile*. New York: Cambridge University Press, 1988.

Andrien, K. L., and L. L. Johnson, eds. *The Political Economy of Spanish America in the Age of Revolution, 1750–1850*. Albuquerque: University of New Mexico Press, 1994.

Barkai, A. *Nazi Economics: Ideology, Theory, and Policy*. New Haven, Conn.: Yale University Press, 1990.

Bethell, L., ed. *The Cambridge History of Latin America*, Vol. III. Cambridge: Cambridge University Press, 1985.

Burns, E. B. *Latin America: A Concise Interpretive History*. Englewood Cliffs, N.J.: Prentice Hall, 1994.

Castenada, J. G. *The Mexico Shock: Its Meaning for the United States*. New York: New Press, 1995.

Child, J. *Geopolitics and Conflict in South America*. New York: Praeger, 1985.

Collier, S., T. Skidmore, and H. Blakemore. *The Cambridge Encyclopedia of Latin America*, 2d ed. Cambridge: Cambridge University Press, 1992.

Conniff, M. L., and T. J. Davis. *Africans in the Americas: A History of the Black Diaspora*. New York: St. Martin's Press, 1994.

Eckes, A. E., Jr. *The United States and the Global Struggle for Minerals*. Austin: University of Texas Press, 1979.

Fitzgerald, E.V.K. "ECLA and the Formation of Latin American Economic Doctrine." In *Latin America in the 1940s: War and Post-war Transitions*, edited by D. Rock. Berkeley: University of California Press, 1994.

Glick, L. A. *Understanding the North American Free Trade Agreement*, 2d ed. Boston: Kluner Law and Taxation Publishers, 1994.

Hansen, R. D. *The Politics of Mexican Development*. Baltimore, Md.: Johns Hopkins University Press, 1971.

Hearden, P. J. *Roosevelt Confronts Hitler: America's Entry into World War II*. DeKalb: Northern Illinois University Press, 1987.

Hirschman, A. O., ed. *Latin American Issues: Essays and Comments*. New York: Twentieth Century Fund, 1961.

Hughlett, L. J., ed. *Industrialization of Latin America*. New York: McGraw-Hill, 1946. (Reprinted, Westport, Conn.: Greenwood Press, 1970).

Humphries, R. A. *Latin America and the Second World War: Volume One, 1939–1942*. London: Athlone, 1981.

_____. *Latin America and the Second World War: Volume Two, 1942–1945.* London: Athlone, 1982.

Jones, W. R. *Minerals in Industry.* Harmondsworth, Middlesex: Penguin, 1950 (Originally published, 1943).

Lowenthal, A. F. *Partners in Conflict: The United States and Latin America in the 1990s.* Baltimore, Md.: Johns Hopkins University Press, 1987.

Lynch, J., ed. *Latin American Revolutions, 1808–1826: Old and New World Origins.* Norman: University of Oklahoma Press, 1994.

MacDonald, G. J., D. L. Nielson, and M. A. Stern. *Latin American Environmental Policy in International Perspective.* Boulder, Colo.: Westview Press, 1996.

Merino, O., and L. A. Newson. "Jesuit Missions in Spanish America: The Aftermath of Expulsion." In *Yearbook, Conference of Latin Americanist Geographers,* 21(1995), edited by D. J. Robinson, Austin: University of Texas Press, 1996, pp. 133–148.

Mitchell, B. R. *British Historical Statistics.* Cambridge: Cambridge University Press, 1988.

_____. *International Historical Statistics: Europe, 1750–1988,* 3d ed. New York: Stockton, 1993.

Naím, M. "Latin America the Morning After." *Foreign Affairs* 74(1995): 45–61.

Nilbo, S. R. *War, Diplomacy and Development: The United States and Mexico, 1938–1954.* Wilmington, Del.: S.R. Books, 1995.

Nystrom, J. W., and N. A. Haverstocky. *The Alliance for Progress: Key to Latin America's Development.* Princeton, N.J.: Van Nostrand, 1966.

Owens, D. J. "Spanish-Portuguese Territorial Rivalry in Colonial Río de la Plata." *Yearbook, Conference of Latin Americanist Geographers,* 19(1993), edited by G. S. Elbow, pp. 15–25. Auburn, AL: Conference of Latin Americanist Geographers.

Park, J. W. *Latin American Underdevelopment: A History of Perspectives in the United States, 1870–1965.* Baton Rouge: Louisiana State University Press, 1995.

Pendle, G. *A History of Latin America.* Harmondsworth, Middlesex: Penguin, 1963.

Rock, D. "Argentina in 1914: The Pampas, the Interior, Buenos Aires. In *The Cambridge History of Latin America,* Vol. 5, *c1870 to 1930,* edited by L. Bethell. Cambridge: Cambridge University Press, 1986.

_____, ed. *Latin America in the 1940s: War and Post-War Transitions.* Berkeley: University of California Press, 1994.

Skeldon, R. *Population Mobility in Developing Countries: A Reinterpretation.* New York: Belhaven, 1990.

Weintraub, S. *NAFTA What Comes Next?* Westport, Conn.: Praeger, 1994.

Wilbanks, T. '"Sustainable Development" in Geographic Perspective.' *Annals of the Association of American Geographers* 84 (1994): 541–556.

Winn, P. *Americas: The Changing Face of Latin America and the Caribbean.* Berkeley: University of California Press, 1992.

Zelinksy, W. "The Hypothesis of the Mobility Transition." *Geographical Review* 61 (1971): 219–249.

Photo Credits

Index